Gaseous Air Pollutants and Plant Metabolism

M. J. Kozioł
and
F. R. Whatley, FRS
Botany School, University of Oxford, Oxford, UK

Butterworths
London Boston Durban Singapore Sydney Toronto Wellington

First published, 1984

© **The several contributors named in the list of contents 1984**

British Library Cataloguing in Publication Data

Gaseous air pollutants and plant metabolism.
  1. Plants, Effects of air pollution—Congresses
  I. Koziol, M. J.   II. Whatley, F.R.
  581.1'33   QK751

    ISBN 0-408-11152-6

Library of Congress Cataloging in Publication Data
Main entry under title:

Gaseous air pollutants and plant metabolism.

    Papers from an international meeting, held in Oxford,
August 2–5, 1982.
    Bibliography: p.
Includes index.
1. Plants, Effect of air pollution on--Congresses.
2. Plants, Effect of gases on--Congresses. 3. Plants--
Metabolism--Congresses. I. Koziol, M. J.
II. Whatley, F. R.
QK751.G34 1983   581.2'4   83-15415
ISBN 0-408-11152-6

Filmset by Mid-County Press, London SW15
Printed in Great Britain at the University Press, Cambridge

Preface

Although many meetings, conferences and symposia on the effects of gaseous pollutants on plants have included papers or sessions on the biochemical interactions of pollutants with plant metabolism, there had not been a symposium devoted primarily to this subject. As the amount of information available on the biochemical interactions of pollutants with various aspects of plant metabolism is increasing steadily, it was thought timely to organize such a symposium. In structuring the meeting, the Organizing Committee was aware of the need to correlate the biochemical aspects with the physiological and ultrastructural ones, to assess the impact of the biochemical responses in terms of effects on crop yields and on the nutritional quality of the crop and to appreciate the importance of these interactions in the context of global pollution problems.

Since it was designed as the first in a proposed series of international symposia, the symposium, of which this volume is the proceedings, consists primarily of review papers; papers which not only summarize the work that has been done but also served to identify those areas meriting further research. In general, a need was identified for more research into the interactions of gaseous pollutants with insect, fungal and viral attacks on crops, into changes in the nutritional quality of crops and into the post-harvest physiology of crops. The importance of suitable diagnostic tests for pollutant resistance for use in plant breeding programmes, for comparative laboratory and field studies and for studies using mixtures of pollutants was also recognized.

International interest in the biochemical interactions of gaseous pollutants with plant metabolism was demonstrated by the participation of delegates from 19 countries. In addition to the papers presented formally during the seven sessions of the symposium, delegates benefited from the many informative posters presented at a Poster Session, held in conjunction with a Trade Exhibition. Both the formal and poster presentations generated interesting and wide-ranging discussions.

The success of the meeting was due ultimately to the delegates in attendance and to the skill and experience of the chairmen of the sessions, especially Professor F. T. Last, Dr T. M. Roberts, Professor V. C. Runeckles, Dr P. J. W. Saunders and Dr A. R. Wellburn.

We thank the staff of St. Catherine's College, Oxford, particularly Miss Valerie Whittock, for the efficient and friendly handling of all domestic needs, and various members of the Oxford Botany School, notably Mr P. G. Turner, Mr K. Burras, Miss S. Mitchell, Miss L. Dyas, Mr J. D. Shelvey, Mr J. Chalmers, Mr P. Connelly and Mrs

D. Wyse. The editorial assistance of Miss B. Beedham is also gratefully acknowledged.
    Finally, on behalf of the Organizing Committee, we should like to thank all the organizations who gave financial assistance to the symposium.

M. J. Kozioł
F. R. Whatley

Organizing committee for the First International Symposium on Gaseous Air Pollutants and Plant Metabolism, held in Oxford, UK, 25 August 1982

Professor F. R. Whatley, FRS, *Botany School, Oxford*
Dr M. J. Kozioł, *Botany School, Oxford*
Dr J. N. B. Bell, *Imperial College, London*
The late Mr D. W. Cowling, *Grassland Research Institute, Hurley*
Professor T. A. Mansfield, *University of Lancaster, Lancaster*
Dr P. J. W. Saunders, *Natural Environment Research Council, Swindon*
Dr T. L. V. Ulbricht, *Agricultural Research Council, London*
Dr A. R. Wellburn, *University of Lancaster, Lancaster*

Acknowledgements

The Organizing Committee gratefully acknowledges the financial assistance given to this Symposium by:

The Agricultural Research Council
An Anonymous Donor
The British Council
The Central Electricity Generating Board
The Commission of the European Communities
Esso Petroleum Company
The London Brick Company Ltd
The Royal Society
Shell International Petroleum Company Ltd

Financial support from the following companies through their participation in the Trade Exhibition is also gratefully acknowledged:

Analysis Automation Ltd, *Southfield House, Eynsham, Oxford*
Baird & Tatlock (London) Ltd, *PO Box 1, Romford, Essex*
Bryans Southern Instruments Ltd, *1 Willow Lane, Mitcham, Surrey*
Delta-T Devices, *128 Low Road, Burwell, Cambridge*
V. A. Howe & Company Ltd, *88 Peterborough Road, London*
Pye Unicam Ltd, *York Street, Cambridge*
Sintrom Electronics Ltd, *14 Arkwright Road, Reading*
Springer-Verlag GmbH & Co. KG, *Postfach 10 52 80, Tiergartenstrasse 17, 6900 Heidelberg 1, FRG*
Techmation Ltd, *58 Edgware Way, Edgware, Middlesex*
Vaisala (UK) Ltd, *11 Billing Road, Northampton*
Waters Associates (Instruments) Ltd, *324 Chester Road, Northwich, Cheshire*

Contents

viii

x   Contents

Abbreviations

As the chapters presented in this book range across various fields, we have endeavoured to keep the number of abbreviations used to a minimum. For convenient reference, we have listed these abbreviations below.

Abbreviations used in the text

9-AA	9-amino-acridine
ACC	1-aminocyclopropane-1-carboxylic acid
BPGA	1,3-bisphosphoglycerate
CDP	cytidine diphosphate
DAD	diaminodurene
DCIP	dichloroindophenol, oxidized
$DCIPH_2$	dichloroindophenol, reduced
DCMU	3-(3',4'-dichlorophenyl)-1,1-dimethylurea
DGDG	digalatosidyl diglyceride
DHAP	dihydroxyacetone phosphate
4',7-DHF	4',7-dihydroxyflavone
DMO	5,5-dimethyloxazolidine-2,4-dione
DPC	diphenyl carbazide
DTE	dithioerythritol
DTT	dithiothreitol
EDU	N-[2-(2-oxo-1-imidazolidinyl)ethyl]-N'-phenylurea
ESR	electron spin resonance spectroscopy
FBP	fructose-1,6-bisphosphate
F6P	fructose-6-phosphate
GDPG	guanosinediphosphoglucose
GDPglucose	guanosinediphosphoglucose
G6P	glucose-6-phosphate
MDA	malondialdehyde
MGDG	monogalactosyl diglyceride
MV	methyl viologen
NMR	nuclear magnetic resonance spectroscopy
$\cdot O_2^-$	superoxide free radical

OAA	oxaloacetic acid
PA	phosphatidic acid
PAN	peroxyacetyl nitrate
PC	phosphatidylcholine
PE	phosphatidylethanolamine
PEP	phosphoenolpyruvate
PG	phosphatidyl glycerol
3-PGA	3-phosphoglycerate
3-PGAL	3-phosphoglyceraldehyde
PI	phosphatidylinositol
PMS	phenazine methosulphate
ppb	parts per billion (parts 10^{-9})
ppm	parts by million (parts 10^{-6})
PS	phosphatidylserine
PSI	photosystem I
PSII	photosystem II
PUSFA	poly-unsaturated fatty acid
r_a	leaf aerodynamic (boundary layer) resistance
r_c	leaf cuticular + surface resistance
r_s	leaf stomatal resistance
r_m	leaf mesophyll resistance
RH	relative humidity
RuBP	ribulose-1,5-bisphosphate
Ru5P	ribulose-5-phosphate
SBP	sedoheptulose-1,7-bisphosphate
S7P	sedoheptulose-7-phosphate
–SH	sulphydryl functional group
TBA	thiobarbituric acid
UDPG	uridinediphosphoglucose
vpd	vapour pressure deficit

Abbreviations of enzymes

Abbreviation	Trivial name	Systematic name
APH	acid phosphatase	orthophosphoric-monoester phosphohydrolase (E.C. 3.1.3.2)
FBPase	fructose-1,6-bisphosphatase	D-fructose-1,6-bisphosphate 1-phosphohydrolase (E.C. 3.1.3.11)
GDH	glutamate dehydrogenase	L-glutamate: NAD$^+$ oxidoreductase (E.C. 1.4.1.2)
GOGAT	glutamate synthase	L-glutamate: NADP$^+$ oxidoreductase (transaminating) (E.C. 1.4.1.13)
GOT	glutamate-oxaloacetate transferase	L-aspartate: 2-oxoglutarate aminotransferase (E.C. 2.6.1.1)
GPD	glyceraldehyde-3-phosphate dehydrogenase	D-glyceraldehyde-3-phosphate: NAD$^+$ oxidoreductase (E.C. 1.2.1.12)
G6PD	glucose-6-phosphate dehydrogenase	D-glucose-6-phosphate: NADP$^+$ 1-oxidoreductase (E.C. 1.1.1.49)
GPT	glutamate pyruvate transferase	L-alanine: 2-oxoglutarate aminotransferase (E.C. 2.6.1.2)
GS	glutamine synthase	L-glutamate: ammonia ligase (ADP-forming) (E.C. 6.3.1.2)
LAP	L-leucine aminopeptidase, aminopeptidase (cytosol)	α-aminoacyl-peptide hydrolase (cytosol) (E.C. 3.4.11.1)
MDH	malate dehydrogenase	L-malate: NAD$^+$ oxidoreductase (E.C. 1.1.1.37)
NADP-GPD	NADP-glyceraldehyde-3-phosphate dehydrogenase	D-glyceraldehyde-3-phosphate: NADP$^+$ oxidoreductase (E.C. 1.2.1.9)
NADP-MD	NADP-malate dehydrogenase, 'malic' enzyme	L-malate: NADP$^+$ oxidoreductase (oxaloacetate-decarboxylating) (E.C. 1.1.1.40)
PEP-Case PEP carboxylase }	phosphoenolpyruvate carboxylase	Orthophosphate: oxaloacetate carboxylyase (E.C. 4.1.1.31)
PFK	phosphofructokinase	ATP: D-fructose-6-phosphate 1-phosphotransferase (E.C. 2.7.1.11)
Pyr., Pi Dikinase	pyruvate, orthophosphate dikinase	ATP: pyruvate, orthophosphate phosphotransferase (E.C. 2.7.9.1)
RuBP-Case RuBP carboxylase }	ribulose-1,6-bisphosphate carboxylase	3-phospho-D-glycerate carboxylyase (E.C. 4.1.1.39)
Ru5P kinase	ribulose-5-phosphate kinase	ATP: D-ribulose-5-phosphate 1-phosphotransferase (E.C. 2.7.1.19)
SBPase	sedoheptulose-1,7-bisphosphatase	D-sedoheptulose-1,7-bisphosphate 1-phosphohydrolase (E.C. 3.1.3.37)
SOD	superoxide dismutase	superoxide: superoxide oxidoreductase (E.C. 1.15.1.1)

Part I

Defining pollution problems

Chapter 1

Air pollution problems in western Europe

J. N. B. Bell
DEPARTMENT OF PURE AND APPLIED BIOLOGY, IMPERIAL COLLEGE AT SILWOOD PARK, ASCOT, UK

Historical

Although air pollution has only become widespread since the Industrial Revolution, it has presented problems on a local scale in western Europe for at least 2000 years. The earliest record of complaints about atmospheric pollution is found in the writings of the Roman poet, Horace, who objected to damage caused by soot on the temples of ancient Rome (Brimblecombe, 1982). Coal smoke probably appeared in London by the beginning of the thirteenth century and subsequently became a sufficiently major nuisance as to provoke a number of medieval statutes aimed at controlling coal combustion (Brimblecombe, 1975). Possibly the first mention of air pollution damage to vegetation is by Evelyn (1661) in *Fumifugium or the Inconvenience of the Aer, and Smoake of London Dissipated*, in which he observes that

'It kills our bees and flowers abroad, suffering nothing in our gardens to bud, display themselves, or ripen; our anemonies and many other choycest flowers, will by no industry be made to blow in London, or the precincts of it, unlesse they be raised on a hot-bed, and governed with extraorindary artifice to accellerate their springing, importing a bitter and ungrateful taste to those few wretched fruits, which never arriving to their desired maturity, seem, like the Apples of Sodome, to fall even unto dust, when they are but touched.'

Despite numerous records of the difficulties involved in growing plants in polluted areas, scientific investigations of this problem did not begin until Stöckhardt (1871) published his account of studies on the 'effects of smoke on spruce (*Picea*) and fir (*Abies*) trees in Germany. Early research into the impact of air pollutants on plants was primarily concerned with obvious acute injury symptoms, but physiological aspects were not entirely neglected, with Oliver (1894) studying effects of London 'fog' on chlorophyll and Cohen and Ruston (1925) examining the influence of sulphur dioxide (SO_2) on photosynthesis.

During the present century, new legislation as well as changing industrial practices and domestic fuel usage have drastically reduced the incidence in western Europe of acute air pollution injury. In the last decade attention has become focused on the ecological impact of acid precipitation, which has tended to overshadow interest in the effect of gaseous pollutants. However, there is abundant evidence that the ambient

3

gaseous air pollutant levels prevailing over large areas of western Europe can produce adverse impacts on plant growth, which in many cases remain undetected in the absence of controlled experiments. This chapter is aimed at providing an overview of current problems in this respect and, as such, is mainly concerned with the ubiquitous primary pollutants, sulphur dioxide (SO_2) and nitrogen oxides (NO_x), and the episodic, but widespread secondary pollutant, ozone (O_3).

Current distribution of major air pollutants in western Europe

Western Europe is characterized by a generally high population density, with large cities and areas of major industrial activity scattered over a substantial proportion of the region. Thus, irrespective of the direction of the prevailing wind, most of western Europe is generally subject to air pollution concentrations above the natural background. Indeed, in the last 20 years changes in the nature of air pollution emissions have generally led to more efficient dispersion which, although alleviating problems close to the source, may have increased the area over which elevated pollutant concentrations occur. In view of these factors, it is not surprising that interest in western Europe is becoming directed increasingly towards understanding the net impact on vegetation of low to moderate concentrations of pollutants both in rural areas and on the fringe of conurbations and industrial locations.

The current distribution of the major phytotoxic primary air pollutants over western Europe has recently been reviewed by Fowler and Cape (1982). They have estimated that approximately 1.3×10^6 ha of agricultural land in the United Kingdom, France, Benelux and West Germany are exposed to an annual mean SO_2 concentration of greater than 0.038 ppm, representing 1.0% of the total area, while about five times this area is subjected to annual means between 0.019 and 0.038 ppm (*Figure 1.1*). Precise knowledge of the distribution of SO_2 outside cities is limited by a lack of rural monitoring sites and *Figure 1.1* is based largely on the modelling exercise by the OECD (1979) on long-range transport of air pollutants, which utilizes 1974 emission data.

Until recently little interest has been shown in pollution by nitrogen oxides in Europe, while in North America most research has been devoted towards understanding its role in the formation of photochemical smogs. Consequently monitoring for NO_x in Europe, even by simple integrating techniques, has been very limited in scope and no clear information is available on historical trends (CENE, 1981). Since regular monitoring of nitric oxide (NO) and nitrogen dioxide (NO_2) with reliable and accurate instrumentation commenced at a number of western European sites over the last decade, it has become clear that both pollutants are more abundant and widely distributed than realized hitherto.

Thus Martin and Barber (1981) recorded a mean annual NO_x concentration of 0.017 ppm at a rural site in central England, which was nearly 50% higher than the mean annual SO_2 concentration (0.012 ppm), with nitrogen dioxide (NO_2) contributing about half of this amount. Measurements at off-street sites in central London have also shown a $SO_2:NO_x$ ratio of 1:1.5, nitric oxide (NO) accounting for a greater proportion of the latter pollutant than in rural areas (Anonymous, 1979; Apling, Rogers and Stevenson, 1981). In The Netherlands mean annual SO_2 and NO_x concentrations of 0.008 ppm and 0.016 ppm, respectively, have been reported for the country as a whole, based on 88 sampling locations (Mooi, 1981). Thus it is apparent that in both rural and urban locations, SO_2 is normally accompanied by greater concentrations of NO_x, on a volume/volume basis. Fowler and Cape (1981) have taken

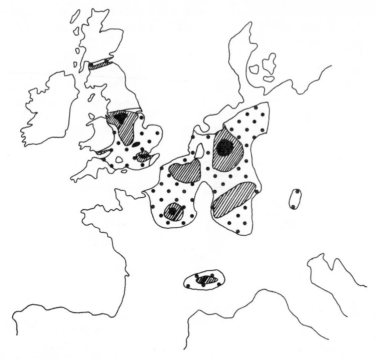

Figure 1.1. Distribution of SO$_2$ over western Europe (based on 1974 emission data). (After Fowler and Cape, 1982.) Annual arithmetic mean SO$_2$ concentration (ppm): ☐ < 0.011; ▦ 0.011–0.019; ▨ 0.019–0.038; ▮ > 0.038

a ratio (v/v) for SO$_2$:NO$_2$ of 1:1.1 as being representative of rural conditions in the United Kingdom and used this to make an approximation of the distribution of NO$_2$ in western Europe. This suggests that the 1% of rural land which is exposed to SO$_2$ concentrations of more than 0.038 ppm, may also experience mean NO$_2$ concentrations of greater than 0.042 ppm.

The third phytotoxic air pollutant which is widely distributed over western Europe is ozone (O$_3$). However, unlike the other two pollutants, it is impossible to make any realistic predictions of the distribution of elevated concentrations, because of its highly episodic nature and the dependence of its formation on specific meteorological conditions, whose incidence varies considerably from year to year. Until systematic monitoring of O$_3$ commenced in western Europe approximately 15 years ago, it was not considered that concentrations could occur above the maximum natural background concentration of about 0.04 ppm. Subsequently it has become apparent that O$_3$ occurs at pollutant concentrations regularly in summertime under suitable meteorological conditions at numerous sites in western Europe.

Table 1.1 shows maximum summertime O$_3$ concentrations recorded at a range of rural, urban and urban fringe locations in south-east England between 1973 and 1980. With the exception of central London in 1980, all sites have experienced O$_3$ concentrations greater than 0.04 ppm in all the years when measurements were made. Marked fluctuations have occurred from year to year, with maximum concentrations being recorded during the hot dry summers of 1975 and 1976. It is clear that high concentrations of O$_3$ are not confined to urban sites: *Table 1.1* shows higher maximum concentrations at rural locations than in the centre of London for five out of eight years. Annual maximum hourly means varied between 0.11 and 0.25 ppm at the rural sites, with up to 30% of days with hourly means greater than 0.08 ppm.

A comparison of O_3 concentrations on the same dates at various European sites, reveals that high concentrations can occur simultaneously in widely separated locations, ranging from Sweden to Italy (*Table 1.2*). This suggests that major O_3 incidents in western Europe may cover a very much larger area than is subjected to SO_2 and NO_2 concentrations greater than 0.038 and 0.042 ppm, respectively. There are insufficient rural O_3 monitoring stations to confirm this, but supporting evidence can

TABLE 1.1. Summer (May–September) ozone concentrations (ppm) at selected sites in south-east England, 1973–1980

Year	O_3 parameter		Rural sites			
		Harwell (Oxfordshire)	Sibton (Suffolk)	Ascot (Berkshire)	Chilworth (Hampshire)
1973	A	0.14	0.14[a]	—	—
	B	16.1%	10.4%[a]	—	—
1974	A	0.12	0.09[a]	—	0.11[a]
	B	11.8%	1.9%[a]	—	2.8%[a]
1975	A	0.18[a]	0.20	—	0.17
	B	23.8%[a]	17.8%	—	21.6%
1976	A	0.25	—	0.14[a]	—
	B	—	—	30.4%[a]	—
1977	A	—	0.11	0.12	—
	B	—	—	3.7%	—
1978	A	—	0.13	0.13	—
	B	—	—	5.8%	—
1979	A	—	0.10	0.14	—
	B	—	2.6%	7.3%	—
1980	A	—	0.06	0.11	—
	B	—	0%	1.4%	—

Year	O_3 parameter		Urban fringe sites			
		Stevenage (Hertfordshire)	Harrow (Greater London)	Teddington (Greater London)	Hainault (Greater London)
1973	A	—	—	—	—
	B	—	—	—	—
1974	A	—	—	—	—
	B	—	—	—	—
1975	A	—	—	0.13	0.12[a]
	B	—	—	15.0%	14.3%[a]
1976	A	0.21	—	0.21	0.18
	B	—	—	34.0%	30.7%[a]
1977	A	0.14	—	—	—
	B	—	—	—	—
1978	A	0.10	—	—	—
	B	—	—	—	—
1979	A	0.11	0.18[a]	—	—
	B	4.3%	4.5%[a]	—	—
1980	A	0.08	0.11[a]	—	—
	B	0.7%	2.6%[a]	—	—

TABLE 1.1 (*continued*)

Year	O_3 parameter	Urban/industrial sites	
		Central London	Canvey Island (Essex)
1973	A	0.14	—
	B	17.2%	—
1974	A	0.16	—
	B	7.6%	—
1975	A	0.15[a]	—
	B	40%[a]	—
1976	A	0.21	—
	B	19.4%	—
1977	A	0.13	0.18
	B	—	—
1978	A	0.15	0.15
	B	6.9%	—
1979	A	0.10	0.09
	B	2.1%	4.1%
1980	A	0.04	0.08
	B	0%	0%

Key:
A — maximum hourly mean concentration ppm.
B — % days with maximum hourly mean O_3 concentrations >0.08 ppm.
[a] Incomplete data set May–September.

References
Derwent *et al.* (1976): Harwell 1973–5, Sibton 1973–5, Chilworth 1974/5, Teddington 1975, Hainault 1975, Central London 1973–5; Apling *et al.* (1977): Harwell 1976, Stevenage 1976; M. R. Ashmore (unpublished): Ascot 1976–80; Ball and Bernard (1978): Hainault 1976, Central London 1976, Teddington 1976; Apling *et al.* (1981a): Sibton 1977/8, Stevenage 1977/8, Canvey Island 1977/8, Central London 1977; Apling *et al.* (1981b): Stevenage 1979/80; Apling *et al.* (1981c): Harrow 1979/80; Apling *et al.* (1981d): Canvey Island 1979/80; Apling *et al.* (1981e): Sibton 1979/80; Apling *et al.* (1981f): Central London 1978–80.

TABLE 1.2. Coincident maximum hourly mean O_3 concentrations (ppm) exceeding 0.08 ppm at western European sites

Date	Location					
	Delft (Netherlands)	Rome (Italy)	Frankfurt (German Federal Republic)	Gothenburg/ Rörvin (Sweden)	London (UK)	Nice (France)
11.8.73	0.12	—	0.13	—	0.09	—
17.8.73	0.19	—	0.13	0.08	0.09	—
20.8.74	0.10	0.11	<0.08	<0.08	0.11	0.12
30.7.75	0.12	0.18	0.09	<0.08	0.12	0.19
3.8.75	<0.08	0.10	0.08	0.12	0.11	0.20
8.8.75	0.12	—	0.09	<0.08	0.15	0.13

(From Guicherit and Van Dop, 1977).

be obtained from large-scale surveys in which *Nicotiana tabacum* cv. Bel-W3, which has a threshold for O_3 injury at a concentration of about 0.05 ppm, is used as a biological indicator. Ashmore, Bell and Reily (1978, 1980) exposed Bel-W3 plants outdoors at 53 mainly rural sites distributed over the British Isles, throughout the summer of 1977 and measured the appearance of O_3 injury on the foliage at weekly intervals. The distribution of the mean of the weekly leaf injury indices, derived from these measurements, is shown in *Figure 1.2*. Injury occurred at all sites, with the exception of those in the extreme north of Scotland. The highest leaf injury indices were recorded in North Wales, Northern Ireland and Central Scotland: these coincided with the areas which experienced the greatest amount of sunshine during summer 1977, showing the importance of the meteorological factor in generating O_3 from its precursors. A similar

Figure 1.2. The distribution of mean O_3-induced leaf injury (%) on *Nicotiana tabacum* cv. Bel-W3 in the British Isles (13.6.77–5.9.77). ☐ 0.01–0.54%; ▨ 0.054–1.06%; ▦ 1.06–1.59%; ▨ 1.59–2.11%; ■ 2.11–2.64%. (Ashmore *et al.*, 1978; Courtesy of Macmillan Journals Ltd)

survey is carried out annually in The Netherlands and *Figure 1.3* also shows the results for 1977, indicating the occurrence of O_3 injury of Bel-W3 plants throughout the country, with the highest concentrations in the coastal regions around Rotterdam and to the north of Amsterdam (NML, 1978).

It is apparent that western Europe currently experiences continuous moderate concentrations of SO_2 and NO_x over 1–5% of agricultural land, with episodes of O_3

Figure 1.3. Distribution of mean O_3-induced leaf injury (%) on *Nicotiana tabacum* cv. Bel-W3 in The Netherlands (6.6.77–28.10.77). (After NML, 1978. Courtesy of Rijksinstituut voor de Volksgezondheid, Bilthoven)

pollution, varying considerably in frequency and intensity, superimposed upon these, but occurring simultaneously as the only pollutant present at potentially phytotoxic levels over much of the less polluted rural areas. In addition, other pollutants, such as fluorides, contribute to the atmospheric mix in the vicinity of specific sources, but their effects are localized.

Effects of moderate levels of ambient air pollution

In view of the large area of western Europe which is exposed either continuously or intermittently to one or more phytotoxic pollutants, there is a clear need for an accurate assessment of the impacts on plants of the different pollutant combinations which occur in rural and urban fringe locations.

Field surveys of crop productivity in areas with different concentrations and mixtures of pollutants are fraught with difficulties in interpretation, due to the confounding effects of climatic and edaphic factors as well as agricultural practices.

However, there is strong circumstantial evidence from field studies that air pollution can reduce plant growth in some rural areas of Europe which are not subjected to strong local point sources. Thus in southern parts of the Pennines, which are located in northern England and are polluted with SO_2 from neighbouring towns and cities, there have been many reports of poor growth of bred pasture grass cultivars on hill farms (Bell and Mudd, 1976). Similarly, attempts by the Forestry Commission to establish conifer plantations in this area have been generally unsuccessful (Lines, 1979). A survey of Scots pine (*Pinus sylvestris*) in the southern Pennines by Farrar, Relton and Rutter (1977) showed a band of countryside, about 50 km wide downwind of Greater Manchester and Merseyside, in which this species was either absent or present in very low abundance (*Figure 1.4*). The occurrence of Scots pine in this area was significantly

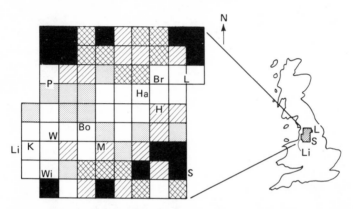

Figure 1.4. Distribution of Scots pine in the southern Pennines. Number of 2 km squares occupied by at least one specimen per 10 km traverse of the 10 km squares shown: ☐0–0.9; ☷1–1.9; ▨2–2.9; ▦3–3.9; ■4–4.9. The towns marked are: Bo, Bolton; Br, Bradford; H, Huddersfield; Ha, Halifax; K, Kirkby; L, Leeds; Li, Liverpool; M, Manchester; P, Preston; S, Sheffield; W, Wigan; Wi, Widnes. (Farrar *et al.*, 1977; Courtesy of Applied Science Publishers Ltd)

negatively correlated with the mean winter SO_2 concentration, the species being absent at sites with concentrations greater than 0.076 ppm. This is in very close agreement with a survey of Knabe (1970) of Scots pine distribution in the Ruhr region of Germany. Considerable caution must be exercised in attributing causality to such correlations, there being many possible reasons for the low abundance of Scots pine in polluted areas (e.g. a lack of interest in planting conifers at such sites), but it is interesting to note that a recent decline in pollution concentrations in the southern Pennines has been accompanied by a marked improvement in the performance of conifer trials (Lines, 1979).

The obvious approach in determining the current impact of air pollution is to carry out controlled experiments in the field in which the performance of plants is compared in ambient and clean air at the same site. Surprisingly, relatively few studies of this type have been carried out in western Europe, most workers preferring to use fumigation systems in which attempts are made, with varying degrees of success, to simulate artificially ambient pollution concentrations.

Table 1.3 shows the results of a selection of pollution exclusion experiments using ambient versus clean air growth chambers in British cities. The earliest experiments of this type were carried out by Bleasdale (1973) in a Manchester suburb 30 years ago,

TABLE 1.3. Effects of ambient air on shoot dry weight of grass species in chamber experiments at urban sites in the United Kingdom

Location	Year	Species	Mean SO_2 concentration (ppm)	Duration (days)	% reduction in ambient cf. clean air	$P<$	Reference
Manchester	1950/51	Lolium	0.066	104	17	0.05	Bleasdale
		perenne	0.051	59	18	0.01	(1973)
Sheffield	1973/74	Lolium	0.026	56	36	0.001	Crittenden
		perenne	0.022	131	20	0.01	and Read
			0.026	86	25	0.001	(1978)
			0.024	116	26	0.001	
		Lolium multiflorum	0.025	56	36	0.001	Crittenden and Read
		Dactylis glomerata	0.017	72	42	0.001	(1979)
Sheffield [a]	1976/77	Lolium	0.017	28	14	0.05	Awang
		perenne	0.011 [b]	28	25	0.05	(1979)
		Lolium	0.017	28	14	0.05	
		multiflorum	0.011 [b]	28	23	0.05	
		Dactylis glomerata	0.017	28	39	0.01	
St Helens	1977–80	Lolium	0.047	240	16%	0.05	Colvill et al.
		perenne	0.036	300	16% increase	NS	(1983)
			0.026	310	6% increase	NS	

[a] Total dry weight.
[b] Chambers subjected to artificial illumination.

before improvements had taken place in British urban air quality. In a series of experiments, reductions in the yield of perennial ryegrass (Lolium perenne) were demonstrated in the presence of ambient air containing mean SO_2 concentrations below 0.07 ppm, compared with plants grown in water-scrubbed clean air. It might be anticipated that such effects would now be absent or minimized under current lower concentrations of urban pollutants. However, recent similar experiments by Crittenden and Read (1978, 1979) in a suburb of Sheffield have clearly indicated a toxic effect of ambient air compared with charcoal filtered air, despite mean SO_2 concentrations lying between 0.015 and 0.030 ppm, with substantial reductions in dry weight taking place. This work has been criticized on the grounds of a lack of treatment replication (IERE, 1981), which could have resulted in environmental differences between the chambers confounding pollution effects. Subsequent investigations by Awang (1979) have used the same chambers, but with the treatments reversed and, in one experiment, similar levels of artificial light employed within the chambers, after covering these with aluminium foil. Again, ambient air containing mean SO_2 concentrations of 0.010–0.020 ppm caused significant growth reductions in several grass species after only 28 days' exposure (Table 1.3).

A major problem inherent in outdoor experiments using closed chambers is that climatic conditions within these are inevitably modified compared with the ambient, in particular temperatures being raised in summertime while light intensities are reduced. These climatic differences in many cases result in growth rates differing from those which would be expected if the plants were grown outdoors at the same site. In the experiments at Sheffield, growth was generally slow, even within the clean air chamber

TABLE 1.4. Effect of light regimen on SO$_2$-induced reductions in mean total dry weight (g) of timothy (*Phleum pratense*)

Light regimen	Dry wt in clean air	Dry wt in SO$_2$[a]	P<
Photosynthetically active radiation 125 μEm^{-2} s^{-1} 12 h day^{-1}	0.030	0.015	0.01
Photosynthetically active radiation 480 μEm^{-2} s^{-1} 16 h day^{-1}	0.725	0.704	NS

NS, Not significant at $P = 0.05$.
[a] 0.012 ppm SO$_2$ for 35 days.
(Davies, 1980).

under summer conditions. There is now good evidence that chronic SO$_2$ injury is greatly enhanced when plants are growing slowly (Cowling and Lockyer, 1978; Bell, Rutter and Relton, 1979; Davies, 1980; Jones and Mansfield, 1982). *Table 1.4* shows the results of an experiment reported by Davies (1980) in which timothy (*Phleum pratense*) seedlings were fumigated with 0.12 ppm SO$_2$ for 35 days under two different light regimens, representing summer (high) and winter (low) intensities, respectively. Under the high light intensity regimen rapid growth took place and no effects of SO$_2$ were observed. In contrast, the plants subjected to the low light intensity regimen grew very slowly and total dry weight was reduced by 50% in the presence of SO$_2$. The mean dry weight of timothy in clean air after 35 days under the low light regimen (0.030 g) is similar to that observed with plants of an approximately equivalent age in Awang's control treatment after 28 days (0.027 g for perennial ryegrass and 0.055 g for Italian ryegrass (*Lolium multiflorum*)), indicating the probability of slow growth enhancing SO$_2$ effects in the Sheffield experiments.

The problems resulting from modification of environmental conditions within closed chambers can be greatly reduced by the use of open-top designs. Colvill *et al.* (1983) have used such a system to determine the effects of the ambient air of St Helens, an industrial town in north-west England, on the growth of perennial ryegrass (*Table 1.3*). In a series of experiments of up to 310 days' duration, between 1977 and 1980, the only reduction in shoot dry weight (16%) was caused by air containing a mean of 0.047 ppm SO$_2$, while no significant effects were detected with mean concentrations of 0.026 and 0.036 ppm. The relatively small impact of these moderately high levels of pollutants may in part be explained by the filtration system in the open-top chamber removing only 50–70% of the ambient SO$_2$, compared with 98–100% efficiency in the other experiments, whose results are described in *Table 1.3*: thus the 'clean air' controls at St Helens were subjected to SO$_2$ concentrations within the range found in the ambient air chambers at Sheffield.

The pollution exclusion experiments described so far have all been carried out within urban areas. Similar investigations have been carried out in an agricultural area, Marston Vale in Bedfordshire, UK, which is subjected to air pollution from several brickworks (*Table 1.5*; Brough, Parry and Whittingham, 1978; Buckenham, Parry and Whittingham, 1982). Barley (*Hordeum vulgare*) grown in open-top chambers ventilated with ambient air showed reductions in grain dry weight, ranging between zero and 49%, compared with plants grown in chambers with a SO$_2$-filtration efficiency of 50–70%. The mean SO$_2$ concentrations recorded at Marston Vale are comparable with those in the Sheffield experiments (*Table 1.4*), as also are the maximum daily mean concentrations (up to 0.083–0.098 ppm). It must be emphasized, however, that the

TABLE 1.5. Effects of ambient air on mean grain dry weight of barley (*Hordeum vulgare*) in a polluted rural area (Marston Vale) in the United Kingdom

Year	Mean CO_2 concentration (ppm)	Maximum daily mean SO_2 concentration (ppm)	Duration (days)	Percentage reduction in dry wt in ambient cf. clean air	Reference
1976	0.023	—	115	49	Brough, Parry and Whittingham (1978)
1979	0.018	0.084	112	37	Buckenham, Parry and Whittingham (1982)
1980	0.020	0.068	146	NS	

NS, Not significant at $P = 0.05$.

results of the Marston Vale experiments cannot be used to predict crop losses over the large area of western Europe exposed to similar mean SO_2 concentrations in view of the presence of gaseous fluorides, also emitted by the brickworks. The difficulties in determining the impact of ambient air pollution in chamber experiments is again shown in this work: identical plots of barley grown outside the chambers generally showed substantially improved productivity compared with the ambient air chamber plants, again raising the possibility of the results being confounded by interactions with microclimate.

Another programme of rural pollution exclusion experiments has been carried out in the United Kingdom, at Ascot, Berkshire, where there are no local point sources of pollutants and SO_2 and NO_x concentrations are generally low. This work is aimed at determining the effect of ambient O_3 pollution on plants and experiments have been conducted in open-top chambers at this site during every summer since 1976. Two clear examples have occurred of visible O_3 injury (*Table 1.6*), in the form of foliar necroses, on the experimental plants at Ascot. In the first example, Ashmore *et al.* (1980) measured a greater amount of such injury on an O_3-sensitive cultivar of the pea (*Pisum sativum*) than on a tolerant cultivar, after a period in 1978 when the O_3 concentration exceeded 0.08 ppm for eight consecutive days, with a maximum hourly mean of 0.14 ppm. The second example occurred in 1981, when similar injury was observed on the leaves of white clover (*Trifolium repens*) and red clover (*T. pratense*) after an episode when the maximum hourly mean O_3 concentration was 0.13 ppm and exceeded 0.07 ppm on four consecutive days. In both the 1978 and 1981 experiments, injury was

TABLE 1.6. Examples of foliar necrosis caused by summertime ozone pollution on plants in open-top chamber experiments at a rural site (Ascot) in the United Kingdom

Year	Species	Maximum hourly mean concentration of O_3 (ppm)	Per cent foliar necrosis	Reference
1978	*Pisum sativum* cv. Kelvedon Wonder (O_3-sensitive)	0.14	1.06%	Ashmore *et al.* (1980)
	P. sativum cv. Onward (O_3-tolerant)		0.05%	
1981	*Trifolium repens* cv. Grassland Huia	0.13	2.81%	M. R. Ashmore (unpublished)
	T. pratense cv. Kuhn		1.14%	

either eliminated or greatly diminished in chambers ventilated with charcoal-filtered air.

An intensive study of the effects of low to moderate ambient air pollution concentrations has been carried out by Rabe and Kreeb (1979) in Stuttgart, Federal Republic of Germany. A range of ornamental, horticultural and agricultural species were grown in closed chambers ventilated with ambient or charcoal-filtered air for periods of four to five weeks, consecutively, from February to October 1976. The authors do not report growth data, but measurements were made of chlorophyll a and protein contents of the foliage (*Table 1.7*). Significant reductions in both these

TABLE 1.7. Physiological effects on plants of ambient air in Stuttgart, Federal Republic of Germany

Species	Growth period	Effects of ambient air cf. clean air		Mean SO_2 concentration (ppm)
		Chlorophyll a	Protein	
Barley (*Hordeum distichon*)	7.2–15.3.76	−7%	−12%	0.049
Tulip (*Tulipa gesnerana*)	21.4–17.5.76	−7%	−8%	0.019
Lucerne, alfalfa (*Medicago sativa*)		NS	−39%	
Broad bean (*Vicia faba*)	17.5–14.6.76	NS	−12%	0.011–0.015
Beet (*Beta vulgaris*)	14.6–12.7.76	−22%	−19%	—
Tomato (*Lycopersicon esculentum*)	9.8–6.9.76	NS	NS	—
Lucerne, alfalfa (*Medicago sativa*)	6.9–4.10.76	−12%	NS	0.015

(Rabe and Kreeb, 1979).
NS, Not significant at $P = 0.05$.

parameters were recorded for all species, except tomato (*Lycopersicon esculentum*), despite spring and summer mean SO_2 concentrations being less than 0.020 ppm. These physiological responses to ambient air pollution not only have implications for the nutritional quality of crops, but also could be expected to result in growth reductions.

The results of the pollution exclusion experiments described in this section suggest that the ambient air of much of rural western Europe contains sufficient concentrations of pollutants, on at least some occasions, to have the potential for substantial reductions in the yield and/or quality of many crop species. However, the importance of the individual pollutants in this respect remains uncertain. In most experiments, only SO_2 has been measured, but in nearly all cases other phytotoxic pollutants, particularly NO_x will also have been present. The contribution of short-term episodes of O_3 pollution during summertime must also be considered. Both Brough, Parry and Whittingham (1978) and Awang (1979) describe experiments in which large growth reductions occurred in the presence of mean SO_2 concentrations between 0.01 and 0.03 ppm during the summer of 1976: it should be noted that this period coincided with the highest O_3 levels ever recorded in the United Kingdom, which probably contributed to the observed effects, although no O_3 measurements were made in either case. Most activated charcoal filtration systems are highly efficient at removing SO_2,

NO_2 and O_3, but do not absorb appreciable quantities of NO. Thus in many pollution exclusion experiments, the 'clean air' chamber may be subjected to the ambient levels of NO, particularly in urban areas and near road-sides, which raises the possibility that, if this gas is producing an injurious effect on its own, the impact of filtration on plant growth will be reduced. A major priority for future research should be the development of differential pollution filtration systems, which can remove any individual phytotoxic gas from the ambient air, thus allowing a separation of the impact of the various pollutants.

Despite the consistent adverse effects of ambient air which have been shown in pollution exclusion experiments, most of these can be criticized on the grounds of possible pollutant/microclimate interactions, resulting from environmental modifications produced by the chambers themselves. Open-top chambers are an improvement in this respect, but invariably show only a partial exclusion of pollution due to penetration of ambient air through the top, particularly under turbulent meteorological conditions. Similar problems experienced in controlled fumigation experiments have stimulated the development of a range of designs of open-air fumigation systems. A parallel development for pollution exclusion experiments, in which the prevailing pollutant mix could be removed continuously by means of a source of clean air without the use of any type of enclosure would be a very welcome development towards understanding the real impact of current ambient levels of phytotoxic gases.

Fumigation experiments with pollutant mixtures

It is now apparent that there is little point in considering single pollutants only when attempting to assess the impact of ambient air pollution in many parts of western Europe. Indeed, as Wellburn (1982) has stated, a case can be made for generally considering the combined effects of SO_2 and NO_2 in this region, rather than treating them as independent phytotoxic agents. A substantial body of information has been built up in recent years on the effects on plants of SO_2/O_3 and SO_2/NO_2 combinations with synergistic, antagonistic and additive effects being demonstrated under different circumstances (Ormrod, 1982). Relatively few fumigation experiments have been performed, however, using mixtures of all three gases.

In view of the episodic occurrence of O_3 singly in otherwise clean rural areas or else superimposed on the continuous SO_2/NO_2 mixtures in the vicinity of conurbations and industrial districts, a priority for research in western Europe lies in designing fumigation experiments to simulate these situations. It is necessary not only to understand the effect of O_3 alone and of SO_2/NO_2 mixtures, but also the impact on plants of a combination of all three pollutants. The results of the very limited number of experiments which have determined the effects of adding O_3 to SO_2/NO_2 mixtures are shown in *Table 1.8*. In all cases an increased response is observed on the addition of O_3 to SO_2/NO_2. A clear synergistic interaction was demonstrated by Fujiwara (1973) on pea, with 18% of the leaf area destroyed in the presence of all three pollutants, in comparison with little or no effects in the O_3 and SO_2/NO_2 treatments. On the other hand, Reinert and Gray (1981) showed only additive effects, with respect to both foliar injury and hypocotyl dry weight reduction in a similar experiment with radish (*Raphanus sativus*). Recent work in The Netherlands (Mooi, personal communication) has demonstrated a remarkable impact on the poplar (*Populus × interamericana*) of adding 0.031 ppm O_3, a concentration within the natural background, to a mixture of

TABLE 1.8. Effects on plants of the addition of O₃ to SO₂/NO₂ mixtures

Species	Gas concentrations (ppm)			Duration	Parameter measured	Effect of pollutant			Reference
	SO_2	NO_2	O_3			$SO_2 + NO_2$	O_3	$SO_2 + NO_2 + O_3$	
Poplar (*Populus × interamericana*) cv. Donk	0.021	0.022	0.031	42 days	Nos. of fallen leaves	+430% cf. clean air	—	+770% cf. clean air	J. Mooi (personal communication)
Pea (*Pisum sativum*)	0.11	0.21	0.11	5 h	% foliar necrosis	0	2%	18%	Fujiwara (1973)
Radish (*Raphanus sativus*) cv. Cherry Belle	0.40	0.40	0.40	6 h	% foliar necrosis Hypocotyl dry wt[a]	11% −11% cf. clean air	45% −20% cf. clean air	57% −35% cf. clean air	Reinert and Gray (1981)

[a] Combined means of two experiments, using 0.2 ppm and 0.4 ppm of SO₂, NO₂ and O₃, respectively; dry weight measured seven days after fumigation.

low concentrations of SO_2 and NO_2 over a period of 42 days: this caused a 44% increase in the extra leaf fall over the control, but it is not known whether a significant interaction took place, because of the absence of a fumigation with O_3 alone.

These experiments demonstrate clearly the urgent requirement for a programme of carefully controlled fumigations using mixtures of SO_2, NO_x and O_3 designed to simulate conditions in western European rural, urban fringe and urban areas, respectively, in order to interpret the results of pollutant exclusion investigations.

Air pollution as a source of plant nutrition

In any assessment of the net impact of air pollution on vegetation in western Europe it is necessary to take into account possible beneficial effects. Both sulphur and nitrogen are essential constituents of proteins and it has been shown that SO_2 and NO_x can act as sources of these nutrients, particularly under conditions of deficiency in the soil, the gases being metabolized along normal pathways (Faller, Herwig and Kühn, 1970; Cowling, Jones and Lockyer, 1973; Anderson and Mansfield, 1979; Yoneyama and Sasakawa, 1979).

The importance of ambient NO_x in plant nutrition in western Europe is uncertain, but probably of little significance compared with agricultural inputs of nitrogen. However, there is good evidence for atmospheric sulphur compounds alleviating sulphur deficiency in the region. Over the last 30 years, substantial increases in the application of nitrogen fertilizers have been accompanied by a progressive reduction in sulphur additions to the land (Bache and Scott, 1979; Saalbach, 1974; Graziano and Rossi, 1978). This has largely been caused by the replacement of ammonium sulphate by high nitrogen content fertilizers, with the result that, in the United Kingdom sulphur inputs have fallen from about 80 kg ha^{-1} per annum to a current average of 20 kg ha^{-1} per annum, a figure which conceals much lower values in some areas (Bache and Scott, 1979). The predicted continuing increase in nitrogen applications has the potential to cause sulphur deficiency in crops, which appears as the nitrogen:sulphur ratio falls below critical values within the range 14–16:1 (Murphy, 1978).

It has been estimated that an average SO_2 concentration of 0.011 ppm, representative of much of rural Britain, will contribute about 38 kg ha^{-1} per annum of sulphur to wheat and grasslands, while the sulphur requirements of the most productive of these crops fall between 12 and 23 kg ha^{-1} per annum (Cowling and Koziol, 1982). Thus, the 23% of western Europe subjected to an annual mean of at least 0.011 ppm SO_2 (Fowler and Cape, 1982) will receive sufficient atmospheric inputs by dry deposition to remove any possibility of sulphur deficiency. Additional inputs of sulphate via wet deposition will further alleviate the problem. About 38% of western Europe is subjected to a total (wet + dry) input of atmospheric sulphur of greater than 15 kg ha^{-1} per annum (OECD, 1981; *Figure 1.5*) and, together with fertilizer additions and release of sulphur by soil mineralization, this would seem sufficient for most crop requirements.

There are, however, records of positive responses by crops to sulphur fertilizers in various parts of western Europe, particularly where atmospheric sulphur concentrations are low. In Ireland the annual total sulphur deposition from the atmosphere is only about 10–12 kg ha^{-1} and a major field experiment was carried out in 1978 to determine the extent of sulphur deficiency in Irish grasslands. Sulphur applications, in the form of gypsum, were made at 36 sites scattered throughout much of the central and southern parts of the country, resulting in yield increases of at least

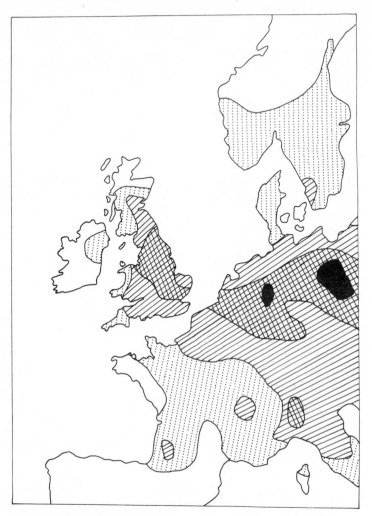

Figure 1.5. Total atmospheric sulphur deposition (kg ha^{-1} per annum) over western Europe: □ < 10; ▨ 10–25; ▧ 25–50; ▩ 50–100; ■ > 100. (CENE, 1981. Courtesy of HMSO, London)

10% at 12 sites, with a maximum of 45% over the growing season (Murphy, 1978, 1979). On this basis, it has been estimated that approximately 30% of the area of productive grassland in the Republic of Ireland may experience some degree of sulphur deficiency (*Figure 1.6*). Elsewhere in the British Isles, the current levels of atmospheric sulphur deposition appear to be sufficient in most places to compensate for any deficiencies in fertilizer applications, although they may be marginal in this respect over much of Scotland and in south-western areas of England and Wales (*Figure 1.1*; McLaren, 1975; Williams, 1975; Scott and Munro, 1979; Roberts, 1981). The extent of sulphur deficient areas in western Europe as a whole remains uncertain, but positive responses to sulphur fertilizers have also been demonstrated in either pot or field trials in south-west France (Delas *et al.*, 1970), the German Federal Republic (Saalbach, 1970, 1974), The Netherlands (Mes and Smilde, 1970), Italy (Casalicchio and Ciafardini, 1977), Norway, Sweden, Spain and Yugoslavia (Meyer, 1977).

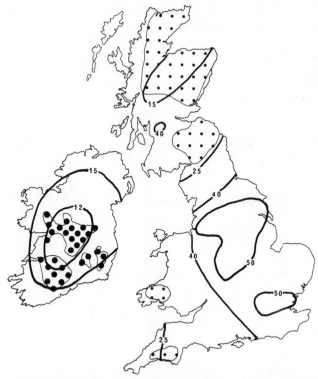

Figure 1.6. Areas of probable and possible sulphur deficiency in the British Isles. ▓ probable deficiency; ▒ possible deficiency; contours indicate total sulphur deposition in kg ha^{-1} per annum. (Modified from Murphy, 1979; Roberts, 1981)

At present the extent and importance of these nutritional effects of atmospheric sulphur compounds in western Europe are not fully understood, but any cost-benefit analysis of programmes to reduce SO_2 emissions must take them into account. However, as Bache and Scott (1979) have pointed out, the cost of adding adequate levels of sulphur to high nitrogen content fertilizers is relatively small, probably representing less than a 10% increase in price.

Future trends in air pollution in western Europe

Improved dispersion of air pollutants from point sources has almost certainly resulted in a reduced impact of gaseous SO_2 on crop growth in western Europe. Any such gains are likely to have been offset by increased emissions of NO_x and the appearance of O_3 as a pollutant, but it is impossible to quantify this effect, due to a lack of historical monitoring data. Predictions of future trends in air pollution injury to plants are difficult to make, in view of uncertainties concerning the level and nature of energy production and economic activity.

The availability and price of various fossil fuels must influence national energy policies, but these will differ between countries and may result in either positive or negative effects on air quality. It is anticipated that in western Europe there will be an overall increase in coal consumption of 37% over the next decade (Anonymous, 1982), but in most countries this will be accompanied by a decline in oil usage. The extent to which fossil fuels may be replaced by nuclear power is subject to political and social pressures which vary in intensity between the nations of western Europe, exacerbating the difficulties in estimating the future significance of gaseous air pollutants in the region. Furthermore, energy demand is related closely to economic activity and thus future air pollution characteristics will also depend to a large extent on the level of recovery from the current recession and the nature of the future European industrial base.

Table 1.9 shows some of the changes in SO_2 and NO_x emissions predicted for different points in time over the next 20 years in various parts of Europe. There is some

TABLE 1.9. Predicted changes in annual European air pollutant emissions

Pollutant	Source	Year	Emission ($\times 10^6$ tonnes)	Reference
Sulphur dioxide	EEC member states	1975	18.0	CONCAWE (1982)
		2000	12.74 (1% a.e.g.r.)[a]	
			13.80 (2% a.e.g.r.)[a]	
			17.80 (2% a.e.g.r. with shortfall in nuclear programme)[a]	
	OECD member states	1974	20.2	OECD (1981)
		1985	25.2	
	United Kingdom	1979	5.26	CENE (1981)
		2000	5.5 (2.7% a.g.g.d.p.)[b]	
			4.7 (2.0% a.g.g.d.p.)[b]	
Nitrogen oxides (as NO_2)	United Kingdom	1980	1.95	CENE (1981)
		2000	2.5 (2.7% a.g.g.d.p.)[b]	
			2.1 (2.0% a.g.g.d.p.)[b]	
		1980	1.95	Barrett *et al.* (1982)
		1990	2.01	
		2000	1.92	

[a] Annual energy growth rate.
[b] Annual growth in gross domestic product.

disagreement between the predictions of different organizations. The OECD (1981) suggested that in the absence of significant improvements in pollution control, there would be a 25% increase in SO_2 emissions by its European member states from 1974 to 1985. In contrast, a recent report by CONCAWE (1982) predicted that within the EEC, SO_2 emissions will fall from the years 1975 to 2000, even with a 2% rise in the annual energy growth rate and a shortfall in the anticipated nuclear programme. NO_x emissions are influenced considerably by the level of motor vehicle usage and so their trends will not necessarily parallel those of SO_2. Thus CENE (1981) have indicated a fall in SO_2 and a rise in NO_x emissions in the United Kingdom up to the year 2000, given a 2% annual growth in gross domestic product. The most recent estimates of NO_x emissions for the United Kingdom, however, show a rise to the year 1990, but a fall to below 1980 values by the end of the century (Barrett *et al.*, 1982).

It could be argued that the current recession in economic activity, particularly in the more polluting heavy industries, might lead to pressures for a relaxation of controls on emissions. Recent developments, however, indicate that such controls may become

TABLE 1.10. EEC limit and guide values (μg m^{-3}) for SO$_2$ and suspended particulates

	Reference period	SO$_2$	Associated value for suspended particulates
Limit values	Year	80 (0.03 ppm)	>40
(median of annual		120 (0.04 ppm)	⩽40
or winter daily	Winter (October–	130 (0.049 ppm)	>60
means)	March)	180 (0.068 ppm)	⩽60
Guide values	Year	40–60	40–60
(mean of daily		(0.015–0.023 ppm)	
means)	24 h	100–150	100–150
		(0.038–0.056 ppm)	

more stringent, accompanied by a greater degree of standardization between the different member states of the EEC. In 1980 the EEC issued a directive on air quality limit and guide values for SO$_2$ and suspended particulates, which prescribes mandatory maximum ground level concentrations of these pollutants (*Table 1.10*; Burhenne, 1981) which must be met throughout the Community by 1983, with an extension for a further ten years to allow the introduction of pollution abatement in difficult cases. The guide values are intended to serve as goals for the long-term protection of health and the environment. The ultimate effect of this legislation will be a reduction in SO$_2$ concentrations in most of the heavily polluted parts of western Europe. This should have either a beneficial or zero effect in other areas, in view of the specific requirement of the directive that its implementation must not lead to a deterioration of air quality anywhere else within the EEC.

Conclusions

Despite increasing interest by most western European countries in air pollution injury to vegetation, the net economic importance of gaseous air pollutants for crop growth in the region is still far from being understood. A recent estimate of 5×10^8 annual loss to agriculture within the 11 European OECD countries has been made on the basis of derived relationships between mean SO$_2$ concentration and the yield of perennial ryegrass, which was assumed to be the most susceptible crop of economic significance (OECD, 1981). However, there is now good evidence that there are other major crop species which are even more sensitive to SO$_2$ (Bell, 1982) and, moreover, the calculation fails to take into account the impact of the other major phytotoxic pollutants. The problems of making estimates of the net impact of air pollution on crop growth are further compounded by a lack of understanding of the beneficial effects, as well as the possibility of interactions with damage caused by insect pests and plant pathogens.

Only when the relative importance of the different pollutants is fully understood, will it be possible to develop realistic air pollution control policies which can be expected to minimize their impact on crop yield. Priority should be given to the establishment of a large-scale programme of field experiments designed to elucidate the effects of SO$_2$, NO, NO$_2$ and O$_3$ in their various mixtures prevailing at different locations in western Europe.

Acknowledgement

I thank Dr M. R. Ashmore for the compilation of the O_3 concentration data.

References

ANDERSON, L. S. and MANSFIELD, T. A. The effects of nitric oxide pollution on the growth of tomato, *Environmental Pollution* **20**, 113–121 (1979)

ANONYMOUS Concentration of some airborne pollutants at various sites in London, *Clean Air* **9**, 61 (1979)

ANONYMOUS European Energy Profile – Western Europe. Issue No. 20 (Ed. J. FEDLER), *Financial Times*, London (1982)

APLING, A. J., SULLIVAN, E. J., WILLIAMS, M. L., BALL, D. J., BERNARD, R. E., DERWENT, R. G., EGGLETON, A. E. J., HAMPTON, L. and WALLER, R. E. Ozone concentrations in south-east England during the summer of 1976, *Nature (London)* **269**, 569–573 (1977)

APLING, A. J., DORLING, T. A., ROGERS, F. S. M., WILLIAMS, M. L. and STEVENSON, K. J. *The Extended National Air Pollution Survey of Gaseous Pollutants. V. Data summaries and intersite comparisons*, Report No. LR374(AP), Warren Spring Laboratory, Stevenage (1981a)

APLING, A. J., DORLING, T. A., LILLEY, K. B., ROGERS, F. S. M. and STEVENSON, K. J. *Survey of Gaseous Air Pollutants at Selected UK Sites. I. Data Digest for Stevenage, Herts., 1978–1980*, Report No. LR397(AP), Warren Spring Laboratory, Stevenage (1981b)

APLING, A. J., DORLING, T. A., LILLEY, K. B., ROGERS, F. S. M. and STEVENSON, K. J. *Survey of Gaseous Air Pollutants at Selected UK Sites. II. Data for Harrow, Middlesex, 1979–1980*, Report No. LR398(AP), Warren Spring Laboratory, Stevenage (1981c)

APLING, A. J., DORLING, T. A., LILLEY, K. B., ROGERS, F. S. M. and STEVENSON, K. J. *Survey of Gaseous Air Pollutants at Selected UK Sites. III. Data Digest for Canvey Island, Essex, 1978–1980*, Report No. LR399(AP), Warren Spring Laboratory, Stevenage (1981d)

APLING, A. J., DORLING, T. A., LILLEY, K. B., ROGERS, F. S. M. and STEVENSON, K. J. *Survey of Gaseous Air Pollutants UK Sites. IV. Data Digest for Sibton, Suffolk, 1979–1980*, Report No. LR400(AP), Warren Spring Laboratory, Stevenage (1981e)

APLING, A. J., DORLING, T. A., LILLEY, K. B., ROGERS, F. S. M. and STEVENSON, K. J. *Survey of Gaseous Air Pollutants at Selected UK Sites. V. Data Digest for Vauxhall Bridge Road, Central London, 1978–1980*, Report No. LR401(AP), Warren Spring Laboratory, Stevenage (1981f)

APLING, A. J., ROGERS, F. S. M. and STEVENSON, K. J. *Ambient Concentrations of Nitric Oxide and Nitrogen Dioxide in the United Kingdom*, Report No. LR396(AP), Warren Spring Laboratory, Stevenage (1981)

ASHMORE, M. R., BELL, J. N. B., DALPRA, C. and RUNECKLES, V. C. Visible injury to crop species by ozone in the United Kingdom, *Environmental Pollution (Series A)* **21**, 209–215 (1980)

ASHMORE, M. R., BELL, J. N. B. and REILY, C. L. A survey of ozone levels in the British Isles using indicator plants, *Nature (London)* **276**, 813–815 (1978)

ASHMORE, M. R., BELL, J. N. B. and REILY, C. L. The distribution of phytotoxic ozone in the British Isles, *Environmental Pollution (Series B)* **1**, 195–216 (1980)

AWANG, M. B. *The Effects of Sulphur Dioxide Pollution on Plant Growth with Special Reference to Trifolium repens*, PhD Thesis, University of Sheffield, Sheffield, UK (1979)

BACHE, B. W. and SCOTT, N. M. Sulphur emissions in relation to sulphur in soils and crops. In *International Symposium on Sulphur Emissions in the Environment*, (Ed. F. F. ROSS), Society for Chemical Industry, London (1979)

BALL, D. J. and BERNARD, R. E. An analysis of photochemical pollution incidents in the Greater London area with particular reference to the summer of 1976, *Atmospheric Environment* **12**, 1391–1402 (1978)

BARRETT, C. F., FOWLER, D., IRWIN, J. G., KALLEND, A. S., MARTIN, A., SCRIVEN, R. A. and TUCK, A. F. *Acidity of Rainfall in the United Kingdom: a Preliminary Report*, Warren Spring Laboratory, Stevenage (1982)

BELL, J. N. B. Sulphur dioxide and the growth of grasses. In *Effects of Gaseous Air Pollution in Agriculture and Horticulture*, (Eds M. H. UNSWORTH and D. P. ORMROD), pp. 225–236, Butterworths, London (1982)

BELL, J. N. B. and MUDD, C. H. Sulphur dioxide resistance in plants: a case study of *Lolium perenne*. In *Effects of Air Pollutants on Plants*, (Ed. T. A. MANSFIELD), pp. 87–103, Cambridge University Press, Cambridge (1976)

BELL, J. N. B., RUTTER, A. J. and RELTON, J. Studies on the effects of low levels of sulphur dioxide on the growth of *Lolium perenne* L., *New Phytologist* **83**, 627–643 (1979)

BLEASDALE, J. K. A. Effect of coal-smoke pollution gases on the growth of ryegrass (*Lolium perenne* L.), *Environmental Pollution* **5**, 275–285 (1973)

BRIMBLECOMBE, P. Industrial air pollution in thirteenth century Britain, *Weather* **30**, 388–396 (1975)

BRIMBLECOMBE, P. Trends in the deposition of sulphate and total solids in London, *The Science of the Total Environment* **22**, 97–103 (1982)

BROUGH, A., PARRY, M. A. and WHITTINGHAM, C. P. The influence of aerial pollution on crop growth, *Chemistry and Industry* **21**, 51–53 (1978)

BUCKENHAM, A. H., PARRY, M. A. J. and WHITTINGHAM, C. P. Effects of aerial pollutants on the growth and yield of spring barley, *Annals of Applied Biology* **100**, 179–187 (1982)

BURHENNE, W. E. (Ed.). *Environmental Law of the European Communities*, Vol. 2. Erich Schmidt Verlag, Berlin (1981)

CASALICCHIO, G. and CIAFARDINI, G. Lo zolfo e l'agricoltura, *L'Italia Agricola* **11**, 81–106 (1977)

CENE *Coal and the Environment* (Commission on Energy and the Environment), Her Majesty's Stationery Office, London (1981)

COHEN, J. B. and RUSTON, A. G. *Smoke: A Study of Town Air*, Edward Arnold, London (1925)

COLVILL, K. E., BELL, R. M., ROBERTS, T. M. and BRADSHAW, A. D. The use of open-top chambers to study the effects of air pollutants, in particular sulphur dioxide, on the growth of ryegrass, *Lolium perenne* L. II. The long-term effect of filtering polluted urban air or adding SO₂ to rural air, *Environmental Pollution (Series A)* **31**, 35–55 (1983)

CONCAWE *SO₂ emission trends and control options in western Europe*, CONCAWE, Den Haag (1982)

COWLING, D. W., JONES, L. H. P. and LOCKYER, D. R. Increased yield through correction of sulphur deficiency in ryegrass exposed to sulphur dioxide, *Nature, London* **243**, 479–480 (1973)

COWLING, D. W. and KOZIOL, M. J. Mineral nutrition and plant response to air pollutants. In *Effects of Gaseous Air Pollution in Agriculture and Horticulture*, (Ed. M. H. UNSWORTH and D. P. ORMROD), pp. 349–375, Butterworths, London (1982)

COWLING, D. W. and LOCKYER, D. R. The effect of SO₂ on *Lolium perenne* L. grown at different levels of sulphur and nitrogen nutrition, *Journal of Experimental Botany* **29**, 257–265 (1978)

CRITTENDEN, P. D. and READ, D. J. The effects of air pollution on plant growth with special reference to sulphur dioxide. II. Growth studies with *Lolium perenne* L, *New Phytologist* **80**, 49–65 (1978)

CRITTENDEN, P. D. and READ, D. J. The effects of air pollution on plant growth with special reference to sulphur dioxide. III. Growth studies with *Lolium multiflorum* Lam. and *Dactylis glomerata* L., *New Phytologist* **83**, 645–651 (1979)

DAVIES, T. Grasses more sensitive to SO₂ pollution in conditions of low irradiance and short days, *Nature* **284**, 483–485 (1980)

DELAS, J., JUSTE, C., TAUZIN, J. and MENET, M. Sulphur deficiency in alfalfa (*Medicago sativa*) in soils of the Charente Area (South West of France). In *Sulphur in Agriculture*, pp. 51–67, An Foras Taluntais, Dublin (1970)

DERWENT, R. G., McINNES, G., STEWART, H. N. M. and WILLIAMS, M. L. *The Occurrence and Significance of Air Pollution by Photochemically Produced Oxidant in the British Isles, 1972–1975*, Report No. LR227(AP), Warren Spring Laboratory, Stevenage (1976)

EVELYN, J. *Fumifugium*, London (1661)

FALLER, N., HERWIG, K. and KÜHN, H. Die Aufnahme von Schwefeldioxyd (³⁵SO₂) aus der Luft: II. Aufnahme, Umbau und Verteilung in der Pflanze, *Plant and Soil* **33**, 283–295 (1970)

FARRAR, J. F., RELTON, J. and RUTTER, A. J. Sulphur dioxide and scarcity of *Pinus sylvestris* in the industrial Pennines, *Environmental Pollution* **14**, 63–68 (1977)

FOWLER, D. and CAPE, J. N. Air pollutants in agriculture and horticulture. In *Effects of Gaseous Air Pollution in Agriculture and Horticulture*, (Eds M. H. UNSWORTH and D. P. ORMROD), pp. 1–26, Butterworths, London (1982)

FUJIWARA, T. Effects of nitrogen oxides on plants, *Kogai To Taisaku* **9**, 253–257 (1973)

GRAZIANO, P. and ROSSI, N. Sulphur in fertilizers in Italy. In *Sulphur in Forages*, (Ed. J. C. BROGAN), pp. 64–74, An Foras Taluntais, Dublin (1978)

GUICHERIT, R. and VAN DOP, H. Photochemical production of ozone in western Europe (1971–1975) and its relation to meteorology, *Atmospheric Environment* **11**, 144–155 (1977)

IERE *Effects of SO₂ and its Derivatives on Health and Ecology*, Vol. 2, Natural Ecosystems, Agriculture and Forestry, International Electric Research Exchange (1981)

JONES, T. and MANSFIELD, T. A. The effect of SO₂ on growth and development of seedlings of *Phleum pratense* under different light and temperature environments, *Environmental Pollution (Series A)* **27**, 57–71 (1982)

KNABE, W. Kiefernwaldverbreiting und Schwefeldioxid- Immissionen in Ruhrgebiet, *Staub, Reinhaltung der Luft* **39**, 32–35 (1970)

LINES, R. Airborne pollutant damage to vegetation — observed damage. *International Symposium on Sulphur Emissions and the Environment*, (Ed. F. F. ROSS), Society for Chemical Industry, London (1979)

MARTIN, A. and BARBER, F. R. Sulphur dioxide, oxides of nitrogen and ozone measured continuously for two years at a rural site, *Atmospheric Environment* **15**, 567–577 (1981)

McLAREN, R. G. Marginal sulphur supplies for grassland herbage in south-east Scotland, *Journal of Agricultural Science, Cambridge* **85**, 571–573 (1975)

MES, A. E. R. and SMILDE, K. W. Sulphur deficiency in The Netherlands, *The Sulphur Institute Journal* **6**, 16–18 (1970)

MEYER, B. *Sulfur, Energy and Environment* Elsevier, Amsterdam (1977)

MOOI, J. Influence of ozone and sulphur dioxide on defoliation and growth of poplars, *Mitteilungen der Forstlichen Bundesversuchsanstait Wien* **137**, 47–51 (1981)

MURPHY, M. D. Responses to sulphur in Irish grassland. In *Sulphur in Forages*, (Ed. J. C. BROGAN), pp. 95–109, An Foras Taluntais, Dublin (1978)

MURPHY, M. D. Much Irish grassland is deficient in sulphur, *Farm and Food Research* **10**, 190–192 (1979)

NML *National Meetnet voor Luchtverontreiniging*, Verslag over de periode 1.10.76–1.10.77. Publication NML-RIV No. 9, Rijksinstituut voor de Volksgezondheid, Bilthoven (1978)

OECD *Programme on the Long Range Transport of Air Pollutants* 2nd edn, OECD, Paris (1979)

OECD *The Costs and Benefits of Sulphur Oxide Control*, OECD, Paris (1981)

OLIVER, F. W. On the effects of urban for upon cultivated plants, *Journal of the Royal Horticultural Society* **13**, 139–151 (1894)

ORMROD, D. P. Air pollutant interactions in mixtures. In *Effects of Gaseous Air Pollution in Agriculture and Horticulture*, (Eds. M. H. UNSWORTH and D. P. ORMROD), pp. 307–331, Butterworths, London (1981)

RABE, R. and KREEB, K. H. Enzyme activities and chlorophyll and protein content in plants as indicators of air pollution, *Environmental Pollution* **19**, 119–137 (1979)

REINERT, R. A. and GRAY, T. N. The response of radish to nitrogen dioxide, sulfur dioxide, and ozone, alone and in combination, *Journal of Environmental Quality* **10**, 240–243 (1981)

ROBERTS, T. M. The effects of stack emissions on agriculture and forestry, *CEGB Research* **12**, 11–24 (1981)

SAALBACH, E. Crop yields in Schleswig-Holstein boosted by sulphur, *The Sulphur Institute Journal* **6**, 14–15 (1970)

SAALBACH, E. The effect of sulphur, magnesium and sodium on yield and quality of agricultural crops, *Pontificia Academiae Scientiarum Scripta Varia* **38**, 541–589 (1974)

SCOTT, N. M. and MUNRO, J. The sulphate status of soils from north Scotland, *Journal of the Science of Food and Agriculture* **30**, 15–20 (1979)

STÖCKHARDT, J. H. Untersuchungen über die schädlichen Einwirkungen des Hütten — und Steinkohlenrauches auf das Wachstum der Pflanzen, insebesondere der Fichten und Tannen, *Tharandter Forstliches Jahrbuch* **21**, 218–254 (1871)

WELLBURN, A. R. Effects of SO_2 and NO_2 on metabolic function. In *Effects of Gaseous Air Pollution in Agriculture and Horticulture*, (Eds. M. H. UNSWORTH and D. P. ORMROD), pp. 169–187, Butterworths, London (1981)

WILLIAMS, C. The distribution of sulphur in the soils and herbage of north-east Pembrokeshire, *Journal of Agricultural Science, Cambridge* **84**, 445–452 (1975)

YONEYAMA, T. and SASAKAWA, H. Transformation of atmospheric NO_2 absorbed in spinach leaves, *Plant and Cell Physiology* **20**, 263–266 (1979)

Chapter 2

Air pollution problems in some central European countries — Czechoslovakia, The German Democratic Republic and Poland

S. Godzik

POLSKA AKADEMIA NAUK, INSTYTUT PODSTAW INŻYNIERII ŚRODOWISKA, ZABRZE, POLAND

Coal, lignite or bituminous, is the major source of energy for both power plants and domestic heating. The most common pollutant in all three countries is sulphur dioxide (SO_2) with an annual emission between 10 and 12 Tg. Other pollutants, like fluorides, organic compounds and heavy metals are of local importance, but can be the cause of serious problems to vegetation in the vicinity of those sources. Until now, the OECD report on the Long Range Transport of Air Pollutants (LRTAP) has been the most extensive study on the emission and dispersion of SO_2 on the continental scale. Applying sector analysis it was found that air masses coming from the direction of these three countries contained higher concentrations of SO_2 and greater amounts of sulphate in precipitation (OECD, 1979). Estimates for SO_2 emission made by Stehlik (1975) are in agreement with the OECD report. All the estimates of emission were based on data from the early 1970s.

The aim of this chapter is to present the latest data published concerning the emission of SO_2 and other gaseous pollutants in Czechoslovakia, the German Democratic Republic and Poland. The effects of these gaseous pollutants on plants, mainly forests, will also be presented.

Air pollutants and their sources

The main source of information on the emission of pollutants is the statistics from fuel consumption. This method has been used in the OECD study concerning SO_2 and NO_x emission; the data obtained for these three countries are summarized in *Table 2.1*. The emission for 1 km^2 has been calculated from these data to provide perhaps a better description of the real situation in these countries.

No substantial improvement in either the quantity or quality of the data available for emission has occurred since the OECD study was completed. Even though some progress has been made in obtaining estimates of emission, the information currently available is limited and rather fragmentary. Compared with the statistics of fossil fuel consumption in the early 1970s, there has been an increase in consumption based on the most recent figures available for 1979 and 1980 (*Table 2.2*). The regional distribution of emission has not changed significantly from that given in the OECD report (1979).

TABLE 2.1. Emission of SO$_2$ in Czechoslovakia, GDR and Poland

	Czechoslovakia	GDR	Poland
Total emission in 10^6 tonnes per annum:			
OECD	2.5	3.4	2.8
Stehlik	2.8	5.0	3.3
Emission calculated for 1 km^2 in tonnes per annum:			
OECD	19.5	31.4	8.9
Stehlik	21.9	46.3	10.5

(After OECD, 1979 and Stehlik, 1975).

TABLE 2.2. Consumption of fossil fuels in Czechoslovakia, GDR and Poland (in 10^6 tonnes per annum)

Country	Coal, bituminous		Coal, lignite		Crude oil products	
	1975	1979	1975	1979	1975	1979
Czechoslovakia	28.1	28.5	86.3	96.3	13.1	15.4
GDR	0.5	—	247.0	256.0	15.9	17.6 [a]
Poland	134.2	165.5	36.0	34.7 [b]	10.3	13.1 [b]

(After Rocznik Statystyczny, 1981 and GUS, 1981)
[a] Data for 1978.
[b] Data for 1980.

TABLE 2.3. Emission of gaseous pollutants in Poland in 1978 according to sources (in 10^3 tonnes per annum)

Pollutant	Total emission	Transportation	Power industry		Domestic heating, etc.
			Utility	Industrial	
SO$_2$	4313	1100	1927	606	680
NO$_x$	1228	435	587	188	18
Hydrocarbons	433	433	—	—	—

(Juda, Nowicki and Jaworski (1980b) as cited in GUS, 1981).

Sulphur dioxide

The annual emission of SO$_2$ is between 10 and 12 Tg (*Table 2.3*), with the largest portion of this, both as total emission and that calculated for each km^2, being emitted in the GDR. Whether Czechoslovakia or Poland is ranked second of the three depends upon the criteria used, that is, total emission or the emission per km^2. More than 60% of the SO^2 emission comes from coal-fired power plants, which represent the largest single source of pollutant. Further increases in the capacity of power plants combined with an increased consumption of coal will lead to a significant increase in the emission of SO$_2$. For Poland, it has been estimated that SO$_2$ emission will reach 7.3 Tg per annum by 1990 and 9.3 Tg per annum by the year 2000. Such significant increases will be caused both by burning bituminous coal of higher sulphur content than now used and by the construction of power plants for burning lignite (Juda, Nowicki and Jaworski, 1980a). No figures for the estimated increases in SO$_2$ emission have been published for Czechoslovakia or the GDR.

Data for the emission of SO_2 for the same country and year differ, depending upon the source of information. Thus for Poland, statistics for the annual emission of SO_2 in 1978 vary from 2.1 to 4.3 Tg (GUS, 1981; Juda, Nowicki and Jaworski, 1980b). A degree of uncertainty must be taken into account when assessing the figures for emission based on fuel consumption and emission factors. Further, as these data are given as averages for the whole country, the concentration of industry in certain areas modifies pollutant distribution within the country.

Nitrogen oxides

The amount of NO_x emitted comes second to SO_2. Estimates for Poland for 1978 are 1.1 Tg NO_x (*Table 2.3*). No data have been published for Czechoslovakia or the GDR. Using the information for fuel consumption (*Table 2.2*) and emission factors (Sittig, 1975; OECD, 1979) it can be estimated that NO_x emissions for Czechoslovakia were 0.9–1.2 Tg per annum and for the GDR 1.5–2.1 Tg per annum. Lack of more detailed information on combustion processes and the amounts of a given fuel used preclude giving definitive values for NO_x emission. These calculated quantities, however, should not be very far from actual emission since the power plants are the major sources of NO_x (*Table 2.3*).

Other pollutants

Two other pollutants, fluorine and ozone, are also potentially important. The quantities of fluorides emitted have been calculated in a similar way to NO_x. Aluminium smelters and the phosphorus fertilizer industry have been taken as the major sources of fluorides. Using emission factors of 2 kg F for each tonne of fertilizer produced and 3 kg F for each tonne of aluminium, the total figure for the emission of F for the three countries is approximately 4.3×10^3 tonnes per annum. The real figures, however, seem to be much higher. Estimates for Poland made in the late 1960s report F emission at 8.5×10^3 tonnes per annum (Paluch and Szalonek, 1970). The only figure for F emission in Czechoslovakia is 330 tonnes per annum, due to an aluminium smelter (Kontriśova, 1980). Coal, which contains small quantities of fluorine, may be an additional source of this pollutant. According to the data available, the emission and effects of F are of local importance.

Long-term measurements on O_3 have been carried out at five locations in the GDR. Annual mean values at the beginning of the monitoring period in 1954 ranged from 13–39 μg O_3 m^{-3}, depending on location. An average rate of increase of 1.3–3.1% was found for the period 1954–1971. At the same monitoring sites, the rates of increase for the period 1972–1977 were 2.5–7.0% with ambient concentrations of 34–61 μg O_3 m^{-3}. The highest 1-h value recorded was 209 μg O_3 m^{-3} (Warmbt, 1979). No other information on O_3 concentrations is available for these countries.

Ground-level concentrations and effects on vegetation

The emission and deposition rates of pollutants vary considerably within each of these countries (*Figures 2.1–2.4*). Only very few values have been published concerning the ground-level concentrations of pollutants on a regional scale (Hanibal and Raab, 1979). To compensate for this lack of data, an indirect method of assessment is proposed based on the injuries sustained by plants, especially forest trees. The area of

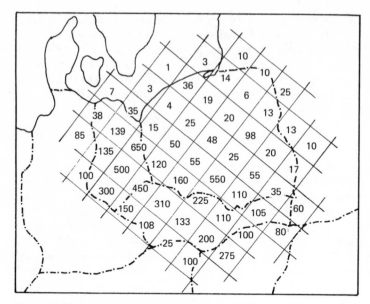

Figure 2.1. Estimated annual emissions of SO_2 (in 10^3 tonnes S) in grid elements with a length of 127 km at 60 °N, based on data from 1973 (OECD, 1979)

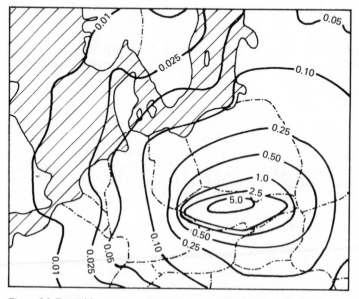

Figure 2.2. Deposition patterns for 1974 due to SO_2 emissions in Czechoslovakia; dry + wet deposition, in g S m^{-2} (OECD, 1979)

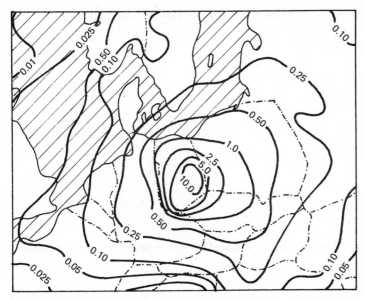

Figure 2.3. Deposition patterns for 1974 due to SO$_2$ emissions in the GDR; dry + wet deposition, in g S m^{-2} (OECD, 1979)

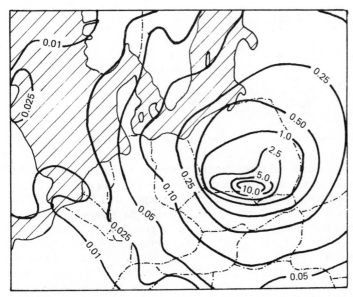

Figure 2.4. Deposition patterns for 1974 due to SO$_2$ emissions in Poland; dry + wet deposition, in g S m^{-2} (OECD, 1979)

TABLE 2.4. Development of injury of forest stands in the North Bohemian region

Year of evaluation	Degree of injury as percent of the total area investigated[a]					Total forest area (ha)
	0	1	2	3	4	
1960	85.3	10.4	3.6	0.7	0.0[b]	176 634
1970	71.3	20.0	7.2	2.2	0.3	173 972
1975	68.4	18.0	10.3	2.5	0.7	171 630
1979	7.9	47.3	23.6	14.0	7.3	167 016

(After Jonaś and Jonaś, 1982)
[a] 0 = no injury
1 = slight injury
2 = medium injury
3 = severe injury
4 = very severe injury.
[b] actual value = 0.03%.

TABLE 2.5. Forest areas injured by air pollutants in Poland in 1980 according to Forest Administration Districts[a] (in ha)

Forest administration district	Location on map (Figure 2.6)	Degree of injury			Total area injured	Area injured as percent of total forest area
		I slight	II medium	III severe		
Białystok	1	—	—	—	—	—
Katowice	2	66 033	94 272	24 695	185 000	31.5
Kraków	3	35 298	705	—	36 003	10.5
Krosno	4	531	317	—	848	0.2
Lublin	5	56 825	4752	522	62 099	13.4
Łódź	6	6839	1670	49	8558	2.3
Olsztyn	7	—	—	—	—	—
Piła	8	—	—	—	—	—
Poznań	9	8208	254	100	8562	2.3
Szczecin	10	735	103	37	875	0.1
Szczecinek	11	—	—	—	—	—
Toruń	12	33 443	6844	68	40 355	6.7
Wrocław	13	16 567	768	768	18 103	3.8
Zielona Góra	14	22 738	2272	—	25 010	6.1
Total		247 217	111 957	26 239	385 413	5.7

(GUS, 1981).
[a] State forests only; an additional 10[5] ha of non-state owned forest was affected by air pollutants.

forests injured by air pollutants in all three countries exceeds 8.94×10^5 ha, with the contribution from each country being: Czechoslovakia, 1.54×10^5 ha and additional areas in the Jizerske Mountains (*Table 2.4* and Vinś, Pospiśil and Kućera, 1982); Poland, 4.8×10^5 ha (*Table 2.5*); and the GDR, 2.6×10^5 ha (Dässler, personal communication). No information on the severity and regional distribution of the forests injured has been published for the GDR. Detailed information is available for the Ore Mountains region, which represents the most severely injured region of Czechoslovakia (*Table 2.4*), and this area is conterminous with the GDR. Other areas sustaining injury, but with no estimate available, are in the Jizerske Mountains, conterminous with Poland (Vinś, Pospiśil and Kućera, 1982). According to Vyskot (1982), one-third of the total forest area of the Czech Republic is endangered by air pollutants; the total area of forests for this republic is 2.6×10^6 ha (Czechoslovak Forestry, 1972).

Both in Czechoslovakia and Poland the areas of highest SO_2 emission and rates of deposition of sulphur compounds (*Figures 2.1–2.4*) coincide with the areas of forests injured (*Tables 2.4* and *2.5, Figure 2.6*). Analysing the development of forest injury in Czechoslovakia from 1960, a dramatic increase in the area of forests injured occurred between 1975 and 1979 (*Table 2.4, Figure 2.5*). The data for Poland are similar. In 1966 the area of forest injured was estimated at about 1.76×10^5 ha, increasing to 2.68×10^5 ha in 1973 and reaching approximately 3.00×10^5 ha in 1977 and

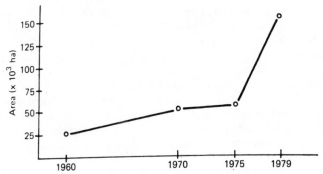

Figure 2.5. Injury forest stands in the Ore Mountains, Czechoslovakia (data from Jonaś and Jonaś, 1982)

Figure 2.6. Location of Forest Administration Districts and of the cities listed in *Table 2.5* that serve as the District Administration Centres (solid circles)

3.85×10^5 ha in 1980. These estimates apply to state forests only; an additional 10^5 ha of forests not owned by the state were injured in 1980 (Wolak, 1980; GUS, 1981). Because of different criteria used to evaluate and classify the severity of injury to forests, it is impossible to analyse more extensively the data for Czechoslovakia and Poland. Some speculation concerning ground-level concentrations of SO_2 can be made on the basis of a study conducted in the early 1960s in the Ore Mountains that correlated measured SO_2 concentrations and the degree of injury observed in the forests (Materna, 1973). During the period of the study, separate mean concentrations of SO_2 were calculated for the summer and winter seasons; ambient concentrations were found to be higher in the winter than in the summer. Although differences were found between monitoring sites, very severe injuries, to the point where entire stands of Norway spruce (*Picea excelsa*) were dying, were found when ambient SO_2 concentrations were in the range of 60–110 μg m^{-3}. Injury ranging from slight to severe was sustained when SO_2 concentrations were 30–40 μg m^{-3} (Materna, 1973). Taking into consideration the importance of site conditions and the modified response of trees to SO_2 on more favourable sites, a concentration factor of 2 would be justified (IUFRO, 1979); i.e. under more favourable edaphic and environmental conditions, it would take twice the ambient concentrations measured in this region to produce the same degree of injury to the stands of spurce. As forests are not uniformly distributed over the regions, the total area receiving elevated concentrations of SO_2 may be several times larger than the area of forests injured. There are no reports on injuries to other plant species on such a large scale other than those reviewed recently by Godzik and Krupa (1982).

Compared with the injuries caused by SO_2, those due to other pollutants are of relatively minor importance. On a local scale, however, the injuries caused by fluoride are serious and the concentrations of F in plant material very high (Paluch and Szalonek, 1970; Börtitz and Reuter, 1977). Although several thousands of hectares of pine forest were killed around a nitrogen fertilizer factory (Gadzikowski, 1980), the cause has been attributed to the emission of fine particles of fertilizer rather than to NO_x.

Final remarks

Because nearly all the power plants burn coal (lignite or bituminous) the most common air pollutant is SO_2 (*Table 2.6*) with these power plants representing the largest single source of this pollutant. Construction of high stacks, commonly accepted, has led to better dispersion of SO_2 and a decrease in the maximum ground level concentrations of

TABLE 2.6. Total emission of SO_2 and NO_x in Czechoslovakia, GDR and Poland (in Tg per annum)

Country	SO_2	NO_x
Czechoslovakia	3.0	0.9–1.2 [a]
GDR	4.0–5.0	1.5–2.1 [a]
Poland	2.1–4.3	1.2

(After GUS, 1981; Hanibal and Raab, 1979; Juda, Nowicki and Jaworski, 1980; Möller, 1982).
[a] Emissions calculated from data for fuel consumption (*Table 2.2*) and emission factors (Sittig, 1975; OECD, 1979).

SO_2. The relatively small distance between the industrial regions of these countries does, however, lead to an overlap of pollutant emission in some areas. The latest published data indicate a permanent trend of increasing consumption of fossil fuels with a corresponding increase in the total emission of pollutants. Thus the amount of SO_2 emitted in these countries is currently estimated at 12 Tg per annum, while emission of NO_x is estimated at 3.6–4.5 Tg per annum (*Table 2.6*).

The injuries to forests from these pollutants, primarily from SO_2, are a serious problem with an area greater than 8.9×10^5 ha being affected, and much larger areas than those reported injured can be potentially affected. A decrease in SO_2 emission is badly needed to save further areas from injury.

Although there are no data for pollutant concentrations on a large scale, the degree of injury sustained by forests can be used as indirect evidence of the occurrence of high ambient concentrations in various regions. In Czechoslovakia and Poland, the areas of highest emission of SO_2 coincide with the areas of forests most severely injured. Other pollutants such as hydrocarbons (from coke plants) and fluorides are of local importance. The significance of NO_x and O_3 cannot be evaluated because of the lack of data.

References

BÖRTITZ, S. and REUTER, F. Untersuchungen über den Fluorgehalt von Blüten in Gebieten mit binengefährdenden Immissionen, *Archiv für Gartenbau* **25**, 247–255 (1977)

Czechoslovak Forestry State Agricultural Publishing House, Prague (1972)

GADZIKOWSKI, R. Oddziaływanie zakładów azotowych w Puławach na środowisko leśne w latach 1967–1978. [Effect of a nitrogen fertilizer factory in Puławach on the forest environment for the period of 1967–1978], *Sylwan* **124**, 17–29 (1980)

Główny Urzad Statystyczny (GUS) Ochrona środowiska i gospodarka wodna. Materiały statystyczne 3. Warzawa. [The Main Statistical Office, 1981, Environment Protection and Water Management, Statistical paper 3, Warsaw] (1981)

GODZIK, S. and KRUPA, S. Effects of sulphur dioxide on the growth and yield of agricultural and horticultural crops. In *Effects of Gaseous Air Pollution in Agriculture and Horticulture*, (Eds. M. H. UNSWORTH and D. P. ORMROD), pp. 247–265, Butterworths, London (1982)

hospodarstvi CSR. Praha [Air Pollution and the Current Problems. Ministry of Agriculture and Water Management, Prague (in Czechoslovakian)] (1979)

IUFRO Resolution on air quality standards for protection of forests, *IUFRO News* **25**, 3 (1979)

JONAS, F. and JONAŚ, F., Jr. [Reclaim of forests soils intoxicated by sulphur compounds in the Ore Mountains (in Czechoslovakian)], *Lesnictvi* **28**, 103–122 (1982)

JUDA, J., NOWICKI, M. and JAWORSKI, W. Stan zanieczyszczenia atmosfery związkami gazowymi na obszarze Polski – stan obecny i perspektywy do roku 2000. [The current status of air pollution by gaseous compounds in Poland – current situation and the perspective for the year 2000.] In *The Influence of the Pollution of Atmosphere on Environment*, pp. 9–15, Polish Scientific and Technical Committee on Management and Protection of Environment, Warsaw (1980a)

JUDA, J., NOWICKI, M. and JAWORSKI, W. Ekspertyza dotycząca emisji zanieczyszczeń powietrza w Polsce w roku 1978. [Expert opinion on air pollution in Poland in the year 1978] as cited in Główny Urząd Statystcyzny (1981)(1980b)

KONTRISOVA, O. [Meadow associations in an area of action of fluorine type emission: Ziar Basin (in Czechoslovakian)]. *Biologicke Prace, Slovenska Akademia Ved.* [Treatises on Biology, Slovak Academy of Sciences] 26/2, 1–159 (1980)

MATERNA, J. Kriterien zur Kennzeichnung einger Immission-einwirkung auf Waldbestände. In *Proceedings of the 3rd International Clean Air Congress*, Düsseldorf, pp. A121–A123 (1973)

MÖLLER, D. Das Verhaltnis von anthropogener zu natürlicher globaler Schwefelemmission, *Zeitschrift für die gesamte Hygiene* **28**, 11–15 (1982)

OECD *The OECD Programme on Long-Range Transport of Air Pollutants*, 2nd edn, Paris (1979)

PALUCH, J. and SZALONEK, I. Die Luftverunreinigung durch Fluor in Polen und ihre toxische Wirkung auf Menschen, Tiere und Pflanzen. Wissenschaftliche Zeitschrift der Humbolt-Universität Berlin, *Mathematisch-naturwissenschaftliche Reihe* **19**, 489–492 (1970)

Rocznik Statystyczny za rok 1981. Główny Urząd Statystyczny. Warzawa. [*Statistical Yearbook for the Year 1981.* The Main Statistical Office, Warsaw.]

SITTIG, M. *Environmental Sources and Emissions Handbook*, Noyes Data Corporation, Park Ridge, New Jersey (USA) (1975)

STEHLIK As cited in Hanibal and Raab (1979)(1975)

VINŠ, B., POSPIŠIL, F. and KUČERA, J. [Evaluation of the development of emission damage in the protected landscape area of the Jizerske Mountains (in Czechoslovakian)], *Lesnictvi* **28**, 87–102 (1982)

VYSKOT, M. Skody na lesich a zivotnim prostredi. [Damage in forests and other living systems (in Czechoslovakian)], *Lesnictvi* **28**, 85–86 (1982)

WARMBT, W. Ergebnisse langjähriger Messungen des bodennahen Ozon in der DDR, *Zeitschrift für Meteorologie* **29**, 24–31 (1979)

WOLAK, J. Wpływ przemysłowych zanieczyszczeń atmosfery na ekosystemy leśne. [Effects of air pollutants on forest ecosystems.] In *The Influence of the Pollution of Atmosphere on Environment*, pp. 213–224, Polish Scientific and Technical Committee on Management and Protection of Environment, Warsaw (1980)

Chapter 3

Defining gaseous pollution problems in North America*

W. W. Heck

AGRICULTURAL RESEARCH SERVICE, US DEPARTMENT OF AGRICULTURE AND
BOTANY DEPARTMENT, NORTH CAROLINA STATE UNIVERSITY, USA

Introduction

The classic approach to a discussion of pollution problems is to focus on sources and emissions. This approach ignores the linkages between pollution sources and plant response, yet such linkages form the basic gaps in defining the extent of the air pollution problem in relation to agricultural production. These concepts were developed by me (Heck, 1982) in my discussion of the air pollution system.

Historically, gaseous pollution problems were associated with the release of pollutants from point sources and the effects of these pollutants on vegetation growing in the vicinity of these sources. Such problems still occur and are the primary reason for our continuing concern for other gaseous emissions (e.g. HF, H_2S, NH_3, HCl, Cl_2) from point sources. Specific information for these pollutants is not included in this chapter because they are considered local problems. Review articles that include discussions of these other pollutants are numerous (Heck and Brandt, 1977; Jacobson and Hill, 1970; Guderian, 1977; Lacasse and Treshow, 1976). The most important of the group, HF, may be present in power plant emissions and thus may be a potential source of concern.

Currently, the three gaseous pollutants of primary concern with respect to effects on vegetation in North America are O_3, SO_2 and NO_2. They are ubiquitous across the continent as a result of mobile source emissions (NO_2 and hydrocarbons) and the increase in the number of power plants with high stacks (SO_2 and NO_2). Ozone results from secondary atmospheric reactions associated with NO_2, hydrocarbons and sunlight. The SO_2 and NO_2 also serve as precursors for acidic deposition. Concentrations of O_3 and NO_2 may show only slight variation when averaged over the growing season for large regions of both the USA and Canada. In recent years, O_3 has been the most intensely studied of these three gases (Heck, Mudd and Miller, 1977; US Environmental Protection Agency, 1978a). Sulfur dioxide was the pollutant of primary concern before 1950 (Heck and Brandt, 1977; Thomas, 1951), but interest in this pollutant has been renewed since 1970 (National Academy of Sciences, 1978). Nitrogen dioxide has not received as much attention since plants are not sensitive to NO_2 at most ambient concentrations (National Academy of Sciences, 1977; US Environmental Protection Agency, 1982a).

* Cooperative investigations of the US Department of Agriculture and the North Carolina State University, Raleigh, NC.

For those interested in various aspects of pollution effects on plants, there are several additional books that should be reviewed (Unsworth and Ormrod, 1982; Mansfield, 1976; Treshow, 1970; Mudd and Kozlowski, 1975). Information on sources, emissions and trends are found in several documents from the US Environmental Protection Agency (1976, 1978a, 1978b, 1980, 1982a, 1982b).

An understanding of the air pollution system (Heck, 1982) permits one to visualize how the sources, emissions, transport, transformations, deposition and monitoring fit into the system and affect plants (*Figure 3.1*). Major factors relevant to the biologist are

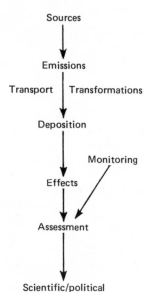

Figure 3.1. The air pollution system from source through effects assessment is shown as a primary component for scientific/political understanding and decision making. This portion of the system feeds into economics, criteria and standards development, and control policies

given in *Table 3.1*. The components of the air pollution system (*Figure 3.1*) are developed to some extent in this chapter. Principal emphasis is given to: the natural background for each pollutant, monitoring principles and concerns, the research protocol needed to study the effects of pollutants on plant metabolism, and the assessment of effects on a regional basis.

The air pollution system

This section addresses aspects of the physical parts of the system that must be understood before a regional or national assessment of the impact of gaseous pollutants on crops can be made. The discussion focuses on major sources and their emissions, transport and transformation of these emissions, and on the monitoring and deposition processes associated with O_3, SO_2 and NO_2. This section concludes with a discussion of plant uptake, which is a linkage process between the physical and biological aspects of the air pollution system.

TABLE 3.1. Major factors to be determined in an assessment of the effects of gaseous pollutants [a]

A. *Sources*
   (1)   Natural versus anthropogenic.
   (2)   Significant sources (coal, oil, natural gas, non-fossil fuels, natural processes).
   (3)   Nature of source (stationary, mobile).

B. *Emissions*
   (1)   Nature of primary gaseous emissions.
   (2)   Relative importance of natural versus anthropogenic emissions.

C. *Atmospheric transport*
   (1)   Modes of gas movement in the atmosphere.
   (2)   Nature of transport (local, regional, global).
   (3)   Residence times (up to days, weeks, months, longer).

D. *Atmospheric transformations*
   (1)   Physical and chemical transformation (type, degree).
   (2)   Participation of natural sources in these transformations.

E. *Monitoring*
   (1)   Rural versus urban.
   (2)   Instrumentation.
   (3)   Biologically important ways to average air-quality data.

F. *Deposition*
   (1)   Gaseous absorption and adsorption.

G. *Biological consequences for plants in terrestrial habitats*
   (1)   Toxicity (low to high, acute to chronic, affecting few to many plants).
   (2)   Likely to accumulate in organisms.
   (3)   Detoxification and/or repair capability (low to high).
   (4)   Type of effect (on health, development, growth, reproduction, phenology, behavior, heredity).

[a] From the National Academy of Sciences (1981) with major modifications.

Sources and emissions

Although we are primarily interested in anthropogenic sources and emissions from such sources, it is imperative that we understand natural emissions and how these relate to anthropogenic emissions. This section highlights the importance of natural and anthropogenic sources, and discusses their implications for plant research.

NATURAL SOURCES/EMISSIONS

All three of the pollutants of primary concern (O_3, SO_2, NO_2) are found naturally, and these background concentrations must be considered when assessing the impact of these gases on crop productivity.

Ozone is thought to come from two principal natural sources (US Environmental Protection Agency, 1978a). First, and most important, is stratospheric intrusion. This phenomenon has been studied, but its overall contribution to background O_3 concentrations has not been adequately assessed. The importance of stratospheric intrusion, in terms of natural background, may be related to storm events, altitude, season of the year, latitude, and other less well understood factors that could make it impossible to determine this natural input of O_3 at any given location or at any given time. An averge range of 0.007–0.020 ppm stratospheric input to tropospheric O_3 concentrations, with hourly maxima of twice these values, was suggested as reasonable by the US Environmental Protection Agency (1978a) for the growing season. It is important that research be conducted to assess the importance of the stratospheric intrusion of O_3. Second, available evidence suggests that biogenic hydrocarbons can participate in the photochemical reaction sequence to produce natural background O_3 (US Environmental Protection Agency, 1978a). These reactions generally cannot be separated from reactions associated with anthropogenic hydrocarbons. Thus, separation must depend on determining emissions and rates of reaction for the biogenic hydrocarbons. There is speculation and some evidence that the terpenes in rural areas may scavenge more O_3 than they generate. Current projections suggest that biogenic sources probably account for less than 10% of the monitored O_3 concentrations.

Nitrogen oxides (NO_x) are produced from all natural combustion sources as well as from biological activity. These sources are responsible for small amounts of NO_2 in the atmosphere with ambient concentration estimates ranging from 0.002–0.005 ppm. The importance of these natural sources is not known, but the consensus is that NO_2 from natural sources does not play a large role in the photochemical reaction sequence of O_3 production (US Environmental Protection Agency, 1982a). These natural background concentrations, no doubt, serve as one source of nutrient nitrogen for plant growth.

Sulfur dioxide is produced from a number of natural sources (National Academy of Sciences, 1978), such as volcanos and fumaroles. Sulfur dioxide is also produced by the oxidation of H_2S and other reduced sulfur compounds that are released into the atmosphere. Thus, some background SO_2 is present, but it has a relatively short half-life in the atmosphere and may be below the detection limit of current analyzers.

The amount that these natural sources contribute to the background concentrations of O_3, SO_2 and NO_2 must be taken into account in the interpretation of data.

ANTHROPOGENIC SOURCES/EMISSIONS

Two primary source classes of O_3, SO_2 and NO_2 (as NO_x) are mobile sources (transportation) and stationary sources (primarily power plants, although smelters are still a major source of SO_2). Any combustion process will add to the amount of hydrocarbon (precursors for O_3 formation). NO_2 and, provided there is a S-containing fuel, SO_2. Ozone, associated with hydrocarbon and NO_2 precursors, is considered a secondary pollutant.

Source and emission data for hydrocarbons and NO_x are shown in *Table 3.2*. Source categories are shown only for 1974, but emission trends cover several years for hydrocarbons (1970–1975), NO_x (1940–1975) and SO_2 (1940–1978). The data show no trends for the hydrocarbons, a major increase in NO_x (primarily in the 1960s) and an increase in SO_2 during the 1960s with a slight trend of reduced emissions since 1970. Trends in SO_2 and NO_x emissions in Canada are similar to those in the USA, but emissions are lower (*Table 3.3*). Current and projected emissions to the year 2000 (by decade) are shown for both Canada and the USA (*Table 3.4*). Data suggest a slight upward trend in emissions from both countries and for both SO_2 and NO_x through the year 2000. These projections are made on the assumption of continued improvement in controls for gaseous pollutants. Thus, the biological scientist can expect no reduction in emissions of SO_2 or NO_x through to the end of this century.

TRANSPORT AND TRANSFORMATION

Plant scientists should understand transport and transformation processes so they can interpret monitoring data in terms of emissions and sources. Problems associated with O_3, SO_2 and NO_2 are not uniformly spaced over the country although O_3 and NO_2 are

TABLE 3.2. Estimate of US total hydrocarbon, NO_x and SO_2 sources and emissions, and emission trends for various time periods[a]

	Emissions (10^6 tonnes per annum)		
	Hydrocarbons (1974)	NO_x (1974)	SO_2 (1978)
Source category			
Transportation	11.3	9.6	0.8
Stationary fuel combustion	1.6	12.1	22.1
Industrial processes	3.3	0.6	4.1
Solid waste	0.9	0.2	0.0
Miscellaneous	12.7	0.2	0.0
Total	29.8	22.7	27.0
Emission trends			
1940		7.1	19.5
1950		9.4	22.0
1960		12.7	21.4
1970	30.7	20.4	29.8
1972	30.9	22.2	
1974	29.8	22.6	
1975	28.1	22.2	26.2
1978			27.0

[a] From US Environmental Protection Agency (1978b, 1980, 1982a); blank spaces indicate that information is lacking.

TABLE 3.3. Estimates of Canadian NO_x and SO_2 sources and emissions[a]

Source category	Emissions (10^6 tonnes per annum)					
	SO_2			NO_x		
	1955	1965	1976	1955	1965	1976
Cu–Ni smelter	2.89	3.90	2.60	—	—	—
Power plants	0.06	0.26	0.61	0.01	0.06	0.21
Other combustion	1.21	1.13	0.88	0.23	0.25	0.05
Transportation	0.08	0.05	0.08	0.32	0.51	1.02
Iron ore processing	0.11	0.16	0.18	—	—	—
Miscellaneous	0.19	1.10	0.95	0.07	0.03	0.19
Total	4.54	6.60	5.30	0.63	0.85	1.47

[a] Linthurst and Altshuller (1983), Chapter Two.

TABLE 3.4. US and Canadian current and projected SO_2 and NO_x emissions[a]

Country	Emissions (10^6 tonnes per annum)					
	SO_2			NO_x		
	1980	1990	2000	1980	1990	2000
USA	24.1	22.8	26.6	19.0	19.5	24.1
Canada	4.2	4.3	4.0	1.7	1.9	2.4
Total	28.3	27.1	30.6	20.7	21.4	26.5

[a] Linthurst and Altshuller (1983), Chapter Two.

more widely spread than SO_2. Concentrations of these pollutants occurring in any location are associated with both emission and transport phenomena.

TRANSPORT

Fifteen years ago air pollution was considered part of urban blight or of large industrial complexes. Studies since 1965 have changed our perception of the problem; air pollution is now recognized as a regional problem with rural as well as urban implications.

It was shown in the late 1960s that O_3, at phytotoxic levels, was present throughout south-western Ohio, not just in urban areas (Heck and Heagle, 1970). Those studies, subsequent monitoring in rural areas, and meteorological studies have elucidated the pervasive nature of the oxidant blanket that cloaks the eastern half of the continent. Long-distance transport of polluted air masses is becoming better understood. Although preliminary back-trajectory studies of air masses rich in O_3, at a particular destination (Heagle, Riordan and Heck, 1979), could show that the air masses had passed over large urban areas during the previous 72 h, additional studies are needed. It was long assumed that O_3 was too reactive to persist in air masses after dark, but a residence time of 0.4 to 90 days is shown in the National Academy of Sciences (1981) report. Measurements in remote areas occasionally show higher values at night than during the day. Most data suggests that anthropogenically-formed O_3 covers the eastern portion of the continent during a large portion of the growing season.

With the increase in high stacks and large megawatt coal-burning power plants, SO_2 is becoming more prevalent in air masses that cover large areas. The north-eastern USA is continuously exposed to low levels of SO_2, much of which originates in the Ohio Valley. Nitrogen oxides and hydrocarbons, from both mobile and stationary sources, are part of these air masses, and thus are widespread over part of eastern North America. As such, they continue to play a role in O_3 formation during long-distance transport.

We do not know how important these pollutants are in the western half of the continent, except in the vicinity of large urban areas or where multiple sources of SO_2 occur. Many of the western power plants are fairly well isolated, burn low-S coal and probably do not pose a major problem.

TRANSFORMATION

Transformation of primary pollutants is initiated as soon as they are emitted. Most NO_x is emitted as NO, but the NO is oxidized to NO_2 in the presence of O_2. Emitted hydrocarbons react with NO_2 in the presence of sunlight, to form O_3. The O_3 builds up after all (or most) NO is oxidized to NO_2. The formation of O_3 is temperature dependent. In urban areas, NO is depleted by mid-morning followed by an increase in O_3 during the late morning and early afternoon. The O_3 concentration may show a 4–6 h plateau before declining in mid-to-late afternoon. In sites downwind of urban areas, the O_3 curve may be offset toward late afternoon. An air mass may move over an urban area, pick up O_3 precursors and produce O_3 far downwind from where the primary pollutants were added. There is continued addition of fresh pollutants to the air mass from sources over which the air mass is moving; without these additions the O_3 level would slowly diminish. Thus, O_3 is continuously formed, used and reformed during long-distance transport.

Both SO_2 and NO_2 are slowly hydrated and oxidized as they move in air masses. The transformation of SO_2 to SO_4^{2-} is estimated to be about $2\% \, h^{-1}$ under average conditions. Conversion of SO_2 to SO_4^{2-} is increased by metal catalysts and O_3. These reactions are continuously occurring in air masses.

Deposition and monitoring

These processes are fundamental to an understanding of air pollution effects on plant systems. Deposition involves the process of pollutant removal from the atmosphere (i.e. uptake by the plant). Monitoring is used to determine the atmospheric concentration of pollutants, usually at ground level. The monitoring data are used as an expression of the plant exposure dose; the data are also used as input for the determination of the amount of pollutant taken up by the plant.

DEPOSITION

Deposition is a physical process whereby the gaseous pollutants are deposited on surfaces. Deposition is the removal (cleansing) mechanism for the atmosphere and involves both wet and dry removal processes. Wet deposition depletes the atmosphere of a certain percentage of the three gases, but has a lesser impact on plant systems. Dry deposition includes gaseous sorption processes of the leaf, plus impaction and settling of particulates.

Gaseous deposition at ground level is the process of primary concern to the plant scientist and is the point of linkage with the atmospheric scientist. Gaseous deposition includes adsorption on plant surfaces, and absorption through stomata or cuticle. Rates of deposition (absorption) are related to flow resistances associated with both external and internal plant factors. There is need to understand further gaseous deposition processes and factors that control the resistances to gas movement into the plant.

MONITORING

Monitoring of pollutant concentration is a critical aspect of measuring gaseous pollutant movement (flux) into plants. To establish the relationship between the atmospheric concentration of a given pollutant and the flux of the pollutant to the plant, an accurate measure of atmospheric concentration is essential. This is often referred to as the exposure concentration for a plant and, over time, is referred to as the exposure dose. In spite of concern for the biological meaning of the exposure dose by some scientists, it is an essential ingredient in the air-pollution system and can be considered the air quality unit that links the atmospheric scientist, the plant scientist and the control officials (the standard setting process). The exposure dose, then, is a measurable manifestation of air quality that lends itself to the concept of a standard and can be used to determine pollutant flux to plants.

Knowing that knowledge of the exposure dose is pivotal to the air pollution system and to an assessment of effects, it is surprising that several factors are still poorly understood. First, the contribution of natural sources to the exposure dose is not understood. Second, the monitoring is inadequate (too few monitoring sites and only scattered data from rural locations) and some problems exist with monitoring techniques. Third, methods of handling and interpreting the monitoring data, so that it is of maximum value to the biologist, are only now being developed.

We must know the contribution of natural background to the ambient concentrations of O_3, SO_2 and NO_2 before reasonable control strategies can be devised and accurate crop loss assessments can be made. The atmospheric chemist is interested in the contribution of natural sources to ambient concentrations of the three gases, but only recently has the presence of natural sources been seriously studied. Currently, it appears that only O_3, of the three gases, receives a significant contribution from natural sources. It is important for the biologist to know whether he can use a single value (or factor) over a large region or whether this is not a reasonable expectation. Current results suggest that a single value may not be feasible (US Environmental Protection Agency, 1978a).

Monitoring was initiated around urban areas primarily for the protection of human health and in the mistaken belief that urban areas were the principal, if not the only, areas with a problem. Monitoring was also done close to point sources, because of concern for human health and to help protect vegetation close to the source. Even after long-distance transport was an accepted phenomenon and high stacks became common, relatively little monitoring was done in rural areas or in areas away from high stacks. Thus, the rural data base is inadequate for all three gases. Additionally, there is some concern for the accuracy of earlier monitoring techniques, especially the wet chemical techniques associated with bubblers. The US Environmental Protection Agency, generally, has developed methods to interpret the data but the concern puts into question some of the long-term trend results that included data prior to 1970. It is essential that monitoring data be collected in a sufficient number of rural areas for the development of county level estimates of exposure dose in both the USA and Canada.

An air-quality data base, using hourly pollutant averages, has been accumulated by the US Environmental Protection Agency. Acceptable long-term averaging techniques that can be used to relate plant effects to pollutant concentrations are being developed and tested by biologists. No one technique is fully accepted. Currently, the averaging times used are 24-h, monthly, and yearly means with some interest in hourly maximum concentrations. These concentration means are easy to understand and use, but they have limited applicability for long-term biological assessment. Many biologists are working on the development of more biologically usable averaging techniques that do not impede the development of reasonable control strategies. It is my belief that the attempts by some plant scientists to use a concentration–time summation dose value (i.e. ppm–h), as a substitute for an averaging technique, are misplaced. Such systems are arbitrary, have little relation to the real world, are impossible to interpret in biological terms, and are impractical for the control official. Likewise, some plant scientists believe the 'uptake' dose (actual amount of pollutant taken into the plant or organism) should be a value to consider. Although the 'uptake' dose is an important value for the plant scientist, it cannot be used by the control official. It is therefore essential that the biologist use the atmospheric monitoring data (exposure dose) and develop biologically meaningful ways to express the pollutant concentration.

Ozone
A need exists to understand the relative contributions of natural versus anthropogenic sources of tropospheric O_3. This is especially important when suggestions are made that 50% or more of the tropospheric O_3 could be from natural sources. The need, then, is to find ways to differentiate the source of O_3 that is being monitored. At present, there may be no location in the USA or much of southern Canada where natural background concentrations can be accurately measured, due to the presence of anthropogenic O_3 precursors.

Ozone monitoring data is archived in the US Environmental Protection Agency's Storage and Retrieval of Aerometric Data (SAROAD) system. This does not include all available data but does include results from over 300 monitoring stations within the USA. These data are primarily from urban sites, but some rural sites are included. In a recent exercise (Office of Technology Assessment, 1982) the SAROAD data base was used to estimate county level averages for the eastern USA. This was done using a Krieging method with some adjustment of the data to remove certain non-representative sites. A primary concern is how well these results represent rural areas when most of these data come from urban sites. Annual O_3 averages for selected rural sites have been reported as high as 0.052–0.074 ppm (US Environmental Protection Agency, 1978a).

Efforts have been made to develop a reasonable concentration–averaging regimen for exposure of annual plants to O_3 in the field (Heagle et al., 1979; Heck et al., 1982). This averaging regimen is based on the diurnal O_3 cycle that has been found at all the National Crop Loss Assessment Network (NCLAN) research sites (Heck et al., 1982). The maximum O_3 concentration on a sunny day spans the 6 to 8 h from mid-to-late morning to mid-to-late afternoon. This also overlaps the mid-to-late morning and early afternoon period of expected maximum plant activity. Thus, a 7-h exposure period (10.00–17.00 h, daylight saving time) was selected and an average across all the days of the exposure period (i.e. the growing season) was obtained. This seasonal 7-h mean daily O_3 concentration includes both a concentration and time concept (Heagle et al., 1979). When it is used with daily 7-h and 24-h mean concentrations, plotted for the season, the three values permit a reasonable biological interpretation of the response of any given crop to O_3. The 7-h daily mean can be used in the development of standards, if relationships can be shown between the 7-h daily and the 7-h seasonal means. In fact, the current 1-h standard can be used if relationships can be developed with the above averaging times. Preliminary work of this type, based on available monitoring data, suggests that the current 1-h standard of 0.12 ppm would equate to a 7-h daily maximum of 0.10–0.11 ppm and an average monthly mean 7-h daily maximum O_3 of 0.06–0.07 ppm (Lokey, Richmond and Jones, 1979). This technique or similar ones could be used to accommodate the needs of both the control official and the biologist. Whatever technique is used, biologists need to develop an averaging format that permits biological interpretation and also lends itself to control needs.

Sulfur dioxide
It appears that background sulfur sources are not a significant source of atmospheric SO_2 and thus the background issue is probably not important. However, the absence of monitoring data for SO_2 in rural areas is more acute than for O_3. Most, if not all, rural monitoring sites are situated close to point sources of SO_2 emissions. Thus, very little monitoring data is available for an assessment of SO_2 concentrations on a regional basis. Sulfur dioxide concentrations are reduced during long-range transport. Thus, we should expect greater variation in SO_2 concentration than for O_3, and a greater dependence on meteorological variables. In considering potential SO_2 problems on vegetation, we must understand the concentration/time/frequency relationships for different situations (e.g. urban–industrial multiple sources, point source low and high stacks, and topographical differences associated with sources). Thus, it may be impossible to develop a single biologically meaningful averaging technique that will apply under all circumstances. There is a need for innovative efforts in the analysis of SO_2 monitoring data.

Nitrogen oxides
Nitrogen dioxide is the component of the NO_x that is of greatest concern to the plant scientist. The natural background of NO_2 appears to be low. Thus, the importance of NO_2 is related more to the location and input of anthropogenic sources. Nitrogen dioxide is reduced during long-distance transport, but some replenishment comes from mobile sources. A minimum of monitoring data is available for NO_2 and only cursory attempts have been made to analyze the data.

Mixtures
An area of need that has not been seriously addressed is the occurrence of these three pollutants in mixtures. Since there is a known interaction between these pollutants for at least some biological systems, serious efforts at data analysis should be directed toward understanding the relationships of these three pollutants in the ambient air. Preliminary information from the Sulfate Regional Experiment (SURE) (Mueller *et al.*, 1980) showed that O_3 tended to peak in mid-afternoon, NO_2 in mid-morning and SO_2 at midday on days with O_3 above 0.05 ppm.

Plant uptake

Pollutant uptake by the plant involves the physical–biological link of the plant leaf. It is at this point that the plant scientist attempts to quantify the amount of pollutant received by the plant.

I have previously discussed the uptake concept (Heck, 1982). Several terms can be used to bridge the gap between the atmospheric and plant scientists, and permit a forum for reasonable discussion. These terms are: exposure dose, uptake dose and effective dose. The exposure dose is the monitored concentration at the plant over some finite time. There is no way to quantify directly the amount received by the plant (uptake dose) in the field using the exposure dose, unless meteorological conditions and the status of the plants are considered. The uptake dose may be defined as that portion of the pollutant absorbed into plant tissues. To understand the relationship between the exposure dose and the uptake dose will require both experimental research and model development. The effective dose is a more elusive concept and requires a knowledge of the physiological resistance mechanism(s) within the leaf. In essence, the effective dose is that amount of pollutant above a threshold amount that can be tolerated by the plant. The effective dose, then, is that portion of the uptake dose that actually induces leaf pathology and/or altered physiology (the stress-causing dose).

Knowledge of exposure dose and estimates of uptake and effective dose are the critical linkage between the atmosphere and the receptor in the air pollution system. Once the pollutant is within the plant system, the plant scientist is responsible for determining its effects on plants and feeding the information back into the air pollution system for scientific/regulatory use (Heck, 1982).

Plant effects

Once the pollutant enters the plant, the plant scientist is faced with a myriad of analytical problems on how to measure the primary effects and how these effects, in turn, affect other parts of the plant. This book is focused on plant metabolism. There is interest in understanding how the outward manifestation of pollutant effects are mediated within the metabolic systems of the plant. This section will address some research needs and approaches that are necessary to develop an understanding of plant effects. This will be followed by a brief discussion of the methodology of assessment, with an example of an ongoing program.

Response of plants — biological organization

This volume addresses the level of biological organization which, in theory, is the level at which all studies should be initiated (Heck, 1973). An overall view of the plant system and the foci within that system where different effects occur is described by Heck (1973). In reality, metabolic studies would never have been initiated without the visible manifestation of effects that first caused scientists to be concerned with the pollution phenomenon. The visual appearance of foliar lesions is normally the starting place for plant stress studies. Scientists first study cause/effect relationships, assess the severity of the effects, translate this to a value judgement, and then try to understand how the effects occurred (mechanistic, metabolic studies).

There is a need to understand the effects of pollutants on biochemical and physiological processes (e.g. net photosynthesis). These studies can often be carried out concurrently with growth and yield studies in whatever exposure facility is being used. A major reason for the few process-oriented studies has been a lack of interest by funding agencies. This has been a source of concern to many plant scientists because a holistic approach is necessary to understand functional changes, develop models and eventually predict plant response to pollutant exposure.

Pollutant stresses on plants are the result of complex physiological reactions. Identifiable cellular changes are often not specific for the pollutant, even if outward symptoms differ. Assuming that plants have resistance mechanisms that differ for each pollutant, cellular changes should be demonstrable as they relate to different mechanisms.

Studies at the biochemical/metabolic level are critical and present a basic challenge to the ingenuity of scientists. The results discussed in this volume attest to both the interest and dedication of those conducting these types of studies. Several major questions that need to be addressed include: what is the primary site of action; what are the chemical species that attack this primary site; and are all other effects secondary to the primary, accounting for symptoms that are similar to those caused by other stresses? Currently, the membrane is considered the most likely primary site of action for O_3 and possibly all gases. However, some evidence suggests that a portion of the pollutant gases can penetrate the membrane and directly affect cell organelles. These concepts need to be tested, and both primary and secondary effects clearly defined.

I have chosen not to discuss needs in other plant effects research areas because they are adequately covered in several references (Heck, 1973; Heck, Taylor and Heggestad, 1973; Heck, 1982). A brief overview and rationale for the types of effects to study is found in Heck (1982). A reasonable progression of studies would be: biochemical processes, physiological processes, growth/biomass, yield and foliar injury. Whatever the progression, the linkages between the biochemical and physiological processes and growth must be addressed, if we are to understand the effects of these pollutants on plants.

Methodology — facilities

The general methodology and facilities currently deemed most appropriate for air pollution research are developed in Heck (1982), but I did not address the special problems associated with biochemical and metabolic systems. The discussion by me (Heck, 1982) covered both facilities and experimental approaches. It encouraged some standardization of methods across laboratories without sacrificing the ability to develop innovative techniques and approaches, as shown in Heck *et al.* (1979). Four basic experimental facilities were discussed briefly and recommended for air-pollution research:

(1) chambers or other devices in the field;
(2) when possible, field studies using ambient conditions without experimental devices;
(3) greenhouse chambers; and
(4) chambers in controlled environment facilities.

Four experimental approaches were also discussed:

(1) modeling;
(2) dose-response;
(3) sensitive versus resistant cultivars; and
(4) acute versus chronic exposures.

When the approach is to study the effects in systems outside the whole plant there are special problems associated with the study of biochemical/metabolic changes induced by gaseous air pollutants. Basically, the study of isolated plant parts is useful to develop hypotheses in terms of the response in the intact plant. These hypotheses must then be tested using the whole plant before firm cause/effect relationships can be established. Until this is done, the whole plant physiologist must continue to work with the 'black box' and continue to hypothesize as to what is happening within the plant.

Biochemical/metabolic studies should be done using controlled environmental growth and exposure facilities. Experimental approaches should eventually use each of the four discussed above (modeling, dose-response, cultivars, acute and chronic exposures). Initial studies should focus on sensitive versus resistant genotypes as a way of identifying a resistance mechanism. Initially acute exposures are useful, but chronic exposure should be instigated as early as possible. Eventually, dose-response and modeling should be considered.

In vitro studies have been difficult to design and interpret. I shall not attempt to specify specific techniques but will comment on problems that must be addressed, if we are to accept results from these types of studies. Working with isolated plant systems requires techniques that are different from those using the intact plant. Yet, the very nature of the changed system is bound to change the chemical responses in the isolated system. The researcher must find some way to link results from these studies to the intact plant.

Much research with SO_2 and NO_2 has utilized SO_3^{2-} and NO_2^- solutions, respectively, to test the effects of SO_2 and NO_2 on isolated plant systems. The problems of choosing reasonable concentrations on the one hand, and on the other of being assured that these ionic species are the active chemical species, have not been critically examined. There may be special properties of SO_2 and NO_2 that trigger the primary plant response. If this is true, how does the researcher expose isolated plant parts to gaseous pollutants with some assurance that plant cells are receiving a pollutant dose? One procedure, used with some success, involves the exposure of tissue or cell suspensions in small cuvettes, but with minimal water. The relative humidity is maintained at 100% and the pollutant concentrations are carefully monitored.

Whatever the decision, the exposure technique is critical to the development of usable data. Research workers in this area should seriously consider new and innovative techniques. Results from these types of studies will eventually be the basic ingredients in the development of mechanistic models which will be used at some level of effects assessment.

Assessment

Assessment is the final step in the study of the effects of air pollutants on crops. It is an iterative process, being continually updated as new information is added that improves assessment capabilities. Assessment should be done at each level of biological organization. In the air pollution system this is the step before an economic appraisal. One assessment of current interest to the US Environmental Protection Agency is the effect on the economically important product of the crop of concern. Thus, assessment, in most people's minds, is equivalent to an assessment of effect on yield and the translation of this value into economic gain or loss.

This chapter will end with a brief overview of the National Crop Loss Assessment Network (NCLAN) (Heck et al., 1982). The NCLAN consists of several research groups cooperating in field work, crop production modeling and economic studies to assess the immediate and long-term economic consequences of the effect of O_3, SO_2, NO_2 and mixtures of them on crop production. The basic aim of NCLAN is to develop dose-response relations for different crops in regions of the USA where they are grown. The data will be used for an economic assessment of crop loss from O_3. SO_2 and NO_2 in the USA. TThe program is also designed to advance the understanding of cause–effect relationships with the intent of developing simulation models. These field studies are designed to provide pollutant dose–crop response data that are as free of artifact as is currently possible using state-of-the-art technology. The basic exposure technique utilizes open-top chambers that permit control of gas(es) around the plant canopy, allowing specific pollution regimes to be imposed on experimental plants, with minimal effect on the crops growing within them. The experimental design uses dose–response as a basic approach. Data analysis uses regression concepts. Results to date suggest an average crop loss in the USA due to gaseous pollutants of 2 to 5% (Heck et al., 1982).

References

GUDERIAN, R. *Air Pollution, Phytotoxicity of acidic gases and its significance in air pollution control, Ecological Studies 22,* Springer-Verlag, Berlin, Germany, 127 pp. (1977)

HEAGLE, A. S., RIORDAN, A. J. and HECK, W. W. Field methods to assess the impact of air pollutants on crop yields. No. 79-46-6. Presented at the *72nd Annual Meeting of the Air Pollution Control Association,* Cincinnati, Ohio. Air Pollution Control Association, Pittsburgh, PA, 24 pp. (1979)

HEAGLE, A. S., PHILBECK, R. B., ROGERS, H. H. and LETCHWORTH, M. B. Dispensing and monitoring ozone in open-top field chambers for plant effects studies, *Phytopathology* **69**, 15–20 (1979)

HECK, W. W. Air pollution and the future of agricultural production. In *Air Pollution Damage to Vegetation,* (Ed. J. A. NAEGELE), pp. 118–129, Advances in Chemistry Series 122. American Chemical Society, Washington, DC (1973)

HECK, W. W. Future directions in air pollution research. In *Effects of Gaseous Air Pollution in Agriculture and Horticulture,* (Eds. M. H. UNSWORTH and D. P. ORMROD), pp. 411–435, Butterworths, London (1982)

HECK, W. W. and BRANDT, C. S. Effects on vegetation: native, crops, forest. In *Air Pollution* (Ed. A. C. STERN), 3rd edn, Vol. 2, pp. 157–229, Academic Press, New York (1977)

. W. and HEAGLE, A. S. Measurement of photochemical air pollution with a sensitive monitoring
ournal of the Air Pollution Control Association **20**, 97–99 (1970)

V. W., KRUPA, S. V. and LINZON, S. N. (Eds) *Handbook of Methodology for the Assessment of Air
on Effects on Vegetation*, Upper Midwest Section, Proceedings, Air Pollution Control Association
lty Conference, April 1978. Air Pollution Control Association, Pittsburgh, PA 392 pp. (1979)

HECK, W. W., MUDD, J. B. and MILLER, R. B. Plants and microorganisms. In *Ozone and Other
Photochemical Oxidants*, pp. 437–585, National Academy of Sciences, Washington, DC (1977)

HECK, W. W., TAYLOR, O. C. and HEGGESTAD, H. E. Air pollution research needs: Herbaceous and
ornamental plants and agriculturally generated pollutants, *Journal of the Air Pollution Control
Association* **23**, 257–266 (1973)

HECK, W. W., TAYLOR, O. C., ADAMS, R., BINGHAM, G., MILLER, J., PRESTON, E. and
WEINSTEIN, L. H. Assessment of crop loss from ozone, *Journal of the Air Pollution Control Association*
32, 353–361 (1982)

JACOBSON, J. S. and HILL, A. C. (Eds) *Recognition of Air Pollution Injury to Vegetation: A Pictorial Atlas*,
Air Pollution Control Association, Pittsburgh, PA (1970)

LACASSE, N. L. AND TRESHOW, M. (Eds) *Diagnosing Vegetation Injury Caused by Air Pollution*, Applied
Science Associates Inc., EPA Contract 68-02-1344. Developed for the Environmental Protection Agency,
146 pp. (1976)

LINTHURST, R. A. and ALTSHULLER, A. P. (Eds) *Critical Assessment Document: the Acidic Deposition
Phenomenon and Its Effects*, Vol. 1, Environmental Protection Agency, Washington, DC, (Draft) (To be
submitted to EPA, March 1983) (1983)

LOKEY, D., RICHMOND, H. and JONES, M. Revision of a Secondary National Ambient Air Quality
Standard — Ozone: A Case Study. Paper No. 79-46-1. Presented at the *72nd Annual Meeting of the Air
Pollution Control Association*, Cincinnati, Ohio Air Pollution Control Association, Pittsburgh, PA (1979)

MANSFIELD, T. A. (Ed.) *Effects of Air Pollutants on Plants. Soc. for Experimental Biology-Seminar Series I.*
Cambridge University Press, Cambridge, England, 209 pp. (1976)

MUDD, J. B. and KOZLOWSKI, T. T. (Eds) *Responses of Plants to Air Pollutants*, Academic Press, NY, 388
pp. (1975)

MUELLER, P. K., HIDY, G. M., WARREN, K., LAVERY, T. F. and BASKETT, R. L. The occurrence of
atmospheric aerosols in the Northwestern United States. In *Annals of The New York Academy of Sciences*
338, 463–482 (1980)

NATIONAL ACADEMY OF SCIENCES *Nitrogen Oxides. Committee on Medical and Biological Effects of
Environmental Pollutants*, National Academy of Sciences, Washington, DC, 333 pp. (1977)

NATIONAL ACADEMY OF SCIENCES *Sulfur Oxides*, Committee on Medical and Biologic Effects of
Environmental Pollutants. National Academy of Sciences, Washington, DC, 209 pp. (1978)

NATIONAL ACADEMY OF SCIENCES *Atmosphere-Biosphere Interactions: Toward a Better
Understanding of the Ecological Consequences of Fossil Fuel Combustion*. Committee of The Atmosphere
and the Biosphere. National Academy of Sciences, Washington, DC, 263 pp. (1981)

OFFICE OF TECHNOLOGY ASSESSMENT. *The Regional Implications of Ozone. Transported Air
Pollutants: An Assessment of Acidic Deposition*. An Interim Draft, July 1982. OTA, Washington, DC (1982)

THOMAS, M. D. Gas damage to plants, *Annual Review of Plant Physiology* **2**, 293–322 (1951)

TRESHOW, M. *Environment and Plant Response*, McGraw-Hill, New York, 422 pp. (1970)

UNSWORTH, M. H. and ORMROD, D. P. (Eds) *Effects of Gaseous Air Pollution in Agriculture and
Horticulture*, Butterworths, London, 522 pp. (1982)

US ENVIRONMENTAL PROTECTION AGENCY *National Air Quality and Emission Trends Report,
1976*. EPA-450/1-77-002. December 1977. Office of Air and Waste Management. Office of Air Qual.
Planning and Standards, Research Triangle Park, NC, 70 pp. (1976)

US ENVIRONMENTAL PROTECTION AGENCY *Air Quality Criteria for Ozone and Other
Photochemical Oxidants*. EPA-600/8-78-004. Office of Research and Development, US Environmental
Protection Agency, Washington, DC, 341 pp. (1978a)

US ENVIRONMENTAL PROTECTION AGENCY *National Air Pollutant Emission Estimates, 1940–
1976*. EPA-450/1-78-003, US Environmental Protection Agency, Research Triangle Park, NC (1978b)

US ENVIRONMENTAL PROTECTION AGENCY *National Air Pollutant Emission Estimates, 1970–
1978*. EPA-450/4-80-002, US Environmental Protection Agency, Office of Air Quality Planning and
Standards, Research Triangle Park, NC (1980)

US ENVIRONMENTAL PROTECTION AGENCY *Air Quality Criteria for Oxides of Nitrogen*. EPA-
600/8-82-026 Environmental Criteria and Assessment Office, Research Triangle Park, NC, 620 pp.
(1982a)

US ENVIRONMENTAL PROTECTION AGENCY *Air Quality Criteria for Particulate Matter and
Sulfur Oxides*. Vols. A, B, C. EPA 600/8-82-029 Environmental Criteria and Assessment Office, Research
Triangle Park, NC (1982b)

Chapter 4

Air pollution problems and the research conducted on the effects of gaseous pollutants on plants in China

Yu Shu-Wen

INSTITUTE OF PLANT PHYSIOLOGY, ACADEMIA SINICA, SHANGHAI, CHINA

Pollution problems

As coal furnishes over 70% of the energy in China, SO_2 is the main air pollutant. The rate of energy utilization of coal (i.e. combustion efficiency) in this country is low, just below 30%, so considerable pollutant is produced during combustion. Consequently, air pollution problems are serious in many cities, especially in the larger ones such as Beijing, Shanghai, Shengyang, Tianjing, Chongqian, Wuhan, Nanjing and Guangzhou. Besides industrial sources, gaseous emissions from household stoves used for both cooking and domestic heating contribute significantly to the air-pollution problem. The pollutants emitted by these stoves with low chimneys and low combustion efficiencies (below 20%) are concentrated at a level within two metres above the ground, where they are particularly harmful to man. In northern China the problem is even more serious, as the period during which domestic heating is necessary is usually much longer than in southern China. Pollutant emissions in winter can be several times the amount in summer (*Table 4.1*). In Beijing, for example, the daily mean concentration of SO_2 during the winter of 1980 was 220 $\mu g \, m^{-3}$, exceeding the permissible limit of 150 $\mu g \, m^{-3}$. According to an estimate made in 1980, the amount of SO_2 discharged into the air in the whole country was 15 million tonnes.

At present there are comparatively few automobiles in China, but the petrol consumed and the exhaust emissions per vehicle are greater than those in the well-developed countries. Thus the concentration of NO_2 from motor vehicle exhaust at the main intersections in certain cities is far in excess of the limit set for health safety (i.e. 100 $\mu g \, NO_2 \, m^{-3}$). In Beijing, for example, the concentration of NO_2 in the downtown section sometimes reached 160–370 $\mu g \, m^{-3}$.

No photochemical smog has, as a rule, been reported around Chinese cities, with the exception of Lanzhou. This city in north-west China is situated in a river basin with mountain ranges on three sides which impede the dispersion of air pollutants. Hydrocarbons and NO_x emitted in large quantities from the petrochemical factories are readily transformed into photochemical oxidants under the intense sunlight the region receives. Ambient concentrations of these pollutants have been sufficient to elicit reports of visible injuries to plants in this region.

Fluoride pollution exists in limited areas in the neighbourhood of iron, steel, aluminium and phosphorus fertilizer works. The case of Baotou is an example of

49

TABLE 4.1. SO$_2$ concentrations in some major cities in China in 1980 (in μg m^{-3}, downtown)

City	Winter	Summer	Average for the year
Beijing	220	50	
Tianjing	350	110	
Lanzhou	117	42	80
Nanjing	136	38	72
Hongzhou	65	42	45

serious, localized injury to plants. In this city in northern China, a large amount of fluoride is emitted from the steel works and smelters, causing air and soil pollution over an area of about 12 000 km^2.

Shanghai is the largest city in China, and it may therefore be useful to look at the conditions in this city to illustrate the pollution situation in China. Industry, commerce and transportation are concentrated in an urban area of 160 km^2, where the population density approaches 38 000 km^{-2} in general but can be as high as 100 000 km^{-2} in certain sections of the city. The annual consumption of coal in the city is about 10 million tonnes, in addition to some 2 million tonnes of petroleum. The use of gas among the inhabitants is limited, being confined to no more than about half the families; coal is still widely and directly used as a domestic fuel. According to the results measured by the Shanghai Environment Monitoring Station, the situation of SO$_2$ pollution in the urban area in the 1970s is presented in *Figure 4.1*. The average SO$_2$ concentration was highest in the commercial area, with the industrial area coming second. The concentrations in both areas are far above the limit set for the average for the year, namely 60 μg m^{-3}. In the late 1970s, SO$_2$ concentrations showed a downward trend in the industrial areas while in residential areas the concentrations remained around the yearly daily mean. The values for the maximum daily means and the median values are also presented.

Figure 4.2 illustrates the trends in NO$_2$ pollution. Except in some commercial areas, NO$_2$ concentrations were within the limits set for the daily mean, namely 100 μg m^{-3}. In general, the trends of SO$_2$ and NO$_2$ pollution did not change markedly in the 1970s. There was, however, a slight decrease in pollution concentrations in 1980 and 1981. The range of daily means recorded in 1980 was 0–750 μg m^{-3} for SO$_2$ and 5–760 μg m^{-3} for NO$_2$, with an average concentration of 126 μg m^{-3} for SO$_2$ and 80 μg m^{-3} for NO$_2$. In 1981, the average concentration of SO$_2$ dropped to 100 μg m^{-3} and that of NO$_2$ to 70 μg m^{-3}. The data presented in *Figure 4.3* were collected at a monitoring station in a residential section. For SO$_2$, the monthly daily means were in excess of the limit set for the yearly average (60 μg m^{-3}) for most months of the year, but did not exceed the standard set for a daily mean (150 μg m^{-3}). For NO$_2$, most of the 90th and 50th percentiles were below the standard set for a daily mean (100 μg m^{-3}).

Pollution research

Since the early 1970s, over 30 research institutions and universities have been conducting fumigation experiments in laboratories as well as field surveys and experiments at polluted sites. Symptoms of acute injury and changes in the anatomy of injured leaves have been studied (Shanghai Institute of Plant Physiology, 1978; Qin,

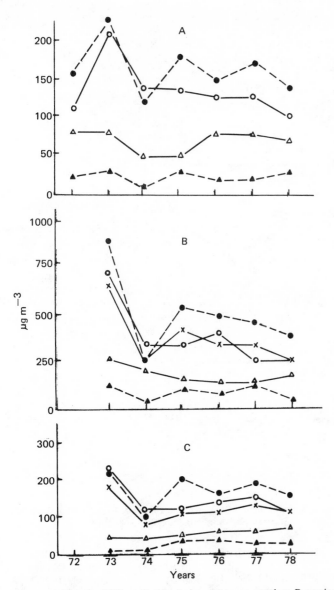

Figure 4.1. Trends of SO$_2$ pollution. A Average concentration; B maximum daily mean; C median level. Commercial area ●---●, industrial area ○——○, residential area △——△, clean area ▲---▲, whole city ×——×

Wu and Wang, 1980), the susceptibility of various plants to several air pollutants has been assessed and a large number of resistant trees and crop plants have been selected (Liu *et al.*, 1980b; Wang *et al.*, 1981a). *Symptoms of Air Pollution Injury in Plants: A Pictorial Atlas* was published in 1981 (Yu and Wang, 1981). Numerous chemical analyses of plant materials have also been performed to determine the ability of various plants to absorb SO$_2$, HF and Cl$_2$, and these data on pollutant concentrations within the plants have been used as indices in biological monitoring.

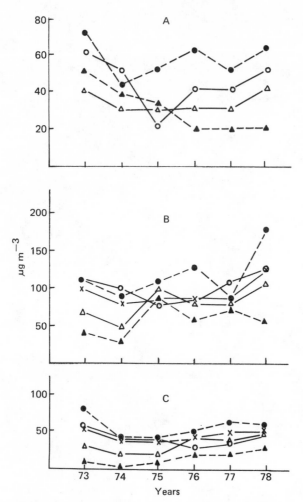

Figure 4.2. Trends of NO₂ pollution. A Average concentration; B maximum daily mean; C median level. Commercial area ●---●, industrial area ○——○, residential area △——△, clean area ▲---▲, whole city ×——×

Concerning biological monitoring, it was found that sesame (*Sesamum orientale*) is a sensitive bio-indicator for both SO₂ and ethylene (Tan *et al.*, 1980), while buckwheat (*Fagopyrum cymosum* Meisn) can be used as a bio-indicator for HF pollution (Jiangsu Institute of Botany, 1978). Indeed, based on the analyses of leaf fluoride content of black poplar (*Populus canadensis* Moench), a map for the distribution of HF pollution in the Baotou region was produced. Wang *et al.* (1981b, 1981c) studied the migration and accumulation of F⁻ in the ecosystem of mulberry silkworms. They found that whenever the fluoride content in the polluted mulberry leaves exceeded 30 ppm, acute injury was induced in the silkworms.

The effects of combinations of gaseous pollutants have also been studied. The results of Tan *et al.* (1980) on the effects of SO₂ and ethylene on sesame showed that 0.23–1.09 ppm ethylene enhanced the injury sustained by exposure to 0.28–0.84 ppm

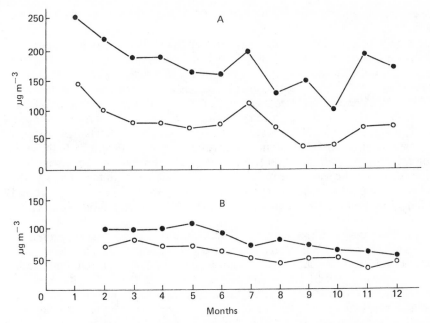

Figure 4.3. SO$_2$ and NO$_2$ concentration in Shanghai (1980). A SO$_2$ ●——● maximum daily mean, ○——○ daily mean. B NO$_2$ ●——● 90th percentile, ○——○ 50th percentile

SO$_2$, but that the presence of SO$_2$ did not affect plant response to ethylene. Bian and Chen (1983) investigated the effects of combinations of SO$_2$ and HF (0.87 ± 0.13 ppm SO$_2$; 0.16 ± 0.07 ppm HF) on buckwheat, and found that these two gaseous pollutants exerted a synergistic effect, but that the fluoride content of the leaves treated with the two pollutants combined was no greater than that in leaves treated with HF alone.

The susceptibility of plants to SO$_2$ has been related to the pH of the leaf sap; at lower pHs plants are more susceptible to injury while higher pHs confer some resistance (Wu, Xu and Li, 1975). Yu *et al.* (1980a) proposed that as the concentrations of the ionic species HSO$_3^-$ and SO$_3^{2-}$ in solution are dependent upon pH, the dominant species at pH 2–5 would be HSO$_3^-$, which is more toxic than SO$_3^{2-}$. Chen (1982) has suggested that the difference between the pH value and the isoelectric point of the leaf cells might be an important factor determining plant resistance to SO$_2$.

Li *et al.* (1980) found, almost simultaneously with two American research groups (Peiser and Yang, 1979; Bressan *et al.*, 1979), that an increased production of ethylene accompanies SO$_2$ injury. However, Li *et al.* (1980) also found that in certain plants such as rice (*Oryza sativa*) and wheat (*Triticum aestivum*) the production of ethylene increased even when exposure concentrations of SO$_2$ were below the threshold causing visible injury. It is evident, therefore, that the production of stress ethylene is not necessarily restricted to wounding. As the extent of SO$_2$ injury increases, ethane is emitted in increasing amounts with a corresponding decrease in the production of ethylene (Li *et al.*, 1980, 1982). More recently, Chou *et al.* (*see* addendum) found that leaves of tobacco (*Nicotiana tabacum*) exposed to 0.2–100 mM HSO$_3^-$ emit propylene and propane in much larger amounts than either ethylene or ethane. In tobacco exposed to 1.8–2.5 ppm SO$_2$, propylene and propane were also emitted. The physiological significance of the production of these 3-carbon gases remains to be explored.

The research group at the Shanghai Institute of Plant Physiology conducted an investigation on the mechanism of SO_2 injury in plants, and proposed that the membrane system of cells was one of the chief sites of interaction with the pollutant. SO_2-induced damage to the plasma membrane results in the loss of differential permeability which can be measured as an increase in the leakage of ions from SO_2-treated cells; the 'leakiness' of the plasma membrane to ions has been proposed as an index of SO_2 injury (Yu et al., 1979; Bian and Chen, 1982). Zhu (1982) has improved a method of measuring electrical resistance of leaves that involves determining the mid-region voltage drop on passing a stable electric current through the leaf tissue. The electrical resistance of the leaf tissue showed a high correlation with the change in the permeability of the plasma membrane, and has been shown to be a more direct and sensitive indicator of membrane damage than the ion-leakage method (Yang et al., 1982).

Membrane lipids may be affected by SO_2. Three kinds of decomposition products of polyunsaturated fatty acids were detected in leaves exposed to SO_2, namely, ethane, thiobarbituric acid-reactive substances (mainly malondialdehyde) and fluorescent substances (Yu et al., 1980b; Liu et al., 1980a; Yu et al., 1981). On the basis of experimental results, Yu et al. (1980b) and Tan et al. (1981) concluded that membrane injury resulted from free radical peroxidation of membrane lipids. In their experiments, Tan et al. (1981) treated plants with five kinds of free-radical scavengers, namely, sodium benzoate, diphenylamine, butylated hydroxytoluene, α-tocopherol (2,6-ditert-butyl-4-hydroxytoluene) and propyl gallate, before exposing plants to SO_2. Plants pretreated with these chemicals sustained less injury, and the production of ethane and the quantities of thiobarbituric acid-reactive and fluorescent substances all decreased correspondingly. Scientists at the Japanese National Institute for Environmental Studies have demonstrated that superoxide dismutase played an important role in the protection of plants against SO_2 injury (Tanaka and Sugahara, 1980; Shimazaki, Sakaki and Sugahara, 1980). Their findings, although representing a different approach, also implicate the participation of free radicals in SO_2 injury to membranes.

With the occurrence of SO_2 injury, not only does the activity of peroxidase increase, but the isoperoxidase zymogram changes as well, showing two new isozyme bands in addition to an intensification of some of the bands (Li et al., 1981). Increased peroxidase activity indicates an increase in the peroxidation processes in the cells, thus providing indirect evidence for a relationship between SO_2 injury and the peroxidation of membrane lipids.

References

BIAN YONG-MEI and CHEN SHU-YUAN Effects of SO_2 and HF singly or in combination on Fagopyrym cymosum, Huanjing Kexue (Journal of Environmental Science) 2, 51–54 (1983)
BIAN YONG-MEI and CHEN SHU-YUAN The effect of SO_2 on membrane permeability of plant cells, Plant Physiol. Communication ,1, 41–45, (1982)
BRESSAN, R. A., LECUREUX, L., WILSON, L. G. and FILNER, P. Emission of ethylene and ethane by leaf tissue exposed to injurious concentrations of sulfur dioxide or bisulfite ion, Plant Physiol. 63, 924–930 (1979)
CHEN RUI-ZHANG Studies on the mechanisms of pollution resistance of plants. I. The relation among pH, isoelectric point, buffering capacity of leaf cells and SO_2 resistance, Plant Physiol. Communication 1, 50–52 (1982)
JIANGSU INSTITUTE OF BOTANY Using gladiolus and Fagopyrum cymosum to detect air pollution of fluorides, Huanjing Kexue 4, 63 (1978)

LI ZHEN-GUO, LIU YU, WU YOU-MEI and YU SHU-WEN Studies on the response and resistance of plants to sulfur dioxide. III. The exposure of plants to SO_2 and stress ethylene production, *Acta Phytophysiologia Sinica* **6(1)**, 47–55 (1980)

LI ZHEN-GUO, WEI PEI-HENG, LIU YU, WU YOU-MEI and YU SHU-WEN Inverse relation between ethylene and ethane production in plants after injuring, *Acta Phytophysiol. Sinica* **8(4)** (in press) (1982)

LI ZHEN-GUO, WU YOU-MEI, LIU YU and YU SHU-WEN Studies on the response and resistance of plants to sulfur dioxide. VII. Effects of SO_2 fumigation on peroxidase of wheat leaves, *Acta Phytophysiologia Sinica* **7(4)**, 363–371 (1981)

LIU YU, LI ZHEN-GUO, WEI PEI-HENG, LI SHAO-LING, WU YOU-MEI and YU SHU-WEN Studies on the response and resistance of plants to sulfur dioxide. IV. Tissue injury and production of ethane in segments of wheat seedlings exposed to HSO_3^-, *Acta Phytophysiologia Sinica* **6(3)**, 307–314 (1980a)

LIU YU-GIN, WU GEN-QIAN, TAN CHANG and YU SHU-WEN On acute injury and relative resistance to sulfur dioxide of trees and shrubs for greening, *Acta Botanica Sinica* **22(3)**, 260–261 (1980b)

PEISER, G. D. and YANG, S. F. Ethylene and ethane production from sulfur dioxide-injured plants, *Plant Physiol.* **63**, 142–145 (1979)

QIN HUI-ZHEN, WU ZHU-JUN and WANG JIA-XI The effects of the harmful gases SO_2 and HF on plant leaf structure, *Acta Botanica Sinica* **22(3)**, 232–235 (1980)

SHANGHAI INSTITUTE OF PLANT PHYSIOLOGY Studies on the response and resistance of plants to sulfur dioxide. I. Acute injury and relative resistance of plants under controlled fumigation condition, *Acta Phytophysiologia Sinica* **4(1)**, 27–37 (1978)

SHIMAZAKI, K., SAKAKI, T. and SUGAHARA, K. Active oxygen participation in chlorophyll destruction and lipid peroxidation in SO_2-fumigated leaves of spinach, *Res. Rep. Natl. Inst. Environ. Stud.* **11**, 91–102 (1980)

TAN CHANG, LI ZHEN-GUO, LIU YU and YU SHU-WEN Sulfur dioxide and ethylene synergism — injury to sesame plants, *Acta Phytophysiologia Sinica* **6(4)**, 431–435 (1980)

TAN CHANG, LIU YU, LI ZHEN-GUO, YU ZI-WEN, YANG WEI-TONG and YU SHU-WEN. Studies on the response and resistance of plants to sulfur dioxide. VIII. Protective effect of free radical-scavengers against SO_2 injury, *Acta Sciential Circumstantial* **1(3)**, 197–206 (1981)

TANAKA, K. and SUGAHARA, K. Role of superoxide dismutase in the defense against SO_2 toxicity and induction of superoxide dismutase with SO_2 fumigation, *Res. Rep. Natl. Int. Environ. Stud.* **11**, 155–164 (1980)

WANG JIA-XI, QIAN DA-FU, LI ZHENG-FANG and GAO XU-PING Selection of resistant, absorptive and sensitive plants to air pollutants, *Bull. Environ. Poll. Ecol.* 95–103 (1981a)

WANG JIA-XI, QIAN DA-FU, LI ZHENG-FANG and GAO XU-PING Effects of fluoride in mulberry leaves on the growth and development of silkworm, *Bull. Environ. Poll. Ecol.* 72–77 (1981b)

WANG JIA-XI, QIAN DA-FU, LI ZHENG-FANG and GAO XU-PING Accumulation and translocation of fluoride in mulberry-silkworm ecosystem, *Bull. Environ. Poll. Ecol.* 78–81 (1981c)

WU ZONG-QING, XU XIAN-YUAN and LI YI Preliminary study on the resistance of trees to gaseous pollutants, *Communication of Science and Technology of Forestry* **7**, 18–20 (1975)

YANG WEI-DONG, TAN CHANG, YU ZI-WEN and YU SHU-WEN Studies on the response and resistance of plants to sulfur dioxide. IX. SO_2 injury and changes of tissue electrical resistance, *Acta Sciential Circumstantial* **2(4)**, 358–363 (1982)

YU SHU-WEN, LIU YU, LI ZHEN-GUO, WU YOU-MEI and YANG HUI-DONG The relation between the resistance of plants to SO_2 and pH of leaf tissue, *Ku Hsueh Tung Bao (Scientia)* **23**, 1097–1098 (in English, **26**, 185–187, 1981) (1980a)

YU SHU-WEN, LIU YU, LI ZHEN-GUO, TAN CHANG, YU ZI-WEN, YANG WEI-TONG and WU YOU-MEI Studies on the mechanism of SO_2 injury in plants, *Ku Hsueh Tung Pao (Scientia)* **24**, 1145–1147 (in English, **26**, 831–834, 1981) (1980b)

YU SHU-WEN, TAN CHENG, YANG WEI-DONG and YU ZI-WEN Studies on the response and resistance of plants to sulfur dioxide. II. Permeability changes of plasma membrane and SO_2 injury, *Acta Phytophysiologia Sinica* **5(4)**, 403–409 (1979)

YU SHU-WEN and WANG JIA-XI (Eds) Symptoms of Air Pollution Injury in Plants — A Pictorial Atlas, *Shanghai Sci. Tech. Pub.* Shanghai (1981)

YU ZI-WEN, TAN CHANG, YANG WEI-DONG and YU SHU-WEN Studies on the response and resistance of plants to sulfur dioxide. VI. Sulfur dioxide injury and the increase of membrane lipid peroxides, *Acta Phytophysiologia Sinica* **7(1)**, 57–65 (1981)

ZHU ZHONG-LING. Measurement of electrical resistance of plant tissue by mid-region voltage drop on passing an electric current, *Plant Physiology Communications* **1**, 54–58 (1982)

Addendum

Evolution of propylene and propane from tobacco leaves exposed to HSO_3^- or SO_2

Chou Yung-Jen, Li Zhen-Guo, Yang Shi-Yung, Yu Zi-Wen and Yu Shu-Wen

INSTITUTE OF PLANT PHYSIOLOGY, ACADEMIA SINICA, CHINA

It has been reported that plants exposed to HSO_3^- or SO_2 and/or subjected to its injurious effects would produce ethylene and ethane. Plants tested included wheat, rice, cayenne pepper, tomato, alfalfa and some curucbitaceous plants. Recently it has been observed in the work of this Institute that in this respect tobacco is quite different from the plants mentioned above. As their leaves reacted with HSO_3^- or SO_2, much propylene and propane were released.

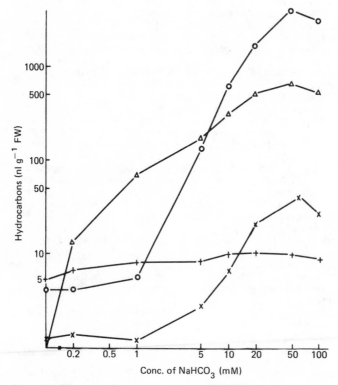

Figure 4.4. Effect of various concentrations of $NaHSO_3$ on evolution of propane, propylene, ethane and ethylene from tobacco leaves. Propane o——o, propylene △——△, ethane × —— ×, ethylene + —— +

When tobacco (*Nicotiana tabacum*) leaves exposed to HSO_3^- within a certain range of concentrations (0.2–50 mM), the formation of propylene and propane increased with increasing concentration (*Figure 4.4*). At lower HSO_3^- concentrations, more propylene was produced than propane, while at higher concentrations much more propane was produced than propylene. The optimum pH for propane production was 6, and for propylene 5–6. Ethylene and ethane were released in small quantities as compared with propylene and propane.

When HSO_3^- was added to the homogenate system of tobacco leaves, propylene and propane were also produced in abundance with the liberation of much ethane. Few hydrocarbon gases were found to be emitted when HSO_3^- was added to the homogenate that had been heated to boiling. However, at a higher concentration of HSO_3^- (50 mM), and after being incubated for a longer period (3 h), propylene and propane were emitted from the boiled system, but in much smaller amounts. This indicated that although their formation is mainly an enzymatic reaction, a non-enzymatic reaction may also be responsible for some of the emission.

Potted plants were exposed to SO_2 in concentrations of 1.8 and 2.5 ppm for 3 h in dynamic fumigation chambers. After fumigation, leaves were sampled and put in flasks, then sealed and incubated for 2 h. Propylene and propane were also found but in small amounts; presumably they were emitted during the course of fumigation. When detached leaves were exposed to SO_2 in sealed vessels at high initial concentrations, propylene was released in large quantities. Propane production increased with increasing SO_2 concentration, but the quantity emitted was much smaller than that of propylene.

The results of the experiments indicated that propane and propylene were the dominant hydrocarbon gases released from tobacco leaves treated with HSO_3^- or SO_2. However, no report on their emission seems to have appeared so far. If the propylene and propane released from tobacco leaves by treatment with HSO_3^- or SO_2 come from the decomposition of polyunsaturated fatty acids just as ethane and pentane do, then the result presented in this chapter provides another proof in favour of the supposition that one mechanism of SO_2 injury is that membrane lipids are being peroxidized.

Chapter 5

Defining pollution problems in the Far East — a case study of Japanese air pollution problems

A. Furukawa

DIVISION OF ENVIRONMENTAL BIOLOGY, NATIONAL INSTITUTE FOR
ENVIRONMENTAL STUDIES, IBARAKI, JAPAN

Introduction

Like most developed countries, Japan has suffered from environmental pollution for a long time. The air and water pollution caused by a copper mine located in Ashio in the 1880s is a well-known historical example of environmental disruption associated with industrialization. However, unlike many other countries, Japan introduced anti-pollution policies about 100 years ago. In 1887, the Osaka Prefecture issued a progressive ordinance in the form of Regulations on Controls of Manufacturing Plants. The purpose of this ordinance was to define the industrial areas so that the disruption to the human environment would be minimized. The anti-pollution measures, which still exist, were the physical transfer in 1895 of the Beshi copper refining plant to Shisaka Island in the Seto Inland Sea, the instigation of regulations to be followed in case of emergency, and the erection of high smokestacks (a 156-m high smokestack was constructed at Hitachi copper mine in 1914).

Although these pollution control measures were progressive at the time, they were insufficient to cope with the increase in industrial activity after the Second World War. Environmental pollution became a serious social problem, and in the 1960s Japan was known world-wide as the 'Polluted Islands'. Almost 20 years have passed since then, and now Japan has partially overcome the environmental pollution problems through continuous efforts to clean up the environment. The Air Pollution Control Law was enacted in 1968 and the Environment Agency established in July 1977. But even in the past two or three years, plant injury caused by air pollution has been reported by scientists, though their findings have not received coverage in the mass media.

The purpose of the present report is to review both the present status of air quality in Japan (based on the *White Paper on Environment* published by the Japan Environment Agency, 1982) and the results of the field survey conducted in the Kantoh district which includes the metropolis of Tokyo (based on the 1975–1981 reports published by the Union of Air Pollution Prevention Association in Kantoh District).

Air pollution problems faced by Japan today

The air pollution monitoring network

In 1965, 15 continuous monitoring stations were set up in the most severely polluted areas of the country. Since then, many monitoring stations have been installed at

TABLE 5.1. Numbers of monitoring stations for air pollution, nationwide, 1980

		SO_2	NO_2	NO	CO	$O_x{}^a$	$NM\text{-}CH^b$	SP^c	Dust	Soot
Monitoring	General	1611	1206	1206	224	953	219	276	1216	1589
stations	Auto	45	260	260	372	48	100	14	60	
Cities	General	617	546	546	185	485	162	131	527	383
	Auto	38	155	155	177	35	70	8	46	

[a] Photochemical oxidants.
[b] Non-methane hydrocarbons.
[c] Suspended particulates.

various locations by local and central governing bodies, and now hundreds of stations continuously monitor air quality (*Table 5.1*). The Ministry of Health and Welfare was responsible for establishing these monitoring stations before the Environment Agency was set up. Monitoring stations are divided into two types according to their purposes. One is the so-called 'general monitoring station' while the other is the 'automobile exhaust monitoring station'. The automobile exhaust monitoring stations are situated near roads or intersections to record the major air pollutants emitted by automobiles. Monitoring networks for air pollution have been consolidated both by the prefectures, which have been entrusted with this responsibility under the Air Pollution Control Law, and by cities that have been authorized by a cabinet order to enforce the Law. Besides these local monitoring stations, 15 national air pollution monitoring stations (general type) and eight environmental background air-monitoring stations have been established (*Figure 5.1*) with the combined purpose of clarifying the state of air pollution caused both by the currently regulated air pollutants and by other substances from a nationwide perspective, and of collecting data for establishing ambient air quality standards and directing further pollution control programmes.

Ambient air quality standards

The air environmental quality standards presently cover SO_2, CO, suspended particulates, NO_2 and photochemical oxidants. The Japanese government makes an effort to achieve and maintain these standards by comprehensive, effective and appropriate pollution controls. The standards which are considered to be the 'desirable levels' are based on current scientific knowledge and will be revised as necessary.

SULPHUR DIOXIDE

Concentrations of SO_2 in the ambient air are determined by the conductimetric method at 1611 general monitoring stations which are distributed in 617 cities throughout the country. The concentration of SO_2 in the atmosphere is restricted by the environmental quality standard to a daily average of hourly values not exceeding 0.04 ppm, with individual hourly values not exceeding 0.1 ppm.

CARBON MONOXIDE

Concentrations of atmospheric CO were recorded at 224 stations in 185 cities in 1980, in addition to 372 automobile exhaust monitoring stations located in 177 cities. Since automobile emissions are the main source of CO, many of the monitoring stations are situated near roadways or intersections. Concentrations of CO are determined by non-

Figure 5.1. National air monitoring network. ● National air pollution monitoring stations installed in major air polluted areas; ○ national background air monitoring stations collecting data for ambient air quality standards

dispersive infrared analysis. To satisfy the air quality standard, the concentration of CO should be less than 10 ppm for the daily average of hourly values, or less than 20 ppm for the average of hourly values taken over eight consecutive hours.

SUSPENDED PARTICULATES

Suspended particulates are defined as those particles with a diameter of 10 μm or less which are suspended in the atmosphere for relatively long periods and which have considerable impact on human health. The air quality standard for suspended particulates states that the daily average of hourly values should not exceed a concentration of 0.10 mg m^{-3}, while hourly values must not exceed 0.20 mg m^{-3}. The concentrations of suspended particulates in the atmosphere may be determined gravimetrically using collection by filtration, by a light-scattering method, by a piezo-electric microbalance method or by a β-ray attentuation method.

NITROGEN DIOXIDE

Data for NO_2 detected by a colorimetric method using the Saltzman reagent were recorded at 1206 general and 260 automobile exhaust monitoring stations in 1980. The air quality standard set for NO_2 restricts daily averages of hourly values to maximum concentrations of 0.04–0.06 ppm.

PHOTOCHEMICAL OXIDANTS

The environmental air-quality standard for photochemical oxidants states that the hourly values should not exceed 0.06 ppm. In 1980, 953 general monitoring stations in 485 cities recorded concentrations of photochemical oxidants. Oxidants are measured using a coulometric method after absorption in 2% neutral-buffered potassium iodide solution. Warnings are issued when the hourly oxidant concentration reaches 0.12 ppm; when concentrations reach or exceed 0.24 ppm, alarms are issued.

Present state of air pollution

SULPHUR DIOXIDE

In the 1960s, the annual average concentrations of SO_2 were approximately 0.06 ppm and visible symptoms of SO_2 injury could be detected on the leaves of many crop plants. However, as a result of the implementation of various anti-pollution measures, such as the establishment of an air quality standard for SO_2 and the steady increase in the number of scrubbers to remove sulphur from stack gases or fuels, the concentration of SO_2 in formerly heavily polluted areas has been markedly reduced to a level well below the standard (*Figure 5.2*). The annual mean concentration of SO_2 monitored continuously from 1965 at 15 general monitoring stations has shown a steady decrease from a peak of 0.059 ppm in 1967 to 0.017 ppm in 1978 and 0.016 ppm in 1980 despite an increase in the consumption of fossil fuels.

CARBON MONOXIDE

Concentrations of CO increased until 1969, thereafter decreasing due to a successive tightening of emission regulations. Between 1970 and 1980, the annual average

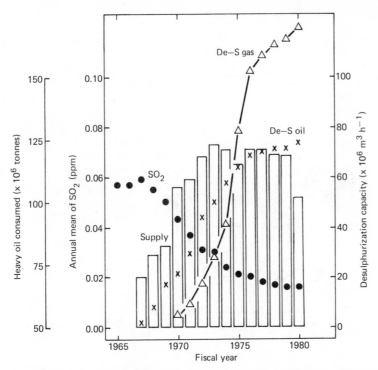

Figure 5.2. Changes in annual average concentrations of sulphur dioxide (●), desulphurization capacities of crude oil (×) and of stack gases (△), and amounts of heavy oil consumed (□), Nationwide, 1965–1980. Concentrations of sulphur dioxide are means from 15 stations. Desulphurization capacities of stack gases are estimated at 0 °C, 1 atm.

concentrations of CO at three automobile exhaust monitoring stations in Tokyo decreased by about 70%. *Figure 5.3* shows the number of monitoring stations for automobile emissions where CO concentrations exceeded the standards, as a percentage of the total number of stations reporting concentrations that satisfied a criterion for validity. To be considered valid for use in emission statistics, data from a monitoring station must consist of at least 6000 hourly samples recorded during a whole year. In 1976, for example, the standard for concentrations of CO was still being exceeded at 32 (or 11.3%) of the 283 automobile exhaust monitoring stations reporting a full year's valid data. In 1980, however, only three (or 0.9%) of 322 automobile exhaust monitoring stations which recorded valid data reported CO concentrations that exceeded the standard, and all 205 general monitoring stations reported CO concentrations that complied with the standard.

SUSPENDED PARTICULATES

Of all the regulated major air pollutants, suspended particulates improved the least. By 1980, the environmental air quality standard for suspended particulates was being achieved at only 79 (or 29.2%) of 271 general monitoring stations which reported more than 6000 hourly records. The problem locations are widely dispersed throughout the country with many stations reporting high concentrations of suspended particulates.

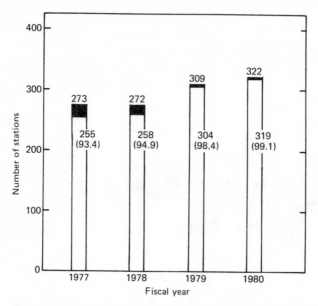

Figure 5.3. Number of automobile exhaust monitoring stations recording carbon monoxide, and stations which complied with the ambient standards. Numbers in parentheses indicate the percentage of stations complying with the standard

NITROGEN DIOXIDE

Since air pollution due to NO_2 is largely affected by automobile emissions in addition to emission from stationary sources, the progress of improvement in atmospheric quality is quite slow compared with SO_2 (*Figure 5.4*). For automobile emissions control, a permissible level of NO_x emission is established in accordance with the Air Pollution Control Law, and a specific emission level is further provided for various types of automobiles based on the above permissible level (*Table 5.2*). The major measures taken to reduce NO_2 emission from the stationary sources include the implementation of combustion techniques designed to reduce the formation of NO_x, fuel conversion and denitrification of stack gases. The improvements achieved in automobile emissions control have been offset somewhat by an increase in the number of motor vehicles and in fuel consumption, so that total nationwide improvements in NO_2 pollution have not been achieved.

PHOTOCHEMICAL OXIDANTS

The number of warnings for photochemical oxidants increased steadily until 1973, then decreased in the following years (*Figure 5.5*). In 1976 and 1977, there were no alarms issued, but three alarms were issued in 1978. Furthermore, the occurrence of typical symptoms of injury by photochemical oxidants was widely reported on many crop plants in various regions, especially in the Kantoh District. Peroxyacetyl nitrate (PAN) is recognized as a photochemical oxidant. As there are no continuous monitoring stations for PAN, it is not known to what extent the air in Japan is polluted by PAN. There are, however, increasing reports of visible injuries typical of PAN being detected on water melon (*Citrullus vulgaris*), peanut (*Arachis hypogaea*), soybean (*Glycine max*),

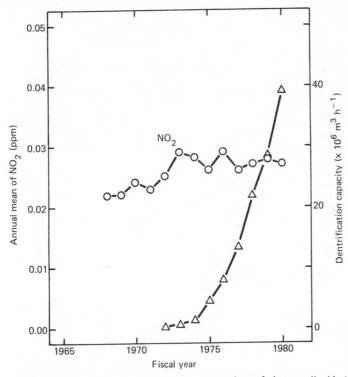

Figure 5.4. Changes in annual average concentrations of nitrogen dioxide (O) and denitrification capacity (△), Nationwide, 1968–80. Nitrogen dioxide concentrations are means from six stations. Denitrification capacities are estimated at 0 °C, 1 atm.

TABLE 5.2. Automobile emission standards

Type of vehicles		Standard		% Reduction	
		Regulation		Regulation	
		I	II	I	II
Normal- and small-sized vehicles using light oil as fuel	Direct injection type	540 ppm	470 ppm	17	18
	Auxiliary chamber type	340 ppm	290 ppm	11	24
Normal- and small-sized vehicles using gasoline or LPG as fuels (except vehicles specially designed to carry ten or less passengers)	Vehicle weight exceeding 2.5 tonnes	1100 ppm	750 ppm	29	52
	Vehicle weight more than 1.7 tonnes but less than 2.5 tonnes	1.2 g km^{-1}	0.9 g km^{-1}	33	50
	Vehicle weight of 1.7 tonnes or less	1.0 g km^{-1}	0.6 g km^{-1}	44	67
Light-weight vehicle (vehicles designed specially to carry passengers, except those vehicles with two-cycle engines)		1.2 g km^{-1}	0.9 g km^{-1}	33	50

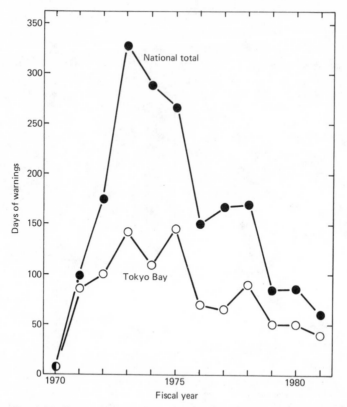

Figure 5.5. Changes in the number of days when photochemical oxidants warnings were issued either nationwide (●) or in the Tokyo Bay area (○). Tokyo Bay area denotes Tokyo, Kanagawa, Chiba, and Saitama prefectures

corn (*Zea mays*) and other plant species. Although the current evidence suggests that the atmospheric environment in urban and suburban districts in Japan is changing, any improvement in air quality is only slight. Most people in Japan believe that air pollution problems have been almost completely overcome, but the present state of the atmospheric environment indicates that their optimism is unfounded. It is true that air pollution due to SO_2 and CO has improved markedly, but photochemical smog, which is one of the most serious forms of air pollution both to man and to vegetation, is not effectively controlled.

NON-REGULATED AIR POLLUTANTS

It is becoming increasingly evident that the air pollutants for which emission controls have been established do not represent all the important factors to be considered as regards air quality. A main product of combustion is NO, which is oxidized to NO_2 in the air. The sources of NO are often stationary, but emission from automobiles is also an important source. Concentrations of NO have been recorded by 1206 general monitoring stations in 546 cities, and by 260 automobile exhaust monitoring stations in 155 cities in 1980. Between 1973 and 1980, the annual average concentration of NO decreased slightly from 0.030 to 0.023 ppm. The concentrations of NO recorded by automobile exhaust monitoring stations are generally higher than those recorded by the general monitoring stations. For example, the annual average concentrations of NO reported by 26 automobile exhaust monitoring stations were 0.099 ppm in 1973 and 0.078 ppm in 1980.

Hydrocarbons are emitted from various stationary and mobile sources. Hydrocarbons are the major precursors in the formation of photochemical oxidants in the atmospheric reactions. Thus, in 1976, the Central Council for Environmental Pollution Control proposed guidelines for ambient concentrations of hydrocarbons to prevent the formation of photochemical oxidants. The new guidelines propose that the average concentrations of non-methane hydrocarbons in the air over any 3 h from 06.00 h should be restricted to a maximum of 0.20–0.31 ppm carbon, depending upon the region.

Compared with other air pollutants, little monitoring is done for hydrocarbons. In 1980, 219 general monitoring stations in 162 cities reported valid data; of these, 74% reported concentrations of hydrocarbons that exceeded the proposed guidelines. Data from automobile exhaust monitoring stations indicate that exhaust gas is indeed a major source of hydrocarbons. For example, the annual average concentrations of non-methane hydrocarbons reported from 17 automobile exhaust monitoring stations was 0.77 ppm in 1977 and 0.63 ppm in 1980 (*Table 5.3*).

TABLE 5.3. Changes in annual average of non-methane hydrocarbons[a]

	Fiscal year		
1977	1978	1979	1980
0.77	0.73	0.66	0.63

[a] Mean value (in ppm) of 17 automobile exhaust monitoring stations.

Impact of photochemical smog in the Kantoh district

The rapid growth of the national economy from the early 1960s brought the Japanese both wealth and an increase in pollution. Pollution by SO_2 increased rapidly as the consumption of petroleum fuels rose steeply during the era of high economic growth. The effort to control the emission of SO_2 from stationary sources has resulted in a marked decrease in SO_2 concentrations in industrial and urban districts. Measures to control the emission of SO_2 and other oxides of sulphur relied on the use of low-sulphur fuels, tall stacks to disperse and hence dilute emissions over a greater area and desulphurization procedures to remove sulphur from fuel oils and stack gases. Since resolving the problem of air pollution by SO_2, the ambient concentrations of photochemical oxidants became the major pollution problem in big cities or industrial areas; Kantoh district has faced such pollution problems for the past ten years.

Kantoh district consists of six prefectures (Ibaraki, Chiba, Saitama, Gunma, Kanagawa and Tochigi) and the Tokyo metropolis. This district contains approximately one-third of the population of Japan; numerous monitoring stations are located in this district to record continuously the ambient concentrations of air pollutants.

In 1975, the agronomists in the Kantoh district began to estimate the effects of air pollutants on vegetation (The Union of Air Pollution Prevention Association in Kantoh District, 1975–1981). Their studies of damage to vegetation focused on

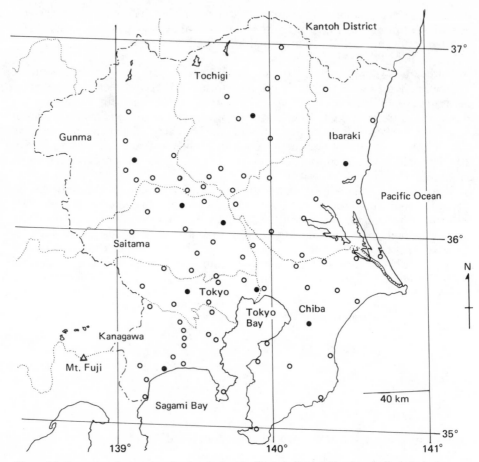

Figure 5.6. Vegetation damage monitoring stations in Kantoh district. Numbers indicate latitude or longitude. ● Stations where cultivation of indicator plants was performed to determine the factors controlling the occurrences of foliar injuries caused by ambient photochemical oxidants

photochemical smog, which reduced both the quality and yield of commercial crops. Damage to vegetation is estimated either by controlled experiments for particular pollutants and crops or by field studies using selected crops. Their main purposes are to estimate the cumulative damage caused by photochemical oxidants and to measure yearly changes in the extent of damage. Eighty-four stations distributed uniformly in the Kantoh district are used to monitor damage to vegetation (*Figure 5.6*).

Plant species used to estimate damage to vegetation include morning glory (*Pharbitis nil* var. Scarlet O'Hara), peanut (*Arachis hypogaea*) and taro (*Colocasia antiquorum* var. *esculenta*). The total number of oxidant warnings issued in the summer season (from June to August) in the Kantoh district has been reduced from 180 in 1973 to 42 in 1981 (*Figure 5.7*). Oxidant concentrations are known to be dependent upon meteorological factors and geographical conditions. It is generally observed that high concentrations of oxidants result largely from local sources during light wind conditions (*Figure 5.8*). Concentrations of oxidants (calculated as dose, ppm × h) at each station for the summer season were plotted, and isopleths drawn to the data points

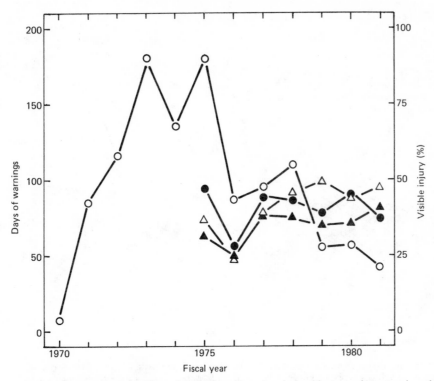

Figure 5.7. Changes in the number of days when photochemical oxidant warnings were issued (○), and visible injuries appeared on morning glory (●), peanut (△) or taro (▲), Kantoh district, 1970–1981

(*Figure 5.9*). The isopleths indicate that the contribution from the Tokyo metropolis where automobile emissions are considered to be higher than in any other region in the Kantoh district does not account for the high oxidant concentrations observed in the local regions.

Because most of the sources of SO_2 are stationary, regional concentrations of SO_2 have been clearly related to meteorological (wind velocity and direction) and geographical (distance from source) conditions (Raynor, Smith and Singer, 1974; Krouse, 1980). The major sources of photochemical oxidants, on the other hand, are mobile, consisting mainly of emission from automobiles. Furthermore, hydrocarbons and NO react in the atmosphere to produce photochemical oxidants. These complications make it difficult to forecast the regional distribution of concentrations of photochemical oxidants.

Trials using plants as indicators of air pollution have been conducted to evaluate the zonal frequencies of the occurrence of phytotoxic concentrations of photochemical oxidants. Tobacco plants (*Nicotiana tabacum*), especially the variety Bel-W3, are widely used as indicator plants in the USA (Heggestad and Menser, 1962). However, in Japan the cultivation of tobacco is restricted by law since the tobacco companies have a monopoly. Among the plants employed as indicators in the Kantoh district, morning glory is preferred because of its high sensitivity to photochemical oxidants and favourable growth habit (Sawada *et al.*, 1972). The results of field surveys and exposure experiments have confirmed the suitability of morning glory as an indicator plant for photochemical oxidants (Nouchi and Aoki, 1979).

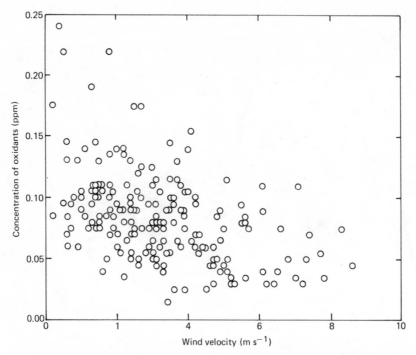

Figure 5.8. The relationship between wind velocity and daily maximum oxidant concentrations in Yokohama. Wind velocity was determined at 9.00 a.m. Data were reported by the Bureau of Environment Protection, Yokohama City, 1976

In Japan, the season for photochemical smog lasts from the end of April to the beginning of October; warnings for photochemical oxidants are issued mostly in July and August. Due to these air pollution situations, the cultivation of bio-indicator plants usually begins in May or June for observations to be made in July and August.

Although the frequencies of oxidant warnings issued in the Kantoh district have decreased over the past seven years, the occurrence of visible injury observed on leaves generally increased (*Figure 5.7*). Furthermore, the occurrence of severe foliar injury on morning glory has spread widely throughout the whole Kantoh district (*Figure 5.10*). The regions where no or only very slight injury could be detected were those facing the Pacific Ocean in south-eastern Chiba and eastern Ibaraki.

In addition to these field surveys, agronomists selected nine stations (one or two stations per prefecture, shown in *Figure 5.7* as solid circles) in the Kantoh district to determine the factors controlling the occurrence of foliar injury in field-grown plants. They cultivated morning glory (var. Scarlet O'Hara), peanut (cv. Dosui) and taro (cv. Chiba-Handachi) at a specified date using the same nutritional conditions and source of seedlings. The observations started in 1976 have been continued to the present.

The effects of dose of oxidant on the degree of foliar necrosis observed on morning glory cultivated in the field are shown in *Figure 5.11*. The threshold concentration of oxidants required for the appearance of foliar necrosis was assumed to be 0.05 ppm, and the oxidant dose values were therefore calculated by integrating only those concentrations greater than 0.05 ppm with time for each observation period (usually

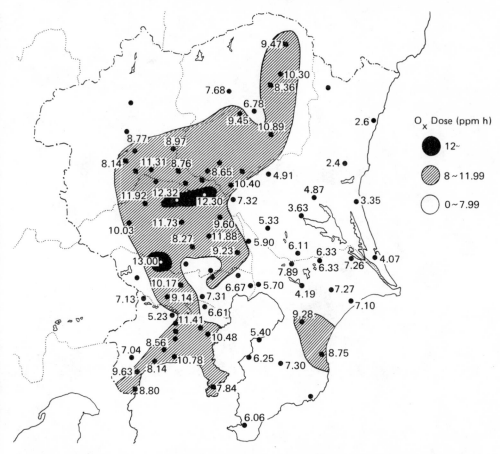

Figure 5.9. Dose isopleths of oxidants concentration in Kantoh district. Numbers indicate the dose (ppm h) of oxidants monitored for one month from July 1st to July 30th in 1981

three to seven days). The degree of foliar injury was assessed by calculating the mean percentages of injured leaves (5% minimum, with increments of 10%).

The results shown in *Figure 5.11* include data from nine different stations. The relation between dose and the degree of foliar injury are represented by linear regressions for three representative years, namely, 1977, 1979 and 1981. There are significant ($P = 0.01$) correlations between these two factors for every regression; thus the trends seem to be real despite the scatter of the points.

It has been well documented that the response of plants to O_3, a major component of that class of pollutants called photochemical oxidants, is influenced by environmental and nutritional factors as well as the age of the leaves (Ting and Heath, 1975). Thus, in a field study, it may be reasonable to assume that plants cultivated in different regions respond quite differently in terms of foliar damage sustained from the ambient concentrations of photochemical oxidants, resulting in the scatter of points observed in *Figure 5.11*.

The trends are different among the years considered. In 1977, the levels of oxidant

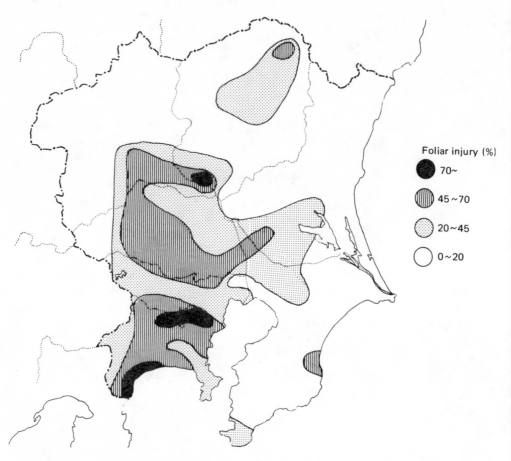

Figure 5.10. Foliar injury isopleths. Foliar injury appearing on morning glory at stations monitoring vegetation damage was plotted and lines of best fit were drawn

dose were high (the maximum was about 2.8 ppm h) and approximately 60% foliar necrosis was observed. In contrast, in 1981 the levels of oxidant dose were lower (the maximum was about 0.8 ppm h), but the degree of foliar injury was comparable with that observed in 1977. The high sensitivity of plants to photochemical oxidants observed in 1981 may be partly due to the growth conditions that year; the climatic conditions were unseasonable at the time of transplanting and plant growth was slower than usual at every monitoring station. Since the sensitivity of plants to oxidants is greatly influenced by growth rate, as discussed by Heck, Dunning and Hindawi (1965), it may be that these plants showing slower growth rates were more susceptible to oxidant injury. However, the results for 1981 showed that taro and peanut, although they grew well, also showed an enhanced sensitivity to the low oxidant dose. The summer of 1980 was rainy and cool. Due to these climatic conditions, plant growth was very poor and the concentrations of photochemical oxidants low, but foliar injuries attributed to photochemical oxidants were still observed.

The foliar symptoms caused by PAN are quite different from those caused by O_3, making it theoretically possible to distinguish between these two photochemical

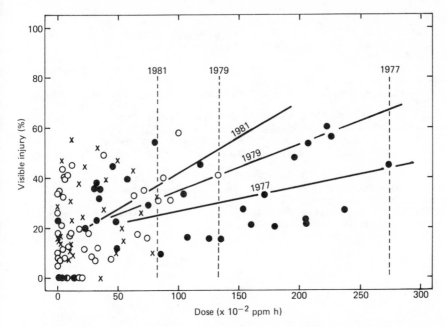

Figure 5.11. Relation between dose of oxidants and the degree of foliar injury appearing on morning glory in 1977 (●), 1979 (○), and 1981 (×) in Kantoh district. ------ indicate the maximum dose detected each year and ——— indicate the linear regression lines for each year as follows; 1977: $y = 16.10 + 0.11 \times (r = 0.541)$, 1979: $y = 15.26 + 0.20 \times (r = 0.422)$, 1981: $y = 10.73 + 0.31 \times (r = 0.425)$, where $y =$ the degree of foliar injuries and $x =$ dose of oxidants

oxidants. In the field, silvering of the lower surfaces of leaves, a typical symptom of PAN injury, was identified; however, the injured areas soon become necrotic making practical diagnosis impossible. The toxicity of PAN is about five times greater than that of O_3, so visible injury could be caused by lower concentrations of this photochemical oxidant. Because of this and of some observation of PAN-type injury in the field, there is the possibility that the increase in injury shown by the bio-indicator plants may be caused by PAN.

Acknowledgements

I thank Dr Y. Matsuoka and Mr T. Takasaki (the Chiba Agricultural Research Center), Mr M. Shinozaki (the Kanagawa Research Center for Environmental Protection) and Mr I. Nouchi (the Tokyo Metropolitan Research Institute for Environmental Protection) for providing the data.

References

HECK, W. W., DUNNING, J. A. and HINDAWI, I. J. Interactions of environmental factors on the sensitivity of plants to air pollution, *Journal of the Air Pollution Control Association* **15**, 511–515 (1965)
HEGGESTAD, H. E. and MENSER, H. A. Leaf-spot sensitive tobacco strain Bel-W_3, a biological indicator of the air pollutant ozone (Abstract), *Phytopathology* **52**, 735 (1962)
JAPAN ENVIRONMENT AGENCY Air pollution: present state and countermeasures. In *White Paper on Environment for 1982*, Japan Environment Agency, Tokyo, pp. 171–226 (1982)

KROUSE, H. R. Sulphur isotopes in our environment. In *Handbook of Environmental Isotope Geochemistry*, (Eds. P. FRITZ and J. C. FONTES), pp. 435–471, Elsevier, Amsterdam (1980)

MATSUOKA, T. Utilization of morning glory as an indicator plant for photochemical oxidants in Japan. In *Proceedings of the 4th International Clean Air Congress*, (Eds. S. K. KASUGA, N. SUZUKI, T. YAMADA, G. KIMURA, K. INAGAKI and K. ONOE), The Japanese Union of Air Pollution Prevention Association, Tokyo, pp. 91–94 (1977)

NOUCHI, I. and AOKI, K. Morning glory as a photochemical oxidant indicator, *Environmental Pollution* **18**, 289–303 (1979)

RAYNOR, G. S., SMITH, M. E. and SINGER, I. A. Meteorological effects of sulfur dioxide concentrations on suburban Long Island, New York, *Atmospheric Environment* **8**, 1305–1320 (1974)

SAWADA, T., KOMEIJI, T., NOUCHI, I., OGUCHI, K. and OHDAIRA, T. Symptoms and distribution of plant damage which was thought to be due to photochemical smog in Tokyo, *Journal of Japan Society of Air Pollution* **7**, 232 (1972)

THE BUREAU OF ENVIRONMENT PROTECTION IN YOKOHAMA CITY Photochemical smog. In the *17th Reports on the Survey of Air Pollution in Yokohama City*, The Bureau of Environment Protection in Yokohama City, Yokohama, pp. 102–145 (1976)

THE UNION OF AIR POLLUTION PROTECTION ASSOCIATION IN KANTOH DISTRICT Study on the effects of photochemical smog on plants (1975–1981)

TING, I. P. and HEATH, R. L. Responses of plants to air pollutant oxidants, *Advances in Agronomy* **27**, 89–121 (1975)

Part II

Pollutants and plant cells

Chapter 6

Permeability of plant cuticles to gaseous air pollutants

K. J. Lendzian

INSTITUT FÜR BOTANIK UND MIKROBIOLOGIE DER TECHNISCHE UNIVERSITÄT MÜNCHEN, FRG

Introduction

The evolution of the cuticle may well have been one of the most crucial factors in the successful colonization of land by bryophyta, pteridophyta and spermatophyta. These plants developed a continuous hydrophobic interphase between their surroundings and their interior so that the loss of water could be reduced to a minimum; but this was in conflict with a second desirable feature, namely the exchange of gases. Generally, the plant achieved a compromise of structure to meet both these requirements. It retains a large surface area, covers it with a cuticle and permits gas exchange through stomata. But areas of a variety of plants are lacking stomata at the adaxial leaf surface or stomata are completely absent, as in the case of the epidermis of the pericarp of many fruits. With respect to astomatous cuticles, there exists the general view that the cuticle limits the exchange of gases. The attempt to quantify the relative impermeability of the cuticle to gases led to diverse speculations such as:

(1) the cuticle is impermeable to gases, or
(2) the cuticle may be significantly permeable.

Conclusive data on the permeability of plant cuticles to gases such as CO_2 and O_2 and for gaseous air pollutants do not so far exist.

A new approach is presented to describe and quantify the permeability to gases of astomatous cuticles from selected plants. The goal of that approach was to determine phenomenological coefficients such as permeability and diffusion coefficients which describe the permeability properties of cuticles and which are directly comparable with the data for cuticular permeability to water. The permeabilities of gaseous air pollutants, CO_2 and O_2, were determined simultaneously to provide a measure for permeation characteristics.

Methods

Astomatous cuticles were isolated from the adaxial leaf surfaces of orange (*Citrus aurantium*) and from the fruits of tomato (*Lycopersicon esculentum*). For that purpose small discs (15–20 mm in diameter) were punched from the leaves and from the pericarp. Care was taken that the cuticles of the discs showed no visible damage. The discs were incubated at 37 °C in a mixture of pectinase, cellulase and sodium azide (2 mM) adjusted to pH 4.0 (Orgell, 1955). After one week, the cuticles were separated from the outer walls of the epidermis. All cuticles were converted into the H^+ form by three successive (10 min) treatments with 1 N HCl. The cuticles were then washed extensively with deionized water and were air-dried. This cuticular material is designated as cuticular membranes (CM). The isolation procedures applied do not change the structure and function of the cuticles.

In a typical experiment a cuticle was mounted between two half cells, one containing a gaseous phase and the other a liquid phase (buffer). The morphological exterior of the cuticle was always in contact with the gaseous phase which contained one of the three gases under investigation. A detailed description of oxygen measurements is given by Lendzian (1982). SO_2 and CO_2 permeation were determined by monitoring the permeation of the tracers $^{35}SO_2$ and $^{14}CO_2$. The permeation apparatus for such experiments is shown in *Figure 6.1*. Water permeability was determined according to Schönherr and Lendzian (1981).

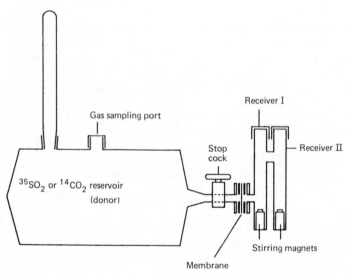

Figure 6.1. Diagram of the apparatus used for SO_2 and CO_2 permeation experiments

Under steady-state conditions (time independence of gas fluxes, no significant changes in the driving force) permeability coefficients (P) were calculated using the equation:

$$P = \frac{M \cdot V}{\Delta t \cdot A \cdot \text{driving force}} \quad (m\ s^{-1})$$

where M is the amount of gas that permeated the membrane of area A (m^2) in the time interval Δt (s), and V (m^3) is the volume of the receiver. The driving force is the gas concentration gradient or the partial pressure gradient between the two half cells.

Results and discussion

Permeability coefficients (P) for different gases and for different cuticles are given in *Table 6.1*. Permeability was always highest for SO_2 and O_2. Carbon dioxide permeability was remarkably lower but exceeded by far the permeability for water at comparable driving forces. With respect to permeation characteristics the SO_2 molecule seems to behave much like the O_2 molecule. Similarities between the molecular structure of SO_2 and CO_2 are not evident.

TABLE 6.1. Permeability coefficients (P) for different cuticles before (CM) and after extraction (MX) of the soluble lipids with chloroform

Plant	P (m s^{-1})			MX/CM
	CM	MX	sol. lipids [a]	
Citrus	1.09×10^{-9} (H_2O)	2.54×10^{-7}	1.09×10^{-9}	233
	5.40×10^{-8} (CO_2)	1.41×10^{-6}	5.62×10^{-8}	26
	2.01×10^{-7} (SO_2)	3.48×10^{-6}	2.13×10^{-7}	17
	3.05×10^{-7} (O_2)	4.65×10^{-6}	3.26×10^{-7}	15
Tomato	2.47×10^{-8} (H_2O)	9.17×10^{-8}	3.38×10^{-8}	3.7
	2.20×10^{-7} (CO_2)	6.64×10^{-7}	3.29×10^{-7}	3.1
	1.10×10^{-6} (O_2)	2.13×10^{-6}	2.27×10^{-6}	1.8
	1.22×10^{-6} (SO_2)	2.15×10^{-6}	2.82×10^{-6}	1.9

[a] Calculated.

P values for SO_2, O_2 and CO_2 are real constants as they do not change when driving forces are altered. This means that all three gases obey Henry's law (proportionality between the partial pressure of a gas and the mole fraction of that gas in a liquid or solid phase). Thus, P values can be used to calculate the actual flux of gases across a cuticular membrane at a given concentration gradient up to a partial pressure of 1 atm. Partial pressures exceeding one atmosphere have not been applied in the experiments described.

The gas permeability of the cuticular membranes investigated was low. The magnitude of hindrance for gas permeation provided by these cuticles can be obtained by comparing the permeability coefficients reported here with those calculated for a water layer of equal thickness and comparable driving force. Cuticular membranes of *Citrus* leaves, for example, were approximately 13 000 (CO_2), 2400 (O_2) and 3500 (SO_2) times less permeable than a water layer. Sulphur dioxide permeation was reduced by a factor of about 2×10^7 compared with permeation through an unstirred air layer of equal thickness. Thus, permeability for gases is strongly reduced but the cuticle is not impermeable for any of the gases investigated.

Plant cuticles are polymer membranes of a heterogeneous composition. In general, they are composed of a cutin polymer matrix and soluble cuticular waxes. As the resistance of the cuticle for water permeation is exclusively determined by the chloroform-extractable waxes (Schönherr, 1976b), the question arose whether such an extraction of the waxes would have any effect on gas permeation. For that purpose cuticular membranes were extracted with chloroform according to Lendzian (1982). After this procedure extracted cuticles (cuticular matrices, MX) were remounted into the gas permeation apparatus and gas permeation was determined. As can be seen in

Table 6.1 extraction of cuticular waxes increased the gas permeability and thus the permeability coefficients. The increase in permeability varied for different cuticles from different plants and for different gases. The effect of extraction was always highest on water permeability and lowest on SO_2 and O_2 permeability. From these data it can be concluded that the soluble waxes of the cuticle play a role in the permeation process. It is unlikely that the barrier for gas permeation is built up by the epicuticular waxes as they do not build up an uniform boundary layer. It is more likely that the waxes embedded within the cuticle form a continuous resistance. Accepting this hypothesis the contribution of the soluble waxes to the overall gas permeability of the cuticles can be estimated by treating the cutin polymer matrix and the soluble lipids as two resistances acting in series:

$$\frac{1}{P(CM)} = \frac{1}{P(MX)} + \frac{1}{P(\text{sol. lipids})}$$

Using this equation permeability coefficients for the permeation of gases through soluble lipids can be calculated (*Table 6.1*). In cuticles of *Citrus* the resistance of the cuticular membrane to water and the three gases investigated was almost completely determined by the resistance contributed by the soluble waxes as P values for CM and soluble waxes were almost identical. In tomato cuticles both the cutin matrix and the soluble lipids contributed to the overall resistance for O_2 and SO_2 permeation.

Gas permeability for *Citrus* cuticles was independent of membrane thickness as the cuticle consists mainly of a cutin polymer matrix (about 97% w/w). The thickness of cuticles from tomato fruits seems to play a minor role for the hinderance of O_2 and SO_2 permeation as about 50% of the resistance was contributed by the soluble waxes which amount to only 4–6% of the total weight.

There are two alternatives for permeation of SO_2 in the cuticle:

(1) SO_2 uses the same pathway as water does, or
(2) SO_2 dissolves in the lipid phase.

In other words, for SO_2 permeability it has to be determined if the cuticle is a porous membrane or a solubility membrane. Schönherr (1976a) has demonstrated that water permeability increases by increasing the pH and thus by increasing the water content of the cuticle. If SO_2 takes the same pathway as water does, the permeability should also increase with increasing pH and in the presence of Na^+ ions. That is not the case. Thus, SO_2 permeates the *Citrus* cuticle via the hydrophobic phase.

To test this hypothesis, the partition coefficient for SO_2 between water and the

TABLE 6.2. Hold-up time (t_0), diffusion coefficient (D), partition coefficient (S_M/S_W) and solubility (S_M) of SO_2 for cuticular membranes (CM) and cuticular polymer matrices (MX) of *Citrus*

Membrane	t_0 (s)	D ($m^2\ s^{-1}$)	S_M/S_W [a]	S_M [b] ($kg\ m^{-3}$)
CM	282	5.32×10^{-15}	688	5.44
MX	198	7.58×10^{-15}	843	6.66

t_0 and D were determined according to Lendzian (1982).

[a] $S_M/S_W = \dfrac{P \cdot x}{D}$; $x = $ thickness $= 3 \times 10^{-6}$ m.

[b] $S_W = $ Henry constant $= 7.9 \times 10^{-3}\ kg\ m^{-3}$ (pH 8.0).

cuticle and the solubility of SO_2 in the membrane have been determined. Sulphur dioxide is 688 times more soluble in the CM and 843 times more soluble in the MX than in water (*Table 6.2*). Introducing the Henry constant, the solubility of SO_2 in the cuticle is calculated to be around 5.44 kg SO_2 m^{-3} (for comparison: 0.32 kg O_2 m^{-3} and 0.022 kg CO_2 m^{-3} membrane). These data clearly show that SO_2 is much more soluble in the cuticle than in water. One has to conclude that SO_2 permeates the Citrus cuticle via the lipophilic phase. All data point to a solubility process with respect to SO_2 permeation. The permeability of other gaseous air pollutants have not been investigated so far.

References

LENDZIAN, K. J. Gas permeability of plant cuticles. Oxygen permeability, *Planta* **155**, 310–315 (1982)

ORGELL, W. H. The isolation of plant cuticle with pectic enzymes, *Plant Physiology* **30**, 78–80 (1955)

SCHÖNHERR, J. Water permeability of isolated cuticular membranes: the effect of pH and cations on diffusion, hydrodynamic permeability and size of polar pores in the cutin matrix, *Planta* **128**, 113–126 (1976a)

SCHÖNHERR, J. Water permeability of isolated cuticular membranes: the effect of cuticular waxes on diffusion of water, *Planta* **131**, 159–164 (1976b)

SCHÖNHERR, J. and LENDZIAN, K. J. A simple and inexpensive method of measuring water permeability of isolated plant cuticular membranes, *Zeitschrift für Pflanzenphysiologie* **102**, 321–327 (1981)

Chapter 7

Uptake and distribution of pollutants in the plant and residence time of active species

S. G. Garsed
IMPERIAL COLLEGE AT SILWOOD PARK, ASCOT, UK

Introduction

In the last ten years we have learned from studies with whole plants the extent to which responses to pollution depend on the concentration(s) of the pollutant(s) and the duration of exposure. We can thus assume that the physiological and biochemical changes associated with the responses will also be concentration-dependent. This chapter provides a framework within which to link our knowledge of the relationship between concentration and injury at the whole plant level with our current understanding of the biochemical effects of pollutants. However, it will become apparent how little we still know of the nature of the chemical reactions of pollutants supplied at concentrations relevant to most polluted areas in the field, so the discussion of particular 'active species' will, of necessity, be rather limited.

The interpretation of plant damage

The relationship between accumulation and injury

With the exception of ozone, peroxyacetyl nitrate (PAN) and ethylene, which do not accumulate in plants to any great extent, exposure of foliage to atmospheric pollutants normally results in an increase in the concentration of the constituent element in plants. Thus SO_2 can increase plant sulphur content three- to fourfold (Katz, 1939), while exposure to HF can increase foliar fluoride by tenfold to one hundredfold (Knabe, 1970). Although many studies of the association between elemental accumulation and plant damage have been made, it has long been recognized that no sound mathematical relationship between the contents of elements in leaves and injury can be described (for SO_2, *see* Katz, 1939; Thomas, Hendricks and Hill, 1950a; for fluorides, *see* Knabe, 1970; Davison and Blakemore, 1976; Davison, 1982; for Cl_2, *see* Brennan, Leone and Daines, 1966).

There are several reasons for this lack of correlation. Firstly, the quantity of pollutant required to cause injury will depend on the flux per unit time. Guderian (1970) has shown for SO_2 that for the same product of *concentration × time*, greater accumulation of sulphur is generally obtained by reducing the concentration and increasing the time, whereas acute plant injury is more severe if the exposure concentration is increased and the time reduced.

Secondly, there is the turnover of pollutant derivatives within the plant. Several processes have been identified, including translocation, dilution by new growth and losses through leaching, gaseous emission or exudation through the roots. Moreover various pollutants may behave differently in all of these respects.

Thirdly, environmental factors will influence the sensitivity of plants to the pollutants. Such interactions are beyond the scope of this chapter and will not be discussed further.

Fourthly, individual metabolic systems within different species may vary in their ability to tolerate the primary pollutant or its products.

Fifthly, individual authors may differ in their views of what constitutes injury, as illustrated by the 'invisible injury' controversy. The groups led by Thomas and by Katz in the 1930s and 1940s concluded that decreases in plant yield resulting from SO_2 pollution occurred only if necrotic lesions were apparent. Their views on the theory that reductions in growth could occur in the absence of visible injury (Stoklasa, 1923) were quite uncompromising: 'No experimental support for this (invisible injury) theory has appeared, but on the contrary a large amount of experimental work has been done which demonstrates that this theory is without foundation' (Thomas, 1951). However, recent work with both ambient versus filtered air experiments (Bleasdale, 1973; Crittenden and Read, 1978; Awang, 1979) and fumigations with single (Taniyama, 1972; Ashenden, 1978; Black and Unsworth, 1979; Bell, Rutter and Relton, 1979) or combined pollutants (Bull and Mansfield, 1974; Ashenden and Williams, 1980) have shown that pollutants can have adverse effects on plants in the absence of visible symptoms.

The problems of defining injury have been underlined further by reports that there is no apparent relationship between sensitivity to acute injury, characterized by visible damage at high gas concentrations, and the effects on growth of lower concentrations over a longer period (Horsman et al., 1979; Ayazloo and Bell, 1981). This work shows not only that estimates of plant sensitivity obtained from experiments involving high concentrations of SO_2 cannot be applied reliably to field situations, but also that there must be more than one mechanism of resistance. Further, maximum sensitivity to acute injury caused by SO_2 apparently occurs under conditions favourable to rapid growth (Thomas, Hendricks and Hill, 1950b), whereas the plants seem to be more sensitive to the long-term effects of the pollutant when growing slowly (Davies, 1980). The fact that resistance to one pollutant cannot necessarily be correlated with resistance to another (e.g. Zimmerman and Hitchcock, 1956, for HF and SO_2) and that even results from greenhouse and field experiments can give different indications of the relative sensitivity of cultivars (Manning and Feder, 1976, for ozone) merely adds to the complications.

The biochemical background to the invisible injury controversy

The groups led by Thomas and by Katz were primarily interested in the effects of acute doses of SO_2. Both groups found that when acute injury occurred, free sulphurous acid could be recovered from the leaves, although the amount declined rapidly after exposure, particularly in the light (Katz, 1939; Thomas, Hendricks and Hill, 1950b). Furthermore there was a direct correlation between leaf injury and the amount of sulphur dioxide or sulphurous acid in the tissues. If low concentrations of SO_2 were given to plants adequately supplied with sulphur, the accumulation was primarily as sulphate with a small increase in organic sulphur. As a result, Thomas, Hendricks and

Hill (1950b) proposed that the effects of SO_2 were local, not systemic and that there was a threshold below which sulphurous acid could be oxidized and rendered less toxic. Above this threshold, free sulphurous acid accumulated and damaged the cells. To support this hypothesis they restated the formula of O'Gara:

$$(c - l)t = k$$

where c is the actual concentration and l is a limiting concentration which can be endured indefinitely.

The idea that the dissolved SO_2 and its dissociation products are the primary agents responsible for acute injury and that the balance between the incoming SO_2 and its destruction is the critical factor in resistance to acute injury is probably now undisputed. Moreover, recent work has shown correlations between resistance to acute injury and both the rates of sulphite oxidation in leaves (Miller and Xerikos, 1979), and the activities of a suphite-oxidizing enzyme (Kondo et al., 1980). Despite the strength of the case for the involvement of sulphite (or sulphurous acid) in acute injury, it is difficult to agree with the concept of an *absolute* threshold for damage since one cannot imagine that all the SO_2 would be oxidized instantaneously on arrival at the cell. Therefore even at low SO_2 concentrations there must be some free sulphurous acid available (perhaps beyond the limits of Thomas' methods of detection) which may cause changes in the cell, even if they are not immediately apparent. To take an example: a single exposure to SO_2 may reduce photosynthetic rate, but if, on removal of the gas, the rate returns to normal, it is concluded that no injury has occurred. However if, on repeated exposure, the ability to recover is lost, it is clear that the plant has in fact suffered some biochemical change from the early exposure which has reduced its capacity to tolerate later exposures (e.g. Puckett et al., 1973). This shows not only the improbability of a *threshold* for damage but also the dangers of short-term measurement of physiological effects.

Although Thomas' and Katz' rigid insistence on a threshold for damage was unfortunate in the light of current evidence, it is also important to realize that their work has suffered from misrepresentation by some recent authors. They recognized that the equations linking injury with sulphurous acid content were applicable only to short-term fumigations and that in the long-term the accumulation of *sulphate* could cause physiological changes, such as a reduction in buffer capacity leading to premature senescence (Thomas, Hendricks and Hill, 1944). However, they did not include such effects in their definition of SO_2 injury, but classified it as 'sulphate toxicity'. Some of the recent work on reductions in plant growth in the absence of visible injury has emphasized the role of senescence, although it is true that in most of the publications effects on yield are attributable, at least in part, to changes in net assimilation rate.

Since the long-term effects of pollutants can be manifested in different ways it is clear that the use of the terms 'chronic injury' is inadequate to cover all plant damage that is not acute, because it fails to differentiate between the long-term effects of sulphate accumulation and the invisible, but probably progressive effects of SO_2 itself at the cellular level. For the remainder of this chapter the terms 'accumulation injury' and 'progressive injury' will be used to differentiate between the long-term effects of sulphate and SO_2, respectively, but it is recognized that both forms of injury probably occur together in the field and may interact. Thus the interpretation of damage at the biochemical level is probably extremely complex.

There is one further point from the invisible injury controversy: Thomas and colleagues performed a large part of their work on a single species, alfalfa (*Medicago sativa*), which is recognized as being extremely sensitive to acute injury. It is possible that alfalfa is so sensitive to acute injury that there is virtually no opportunity for any other injury to occur first. Moreover, it is a fast-growing species which may readily incorporate the absorbed sulphur into organic forms, and may therefore be insensitive to accumulation injury.

This brief review of the invisible injury controversy highlights some of the problems involved in defining:

(1) what a particular pollutant exposure represents in terms of the nature of the active species at the chemical level;
(2) what we mean by injury;
(3) how far results from short-term experiments may be applicable to plants receiving prolonged exposures in the field; and
(4) how far results from one species may be applicable to another.

The rest of the chapter will concentrate primarily on the first three subjects.

Analysis of plant response

To analyse factors causing plant injury, a framework is required. One such is described by Levitt (1972). By a simple analogy to the laws of moments, a stress produces a strain proportional to the initial stress. Such a strain may be either elastic (recoverable) or plastic (irrecoverable) and may result from the primary stress itself, or a secondary stress induced by the initial strain (*Figure 7.1*).

Correspondingly there are various mechanisms of resistance (*Figure 7.2*). Thus the plant may either avoid the stress (through stomatal closure in the case of pollutant), avoid the strain at the physiological level, or tolerate any strains which may occur. In this chapter the terms 'avoidance', 'tolerance', 'stress' and 'strain' will be used *sensu* Levitt. However, in the present context the distinction between plastic and elastic strains seems to be more appropriate as a theoretical concept than of practical use, so no attempt will be made to distinguish between the terms in the examination of pollutant effects. An example was given earlier in which the capacity for photosynthesis to recover from intermittent exposure to SO_2 decreased as the number of exposures increased. Is it really possible to indicate the point at which the pollutant effects become irreversible in the same way as leaves can reach a permanent wilting point when exposed to drought?

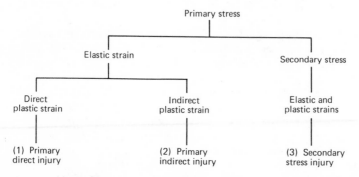

Figure 7.1. Kinds of stress injury (after Levitt, 1972)

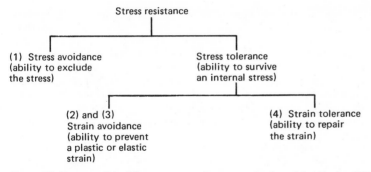

Figure 7.2. Four possible different stress resistance mechanisms (after Levitt, 1972)

Pollutant flux and the role of mesophyll resistance in avoidance

With modern methods of gas analysis it is possible to separate the resistances to pollutant flux into various components: aerodynamic (leaf boundary layer) resistance (r_a); cuticular plus surface resistance (r_c); stomatal resistance (r_s) and mesophyll resistance (r_m). While the nature of the first three resistances are well understood and have been discussed comprehensively by Unsworth, Biscoe and Black (1976), the fourth, mesophyll resistance, is more of a measurable barrier to movement than any physiologically-definable mechanism. It is clear, however, that it will play a major role in determining the nature of the active species reaching the sensitive parts of the cell, and thus in the avoidance of cellular damage. There are three pieces of evidence which point to mesophyll resistance having a role in avoidance:

(1) Different gases are not necessarily taken up in similar proportion by different species (*Table 7.1*). Thus there is a degree of selectivity in the amount of each gas taken up;

(2) Differences in pollutant flux between plant species may be much greater than can be explained solely by differences in stomatal resistance (e.g. Klein *et al.*, 1978);

TABLE 7.1. Relative accumulation of sulphur, chloride and fluoride after exposure of six species to SO$_2$ (177 ppb for 421 h), HF (1.93 ppb for 421 h) and HCl (209 ppb for 319 h)

Species	S accumul. / S supply	Cl accumul. / Cl supply	F accumul. / F supply	Rel. Cl accumul. (S = 100)	Rel. F accumul. (S = 100)
Rye (*Secale cereale*)	1.3	8.1	13.6	620	1050
Barley (*Hordeum vulgare*)	2.3	8.6	12.0	370	520
Ponderosa pine (*Pinus ponderosa*)	0.51	0.69	4.0	135	784
Norway spruce (*Picea abies*)	0.63	0.68	4.3	130	680
Strawberry (*Fragaria chiloensis*)	1.2	1.3	16.5	110	1370
Chrysanthemum indicum	4.0	5.18	29.8	130	740
Mean	1.66	4.1	13.4	249	857

After Guderian (1977).

(3) Rates of uptake of individual gases cannot necessarily be predicted from purely physical parameters, e.g. solubility.

It can be assumed that the initial sink for most of the gas molecules arriving at the plant is water; either externally on the cuticle or on the wet cell walls inside the leaf. It is thus obvious that the solubility of the gas must be an important factor in the flux equation. *Table 7.2* contrasts the solubilities of a range of gases in water at 20 °C with their rates of uptake by a standardized plant canopy. It is clear that there is some relationship between the two, but while the solubilities are over a range of 10^4, the uptake rates vary only by 10^2.

TABLE 7.2. **Comparison of solubility and uptake data for various gases**

Gas	[a] Solubility in water at 20°C, 1 atm mmol cm^{-3}	[b] Steady-state uptake by alfalfa canopy (μl ppb^{-1} min^{-1} m^{-2})
CO	0.0010	No data
NO	0.0014	10
O_3	0.012	100
CO_2	0.039	11
SO_2	1.6	170
HF	18	220
NO_2	Highly soluble but decomposes	120

[a] After Bennett, Hill and Gates (1973).
[b] After Bennett and Hill (1973).

Recent work by Taylor, McLaughlin and Shriner (1982) in which the uptake of a range of sulphur gases of varying solubility in water is compared with uptake by foliage shows an approximately linear relationship between unit leaf flux and $\log_{10}$ solubility. However, there seem to be differences in the values for individual gases which are not explained entirely in terms of solubility. Thus we must consider r_m in terms of reactions in water and of other factors.

Influence of solubility and dissociation

The solubility of pollutant gases depends largely on their ability to dissociate in water. Both SO_2 and NO_2 dissociate in a complex manner. For NO_2, most authors do not quote solubility figures, although they state that it is highly soluble but decomposes. Reactions of SO_2 in water on the other hand have been studied extensively. Vass and Ingram (1949) demonstrated the pH-dependent nature of the dissociation:

$$SO_2 + H_2O \rightleftharpoons \text{'}H_2SO_3\text{'} \rightleftharpoons HSO_3^- + H^+ \rightleftharpoons SO_3^{2-} + 2H^+$$

The percentage distribution of the sulphur species over a range of pH values is shown in *Figure 7.3*. At high pH the sulphite ion predominates; around pH 4–5, the bisulphite ion; whereas below pH 4 there is an increasing proportion of sulphurous acid. 'H_2SO_3' is the theoretical product of dissolved SO_2 at low pH, but it has never been found and it is probable that SO_2 exists in solution as undissociated molecules in a loose hydrate (Falk and Giguère, 1958). The fact that the solubility of SO_2 in water declines below pH 4 with a curve very similar to that for the increase of 'H_2SO_3' (Brimblecombe and

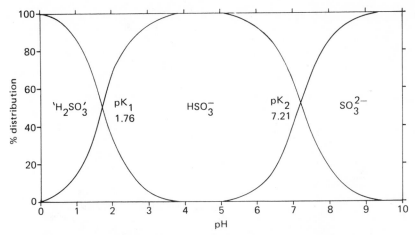

Figure 7.3. Sulphurous acid distribution diagram (after Puckett *et al.*, 1973)

Spedding, 1972) would support this view (*Figure 7.4*). Davison (1982) has shown remarkably similar curves for the relationship between the increase in the proportion of hydrofluoric acid in solution with declining pH, and the associated decrease in the solubility of fluoride.

Hocking and Hocking (1977) conclude that the major form of dissolved SO_2 at physiological pH values is HSO_3^- with a relatively small (but probably important) proportion of undissociated SO_2. Furthermore, they show that the solubility of SO_2 and consequently the equilibrium between atmospheric and aqueous concentrations is dependent on temperature. Thus if purely physical factors were the only ones affecting flux, gas uptake should be dependent on external concentration, temperature and the pH of the water.

Unfortunately it is not always possible to extrapolate values for the solubility of gases in distilled water to their dissolution in internal or external plant water; for instance, Brimblecombe (1978) has shown that dew can contain more dissolved SO_2 than would be expected from experiments with distilled water. When SO_2 dissolves in water, the resultant decrease in pH reduces the capacity to retain SO_2 in solution. However, water on plant surfaces does not necessarily behave like distilled water because of the presence of metabolites (leached from the cells) which may neutralize the absorbed gas, thereby preventing the fall in pH and maintaining the capacity of the water to absorb SO_2. Under certain conditions, quite high concentrations of neutralizing compounds may occur. For instance, Rowlatt, Crawford and Unsworth (1978) have shown that the pH of throughfall below senescent leaves (those most susceptible to leaching) may be approximately two units greater than that of the incident rain.

Factors independent of solubility

The fact that gas flux is not dependent only on solubility, even when the differences between distilled water and that on plant surfaces are taken into account, implies the operation of a sink mechanism to maintain a concentration gradient between the atmospheric and aqueous forms. For NO and SO_2 such a gradient may occur through oxidation to NO_2 and SO_4^{2-} respectively. The factors responsible for maintaining the

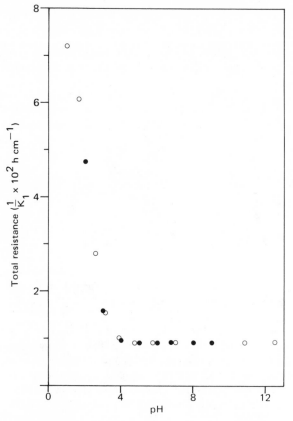

Figure 7.4. Effect of pH on resistance to SO_2 solution. ● Calculated values (Liss, 1971); ○ experimental values (after Brimblecombe and Spedding, 1972)

flux of NO are discussed by Mansfield and Freer-Smith (1981). These authors consider some of the reasons why fluxes of NO are only two- to threefold less than those of NO_2, despite the large differences in solubility (Law and Mansfield, 1982; *see also* Bengtson *et al.*, 1980). They suggest that the greater solubility of NO in xylem sap than in distilled water (Anderson and Mansfield, 1979) may be a contributory factor. The xylem sap, which is continuous with the water on the external surfaces of the leaf mesophyll cells, may contain sufficient solutes to increase the solubility of NO in a manner comparable to that described above for SO_2. However, such increased solubility cannot account on its own for more than a part of the discrepancy, and the authors consider that oxidation of NO, which occurs both in the air and in water, is the major influence on the flux of NO. However, Wellburn (this volume) has found that NO and NO_2 act differently on enzyme systems, indicating that oxidation to NO_2 is not a prerequisite for the absorption of NO. This implies that a strong physiological sink must exist to maintain the concentration gradient, given the low solubility of NO. Cope and Spedding (1982) compared the fluxes of SO_2 and H_2S to vegetation and found that uptake of H_2S in vegetation was nearly comparable to that of SO_2, despite its much lower solubility in water and suggested that a metabolic sink was responsible for maintaining the concentration gradient for H_2S.

Despite the strong evidence for the involvement of an active process in the movement of pollutants across membranes, the precise physiological mechanisms remain unknown. It is known that one gas may affect the uptake of another (Elkiey and Ormrod, 1981a), and that for ozone r_m can depend strongly on the concentration of O_3 and the duration of exposure (Tingey and Taylor, 1982). An increase in r_m with increasing NO_2 concentrations (all above 1 ppm) and time has been described by Srivastava, Jolliffe and Runeckles (1975). The fact that comparable increases in r_m occurred when intermittent exposures were given led the authors to conclude that the increase in mesophyll resistance was a result of physiological changes in the membrane rather than a weakening of the concentration gradient resulting from the accumulation of NO_2. Such an interpretation is consistent with our knowledge that membranes are primary sites of attack at high pollutant concentrations. Thus an incoming gas might influence avoidance at the cellular level through an effect on mesophyll resistance, just as it may influence avoidance at the whole-plant level through effects on stomatal resistance. As yet, however, we have no information on the extent to which long-term exposure to low concentrations of pollutants can modify avoidance at the cellular level.

Fate of externally-sorbed pollutants and the role of cuticular resistance in avoidance

Pollutants deposited onto wet leaf surfaces

Besides the pollutants absorbed inside the plant, a proportion may be deposited on the external surfaces of the leaves. In the case of SO_2 some deposition occurs on dry leaves, but the rate increases rapidly in the presence of surface moisture. Fowler and Unsworth (1974) have shown that the surface resistance (r_c) decreases by more than an order of magnitude when dew is present on a wheat crop. Garland and Branson (1977) estimate that in lowland Britain foliage is wet about 30% of the time, through a combination of dew and rain. Not only is the wet foliage an efficient sink for pollutants such as SO_2 and NO_2, but the rain itself can contain appreciable quantities of pollutants, depending on the atmospheric concentrations (Davies, 1979; Davison, 1982). In the case of sulphur pollutants, concentrations of sulphite and sulphate as high as 0.15 mM and 1.8 mM respectively have been measured in the rain in Manchester (Ferguson, Lee and Bell, 1978).

While the stomatal pathway is undoubtedly the major route for pollutant uptake during conditions favourable to growth, it is not a continuously-available pathway, unlike that across the cuticle. Fowler (1980) suggests that over the year as a whole the stomatal and cuticular pathways for SO_2 to perennial grasses may be of similar magnitude. Since the former predominates in the summer, it will be in the winter, when atmospheric SO_2 concentrations are highest, the foliage frequently wet, and the stomata closed for long periods, that movement across the cuticle will be the major route for SO_2 entering the plant. Studies with gaseous SO_2 and dry foliage have shown that climatic conditions typical of winter (low irradiance and short days), which lead to slow growth even in absence of pollutants, also greatly increase the adverse effects of SO_2 on growth (Davies, 1980). Moreover, it has recently been shown by Elkiey and Ormrod (1981b) that both SO_2 and NO_2 cause greater damage to timothy (*Phleum pratense*) foliage which has been misted, compared with dry foliage, so it would be expected that a combination of wet foliage and poor growing conditions would have severe effects on grass productivity in winter. The absence of a definitive investigation of this question makes the true effect of overwinter exposures of crops difficult to estimate.

There are numerous accounts of the absorption from droplets of pollutants and their products, including sulphite (Garsed and Mochrie, 1980) suphate (Biddulph, Cory and Biddulph, 1956) and fluorides (reviewed by Davison, 1982), but the factors limiting the process are hardly understood. Chamel and Garrec (1977) found very low rates of fluoride movement across isolated pear (*Pyrus communis*) leaf cuticles despite the high concentration gradient employed. However, they took care to use only undamaged cuticles and suggested that higher rates might be obtained under natural conditions when the cuticles are more likely to be damaged.

Pollutants deposited onto dry leaf surfaces

Most information concerning the behaviour of pollutants deposited onto dry leaf surfaces refers to SO_2. A certain amount of the gas initially deposited can be removed subsequently by washing (Garsed and Read, 1977a, 1977b; Raybould, Unsworth and Gregory, 1977). However, the proportion which can be removed appears to depend on the interval between the fumigation and the washing. Garland and Branson (1977) were able to wash from pine shoots 70% of the total $^{35}SO_2$ after 0.05 min, but less than 10% after 20 min. No definitive explanation for this observation has yet been provided, although Garland and Branson suggested that either the SO_2 was re-emitted to the atmosphere or immobilized in or on the leaf. Sulphur dioxide has been shown to accumulate at specific sites in the epidermis, particularly the guard cells (Weigl and Ziegler, 1962). Whether these sites of accumulation represent sites of absorption or correspond to the positions of ectodesmata (Schönherr and Bukovac, 1970; Franke, 1971) has yet to be determined. It is also uncertain whether any of the sulphur deposited on a dry leaf may subsequently penetrate without rewetting, or even the extent to which rewetting may allow the sulphur to re-enter solution and subsequently be absorbed.

Reactions in the cell

The interpretation of biochemical data

The first problem in discussing pollutant metabolism is that in most cases it is impossible to determine the quantity or precise nature of the active species which enter the cell. In the case of SO_2, not only may oxidation occur on the cell wall, but the individual chemical species may cross the plasmalemma at different rates (Puckett *et al.*, 1973) and once in the cell have different affinities for particular sites. For example, Ziegler (1977) found that ^{35}S accumulated in chloroplast lamellae to a much greater extent when the label was supplied as sulphite, than when supplied as sulphate. Furthermore the individual reactions of the chemical species may depend on the concentration of pollutants supplied. Thus several workers have described the formation of α-hydroxysulphonates by direct addition reactions between SO_2 and aldehydes, but these studies have always involved the use of very high concentrations of SO_2 (1% in one case). Ziegler (1975) points out that ionic reactions leading to sulphonates predominate at high SO_2 concentrations, but oxygen-dependent free-radical reactions prevail at low concentrations, and it is now generally accepted that the formation of sulphonates is negligible at concentrations of SO_2 representative of ambient conditions. Furthermore, Pierre (1977) has shown that enzymes responded differently to sulphite and α-hydroxysulphonate indicating that the two sulphur compounds act in a different manner at the cellular level. In the same paper Pierre illustrated the degree to which studies of the biochemical effects of sulphur compounds *in vivo* and *in vitro* may yield different results. Similarly Srinivasan and Rao (1980) found a strong effect of fluoride on nitrate reduction *in vivo*, but not *in vitro*. It appears that NADH generation rather than nitrate reduction *per se* was the sensitive process in this case.

Even when the flux to a cell is known the response of a particular organelle may differ because of compartmentation, both in the physical sense, as in the case of membrane-bound organelles, and also in a physiological sense, as in the case of isozymes. Pahlich (1973) illustrated this latter point by demonstrating the inhibition of mitochondrial, but not cytoplasmic, glutamate-oxaloacetate transferase (GOT) by SO_2.

The nature of pollution damage

Pollutant-induced strains may result not only from the direct effect of the primary stress, but also through secondary stresses. A major source of secondary stress with some pollutants is believed to be free radicals and there is strong evidence that free radical scavengers may protect plants from injury. Since the nature of the free radical reactions is discussed elsewhere in this book, this chapter will concentrate on the nature of direct strains.

Once inside the plasmalemma, the pollutant may be able to interfere with the normal metabolic processes of the cell. The pollutants can be divided into three classes, depending on their likely metabolic fate:

(1) non-accumulating, e.g. ozone and ethylene;
(2) accumulating and likely to become involved in 'normal' metabolism, largely S- and N-containing compounds;
(3) accumulating, but not likely to be involved in 'normal' metabolism, e.g. HF, HCl, Cl_2.

The discussion will concentrate largely on the pollutants in category (2) because they are the most widely studied and probably also the most complex as a result of the variety of their possible oxidation states. They are already present in the cell in quantity as a result of absorption by the roots. However, in the case of products of sulphur dioxide it is likely that a substantial proportion of the soluble forms are located in the vacuole and therefore are relatively unavailable to cytoplasmic or organelle-bound enzymes. Smith (1981) reports that in cultured tobacco cells, incoming sulphate labels cytoplasmic sulphate pools within 1 h and the amino acid pools within 6 h. The excess sulphate and amino acids were transported to the vacuole. Thus the likelihood exists that the incoming metabolites which cross the plasmalemma into the cytoplasm may be used in preference to the endogenous materials. Kylin (1960) found preferential use of incoming $^{35}SO_4^{2-}$ over endogenous SO_4^{2-} in deseeded wheat plants, and several authors have found rapid incorporation of the products of $^{35}SO_2$ into soluble organic or insoluble forms (e.g. Faller, Herwig and Kuhn, 1970; Ayazloo, Garsed and Bell, 1982). This preferential use of incoming metabolites indicates that care is required in the interpretation of data from pulse-feeding experiments, particularly since the evidence from long-term fumigations points to the accumulation of inorganic rather than the organic sulphur found initially. *Table 7.3* illustrates some results of studies with $^{35}SO_2$. With high concentrations of SO_2, most radioactivity is found as sulphate, with only a small proportion of organic material. A similar effect is found in darkness

TABLE 7.3. Effect of carrier concentration on the percentage distribution of ^{35}S in the third leaf of sunflower (*Helianthus annuus*) **five days after exposure to SO_2**

Carrier SO_2 added (ppb)	Fraction		
	Sulphate	Acid-insoluble	Acid-soluble
0	10	86	4
75	31	63	6
187	53	44	3
373	76	22	2
560	76	22	2

Selected data from Faller, Herwig and Kuhn (1970).

(Garsed and Read, 1977b). At lower concentrations there is substantial incorporation of radioactivity into organic material. However, all the long-term studies of the increase in sulphur after fumigation at low concentrations show that accumulation occurs primarily as sulphate, with a relatively small increase in the organic fraction (e.g. Thomas, Hendricks and Hill, 1944; Keller and Jäger, 1980). The exact degree of enrichment of the organic and inorganic fractions depends on many factors including nitrogen and potassium nutrition (Jäger and Klein, 1976) or sulphur nutrition alone (Thomas, Hendricks and Hill, 1944). Most of the excess soluble organic sulphur accumulates in the thiol compounds glutathione and cysteine; Grill and Esterbauer (1973) and Grill, Esterbauer and Klosch (1979) report consistently higher levels of these two compounds in leaves collected from polluted sites than from control sites. There is also a small increase in protein —SH (Grill *et al.*, 1980).

These results suggest that at the beginning of exposure to low concentrations of SO_2 a substantial proportion of the incoming gas may follow the normal reduction pathways for sulphur compounds, including the formation of insoluble materials and thiols (*see Table 7.3*). But why, then, does sulphate accumulate in the long term? Does the reduction pathway become saturated so that less of the incoming SO_2 is reduced and accumulates as sulphate in the cell, as is the case with high concentrations, or does the formation of sulphate result from the extensive desulphuration of amino compounds? Two pieces of evidence suggest that the latter process may be involved:

(1) ^{35}S-cysteine can be degraded rapidly to yield $^{35}SO_4^{2-}$ (Harrington and Smith, 1980; Filner *et al.*, this volume).

(2) Plants exposed to H_2S pollution subsequently accumulate sulphate in the vacuole (Steubing and Jäger, 1978). It is important to note that the effects of H_2S on enzymes were different from those of SO_2, and thus not attributable to the accumulation of SO_4^{2-}.

It was also notable that the low concentrations of H_2S used by Steubing and Jäger caused appreciable stimulation of growth, an effect also reported by Thompson and Kats (1978). In view of the suggestion by Hammett (1930), that cell division is controlled by the balance between incompletely oxidized and reduced sulphur radicals, an idea used by Bleasdale (1973) to explain the effects of coal smoke on the growth of ryegrass, studies of the metabolism of low concentration of both SO_2 and H_2S after various intervals of pollutant exposure are particularly desirable.

Nitrogen oxides are the other major pollutants which can become incorporated into metabolic pathways.

The simplified series of reactions for NO_2 is:

$$NO_2 \rightarrow \text{nitrous acid} \xrightarrow{\text{Nitrite reductase}} NH_3 \xrightarrow{\text{GS/GOGAT}} \text{amino acids}$$

Since nitrous acid is known to be toxic (*see* Wellburn, 1982), its removal through metabolism will be of benefit to the plant. *Table 7.4* shows some results from studies with $^{15}NO_2$. It is clear that metabolism of NO_2 is extremely rapid since incorporation into organic material is virtually complete within a short time of exposure. In contrast to studies with $^{35}SO_2$ there is no accumulation of nitrate comparable to that of sulphate. However, one similarity to the SO_2 data suggested in the limited figures available is the large difference in the proportions of insoluble material. With the low concentration and longer duration of exposure in the experiments of Rogers, Campbell and Volk (1979), 66% of the organic material was insoluble, whereas Kaji *et al.* (1980) found that less than 20% was insoluble. The extent to which such differences are a result of concentration, duration or both is a question which needs resolution.

The rate of nitrous acid metabolism will depend largely on the activity of nitrite reductase, the level of which is well known to be increased by exposure to NO_2. The effects of high NO_2 concentrations on nitrite reductase activity in the short term seem to depend on species. Yoneyama *et al.* (1979) have shown that maize (*Zea mays*), which is relatively resistant to acute injury by NO_2, has no detectable nitrite accumulation and has a rapid induction of nitrite reductase. The intermediately-sensitive sunflower (*Helianthus annuus*) also had rapid induction of nitrite reductase but still accumulated nitrite, while most sensitive species, such as the bean (*Phaseolus vulgaris*), accumulated nitrite and showed little increase in nitrite reductase. The importance of nitrite reductase in dealing with the chemical products of NO_2 in the long-term metabolism of NO_x has been shown by Wellburn *et al.* (1981), who found that when SO_2 was present at the same time as NO_2, the induction of nitrite reductase was inhibited. They proposed that the more-than-additive inhibition of growth when SO_2 and NO_2 were present together (Ashenden and Mansfield, 1978; Ashenden, 1979) was due to the resulting accumulation of nitrite. Although it seems possible to explain the more-than-additive effects of SO_2 and NO_2 at low concentrations there is evidence that as concentrations of NO_2 and SO_2 increase, their influence on CO_2 exchange changes from more-than-additive to less-than-additive (White, Hill and Bennett, 1974). This suggests that at higher concentrations there might be reactions, influencing plant response, other than those involving nitrite reductase.

Movement and cycling of pollutants and their products

The non-metabolized pollutants, such as fluoride and chloride, are probably translocated largely in the xylem, since it is characteristic of such elements that they accumulate in the leaf margins. However, some translocation of fluoride must also occur in the phloem, since there is evidence also for substantial accumulation in the roots. The possible chemical nature of the translocated fluoride is discussed by Weinstein and Alscher-Herman (1982).

The metabolized pollutants SO_2 and NO_2 can be incorporated rapidly into organic molecules as already discussed. The principal organic compounds produced, the amino compounds glutamine and asparagine in the case of NO_x, and glutathione and cysteine in the case of SO_2, are the major organic forms in which nitrogen and sulphur are

96

TABLE 7.4. Some values for percentage distribution of ^{15}N after exposure of plants to $^{15}NO_2$

Species	NO_2 concentration (ppb)	Duration	Irradiance	Percentage distribution of ^{15}N			Reference
				Insoluble organic	Soluble organic	Inorganic	
Bean (*Phaseolus vulgaris*)	320	3 h	660 μE m^{-2} s^{-1}	63	34.5	2.5	Rogers, Campbell and Volk (1979)
Sunflower (*Helianthus annuus*)[a]	4000–6000	20 min	30 k lux / Dark	17 / 11	83 / 86	1 / 14	Kaji *et al.* (1980)
Spinach (*Spinacea oleracea*)[a]	4000–6000	20 min	30 k lux / Dark	16 / 7	84 / 93	1 / 1.5	
Spinach (*Spinacea oleracea*)	4000	2.5 h	30 k lux	Amino acids only measured 99.4		0.62	Yoneyama and Sasakawa (1979)

[a] The tabulated values of Kaji *et al.* do not take into account that approximately 40% of the ^{15}N was in a fraction labelled 'other'. Moreover the totalled percentages all exceed 100%.

translocated under normal conditions (Pate, 1980; Rennenberg, Schmitz and Bergmann, 1979). Because pollutants affect the levels of phloem-mobile metabolites in plants, they may modify their susceptibility to insect attack (Alstad, Edmunds and Weinstein, 1982). Nitrogen and sulphur pollutants in particular increase the concentrations of those amino acids most necessary to phloem-feeders like aphids and leafhoppers and increased levels of aphid infestation have been shown to occur in polluted regions (e.g. Villemant, 1981).

The metabolic products of SO_2 are transported primarily to regions of high metabolic activity, particularly the stem apices, and show patterns of movement with regard to leaf age similar to those for $^{14}CO_2$ assimilates (Garsed and Read, 1974). It is probable that the products of NO_x behave similarly, although our information on patterns of movement and chemical forms of translocated $^{15}NO_x$ is limited. In the case of SO_2, the known mobility of sulphite in the phloem (Garsed and Mochrie, 1980) raises the question of the likely effects on morphogenesis of the sulphur accumulated in the growing points, since Hammett (1930) proposed that cell division is controlled by the balance between incompletely oxidized and reduced sulphur radicals. The observations of suppressed lateral branching in trees exposed to SO_2 and NO_x mixtures (Freer-Smith, personal communication) now require examination at the biochemical level.

The patterns of translocation are greatly influenced by the relative physiological ages of the leaves, so an understanding of the cycling and losses of pollutants as plants age is an essential requirement for interpreting the long-term pollutant balance of the plant. Some idea of the complexity involved can be seen from the work of Raybould, Unsworth and Gregory (1977), who studied the origins of sulphur collected below a developing wheat canopy. They found that while leaching of internal materials was the major source of the sulphur collected below the crop, it accounted for only 30% of the total sulphur lost as the tops aged. The sulphur lost from the tops had not accumulated in the roots because they also had a decrease in sulphur content. The authors suggested that the losses occurred either through volatilization, or through leakage from the roots into the soil, a process frequently observed during experiments with $^{35}SO_2$ (e.g. Garsed and Read, 1977b), but they did not look further at the precise mechanisms involved. The re-emission of gases has been studied largely as a short-term process. It is well known that H_2S is emitted from plants after exposure to SO_2 (e.g. De Cormis, 1968). The process is dependent on light and on the concentration of pollutant supplied, so most authors interpret the process, perhaps somewhat teleologically, as a means for the plant to remove excess sulphur, i.e. an avoidance mechanism. De Cormis (1968) was unable to find H_2S emission in the absence of SO_2 pollution, but the work of Raybould, Unsworth and Gregory (1977) suggests that studies of emission from plants growing at ambient levels of pollution and normal levels of soil sulphate would be desirable (some recent findings of Filner et al. on this subject appear elsewhere in this book). Unpublished work by Bell and Clough showed that after exposure to low concentrations of $^{35}SO_3^{2-}$ (20–200 ppb) approximately 0.1% of the initial radioactivity could be recovered as SO_2 and 0.01% as H_2S within a few hours. It is not clear whether the SO_2 was re-emitted from inside or outside the lamina, but it is possible that the former site may have been involved, since Wilson, Bressan and Filner (1978) were able to detect gaseous SO_2 after immersion of petioles in sulphite solutions.

It is not only the metabolized pollutants which can be re-emitted from plants. Davison and Blakemore (1976) found substantial fluctuations in the fluoride contents of swards, and recent work has shown that the losses were due more to gaseous emission than to rainfall or dilution by new growth (Davison, 1982). The behaviour of fluorides is now being shown to be far more complex than had previously been supposed, since they do not merely accumulate in the leaf margins until leaf death, but are actively cycled like most other elements.

Little is known of the cycling or re-emission of nitrogenous compounds after exposure to pollutants. Both NO and NO_2 have been found to be emitted from herbicide-treated soybean plants (Klepper, 1979) but the extent of emission under non-pathological conditions is less certain. In view of the importance of pollutant cycling within, and losses from, the plant in the avoidance of accumulation injury, remarkably little work has been done to further our understanding of the processes involved.

Analysis of dose-response in relation to biochemistry

The dependence of the type of injury on the pollutant flux and duration of exposure obviously causes problems in the interpretation of biochemical data, but it also presents an opportunity to study the mechanisms of action of the pollutants through comparison of plant responses to different exposures. A particularly important biochemical question is the nature of the three types of injury described previously. So far, comparisons of the effects of high and low concentrations of a pollutant on genotypes known to show differential sensitivity to acute and long-term injury, have been confined to SO_2 (Horsman et al., 1979; Ayazloo and Bell, 1981). Acute injury can be explained by the short-term effects of high concentrations of SO_2, but how much of the long-term injury can be attributed to accumulation and how much to progressive damage by SO_2 before it is oxidized to sulphate? Comparisons of the relationship between the long-term and short-term effects of various other pollutants may help to answer this question. For instance, NO_x shows several oxidation states but the evidence suggests that there is less capacity for the accumulation of inorganic compounds than is the case with sulphur dioxide. With fluoride there is a high capacity for progressive injury because of the reactivity of HF, and this is accompanied by substantial accumulation, probably in the form of complexes (Weinstein and Alscher-Herman, 1982). Ozone, on the other hand, has a high capacity for progressive injury, but does not accumulate.

Another approach to the question of accumulation injury would be to compare the biochemical responses of halophytes with mesophytes during long-term exposure to both accumulating and non-accumulating pollutants. Thomas, Hendricks and Hill (1950a) noted in their survey of the sulphur content of vegetation that many halophytes could, without apparent injury, absorb quantities of sulphate much higher than those found in normal plants.

The work of Ayazloo and Bell (1981) on the relationship between sensitivity to acute injury caused by short-term exposure, and injury represented by growth reduction as a consequence of long-term exposure to SO_2, presents a paradox which may illustrate some of the problems of interpreting long-term injury. These authors found that within grass populations collected from Helmshore (a moderately polluted site in northern Britain), tolerance to both types of injury had occurred but not necessarily together in the same plant. It is easy to see how tolerance to the type of injury causing growth reductions has occurred; the sensitive plants simply grew less rapidly and were thus less competitive. But how did resistance to acute injury occur in genotypes that were not resistant to growth reductions, since SO_2 levels of Helmshore rarely appear high enough to produce sufficient acute injury to exert a selective pressure on its own? Since a *gradual* process of selection through growth reduction dependent solely on SO_2

cannot explain the resistance to acute injury, it is possible that a strain develops, progressively, via a mechanism related to acute injury. The strain would then exert its selective pressure not on its own, but through interactions with another environmental factor. Frost (Davison and Bailey, 1982) and perhaps water stress are two of the likely candidates for such a factor since sensitivity to both is known to be increased by SO_2, and both may cause irreparable damage within a short time. Mansfield and Freer-Smith (1981) have an alternative explanation. They suggest that NO_x is the interacting agent and that the activity of the enzyme nitrite reductase is depressed by SO_2 in plants sensitive to acute injury which leads to the accumulation of nitrite in toxic concentrations. It is recognized that other authors may put different interpretations on the work of Ayazloo and Bell, but the importance of the study is that it demonstrates most convincingly the principle that there is more than one mechanism of long-term injury.

Conclusions

In trying to discuss the nature of the active chemical species derived from atmospheric pollutants that can influence plants, we have been drawn away from a simple catalogue of chemicals into a consideration of the concept of injury, with frequent references to the role of pollutant flux in determining these species. If we are attempting ultimately to determine biochemical responses and then apply our knowledge to the improvement of crops grown in polluted areas, then we must work under conditions relevant to those areas. Just as high-concentration fumigations elicit a totally different whole-plant response from low concentrations, as illustrated by the invisible injury controversy, so also do they elicit totally different biochemical responses through the formation of different chemical species. We have suffered in the past because biochemical studies have rarely been related to growth and *vice versa*. The success of the collaboration at Lancaster between the plant physiologists led by Professor Mansfield and the biochemists under Dr Wellburn in identifying the SO_2/NO_2 interaction under fairly 'realistic' conditions indicates a valuable way forward.

References

ALSTAD, D. N., EDMUNDS, G. F. and WEINSTEIN, L. H. Effects of air pollutants on insect populations, *Annual Review of Entomology* **27**, 369–384 (1982)

ANDERSON, L. S. and MANSFIELD, T. A. The effects of nitric oxide pollution on the growth of tomato, *Environmental Pollution* **20**, 113–121 (1979)

ASHENDEN, T. W. Growth reductions in cocksfoot (*Dactylis glomerata* L.) as a result of SO_2 pollution, *Environmental Pollution* **15**, 161–166 (1978)

ASHENDEN, T. W. The effects of long-term exposures to SO_2 and NO_2 pollution on the growth of *Dactylis glomerata* and *Poa pratensis*, *Environmental Pollution* **18**, 249–258 (1979)

ASHENDEN, T. W. and MANSFIELD, T. A. Extreme pollution sensitivity of grasses when SO_2 and NO_2 are present in the atmosphere together, *Nature (London)* **273**, 142–143 (1978)

ASHENDEN, T. W. and WILLIAMS, I. A. D. Growth reductions in *Lolium multiflorum* Lam. and *Phleum pratense* L. as a result of SO_2 and NO_2 pollution, *Environmental Pollution (Series A)* **21**, 131–139 (1980)

AWANG, M. B. *The effects of sulphur dioxide pollution on plant growth with special reference to Trifolium repens*, PhD Thesis, University of Sheffield, Sheffield, UK (1979)

AYAZLOO, M. and BELL, J. N. B. Studies on the tolerance to sulphur dioxide of grass populations in polluted areas. I. Identification of tolerant populations, *New Phytologist* **88**, 203–222 (1981)

AYAZLOO, M., GARSED, S. G. and BELL, J. N. B. Studies on the tolerance to sulphur dioxide of grass populations in polluted areas. II. Morphological and physiological investigations, *New Phytologist* **90**, 109–126 (1982)

BELL, J. N. B., RUTTER, A. J. and RELTON, J. Studies on the effects of low levels of sulphur dioxide on the growth of *Lolium perenne* L., *New Phytologist* **83**, 627–643 (1979)

BENGTSON, G., BOSTROM, C. A., GREENFELT, P., SKARBY, L. and TROEN, E. Deposition of nitrogen oxides to Scots pine (*Pinus sylvestris* L.). In *Proceedings of the International Conference on The Ecological Impact of Acid Precipitation*, (Eds. D. DRABLØS and A. TOLLAN), pp. 154–155, SNSF, Oslo (1980)

BENNETT, J. H. and HILL, A. C. Absorption of gaseous air pollutants by a standardized plant canopy, *Jounal of the Air Pollution Control Association* **23**, 203–206 (1973)

BENNETT, J. H., HILL, A. C. and GATES, D. M. A model for gaseous pollutant sorption by leaves, *Journal of the Air Pollution Control Association* **23**, 957–962 (1973)

BIDDULPH, O., CORY, R. and BIDDULPH, S. The absorption and translocation of sulfur in red kidney bean, *Plant Physiology* **31**, 28–33 (1956)

BLACK, V. J. and UNSWORTH, M. H. Effects of low concentrations of sulphur dioxide on net photosynthesis and dark respiration of *Vicia faba*, *Journal of Experimental Botany* **30**, 473–483 (1979)

BLEASDALE, J. K. A. Effects of coal-smoke pollution on the growth of ryegrass (*Lolium perenne* L.), *Environmental Pollution* **5**, 275–285 (1973)

BRENNAN, E., LEONE, I. N. and DAINES, R. H. Response of pine trees to chlorine in the atmosphere, *Forest Science* **12**, 386–390 (1966)

BRIMBLECOMBE, P. 'Dew' as a sink for sulphur dioxide, *Tellus* **30**, 151–157 (1978)

BRIMBLECOMBE, P. and SPEDDING, D. J. Rate of solution of gaseous sulphur dioxide at atmospheric concentrations, *Nature* **236**, 225 (1972)

BULL, J. N. and MANSFIELD, T. A. Photosynthesis in leaves exposed to SO_2 and NO_2, *Nature* **250**, 443–444 (1974)

CHAMEL, A. and GARREC, J. P. Penetration of fluorine through isolated pear leaf cuticles, *Environmental Pollution* **12**, 307–310 (1977)

COPE, D. M. and SPEDDING, D. J. Hydrogen sulphide uptake by vegetation, *Atmospheric Environment* **16**, 349–353 (1982)

CRITTENDEN, P. D. and READ, D. J. The effects of air pollution on plant growth with special reference to sulphur dioxide. II. Growth studies with *Lolium perenne* L., *New Phytologist* **80**, 49–62 (1978)

DAVIES, T. Grasses more sensitive to SO_2 pollution in conditions of low irradiance and short days, *Nature* **284**, 483–485 (1980)

DAVIES, T. D. Dissolved sulphur dioxide and sulphate in urban and rural precipitation (Norfolk, UK), *Atmospheric Environment* **13**, 1275–1285 (1979)

DAVISON, A. The effects of fluorides on plant growth and forage quality. In *Effects of Gaseous Air Pollution in Agriculture and Horticulture*, (Eds. M. H. UNSWORTH and D. P. ORMROD), pp. 267–291, Butterworths, London (1982)

DAVISON, A. W. and BAILEY, I. F. SO_2 pollution reduces the freezing resistance of ryegrass, *Nature* **297**, 400–402 (1982)

DAVISON, A. W. and BLAKEMORE, J. Factors determining fluoride accumulation in forage. In *Effects of Air Pollutants on Plants*, (Ed. T. A. MANSFIELD), pp. 17–30, Cambridge University Press, Cambridge (1976)

DE CORMIS, L. Dégagement d'hydrogène sulfuré par des plantes soumises à une atmosphère contenant de l'anhydride sulfureux, *Compte Rendu de l'Academie de Sciences, Série D* **266**, 683–685 (1968)

ELKIEY, T. and ORMROD, D. P. Absorption of ozone, sulphur dioxide and nitrogen dioxide by petunia plants, *Environmental and Experimental Botany* **21**, 63–70 (1981a)

ELKIEY, T. and ORMROD, D. P. Sulphur and nitrogen nutrition and misting effects on the response of bluegrass to ozone, sulphur dioxide, nitrogen dioxide or their mixture, *Water, Air and Soil Pollution* **16**, 177–186 (1981b)

FALK, M. and GIGUÈRE, P. A. On the nature of sulphurous acid, *Canadian Journal of Chemistry* **36**, 1121–1125 (1958)

FALLER, N., HERWIG, K. and KUHN, H. Die Aufnahme von Schwefeldioxyd ($S^{35}O_2$) aus der Luft. II. Aufnahme, Umbau and Verteilung in der Pflanze, *Plant and Soil* **33**, 283–295 (1970)

FERGUSON, P., LEE, J. A. and BELL, J. N. B. The effects of sulphur pollutants on the growth of *Sphagnum* species, *Environmental Pollution* **16**, 151–162 (1978)

FOWLER, D. Removal of sulphur and nitrogen compounds from the atmosphere in rain and by dry deposition. In *Proceedings of the International Conference on The Ecological Impact of Acid Precipitation*, (Eds. D. DRABLØS and A. TOLLAN), pp. 22–32, SNSF, Oslo (1980)

FOWLER, D. and UNSWORTH, M. H. Dry deposition of sulphur dioxide on wheat, *Nature* **249**, 389–390 (1974)

FRANKE, W. The entry of residues into plants via ectodesmata (ectocythodes), *Residue Reviews* **38**, 81–115 (1971)

GARLAND, J. A. and BRANSON, J. R. The deposition of sulphur dioxide to pine forest assessed by a radioactive tracer method, *Tellus* **29**, 445–454 (1977)

GARSED, S. G. and MOCHRIE, A. Translocation of sulphite in *Vicia faba* L., *New Phytologist* **84**, 421–428 (1980)

GARSED, S. G. and READ, D. J. The uptake and translocation of $^{35}SO_2$ in soy-bean *Glycine max* var. *biloxi*, *New Phytologist* **73**, 299–307 (1974)

GARSED, S. G. and READ, D. J. The uptake and metabolism of $^{35}SO_2$ in plants of differing sensitivity to sulphur dioxide, *Environmental Pollution* **13**, 173–186 (1977a)

GARSED, S. G. and READ, D. J. Sulphur dioxide metabolism in *Glycine max* var. *biloxi*. II. Biochemical distribution of $^{35}SO_2$ products, *New Phytologist* **78**, 583–592 (1977b)

GRILL, D. and ESTERBAUER, H. Cystein und Glutathion in gesunden und SO₂-geschädigten Fichtennadeln, *European Journal of Forest Pathology* **3**, 65–71 (1973)

GRILL, D., ESTERBAUER, H. and KLOSCH, U. Effect of sulphur dioxide on glutathione in leaves of plants, *Environmental Pollution* **13**, 187–194 (1979)

GRILL, D., ESTERBAUER, H., SCHARNER, M. and FELGITSCH, C. H. Effect of sulfur dioxide on protein-SH in needles of *Picea abies*, *European Journal of Forest Pathology* **10**, 263–267 (1980)

GUDERIAN, R. Untersuchungen über quantitative Beziehungen zwischen dem Schwefelgehalt von Pflanzen und dem Schwefeldioxid gehalt der Luft. I. *Zeitschrift für Pflanzenkrankheit* **77**, 200–220 (1970)

GUDERIAN, R. *Air Pollution. Phytotoxicity of Acidic Gases and its Significance in Air Pollution Control*, Springer-Verlag, Berlin (1977)

HAMMETT, F. S. The natural chemical regulation of growth by increase in cell number, *Proceedings of the American Philosophical Society* **69**, 217–223 (1930)

HARRINGTON, H. M. and SMITH, I. K. Cysteine metabolism in cultured tobacco cells, *Plant Physiology* **65**, 151–155 (1980)

HOCKING, D. and HOCKING, M. B. Equilibrium solubility of trace atmospheric sulphur dioxide in water and its bearing on air pollution injury to plants, *Environmental Pollution* **9**, 57–64 (1977)

HORSMAN, D. C., ROBERTS, T. M., LAMBERT, M. and BRADSHAW, A. D. Studies on the effect of sulphur dioxide in perennial ryegrass (*Lolium ferenne* L.). I. Characteristics of fumigation system and preliminary experiments, *Journal of Experimental Botany* **30**, 485–493 (1979)

JÄGER, H.-J. and KLEIN, H. Modellversuche zum Einfluss der Nährstoffversorgung auf die SO₂-Empfindlichkeit von Pflanzen, *European Journal of Forest Pathology* **6**, 347–354 (1976)

KAJI, M., YONEYAMA, T., TOTSUKA, T. and IWAKI, H. Absorption of atmospheric NO₂ by plants and soils. VI. Transformation of NO₂ absorbed in the leaves and transfer of the nitrogen through the plants. In *Studies on the Effect of Air Pollutants on Plants and Mechanisms of Phytoxicity*. Research Report from the National Institute for Environmental Studies, Ibaraki, Japan **11**, 51–58 (1980)

KATZ, M. (Ed.) *Effect of sulphur dioxide on vegetation*. National Research Council of Canada. Report No. 815 (1939)

KELLER, Th. and JÄGER, H.-J. Der Einfluss bodenbürtiger Sulfationen auf den Schwefelgehalt SO₂-begaster Assimilationsorgane von Waldbaumarten, *Angewandte Botanik* **54**, 77–89 (1980)

KLEIN, H., JÄGER, H.-J., DOMES, W. and WONG, C. H. Mechanisms contributing to differential sensitivities of plants to SO₂, *Oecologia* **33**, 203–208 (1978)

KLEPPER, L. Nitric oxide (NO) and nitrogen dioxide (NO₂) emissions from herbicide-treated soybean plants, *Atmospheric Environment* **13**, 537–542 (1979)

KNABE, W. Natural loss of fluorine from needles of Norway spruce (*Picea abies* Karst), *Staub* (English Edition) **30**, 29–32 (1970)

KONDO, N., AKIYAMA, Y., FUJIWARA, M. and SUGAHARA, K. Sulfite oxidising activities in plants. In *Studies on the Effects of Air Pollutants on Plants and Mechanisms of Phytotoxicity*. Research Report from the National Institute for Environmental Studies, Ibaraki, Japan **11**, 137–150 (1980)

KYLIN, A. The incorporation of radio-sulphur from external sulphate into different sulphur fractions of isolated leaves, *Physiologia Plantarum* **13**, 366–379 (1960)

LAW, R. M. and MANSFIELD, T. A. Oxides of nitrogen in the greenhouse atmosphere. In *Effect of Gaseous Air Pollutants in Agriculture and Horticulture*, (Eds. M. H. UNSWORTH and D. P. ORMROD), pp. 93–112, Butterworths, London (1982)

LEVITT, J. *Responses of plants to environmental stresses*. Academic Press, New York (1972)

LISS, P. S. Exchange of SO₂ between the atmsophere and natural waters, *Nature* **233**, 327–329 (1971)

MANNING, W. J. and FEDER, W. A. Effects of ozone on economic plants. In *Effects of Air Pollutants on Plants*, (Ed. T. A. MANSFIELD), pp. 47–60, Cambridge University Press, Cambridge (1976)

MANSFIELD, T. A. and FREER-SMITH, P. H. Effects of urban air pollution on plant growth, *Biological Reviews* **56**, 343–368 (1981)

MILLER, J. E. and XERIKOS, P. B. Residence time of sulphite in SO₂ 'sensitive' and 'tolerant' soybean cultivars, *Environmental Pollution* **18**, 259–264 (1979)

PAHLICH, E. Über den Hemm-Mechanismus mitochondrialer Glutamat-Oxalacetat Transaminase in SO$_2$-begasten Erbsen, *Planta* **110**, 267–278 (1973)

PATE, J. S. Transport and partitioning of nitrogenous solutes, *Annual Review of Plant Physiology* **31**, 313–340 (1980)

PIERRE, M. Action du SO$_2$ sur le métabolisme intermédiaire. II. Effet de doses subnécrotiques de SO$_2$ sur les enzymes de feuilles de Haricot, *Physiologie Végétale* **15**, 195–205 (1977)

PUCKETT, K. J., NIEBOER, E., FLORA, W. P. and RICHARDSON, D. H. S. Sulphur dioxide: its effect on photosynthetic ^{14}C fixation in lichens and suggested mechanisms of phytotoxicity, *New Phytologist* **72**, 141–154 (1973)

RAYBOULD, C. C., UNSWORTH, M. H. and GREGORY, P. J. Sources of sulphur collected below a wheat canopy, *Nature* **267**, 146–147 (1977)

RENNENBERG, H., SCHMITZ, K. and BERGMANN, L. Long-distance transport of sulfur in *Nicotiana tabacum*, *Planta* **147**, 57–62 (1979)

ROGERS, H. H., CAMPBELL, J. C. and VOLK, R. J. Nitrogen-15 dioxide uptake and incorporation by *Phaseolus vulgaris* (L.), *Science* **206**, 333–335 (1979)

ROWLATT, S., CRAWFORD, D. V. and UNSWORTH, M. H. Sulphur cycle in wheat and other farm crops. In *Sulphur in Forages*, (Ed. J. C. BROGAN), pp. 1–14, An Foras Tulantais, Dublin (1978)

SCHÖNHERR, J. and BUKOVAC, M. J. Preferential polar pathways in the cuticle and their relationship to ectodesmata, *Planta* **92**, 189–201 (1970)

SMITH, I. K. Compartmentation of sulfur metabolites in tobacco cells. Use of efflux analysis, *Plant Physiology* **68**, 937–940 (1981)

SRINIVASAN and RAO, C. S. R. Effect of fluoride on *in vivo* nitrate reduction in rice leaves (*Oryza sativa* L.), *Experientia* **36**, 634–635 (1980)

SRIVASTAVA, H. S., JOLLIFFE, P. A. and RUNECKLES, V. C. Inhibition of gas exchange in bean leaves by NO$_2$, *Canadian Journal of Botany* **53**, 466–474 (1975)

STEUBING, L. and JÄGER, H.-J. Ökophysiologische-biochemische Wirkung von H$_2$S auf *Pisum sativum* L., *Angewandte Botanik* **52**, 137–147 (1978)

STOKLASA, J. *Die Beschädigungen der Vegetation durch Rauchgase und Fabriksexhalatonen*. Urban und Schwarzenberg, Berlin (1923)

TANIYAMA, T. The symptoms and mechanism of sulphur dioxide injury in crop plants, *Bulletin of the Faculty of Agriculture*, Mie University, Tsu, Japan **44**, 11–130. Central Electricity Generating Board Translation 7430 (1972)

TAYLOR, G. E., McLAUGHLIN, S. B. and SHRINER, D. S. Effective pollutant dose. In *Effects of Gaseous Air Pollution in Agriculture and Horticulture*, (Eds. M. H. UNSWORTH and D. P. ORMROD), pp. 458–460, Butterworths, London (1982)

THOMAS, M. D. Gas damage to plants, *Annual Review of Plant Physiology* **2**, 293–322 (1951)

THOMAS, M. D., HENDRICKS, R. H. and HILL, G. R. Some chemical reactions of sulphur dioxide after absorption by alfalfa and sugar beets, *Plant Physiology* **19**, 212–226 (1944)

THOMAS, M. D., HENDRICKS, R. H. and HILL, G. R. Sulfur content of vegetation, *Soil Science* **70**, 9–18 (1950a)

THOMAS, M. D., HENDRICKS, R. H. and HILL, G. R. Sulfur metabolism of plants. Effect of sulfur dioxide on vegetation, *Industrial and Engineering Chemistry* **42**, 2231–2235 (1950b)

THOMPSON, C. R. and KATS, G. Effects of continuous H$_2$S fumigation on crop and forest plants, *Environmental Science and Technology* **12**, 550–554 (1978)

TINGEY, D. T. and TAYLOR, G. E. Variation in plant response to ozone: A conceptual model of physiological events. In *Effects of Gaseous Air Pollution in Agriculture and Horticulture*, (Eds. M. H. UNSWORTH and D. P. ORMROD), pp. 113–138, Butterworths, London (1982)

UNSWORTH, M. H., BISCOE, P. V. and BLACK, V. Analysis of gas exchange between plants and polluted atmospheres. In *Effects of Air Pollutants on Plants*, (Ed. T. A. MANSFIELD), pp. 5–16, Cambridge University Press, Cambridge (1976)

VASS, K. and INGRAM, M. Preservation of fruit juices with less sulphur dioxide, *Food Manufacture* **24**, 414 (1949)

VILLEMANT, C. Influence de la pollution atmosphérique sur les populations d'aphides du pin sylvestre en Forêt de Roumare (Seine Maritime), *Environmental Pollution (Series A)* **24**, 245–262 (1981)

WEIGL, J. and ZIEGLER, H. Die räumliche Verteilung von ^{35}S und die Art der markierten Verbindungen in Spinatblättern nach begasung mit ^{35}SO$_2$, *Planta* **58**, 435–447 (1962)

WEINSTEIN, L. H. and ALSCHER-HERMAN, R. Physiological responses of plants to fluorine. In *Effects of Gaseous Air Pollution in Agriculture and Horticulture*, (Eds. M. H. UNSWORTH and D. P. ORMROD), pp. 139–167, Butterworths, London (1982)

WELLBURN, A. R. Effects of SO$_2$ and NO$_2$ on metabolic function. In *Effects of Gaseous Air Pollution in Agriculture and Horticulture*, (Eds. M. H. UNSWORTH and D. P. ORMROD), pp. 169–187, Butterworths, London (1982)

WELLBURN, A. R., HIGGINSON, C., ROBINSON, D. and WALMSLEY, C. Biochemical explanations of more than additive inhibitory effects of low atmospheric levels of sulphur dioxide plus nitrogen dioxide upon plants, *New Phytologist* **88**, 223–237 (1981)

WHITE, K. L., HILL, A. C. and BENNETT, J. H. Synergistic inhibition of apparent photosynthesis rate of alfalfa by combinations of sulfur dioxide and nitrogen dioxide, *Environmental Science and Technology* **8**, 574–576 (1974)

WILSON, L. G., BRESSAN, R. A. and FILNER, P. Light dependent emission of hydrogen sulfide from plants, *Plant Physiology* **61**, 184–189 (1978)

YONEYAMA, T. and SASAKAWA, H. Transformation of atmospheric NO_2 absorbed in spinach leaves, *Plant and Cell Physiology* **20**, 263–266 (1979)

YONEYAMA, T., SASAKAWA, H., ISHIZUKA, S. and TOTSUKA, T. Absorption of atmospheric NO_2 by plants and soils. II. Nitrite accumulation, nitrite reductase activity and diurnal change in NO_2 absorption of leaves, *Soil Science and Plant Nutrition* **25**, 267–275 (1979)

ZIEGLER, I. The effect of SO_2 pollution on plant metabolism *Residue Reviews* **56**, 79–105 (1975)

ZIEGLER, I. Subcellar distribution of ^{35}S-sulfur in spinach leaves after application of $^{35}SO_4^{2-}$, $^{35}SO_3^{2-}$ and $^{35}SO_2$, *Planta* **135**, 25–32 (1977)

ZIMMERMAN, P. W. and HITCHCOCK, A. E. Susceptibility of plants to hydrofluoric acid and sulphur dioxide gases. *Contributions from the Boyce Thompson Institute for Plant Research* **18**, 263–279 (1956)

Chapter 8

Pollutants and plant cells: effects on membranes

J. B. Mudd*, S. K. Banerjee†, M. M. Dooley‡ and K. L. Knight§

DEPARTMENT OF BIOCHEMISTRY AND STATEWIDE AIR POLLUTION RESEARCH
CENTER, UNIVERSITY OF CALIFORNIA, USA

Introduction

The plant is accustomed to the uptake and evolution of carbon dioxide and oxygen that
are generated and consumed in the course of metabolism. The uptake and destiny of
pollutant gases such as sulfur dioxide, oxides of nitrogen, hydrogen fluoride,
peroxyacylnitrates, and ozone are a function firstly of their concentration and water
solubility, and secondly of their reactivity. A very reactive molecule such as ozone may
be expected to oxidize the compounds with which it first comes in contact, such as those
of the plasma membrane or even the cell wall, but in the case of oxides of nitrogen and
sulfur dioxide, the hydration products are to some extent normal metabolites and can be
dealt with by the metabolic machinery of the cell. Even in cases where the pollutant gas
affects membrane permeability we must try to determine whether this is a direct effect
or a consequence of intracellular reactions.

In this chapter we have only briefly reviewed the possible effects on membranes of
sulfur dioxide, oxides of nitrogen and peroxyacetylnitrate, and have discussed ozone
more extensively as it is the most likely candidate for producing direct effects on the
plasma membrane.

Sulfur dioxide

There are several reports of the reaction of sulfur dioxide or sulfite with functional
groups important in proteins. Yang (1973) described the oxidation of tryptophan in the
presence of Mn^{2+} and sulfite; oxidation was dependent on aerobic conditions and was
inhibited by superoxide dismutase (SOD). Eickenroht, Gause and Rowlands (1975)
reported the reaction of SO_2 with proteins, concentrating on the quenching of
tryptophan fluorescence. This work was expanded by Gause and Rowlands (1975) who

* Present address: ARCO Plant Cell Research Institute, 6560 Trinity Court, Dublin, California 94566
USA.
† Present address: Department of Biochemistry, University College of Science, Calcutta-700019, India.
‡ Present address: Department of Chemistry, Louisiana State University, Baton Rouge, Louisiana
70803 USA.
§ Present address: Department of Biological Chemistry, University of California, Los Angeles,
California 90024 USA.

exposed lymphocytes to SO_2 and concluded from the behavior of spin-labelled compounds inserted in the membrane that protein aggregates were formed. Another well known reaction of sulfite with proteins is the cleavage of disulfide bonds

$$RSSR^1 + SO_3^{2-} \rightarrow RSSO_3^- + R^1S^-$$

In spite of these potential reactions with proteins of the plasma membrane, the preponderance of evidence suggests that SO_2 penetrates the cell where it affects various metabolic pathways. Pahlich *et al.* (1972) reported activation of glutamate dehydrogenase by SO_2 while glutamine synthetase was unaffected. These results have been verified and extended by Wellburn *et al.* (1976) who found stimulation of glutamate dehydrogenase (GDH), glutamate oxaloacetate transaminase (GOT) and glutamate pyruvate transaminase (GPT) after exposure to 1 ppm SO_2. These changes may be found in tissues which are combatting the deleterious effects of SO_2, so that responses are made to elevated amounts of sulfite and sulfate. When visible injury is detected, it seems likely that superoxide is an intermediate in the toxic reactions. Tanaka and Sugahara (1980) have shown that high SOD content in leaves is correlated with resistance to SO_2 damage and that the inactivation of SOD by diethyldithio-carbamate increased the susceptibility to SO_2. The toxicity of SO_2 is associated with the production of malonaldehyde, a product of lipid oxidation (Shimazaki, Sakaki and Sugahara, 1980), which could be considered evidence of the reaction of SO_2 with membrane lipids. However, the effects on chlorophyll and carotenoids indicate that SO_2 has penetrated the cell as far as the chloroplast and that lipid oxidation may be a late event in the sequence of toxic reactions.

Oxides of nitrogen

The reaction of nitrogen dioxide with unsaturated compounds including lipids of biological membranes has been well studied. Two types of reaction have been observed:

(1) addition of NO_2 at the double bond; reversal of this addition causes isomerization of the double bond from *cis* to *trans*
(2) abstraction of hydrogen at methylene groups interrupting double bond systems.

$$NO_2 + 18{:}2^{9,12} \rightarrow \underline{\qquad}\diagup\diagdown\underline{\qquad}\diagup_{NO_2} + \underline{\qquad}\diagup\diagdown\underline{\qquad}\diagup + HONO$$

In both cases peroxidation of the fatty acid is initiated. Felmeister, Amanat and Weiner (1970) have exposed monomolecular films of phospholipid to NO_2 and found that under constant surface pressure the area of the film increased. This would be consistent with derivatizations as indicated above. Exposure of monomolecular films of cholesterol to NO_2 results in a decrease of the surface area because the cholesterol nitrate formed was desorbed from the surface layer (Kamel, Weiner and Felmeister, 1971).

These results imply that reactions with NO_2 on this lipid bilayer could have serious consequences. It is debatable, however, that these reactions take place *in vivo*. Csallany and Ayaz (1978) found that long-term exposure of mice to NO_2 did not increase the concentrations of lipofuschin, a pigment characteristic of lipid peroxidation. In plant tissue there is no evidence of lipid peroxidation or any other effect of NO_2 on membranes. On the contrary, the effects of exposure to NO_2 can be interpreted as a response to elevated NO_2^- and NO_3^- concentrations inside the cell:

$$2NO_2 + H_2O \rightarrow NO_2^- + NO_3^- + 2H^+$$

Lee and Schwartz (1981) have pointed out that this reaction is dependent on the concentration of NO_2 and its solubility; some reactions of NO_2 within the cell may be attributed to solvated NO_2. Zeevart (1976) reported that several species of plants exposed to NO_2 sustained more damge in darkness than in light, which is consistent with the reduction of NO_3^- to NO_2^- and then to NH_4^+ by NADPH generated photosynthetically. Indeed, Wellburn, Wilson and Aldridge (1980) showed that exposure of tomato (*Lycopersicon esculentum*) to NO resulted in increased activities of nitrite reductase, GDH, GPT and GOT.

Peroxyacetylnitrate (PAN)

A number of model reactions of peroxyacetyl nitrate have been studied. Thiol groups are very reactive and can be either oxidized or acetylated (Mudd, 1966); such reactions could affect membrane proteins. PAN also reacts with olefins to form epoxides (Darnall and Pitts, 1970), but it is not known whether this reaction takes place in biologically important lipids, or what the consequence would be. At the moment, the best hypothesis of reaction of PAN with biological membranes would be the oxidation of the sulfhydryl groups of proteins. If PAN can penetrate into the cytoplasmic compartment, other reactions become relevant, such as the oxidation of low molecular weight sulfhydryl compounds (Leh and Mudd, 1974a) and the oxidation of reduced nicotinamide coenzymes (Mudd and Dugger, 1963).

Ozone

The permeability of plant membranes is definitely changed as a result of exposure to ozone. Heath and Frederick (1979) found that the efflux of ^{86}Rb from algal (*Chlorella*) cells was much more rapid after exposure to 115 ppm O_3. Similarly, the permeability to ^{86}Rb could be measured as an increased uptake after exposure to O_3. Evans and Ting (1973) exposed discs of leaves of pinto bean (*Phaseolus vulgaris*) to 0.5 ppm O_3 and observed a greater leakage of [86]Rb. The uptake of D-glucose and 2-deoxyglucose by leaf discs from pinto bean also was stimulated by exposure to O_3 (Perchorowicz and Ting, 1974). This effect was not noticeable immediately after the exposure to O_3 (in fact some inhibition to uptake was observed) but was pronounced one day after the exposure. In the case of D-glucose, there was an increase in the proportion of radioactive label in the insoluble fraction of the cell as a consequence of exposure to O_3, indicating some unspecified change in the metabolism of the glucose which had been taken up. Further study of these phenomena by Sutton and Ting (1977) revealed that the stimulation of uptake of ^{86}Rb or 2-deoxyglucose was greatest one day after exposure of the plant material to 0.5 ppm O_3, after which there was a tendency for the exposed plants to recover to the rates of uptake observed in the control plants.

These effects of O_3 on permeability are not restricted to the plasma membrane. Nobel and Wang (1973) exposed isolated chloroplasts to 50 ppm O_3 and found that the permeability to *meso*-erythritol and to glycerol increased. It should be noted that the exposure to O_3 also inhibited photosynthetic phosphorylation of the chloroplasts, but it is not clear whether this inhibition is a consequence of the effects on permeability or is independent of these effects.

The chemical basis for the effects of O_3 on permeability is harder to explain. Many authors have pointed to the supposed susceptibility of the unsaturated fatty acids of the membrane to oxidation. Teige, McManus and Mudd (1974) found that liposomes

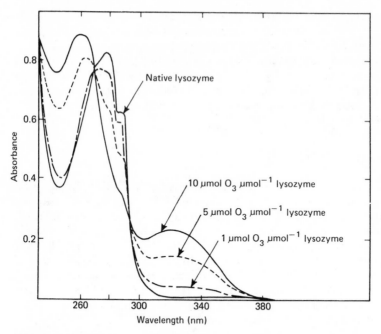

Figure 8.1. Ultraviolet spectra of native and O_3-oxidized lysozyme. Lysozyme (10 mg/ml) in 0.1 M phosphate buffer (pH 7) was exposed to ozone delivered at 1.9 μmol min^{-1} for varying lengths of time and the UV spectra were recorded

prepared from phosphatidylcholine were susceptible to oxidation by O_3. It was demonstrated in this paper and in the papers of Heath and Tappel (1976) and Heath (1978) that the oxidation of fatty acids was by direct ozonolysis rather than by lipid peroxidation. These results, however, may not represent the situation in biological membranes in which protein plays an important role.

The question, then, is whether the protein or the lipid of the membrane is oxidized preferentially. This dilemma is illustrated by the results of Kindya and Chan (1976), Chan, Kindya and Kesner (1977), and Kesner, Kindya and Chan (1979). Their experiments showed the sodium–potassium ATPase of the erythrocyte membrane was inhibited after exposure to O_3, and concluded that the sequence of events was lipid peroxidation followed by inhibition of enzyme activity by the products of lipid peroxidation. This conclusion does not allow for any direct reaction of O_3 with amino acid residues of the proteins of the membrane.

We would like to emphasize the susceptibility of proteins to inactivation by O_3 and to comment on the relative susceptibility of proteins and lipids to O_3. As examples of the effects of O_3 on proteins, the susceptibility of lysozyme, glyceraldehyde-3-phosphate dehydrogenase, and glycophorin to oxidation by O_3 have been studied, and in the case of glycophorin, the effect of imbedding the protein in a phospholipid bilayer on the specificity of oxidation has been studied.

Lysozyme

The oxidation of lysozyme by O_3 has been studied by Previero, Coletti-Previero and Jolles (1967), by Kuroda, Sakiyama and Narita (1975) and by Leh and Mudd (1974b).

Although there was general agreement that tryptophan was oxidized, there was some discrepancy in the site of its oxidation; Previero, Coletti-Previero and Jolles (1967) and Leh and Mudd (1974b) concluded tryptophan residues 108 and 111 were oxidized while Kuroda, Sakiyama and Narita (1975) concluded that residue(s) 62(63) were oxidized. Dooley and Mudd (1982) have varied the reaction conditions in order to obtain maximum specificity of oxidation and now agree with Kuroda, Sakiyama and Narita (1975) that under conditions of high protein concentration and low O_3 concentration, tryptophan residues 62(63) are specifically oxidized. The oxidation of tryptophan in lysozyme can readily be followed by its conversion to N-formylkynurenine by measurement of the ultraviolet absorbance spectrum (*Figure 8.1*). The site of oxidation can be determined by degradation of the protein by trypsin and separation of the tryptic fragments (*Figures 8.2* and *8.3*). The fragment containing tryptophans 62 and 63 is the only one showing the characteristic absorbance spectrum of N-formylkynurenine. Under certain conditions this modified fragment chromatographs with the unmodified fragment containing tryptophans 108 and 111 and can therefore confuse the interpretation of the result. Although methionine is oxidized in these experiments, it does not change the catalytic activity of the enzyme. The inactivation of biological activity is related to oxidation of residues which are either directly involved in the chemical transformation or involved in the binding (or orientation) of the substrate. Previous knowledge of the mechanism of action of lysozyme allows us to conclude that the inactivation of lysozyme by O_3 is an effect on substrate binding rather than on the catalytic site.

Glyceraldehyde-3-phosphate dehydrogenase

Results previously obtained had demonstrated the susceptibility of glyceraldehyde-3-phosphate dehydrogenase to inactivation by O_3 (Freeman, Sharman and Mudd, 1979).

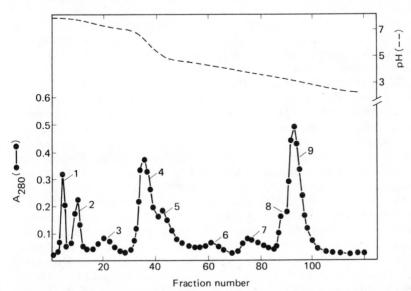

Figure 8.2. Separation of the tryptic peptides of native lysozyme. Native lysozyme was digested with trypsin and the peptides were separated by chromatography on Dowex 1: 3.0 ml fractions were collected and analyzed for absorbance at 280 nm and for pH. Peaks 1, 4, 8, and 9 were shown to contain tryptophan by measuring the absorbance of ultraviolet light

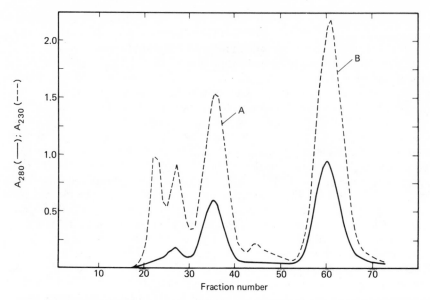

Figure 8.3. Sephadex chromatography of peaks 8 and 9 from Dowex 1 chromatography of trypsin-digested native lysozyme. Peaks 8 and 9 (*Figure 8.2*) were pooled, lyophilized, redissolved, and chromatographed on Sephadex G-25. Fraction size was 1.8 ml. Peaks A and B were both shown to contain tryptophan by measuring the absorbance of ultraviolet light. Amino acid analysis of peptides B and A show them to contain tryptophans 62, 63 and 108, 111 respectively. When the lysozyme has been treated with ozone, only peptide B is found to contain tryptophan oxidation products

The protein has been thoroughly studied by enzymologists, which makes the effects of a new variable, O_3, easier to interpret. It is known that the enzyme consists of four identical subunits and that each subunit has four sulfhydryl groups. One of these sulfhydryl groups is particularly reactive with derivatizing reagents such as *p*-mercuribenzoate. This reactive sulfhydryl group is intimately involved in the catalytic activity because it forms a covalent link with the substrate.

Ozone readily inactivated pure glyceraldehyde-3-phosphate dehydrogenase, and both fast-reacting and slow-reacting sulfhydryls were oxidized. Measurement of the ultraviolet absorbance spectrum of the oxidized protein also showed that tryptophan was converted to N-formylkynurenine (*Figure 8.4*). Analysis of oxidation of the amino acid residues showed that in the early stages of exposure to O_3, the fast-reacting sulfhydryls were most susceptible, but with time both slow-reacting sulfhydryls and tryptophan residues were more oxidized than the fast-reacting sulfhydryls (*Figure 8.5*). Examination of the ozonized protein by amino acid analysis showed that even histidine and methionine residues were oxidized (Knight, 1981). In the face of multiple oxidations is it possible to determine a specific cause of the inactivation by ozone?

Glyceraldehyde-3-phosphate dehyrogenase can be derivatized and inactivated by sodium dithionate in stoichiometric amounts: only the sulfhydryl of the active site is derivatized. Activity can be completely recovered by removal of the blockage using dithiothreitol (*Figure 8.6*). When the derivatized enzyme was exposed to O_3 and the blockage subsequently removed, all of the enzymic activity was recovered. Amino acid analysis of the derivatized and ozonized enzyme showed that slow-reacting sulfhydryls, tryptophan and histidine residues had been oxidized. It appears, therefore, that oxidation of these residues does not affect the catalytic activity (*Figure 8.7*).

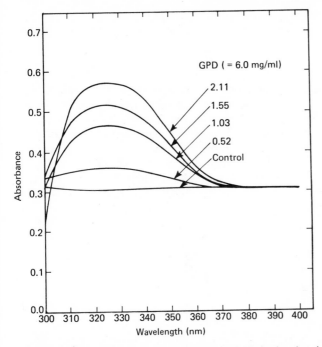

Figure 8.4. Effect of ozone on the absorbance of GPD in the ultraviolet region. Enzyme was pre-treated with 5 mM DTT and 2.0 mM NAD. Enzyme concentration in the reaction vessel was 1.20 mg/ml. Following reaction with ozone each sample was concentrated to 6.0 mg/ml using an Amicon minicon B-15 concentrator. Difference spectra were taken with a Cary 15 spectrophotometer using a 1 cm pathlength quartz cuvette, using untreated enzyme as the reference

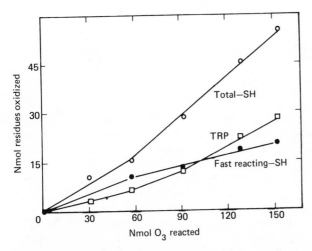

Figure 8.5. Oxidation of enzyme tryptophan and —SH groups by ozone. Enzyme concentration in the reaction vessel was 0.74 mg/ml in 50 mM phosphate, pH 8.0. Ozone flow = 31.5 nmol/min. Tryptophan oxidation was assayed by spectrophotometric determination of N-formylkynurenine

These results show that the oxidation of amino acids in proteins is relatively non-specific and that many residues can be oxidized without loss of biological activity.

Glycophorin

The best characterized transmembrane protein is glycophorin from the erythrocyte membrane. The primary amino acid sequence has been determined, and it is known that the amino acids imbedded in the lipid bilayer are residues 72–92. The protein contains neither cysteine nor tryptophan residues, but the location of two residues of methionine make glycophorin particularly suitable for a study of its reaction with O_3. Methionine at residue 8 would normally be exposed exterior to the cellular membrane whereas methionine 81 would normally be within the lipid bilayer. We therefore investigated whether residue 81 is protected from oxidation by O_3 and whether methionine 8 is more readily oxidized than fatty acids of the lipid bilayer.

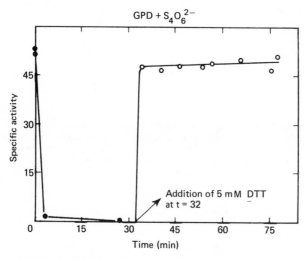

Figure 8.6. Reversible inhibition of GPD activity with $S_4O_6^{2-}$. 27.4 μM enzyme monomer was incubated with 29.6 μM $S_4O_6^{2-}$ at 4°C in 50 mM phosphate, pH 8.0. When inhibition was complete DTT was added to a final concentration of 5.0 mM. Enzyme was then assayed periodically to check for recovery of activity

Glycophorin can be isolated as a water soluble protein from erythrocytes and in this form exposure to O_3 oxidizes both methionine residues. Cyanogen bromide cleaves polypeptide chains at methionine residues. In the native protein cleavage by cyanogen bromide yields three fragments. Ozonization of methionine generates methionine sulfoxide which is not susceptible to cleavage by cyanogen bromide, and the ozone-treated protein remains as one polypeptide chain. When glycophorin is incorporated into spherical phospholipid vesicles, the orientation of the protein is the same as in the erythrocyte: the amino terminus is external and the carboxy terminus is internal. The protein-lipid mixture can now be exposed to O_3 and after the exposure the protein and the lipid can be separated and analyzed separately. The results of amino acid analysis show that only one of the methionine residues is oxidized and cyanogen bromide cleavage yields only two fragments which can be analyzed to show that methionine 8

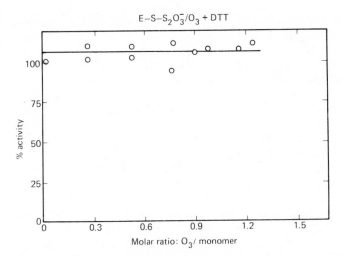

Figure 8.7. Protection of GPD against inactivation with low levels of ozone by derivatization with $S_4O_6^{-2}$. 1.7 μmol enzyme monomer were derivatized with 2.0 μmol $S_4O_6^{2-}$. Concentration of derivatized enzyme in the reaction vessel was 0.90 mg/ml. Ozone flow = 25.5 nmol/min. Following reaction with ozone aliquots of enzyme were incubated for 40 min with 5 mM DTT prior to activity assay

has been oxidized to methionine sulfoxide. Analysis of the lipid shows very little oxidation of the unsaturated fatty acids. These results are shown in Figure 8.8.

We conclude that the amino acids of polypeptide chains external to the membrane are susceptible to oxidation by O_3 because of their accessibility and inherent susceptibility to oxidation. The double bonds of fatty acids, while inherently

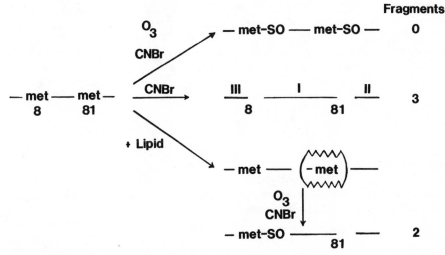

Figure 8.8. Schematic representation of the reaction of ozone with glycophorin. When untreated glycophorin is reacted with cyanogen bromide, both methionine residues react and three peptides are formed. Treatment of glycophorin in aqueous solution with ozone oxidizes both methionine residues to methionine sulfoxide: there is no cleavage on subsequent treatment with cyanogen bromide. If the glycophorin is imbedded in a phospholipid bilayer, methionine 81 is protected but residue 8 is oxidized to methionine sulfoxide. Subsequent treatment with cyanogen bromide yields two fragments

susceptible to oxidation by O_3 are not readily accessible to O_3 and are therefore less oxidized. Goldstein and McDonagh (1975) have reported that the native fluorescence of tryptophan in erythrocyte membranes is eliminated by exposure to O_3. It appears that the tryptophan oxidized in these experiments was external to the membrane because the amount of tryptophan in the membrane determined by amino acid analysis was not greatly decreased. Apparently the largest proportion of tryptophan was in an environment which quenched fluorescence and was not accessible to oxidation by O_3. This result is consonant with our results with glycophorin in lipid vesicles and we concur with Goldstein and McDonagh (1975) that '...ozone has a predilection for more hydrophilically located tryptophan molecules'.

Would an effect of ozone on membrane proteins be sufficient to explain the observed changes in membrane permeability? Reagents which react with sulfhydryl groups of membranes do inhibit phosphate transport in *Chlorella* (Jeanjean, Hourmant and Ducet, 1975). It seems unlikely that this is the only effect of O_3 on the cell. Some fraction of the administered O_3 probably penetrates the cell and exerts an effect there. Carnahan, Jenner and Wat (1978) have shown that N-[2(-oxo-1-imidazolidinyl) ethyl]N'-phenylurea protects against O_3 injury, and Lee and Bennett (1982) demonstrated that the content of SOD increased in the leaves of protected plants. It appears that these effects occur at the intracellular level. If superoxide production is a consequence of exposure to O_3 one might expect it to have deleterious effects in many other places as well as in the plasma membrane.

Conclusions

Although the plasma membrane appears to be a likely point of weakness to damage by air pollutants, purely on the grounds of accessibility, the evidence indicates this is certainly not the only point of attack. In the case of sulfur dioxide and oxides of nitrogen, the gases clearly have access to the cytosol where the metabolism of the cell is modified in response to elevated concentrations of the ionic species. If the concentration of anions exceeds the metabolic capacity of the cell, deleterious reactions ensue which may include damage to membranes. Ozone clearly affects the permeability of the plasma membrane, but this is by no means the only effect. A fraction of the O_3 appears to penetrate the cell and react with intracellular components such as glutathione and functional groups of proteins.

Acknowledgement

Research in the authors' laboratory was supported by grant ES-917 from the National Institute of Environmental Health Sciences.

References

CARNAHAN, J. E., JENNER, E. L. and WAT, E. K. W. Prevention of ozone injury of plants by a new protectant chemical, *Phytopathology* **68**, 1225–1229 (1978)
CHAN, P. C., KINDYA, R. J. and KESNER, L. Studies on the mechanism of ozone inactivation of erythrocyte membrane (Na$^+$ +K$^+$)-activated ATPase, *Journal of Biological Chemistry* **252**, 8537–8541 (1977)

CSALLANY, A. S. and AYAZ, K. L. Long-term NO_2 exposure of mice in the presence and absence of vitamin E.I. Effect on body weights and lipofuscin in pigments, *Archives of Environmental Health* November/December, 285–291 (1978)

DARNALL, K. R. and PITTS, J. N. Peroxyacetyl nitrate. A novel reagent for oxidation of organic compounds, *Chemical Communications* 1305–1306 (1970)

DOOLEY, M. M. and MUDD, J. B. Reaction of ozone with lysozyme under different exposure conditions, *Archives of Biochemistry and Biophysics* **218**, 459–471 (1982)

EICKENROHT, E. Y., GAUSE, E. M. and ROWLANDS, J. R. The interaction of SO_2 with proteins, *Environmental Letters* **9**, 265–277 (1975)

EVANS, L. S. and TING, I. P. Ozone-induced membrane permeability changes, *American Journal of Botany* **60**, 155–162 (1973)

FELMEISTER, A., AMANAT, M. and WEINER, N. D. Interactions of gaseous air pollutants with egg lecithin and phosphatidyl ethanolamine monomolecular films, *Atmospheric Environment* **4**, 311–319 (1970)

FREEMAN, B. A., SHARMAN, M. C. and MUDD, J. B. Reaction of ozone with phospholipid vesicles and human erythrocyte ghosts, *Archives of Biochemistry and Biophysics* **197**, 264–272 (1979)

GAUSE, E. M. and ROWLANDS, J. R. Effects of sulfur dioxide and bisulfite ion upon human lymphocyte membranes, *Environmental Letters* **9**, 293–305 (1975)

GOLDSTEIN, B. D. and McDONAGH, E. M. Effect of ozone on cell membrane protein fluorescence. I. *In vitro* studies utilizing the red cell membrane, *Environmental Research* **9**, 179–186 (1975)

HEATH, R. L. The reaction stoichiometry between ozone and unsaturated fatty acids in an aqueous environment, *Chemistry and Physics of Lipids* **22**, 25–37 (1978)

HEATH, R. L. and FREDERICK, P. E. Ozone alteration of membrane permeability in *Chlorella*. I. Permeability of potassium ion as measured by [86]rubidium tracer, *Plant Physiology* **64**, 455–459 (1979)

HEATH, R. L. and TAPPEL, L. A. A new sensitive assay for the measurement of hydroperoxides, *Analytical Biochemistry* **76**, 184–191 (1976)

JEANJEAN, R., HOURMANT, A. and DUCET, G. Effet des inhibiteurs de groupes SH sur le transport du phosphate chez *Chlorella pyrenoidosa*, *Biochimie* **57**, 383–390 (1975)

KAMEL, A. M., WEINER, N. D. and FELMEISTER, A. Identification of cholesteryl nitrate as a product of the reaction between NO_2 and cholesterol monomolecular films. In *Chemistry and Physics of Lipids* **6**, 225–234 (1971)

KESNER, L., KINDYA, R. J. and CHAN, P. C. Inhibition of erythrocyte membrane ($Na^+ + K^+$)-activated ATPase by ozone-treated phospholipids, *Journal of Biological Chemistry* **254**, 2705–2709 (1979)

KINDYA, R. J. and CHAN, P. C. Effect of ozone on erythrocyte membrane adenosine triphosphatase, *Biochimica et Biophysica Acta* **429**, 608–615 (1976)

KNIGHT, K. L. *The reaction of ozone with glyceraldehyde-3-phosphate dehydrogenase.* PhD dissertation, University of California, Riverside (1981)

KURODA, M., SAKIYAMA, F. and NARITA, K. Oxidation of tryptophan in lysozyme by ozone in aqueous solution, *Journal of Biochemistry* **78**, 641–651 (1975)

LEE, E. H. and BENNETT, J. H. Superoxide dismutase. A possible protective enzyme against ozone injury in snap beans (*Phaseolus vulgaris* L.), *Plant Physiology* **69**, 1444–1449 (1982)

LEE, Y.-N. and SCHWARTZ, S. E. Reaction kinetics of nitrogen dioxide with liquid water at low partial pressure, *Journal of Physical Chemistry* **85**, 840–848 (1981)

LEH, F. and MUDD, J. B. Reaction of peroxyacetyl nitrate with cysteine, cystine, methionine, lipoic acid, papain, and lysozyme, *Archives of Biochemistry and Biophysics* **161**, 216–221 (1974a)

LEH, F. and MUDD, J. B. *Reaction of ozone with lysozyme*. ACS Symposium Series **3**, 22–39 (1974b)

MUDD, J. B. Reaction of peroxyacetyl nitrate with glutathione, *Journal of Biological Chemistry* **241**, 4077–4080 (1966)

MUDD, J. B. and DUGGER, W. M. Jr. The oxidation of reduced pyridine nucleotides by peroxyacyl nitrates, *Archives of Biochemistry and Biophysics* **102**, 52–58 (1963)

NOBEL, P. S. and WANG, C.-T. Ozone increases the permeability of isolated pea chloroplasts, *Archives of Biochemistry and Biophysics* **157**, 388–394 (1973)

PAHLICH, E., JÄGER, H.-J. and STEUBING, L. Beeinflussung der Aktivitaten von Glutamat-dehydrogenase und Glutaminsynthetase aus Erbsenkeimlingen durch SO_2, *Angewandte Botanik* **46**, 183–197 (1972)

PERCHOROWICZ, J. T. and TING, I. P. Ozone effects on plant cell permeability, *American Journal of Botany* **61**, 787–793 (1974)

PREVIERO, A., COLETTI-PREVIERO, M.-A. and JOLLES, P. Localization of non-essential tryptophan residues for the biological activity of lysozyme, *Journal of Molecular Biology* **24**, 261–268 (1967)

SHIMAZAKI, K.-I., SAKAKI, T. and SUGAHARA, K. Active oxygen participation in chlorophyll destruction and lipid peroxidation in SO_2-fumigated leaves of spinach, *Research Reports of the National Institute of Environmental Studies (Japan)* **11**, 91–101 (1980)

SUTTON, R. and TING, I. P. Evidence for the repair of ozone-induced membrane injury, *American Journal of Botany* **64**, 404–411 (1977)

TANAKA, K. and SUGAHARA, K. Role of superoxide dismutase in defense against SO_2 toxicity and an increase in superoxide dismutase activity with SO_2 fumigation, *Plant and Cell Physiology* **21**, 601–611 (1980)

TEIGE, B., McMANUS, T. T. and MUDD, J. B. Reaction of ozone with phosphatidylcholine liposomes and the lytic effect of products on red blood cells, *Chemistry and Physics of Lipids* **12**, 153–171 (1974)

WELLBURN, A. R., CAPRON, T. M., CHAN, H.-S. and HORSMAN, D. C. Biochemical effects of atmospheric pollutants on plants. In *Effects of Air Pollutants on Plants*, (Ed. T. A. MANSFIELD), pp. 105–114, Cambridge University Press, Cambridge (1976)

WELLBURN, A. R., WILSON, J. and ALDRIDGE, P. H. Biochemical responses of plants to nitric oxide polluted atmospheres, *Environmental Pollution* **22**, 219–228 (1980)

YANG, S. F. Destruction of tryptophan during the aerobic oxidation of sulfite ions, *Environmental Research* **6**, 395–402 (1973)

ZEEVART, A. J. Some effects of fumigating plants for short periods with NO_2, *Environmental Pollution* **11**, 97–108 (1976)

Chapter 9

Effects of various gaseous pollutants on plant cell ultrastructure

Satu Huttunen

KASVITIETEEN LAITOS, OULUN YLIOPISTO, OULU, FINLAND

and

Sirkka Soikkeli

EKOLOGINEN YMPÄRISTÖHYGIENIA, KUOPION KORKEAKOULU, KUOPIO, FINLAND

Introduction

The effects of various gaseous pollutants on plant cell ultrastucture have been studied in several plant species (Thomson, Dugger and Palmer, 1966; Fischer, 1967; Wellburn, Majernik and Wellburn, 1972; Fischer, Kramer and Ziegler, 1973; Godzik and Knabe, 1973; Malhotra, 1976; Soikkeli and Tuovinen, 1979). The chloroplasts of mesophyll cells have been the most common objects of study, and injuries of the chloroplast envelopes are often the first signs of pollutant damage at the ultrastructural level (Godzik and Sassen, 1974). An increase in the staining density of the envelopes and the accumulation of electron-dense material between the two membranes have been reported as some of the first changes in chloroplasts caused by exposure of plants to O_3 (Thomson, Dugger and Palmer, 1966; Thomson, Nagahashi and Platt, 1974; Swanson, Thomson and Mudd, 1973; Pell and Weissberger, 1976). The swelling and other changes in the structure of the chloroplast thylakoids in plants exposed to air pollutants is often accompanied by a granulation of the stroma. These ultrastructural changes can be related to some extent to changes in the photosynthetic capacity of the mesophyll cells (Wellburn, Majernik and Wellburn, 1972; Malhotra, 1976).

Observations on the changes in the ultrastructure of other organelles are few. Mitochondrial injuries are usually described only during the later stages of cell injury. The changes described in the cytoplasm are also very few and include, at the first signs of injury, the appearance of small electron-dense particles and/or lipid-like material in the cytoplasm of spruce (*Picea*) and pine (*Pinus*) needles. Lipid droplets have been found in the cytoplasm of HF fumigated plants, and these changes might indicate disturbances in lipid metabolism (Wei and Miller, 1972; Shimazaki, Sakaki and Sugahara, 1980; Soikkeli, 1981a, 1981b). Initial changes in the cytoplasm also include an increase in the amount of endoplasmic reticulum and vacuolization in the cells (Wei and Miller, 1972; Soikkeli and Tuovinen, 1979). The granular appearance of the cytoplasm caused by SO_2, HF, or both has usually been reported as one of the later events of cell damage (Engelbrecht and Louw, 1973; Młodzianowski and Białobok, 1977). A decrease in the number of polysomes in the cytoplasm has been reported in cells of soybeans (*Glycine max*) showing visible foliar injury after exposure to HF (Wei and Miller, 1972).

117

Very little is known about the effects of pollutants on cells other than those of the leaf mesophyll. Changes in the structure of the plant epidermis have been reported. It was observed that white clover (*Trifolium repens*) (Sharma and Butler, 1973), red clover (*Trifolium pratense*) (Sharma and Butler, 1975) and kudzu (*Pueraria lobata*) (Sharma, Chandler and Salemi, 1980) growing in polluted environments had a lower stomatal frequency and showed an increase in the number and length of trichomes on leaves compared with plants growing in relatively unpolluted areas. Leaves of sweet gum (*Liquidambar styraciflua*) from a polluted environment also showed an increase in the number of trichomes on the lower epidermis, but the number of stomata was found to be greater than that in leaves from trees growing in a cleaner area (Sharma and Tyree, 1973). Surface injuries and structural changes have also been reported in leaves of other plant species, one of the common changes being a decrease in the hirsuteness so that a smoother cell surface is seen when compared with leaves from a cleaner environment (Huttunen, 1979); an example of this is presented in *Figure 9.1*. Small folds, present in the outer epidermal cell walls of normal leaves of horse chestnut (*Aesculus hippocastanum*) were absent in leaves sampled from polluted trees. Stomata of leaves from polluted areas also had an abnormal appearance, and the epidermal cells were differently shaped, had fewer folds and presented a smoother cell surface compared with control leaves (Godzik and Sassen, 1978). The adaxial surface of cowberry (*Vaccinium vitis-idaea*) leaves exposed to pollution from traffic also showed a smoother surface structure than control leaves (Mikkonen and Huttunen, 1981). In mosses, injuries observed in the surface cell structures include the loss of turgidity and alterations in the morphology of the cell surface (Huttunen, Karhu and Kallio, 1981), while in lichens growing in polluted areas an uncharacteristic development of white pruinose lobe tips may be seen, as observed on *Parmelia bolliana* (Will-Wolf, 1980). Le Blanc and Rao (1973) also found that the outer cortices of other lichens (*Parmelia sulcata* and *Physica millegrana*) were covered in an abundance of waxy material. Holopainen (1981) investigated the effects of air pollution on the ultrastructure of the epiphytic lichens *Hypogymnia physodes* and *Alectoria capillaris* and found that in comparison with controls, the chloroplasts from the pollutant-exposed phycobionts appeared more rounded, the pyrenoglobuli and cytoplasmic storage bodies were smaller than normal and the number of polyphosphate bodies had increased. In the pollutant-exposed mycobionts storage droplets were very small or absent entirely and many vacuoles and dark inclusions appeared in the hyphae in contrast to control mycobionts.

Some recent studies on the effects of air pollutants on cell ultrastructure have aimed to clarify the injuries sustained by conifer needles under field conditons, and hence are concerned with the long-term effects of air pollutants (Soikkeli and Tuovinen, 1979; Soikkeli, 1981a, 1981b). An understanding of the normal seasonal changes in needles provides a basis for understanding pollutant-induced changes over long periods (Chabot and Chabot, 1975; Soikkeli, 1978). Energy for the growth of new needles in several conifers is derived in part from the photosynthate stored in the previous year's needles. Following their expansion, the current needles then assume the major role of supplying carbon to the tree for growth, maintenance processes and storage. The older needles contribute progressively less to the carbon economy of the tree as they age. The cause of this physiological progression is likely to be closely related to the developmental and seasonal changes in the cell structures. The effects of air pollutants, then, are superimposed upon these seasonal and developmental changes, thereby restricting comparisons from polluted and clean environments to needles of similar developmental stages.

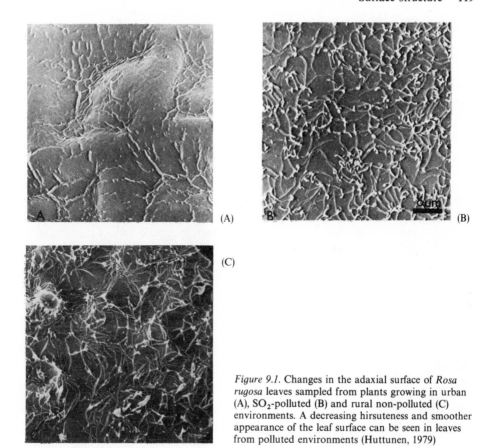

Figure 9.1. Changes in the adaxial surface of *Rosa rugosa* leaves sampled from plants growing in urban (A), SO_2-polluted (B) and rural non-polluted (C) environments. A decreasing hirsuteness and smoother appearance of the leaf surface can be seen in leaves from polluted environments (Huttunen, 1979)

Surface structure

Protection against wind, desiccation, cold, frost, plant pathogens, insects and air pollution are among the important functions of the plant surface. In higher plants, the leaf surface is covered with a cuticle which consists of the polymer, cutin, embedded in waxes. In protecting against desiccation, the cuticle also impedes gaseous diffusion into and out of the leaf, and thus represents the main barrier to the entry of air pollutants. In lower plants, the cuticular structure is not developed so that the cell wall is the only protective structure (Kolattukudy, Espelie and Soliday, 1981). The structure, chemistry and biochemistry of the natural waxes in the cuticle have been widely studied, and an introduction to their composition can be found in Martin and Juniper (1970), Kolattukudy (1976, 1980) and Kolattukudy, Espelie and Soliday (1981). The thickness of the cuticle is not important, but its architecture and structural resistance to weathering — which help to maintain the integrity of the cuticle — are ecophysiologically important.

The effects of airborne pollutants on the epicuticular wax structures have been investigated, especially in conifer needles (Grill, 1973; Cape and Fowler, 1981;

Huttunen and Laine, 1981, 1983; Huttunen, Mäkelä and Laine, 1982), but broad-leaved trees (Godzik and Sassen, 1978; Espelie, Davis and Kolattukudy, 1980), dwarf shrubs (Mikkonen and Huttunen, 1981) and mosses (Huttunen, Karhu and Kallio, 1981) have been studied as well. The most obvious effect of air pollutants is the erosion of the epicuticular wax structures around the stomata of conifer needles (Cape and Fowler, 1981; Huttunen and Laine, 1981, 1983). The normal weathering of the wax of needles takes place over several years; in Scots pine (*Pinus sylvestris*) needles in northern Finland, this erosion typically takes five to six years, and in Norway spruce (*Picea abies*) the weathering period is even longer (Huttunen and Laine, 1982). In a study of Scots pine needles, the erosion of surface wax was enhanced in trees growing in polluted areas (*Figure 9.2*; Huttunen and Laine, 1983). As conifer needles are especially sensitive to water loss during late winter, despite stomata being closed (Leyton and Juniper, 1963; Huttunen, Havas and Laine, 1981), the pollutant-enhanced erosion of the protective cuticle can have important consequences.

The carbon fraction of the wax structure of developing pine needles is characterized by consisting primarily of low molecular weight waxes in the range of C_{14} and C_{26}, with most waxes being below C_{19}. Later during the winter the lowest molecular waxes, i.e. between C_{14} and C_{17}, disappear. The first sign of wax erosion near the stomata of Scots pine is the fusion of the wax fibres; this typically occurs in August, about one or two months after needle flushing. The fusion of the wax fibres is accompanied by the erosion of the low molecular waxes and the *de novo* synthesis of higher molecular waxes. At this time, an increase in the amount of C_{23} epicuticular wax can also be observed (Huttunen, Mäkelä and Laine, 1982). During the summer, the amount of wax in the current year's needles of Scots pine was found to be irregular due to climatic and other environmental conditions, and considerable variation among pines was observed (Huttunen, Mäkelä and Laine, 1982). Nevertheless, when the amounts of wax in needles from trees in rural and urban environments were compared, samples from the rural areas were found to contain more wax in almost all cases.

Injuries to the resin ducts are among the first symptoms of pollution damage to needle tissues, the general response being an initial hypertrophy of the epithelial cells

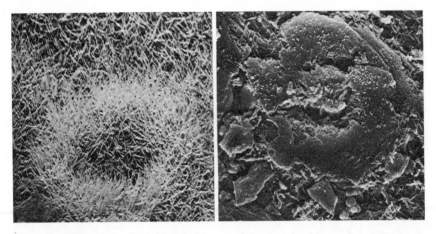

Figure 9.2. The structure of epicuticular wax around the stomata of Scots pine needles in winter. On the left, the normal fibre structure of waxes under winter conditions; on the right, weathered epicuticular wax without crystalloid structure of needles growing in polluted air. (Magnification × 2000 reduced to 80%)

leading to occlusion of the resin ducts (Stewart, Treshow and Harner, 1973; Huttunen, 1975, 1978). During winter, exposure to air pollutants resulted in the total collapse of the epithelial cells of the resin ducts of Scots pine needles, although the needles retained a green and healthy-looking outer appearance (Huttunen, 1975, 1978). These needles showed an accumulation of sulphur compounds, especially in the transfusion tissues; a granulation of the transfusion cells was also observed.

The effects of air pollutants on the surface structures of lower plants have been studied in transplantation experiments with mosses. Disturbances in structure such as the loss of normal cell turgidity and the disappearance of the mamillae of surface cells and of the wrinkled surface structure were observed before any visible symptoms of damage. The changes in ultrastructure revealed by electron microscopy could be better related to disturbances in metabolism and in photosynthesis than to those changes seen with conventional light microscopy. Changes in the surface structure, namely the change in the microenvironment caused by the disappearance of the mamillae and the development of a dried, wrinkled cell structure, were found to correlate with changes in the nitrogenase activity of the free-living blue-green algae on moss leaves (Huttunen, Karhu and Kallio, 1981).

Mesophyll cells and chloroplasts

In most studies, the development of ultrastructural injury is described in actively metabolizing tissues (Kärenlampi and Soikkeli, 1980). The most often observed changes in the features of chloroplasts are:

(1) a swelling, rounding or decrease in the size of the chloroplasts (Fischer, Kramer and Ziegler, 1973; Engelbrecht and Louw, 1973; Swanson, Thomson and Mudd, 1973; Soikkeli and Tuovinen, 1979);
(2) an apparent doubling of the chloroplast envelopes caused by the invagination of the inner membrane (Godzik and Knabe, 1973);
(3) a stretching of the chloroplast envelope (Swanson, Thomson and Mudd, 1973; Soikkeli, 1981b; Soikkeli and Tuovinen, 1979);
(4) an increase in the staining density of the envelopes and the accumulation of electron-dense material between the two membranes (Swanson, Thomson and Mudd, 1973; Pell and Weissberger, 1976);
(5) a swelling of the chloroplast thylakoids (Wellburn, Majernik and Wellburn, 1972; Fischer, Kramer and Ziegler, 1973; Godzik and Knabe, 1973; Młodzianowski and Białobok, 1977; Białobok, 1978);
(6) a reduction of grana lamellae (Godzik and Sassen, 1974; Soikkeli and Tuovinen, 1979); and
(7) a granulation of the chloroplast stroma (Thomson, Dugger and Palmer, 1966; Fischer, Kramer and Ziegler, 1973; Młodzianowski and Białobok, 1977).

Swelling and curling of thylakoids have both been observed in chloroplasts of plants exposed to NO_x or to HF (Soikkeli and Tuovinen, 1979; Soikkeli, 1981a, 1981b). The curling of thylakoids in the chloroplasts of conifer needles might be caused by the combined action of frost and air pollutants (Soikkeli, 1981b). Although the swelling of thylakoids has also been described for other pollutants such as SO_2 and NO_2 (Wellburn, Majernik and Wellburn, 1972; Malhotra, 1976; Soikkeli, 1981c), it cannot be considered a specific reaction to a given pollutant but rather as a general symptom of acute injury.

Injuries characterized by a reduction in the number of grana and the appearance of light-coloured plastoglobuli resembling lipid material were found to be especially frequent in trees within urban areas, and in trees in the periphery of forests bordering urban areas, where the annual mean concentration of SO_2 was greater than 60 μg m^{-3} (Soikkeli and Tuovinen, 1979). Even where the annual mean concentration did not exceed 40 μg SO_2 m^{-3}, needle damage, although mild, could be detected at the light microscope level in trees from urban forests (Huttunen et al., 1979; Soikkeli, 1981b; Huttunen, Karhu and Pakonen, 1982).

Fibrillar and crystalloid structures have been observed in the chloroplast stroma by many authors, especially in O_3-treated material (Thomson, Dugger and Palmer, 1966; Thomson, Nagahashi and Platt, 1974; Thomson and Swanson, 1972). These structures were found at mild stages of O_3 injury, and frequently accompanied injury to the chloroplast envelopes. Crystalline bodies were also found in SO_2 and NO_2 treated plants (Godzik and Sassen, 1974; Dolzmann and Ullrich, 1966). Fumigation with NO_2 was also found to give rise to tubular protrusions from the chloroplast envelopes which were closely associated with the mitochondria (Lopata and Ullrich, 1975).

Soikkeli (1978, 1980) studied the seasonal changes in the ultrastructure of Norway spruce (Picea abies) and Scots pine (Pinus sylvestris) needles not affected by air pollutants to provide a basis for assessing pollutant-induced changes in ultrastructure. The ultrastructure of the needles and the seasonal changes in it were similar in the two conifers. Most of the mesophyll cells appear to be functionally mature at the time when the needles emerge from the bud scales. The current year's needles were shown to have achieved a mature cell structure in August; the chloroplasts form a single layer near the cell walls and each chloroplast contains a single, large starch grain.

In autumn, Soikkeli (1980) observed various changes in the ultrastructure of the needles:

(1) the starch grains disappear, signifying a shift in the starch-sugar equilibrium towards soluble sugars, which are important to the frost hardiness of the needles;
(2) the chloroplasts collect together in one corner of the cell in spruce needles, and in several corners of the cell in pine needles;
(3) the cytoplasm becomes net-like and rich in ribosomes; and
(4) lipid droplets appear in abundance in the cytoplasm.

Mild injuries induced by air pollutants could be seen in the needle samples collected in autumn (October). Two types of injury were identified. In the first, the number of grana in the chloroplasts was reduced and the number of plastoglobuli increased. In the second type of injury, the plastoglobuli in the cytoplasm were lightening and the amount of lipid-like material accumulating (Figure 9.3). These changes were found to be symptoms of chronic injury in conifer needles (Soikkeli, 1981c). This injury to the current year's needles was found to be similar to the injury sustained by older spruce needles (Soikkeli and Tuovinen, 1979; Soikkeli and Paakkunainen, 1981).

Although chloroplasts have been reported to disorganize to some degree in both spruce (Senser, Schötz and Beck, 1975) and pine (Martin and Öquist, 1979) needles during the winter, Soikkeli (1978, 1980) observed that all the cell organelles of both spruce and pine needles remained intact throughout the year in needles collected from non-polluted areas. Towards the end of winter, both types of injury discussed above could be observed in older needles from polluted areas. In most cases, one type of injury was dominant with the second type occurring additionally. In general, the degree of injury was more severe in older than in younger needles (Soikkeli, 1981c).

Fairly early in the spring (March, April or May, depending upon the geographic

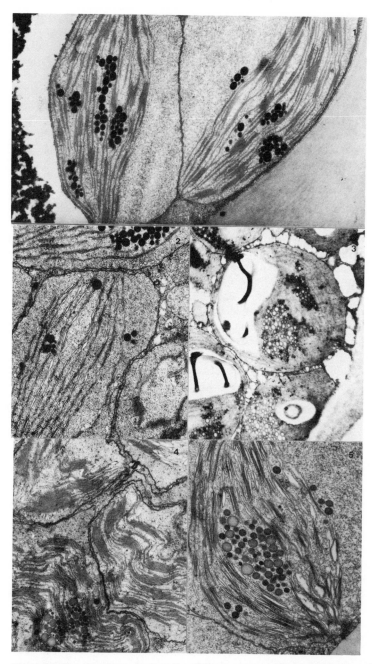

Figure 9.3. The ultrastructure of needles of Norway spruce (*Picea abies* L. Karst.). (1) The chloroplast of a spruce needle from a clean area in winter. (2) The decrease in the number of grana lamellae resulting from sulphur compounds pollution in winter. (3) The increasing lightness in the colour of the plasto-globuli and the accumulation of lipid-like droplets in needles under sulphur compounds pollution in spring. (4) The curling of lamellae in chloroplasts under $SO_2 + NO_x$ pollution in winter. (5) The swelling of lamellae under $SO_2 + NO_x$ pollution in winter. (Magnification in all figures, $\times 20\,000$ reduced to two-thirds: micrographs by S. Soikkeli)

location), the ultrastructure of the needles changes to correspond to summer conditions again; the chloroplasts are near the cell walls and contain starch, the cytoplasm is no longer net-like in appearance and is rich in polysomes, indicating active cellular metabolism (Soikkeli, 1980). By August, the initial stages of pollutant injury, i.e. the slight swelling of the stroma lamellae, were distinguished in the current year's needles. With time, the swelling spread to the grana thylakoids in the green needles (Soikkeli and Paakkunainen, 1981), and in October Soikkeli (1981b) observed in the current year's needles a reduction in the number of grana and the appearance of the second type of injury identified in conifer needles, namely an increase in the amount of the lipid-like material in the cytoplasm.

Godzik and Knabe (1973) analysed the needles of Scots pine in the Ruhr area and found that all the samples showed chloroplast injuries; similar observations were made on the Weymouth pine (*Pinus strobus*) by Parker and Philpott (1961) and by Harris (1971). Needles of European larch (*Larix decidua*) exposed to air pollutants have been observed to display swelling and degradation of the chloroplast thylakoids, granulation of the cytoplasm and the plastid matrix, degradation of ribosomes and the endoplasmic reticulum, agglutination of chromatin and cell plasmolysis (Młodzianowski and Białobok, 1977). Two types of chloroplast degradation were observed. In pollutant-tolerant individuals, the stroma remained unchanged and the thylakoids were the first to degenerate, while in the second type of chloroplast degradation, found in pollutant-susceptible individuals, the thylakoids were more stable and the stroma degenerated first. Such types of plastid degradation have not been observed in the mesophyll cells of Scots pine (Młodzianowski and Młodzianowska, 1980).

The information obtained so far is insufficient to explain all the seasonal changes and their interactions with gaseous air pollutants, but ultrastructural changes increase in severity with increasing age of the needles.

Conclusions

The information on pollutant-induced changes in morphology obtained at different levels of organization by using either light or electron microscopy should ideally be related to plant biochemical and physiological studies. Mansfield and Freer-Smith (this volume) have discussed the physiological role of stomata in pollutant stress avoidance. The morphological changes seen in the epidermis of plants may represent various reactions to pollutant stress by resistant or sensitive species. For example, lower stomatal frequencies and increased length and number of trichomes as observed in clover (*Trifolium*) and kudzu (*Pueraria*) growing in polluted environments may be adaptations aimed at stress avoidance. The decrease in stomatal frequency would represent a morphological/physiological restriction to the entry of the pollutant to leaf tissues while an increase in the number and length of trichomes alters the surface of the leaf and would be expected to increase the aerodynamic resistance (r_a) to the transfer of pollutants from the atmosphere to the leaf surface. A decrease in the number of trichomes, on the other hand, might represent the reaction of a sensitive species to a pollutant stress. Morphological studies on plants growing in, and adapted to, polluted environments can augment the information obtained in laboratory experiments on the mechanisms of pollutant avoidance/tolerance and on the response of resistant and sensitive species to this environmental stress.

Perhaps the greatest potential for understanding the resistance or sensitivity of plants to air pollution and for elucidating the mechanisms of stress metabolism comes from the combination of electron microscopy with biochemical and histochemical investigations. Wellburn, Majernik and Wellburn (1972) and Malhotra (1976) have been able to relate to some extent the pollutant-induced changes in the ultrastructure of chloroplasts to changes in the photosynthetic capacity of mesophyll cells, and Parry and Whittingham (this volume) have argued that changes in the chloroplast stroma, such as the granulation observed in response to air pollutants and other stresses, would be expected to reflect changes in the structure and catalytic activity of enzymes of C_3 photosynthesis. Ultrastructural studies have shown that exposure to air pollutants can also disrupt the structure of the thylakoids and grana within the chloroplasts, and such disruptions are likely to have important consequences on the activities of PSI and PSII (see Sugahara, this volume) and on the light-modulated enzymes of photosynthesis (see Alscher-Herman, this volume).

When studying the possible effects of very low concentrations of pollutants on evergreen plants over prolonged periods, the most sensitive and informative methods should be used. Among these methods, electron microscopy, biochemical investigations of lipid metabolism and of enzymatic changes characteristic of stress metabolism are the most useful. The information based on very few cells in electron microscopy must be generalized to the tissue level with the aid of light microscopy.

References

BIAŁOBOK, S. Studies on the effect of sulphur dioxide and ozone on the respiration and assimilation of trees and shrubs in order to select individuals resistant to the action of these gases, *Polish Academy of Sciences, Fourth Annual Report*, 100 pp. (1978)

CAPE, N. and FOWLER, D. Changes in epicuticular wax of *Pinus sylvestris* exposed to polluted air, *Silva Fennica* 15, 457–458 (1981)

CHABOT, J. F. and CHABOT, B. F. Developmental and seasonal patterns of mesophyll ultrastructure in *Abies balsamea*, *Canadian Journal of Botany* 53, 295–304 (1975)

DOLZMANN, P. and ULLRICH, H. Einige Beobachtungen über Beziehungen zwischen Chloroplasten und Mitochondrien im Palisadenparenchym von *Phaseolus vulgaris*, *Zeitschrift für Pflanzenphysiologie* 55, 165–180 (1966)

ENGELBRECHT, H. P. and LOUW, C. W. Hydrogen fluoride injury in sugar-cane: some ultrastructural changes. In *Proceedings of the Third International Clean Air Congress*, VDI-Verlag GmbH, Düsseldorf, A 157–159 (1973)

ESPELIE, K. E., DAVIS, R. W. and KOLATTUKUDY, P. E. Composition, ultrastructure and function of the cutin- and suberin-containing layers in the leaf, fruit peel juice sac and inner seed coat of grapefruit (*Citrus papradisi* Macfed.), *Planta* 149–151 (1980)

FISCHER, F., KRAMER, D. and ZIELGER, H. Elektronenmikroskopische Untersuchungen SO_2-begaster Blätter von *Vicia faba* L. Beobachtungen am Chloroplasten mit akuter Schädigung, *Protoplasma* 76, 83–96 (1973)

FISCHER, K. *Cytologische und physiologische Wirkungen von SO_2 auf höhere Pflanzen*. Dissertation, Darmstadt (1967)

GODZIK, S. and KNABE, W. Vergleichende elektronenmikroskopische Untersuchungen der Feinstruktur von Chloroplasten einiger *Pinus-Arten* aus den Industriegebieten an der Ruhr und in Oberschlesien. In *Proceedings of the Third International Clean Air Congress*, VDI-Verlag GmbH, Düsseldorf, A 164–170 (1973)

GODZIK, S. and SASSEN, M. M. A. Einwirkung von SO_2 auf die Feinstruktur der Chloroplasten von *Phaseolus vulgaris*, *Phytoplathologische Zeitschrift* 79, 155–159 (1974)

GODZIK, S. and SASSEN, M. M. A. A scanning electron microscope examination of *Aesculus hippocastanum* L. leaves from control and air polluted areas, *Environmental Pollution* 17, 13–18 (1978)

GRILL, D. Rasterelektronenmikroskopische Untersuchungen an SO_2-belasteten Fichtennadeln, *Phytopathologische Zeitschrift* 78, 75–80 (1973)

HARRIS, W. M. Ultrastructural observations on the mesophyll cells of pine leaves, *Canadian Journal of Botany* 49, 1107–1109 (1971)

HOLOPAINEN, T. Alterations in the ultrastructure of epiphytic lichens *Hypogymnia physodes* and *Alectoria capillaris* caused by air pollution, *Silva Fennica* 15, 469–474 (1981)

HUTTUNEN, SAIJA. Ilman epäpuhtauksien vaikutus eräisiin koristekasveihin Kokkolassa [*The effects of air pollutants on garden species at Kokkola, northern Finland*]. *Pro gradu. Helsingin yliopiston puutarhatieteen laitos.* 80s (1979)

HUTTUNEN, SATU. The influence of air pollution on the forest vegetation around Oulu, *Acta Universitatis Ouluensis A* **33**, *Biologia* **2**, 78 pp. (1975)

HUTTUNEN, S. The effects of air pollution on provenances of Scots pine and Norway spruce in northern Finland, *Silva Fennica* **12**, 1–16 (1979)

HUTTUNEN, S., HAVAS, P. and LAINE, K. Effects of air pollutants on the wintertime water economy of the Scots pine, *Pinus sylvestris, Holarc. Ecol.* **4**, 94–101 (1981)

HUTTUNEN, S., KARHU, M. and KALLIO, S. The effect of air pollution on transplanted mosses, *Silva Fennica* **15**, 495–504 (1981)

HUTTUNEN, S., KARHU, M. and PAKONEN, T. Integrated study of microscopic injuries after stress development in *Pinus sylvestris*. In *Abstracts, XII International Meeting for Specialists in Air Pollution Damages in Forests*, August 23–29, Oulu, Finland (1982)

HUTTUNEN, S. and LAINE, K. The structure of pine needle surface (*Pinus sylvestris* L.) and the deposition of air borne pollutants, *Archiwum Ochrony Srodowiska* **2–4**, 29–38 (1981)

HUTTUNEN, S. and LAINE, K. Effects of air-borne pollutants on the surface wax structure of *Pinus sylvestris* L needles, *Annales Botanici Fennici* **20**, 79–86 (1983)

HUTTUNEN, S., MÄKELÄ, M. and LAINE, K. Quality and structure of surface waxes in pine needles and long-term effects of air pollutants (in preparation) (1982)

HUTTUNEN, S., MANNINEN, J., LAINE, K., FORSTEN, P. PAKONEN, T. and TÖRMÄLEHTO, H. Rikkiyhdisteiden leviäminen ja vaikutus kasvillisuuteen Porissa v. 1979 [*Dispersion and effects of airborne sulphur compounds on vegetation at Pori in 1979*], *Porin kaupungin tutimuksia* **32**, 1979, 38 s (1979)

KÄRENLAMPI, L. and SOIKKELI, S. Morphological and fine structural effects of different pollutants on plants: development and problems of research. In *Papers Presented to the Symposium on the Effects of Airborne Pollution on Vegetation, United Nations*, Warsaw, Poland, 1980, pp. 92–99 (1980)

KOLATTUKUDY, P. E. (Ed.) *Chemistry and Biochemistry of Natural Waxes.* Elsevier, Amsterdam (1976)

KOLATTUKUDY, P. E. Cutin, suberin and waxes. In *The Biochemistry of Plants*, Vol. 4, *Lipids: structure and function*, (Eds. P. K. STUMPF and E. E. CONN), pp. 571–645, Academic Press, New York (1980)

KOLATTUKUDY, P. E., ESPELIE, K. E. and SOLIDAY, C. L. Hydrophobic layers attached to cell walls, cutin, suberin and associated waxes. In *Plant Carbohydrates II. Extracellular carbohydrates*, (Eds. W. TANNER and F. A. LOEWUS), pp. 225–254, *Encyclopedia of Plant Physiology, New Series*, **Vol. 13B**, Springer-Verlag, Berlin (1981)

LE BLANC, F. and RAO, D. N. Effects of sulphur dioxide on lichen and moss transplants, *Ecology* **53**, 612–617 (1973)

LEYTON, L. and JUNIPER, B. E. Cuticle structure and water relations of pine needles, *Nature* **198**, 770–771 (1963)

LOPATA, W.-D. and ULLRICH, H. Untersuchungen zu stofflichen und strukturellen Veränderungen an Pflanzen unter NO_2-Einfluss. *Staub.-Reinhalt, Luft* **35**, 196–200 (1975)

MALHOTRA, S. S. Effects of sulphur dioxide on biochemical activity and ultrastructural organization of pine needle chloroplasts, *New Phytologist* **76**, 239–245 (1976)

MARTIN, B. and ÖQUIST, G. Seasonal experimentally induced changes in the ultrastructure of *Pinus sylvestris, Physiologia Plantarum* **46**, 42–49 (1979)

MARTIN, J. T. and JUNIPER, B. E. *The Cuticles of Plants*, Edward Arnold Ltd, London (1970)

MIKKONEN, H. and HUTTUNEN, S. Dwarf shrubs as bioindicators, *Silva Fennica* **15**, 475–480 (1981)

MŁODZIANOWSKI, F. and BIAŁOBOK, S. The effect of sulphur dioxide on ultrastructural organization of Larch needles, *Acta Societatis botanicorum Polaniae* **46**, 629–634 (1977)

MŁODZIANOWSKI, F. and MŁODZIANOWSKA, L. *Cytochemical localization of enzyme activity in Scots pine needles treated with SO_2*, Polish Academy of Sciences, Institute of Dendrology Publication Pl-Fs-74: 79–81 (1980)

PARKER, J. and PHILPOTT, D. E. An electron microscope study of chloroplast condition in summer and winter in *Pinus strobus, Protoplasma* **53**, 575–583 (1961)

PELL, E. J. and WEISSBERGER, W. C. Histopathological characterization of ozone injury to soybean foliage, *Phytopathology* **66**, 856–861 (1976)

SENSER, M., SCHÖTZ, F. and BECK, E. Seasonal changes in structure and function of spruce chloroplasts, *Planta* **126**, 1–10 (1975)

SHARMA, G. K. and BUTLER, J. Leaf cuticular variations in *Trifolium repens* L. as indicators of environmental pollution, *Environmental Pollution* **5**, 287–293 (1973)

SHARMA, G. K. and BUTLER, J. Environmental pollution: leaf cuticular patterns in *Trifolium pratense* L., *Annals of Botany* **39**, 1087–1090 (1975)

SHARMA, G. K., CHANDLER, C. and SALEMI, L. Environmental pollution and leaf cuticular variation in kudzu (*Pueraria lobata* Willd.), *Annals of Botany* **45**, 77–80 (1980)

SHARMA, G. K. and TYREE, J. Geographic leaf cuticular and gross morphological variations in *Liquidambar styraciflua* L. and their possible relationship to environmental pollution, *Botanical Gazette* **134**, 179–184 (1973)

SHIMAZAKI, K., SAKAKI, T. and SUGAHARA, K. Active oxygen participation in chlorophyll destruction and lipid peroxidation in SO_2-fumigated leaves of spinach. In *Studies on the Effects of Air Pollutants on Plants and Mechanisms of Phytotoxicity*. Research Report from the National Institute for Environmental Studies, Ibaraki, Japan, No. 11: 91–101 (1980)

SOIKKELI, S. Seasonal changes in mesophyll ultrastructure of needles of Norway spruce (*Picea abies*), *Canadian Journal of Botany* **56**, 1932–1940 (1978)

SOIKKELI, S. Ultrastructure of the mesophyll in Scots pine and Norway spruce: seasonal variation and molarity of the fixative buffer, *Protoplasma* **103**, 241–252 (1980)

SOIKKELI, S. Comparison of cytological injuries in conifer needles from several polluted industrial environments in Finland, *Annales Botanici Fennici* **18**, 47–61 (1981a)

SOIKKELI, S. The effects of chronic urban pollution on the inner structure of Norway spruce needles, *Savonia* **4**, 1–10 (1981b)

SOIKKELI, S. A review of the structural effects of air pollution on mesophyll tissue of plants at light and transmission electron microscope level, *Savonia* **4**, 11–34 (1981c)

SOIKKELI, S. and PAAKKUNAINEN, T. The effect of air pollution on the ultrastructure of the developing and current year needles of Norway spruce. *Mitteilungen der forstlichen Bundesversuchsanstalt, Wien*, **137/II**, 159–163 (1981)

SOIKKELI, S. and TUOVINEN, T. Damage in mesophyll ultrastructure of needles of Norway spruce in two industrial environments in central Finland, *Annales Botanici Fennici* **16**, 50–64 (1979)

STEWART, D., TREESHOW, M. and HARNER, F. M. Pathological anatomy of conifer needle necrosis, *Canadian Journal of Botany* **51**, 983–988 (1973)

SWANSON, E. S., THOMSON, W. W. and MUDD, J. B. The effect of ozone on leaf cell membranes, *Canadian Journal of Botany* **51**, 1213–1219 (1973)

THOMSON, W. W., DUGGER, W. M., Jr. and PALMER, R. L. Effects of ozone on the fine structure of the palisade parenchyma cells of bean leaves, *Canadian Journal of Botany* **44**, 1677–1682 (1966)

THOMSON, W. W., NAGAHASHI, J. and PLATT, K. Further observations on the effects of ozone on the ultrastructure of leaf tissue. In *Air Pollution Effects on Plant Growth*, (Ed. M. DUGGER), pp. 83–93, American Chemical Society Symposium Series, No. 3. The American Chemical Society, Washington, DC (1974)

THOMSON, W. W. and SWANSON, E. S. Some effects of oxidant air pollutants (ozone and peroxyacetyl nitrate) on the ultrastructure of leaf tissue, *Proceedings of the Electron Microscopy Society of America* **30**, 360–361 (1972)

WEI, L. L. and MILLER, G. W. Effect of HF on the fine structure of mesophyll cells from *Glycine max* Merr, *Fluoride* **5**, 67–73 (1972)

WELLBURN, A. R., MAJERNIK, O. and WELLBURN, F. A. Effects of SO_2 and NO_2 polluted air upon the ultrastructure of chloroplasts, *Environmental Pollution* **3**, 37–49 (1972)

WILL-WOLF, S. Effects of a 'clean' coal-fired power generating station on four common Wisconsin lichen species, *The Bryologist* **83**, 296–300 (1980)

Part III

Effects on light reactions

Chapter 10

The role of stomata in resistance mechanisms

T. A. Mansfield and P. H. Freer-Smith
DEPARTMENT OF BIOLOGICAL SCIENCES, UNIVERSITY OF LANCASTER, UK

Conceptual approach to resistance mechanisms

One of the major objectives in the analysis of effects of air pollution at the physiological and biochemical levels is the development of an understanding of what determines resistance and susceptibility. In the well-established terminology used by those who study a variety of stress phenomena in plants, resistance can be the result of mechanisms of two kinds, namely *stress avoidance* or *stress tolerance* (Levitt, 1972; Taylor, 1978). In stress avoidance the plant is able to exclude the agent concerned, whereas in stress tolerance the agent is able to act internally but without serious consequences on the functioning of cells or tissues. Stress tolerance is conveniently subdivided into *strain avoidance* and *strain tolerance*, where *strain* represents a harmful change induced by the stress-inducing agent. Strain tolerance implies an effect on normal functioning (but not to the extent that survival is affected) and strain avoidance is achieved when normal functioning of a process continues even though there may be energetic costs involved.

Taylor (1978) argued convincingly for the adoption of this terminology by those interested in the factors involved in plant resistance to air pollutants. The subdivision of stress tolerance into strain avoidance and tolerance may, however, often be difficult because of our limited understanding, but the main division of resistance into stress avoidance and tolerance does involve some mechanisms which are beginning to be understood, namely those involving stomatal movements. The content of this contribution is not directly concerned with plant metabolism, but the physiological issues raised cannot be ignored by those whose main interests are the effects of pollutants at the cellular level.

The role of stomata in stress avoidance

Stomata play an indispensable part in the mechanisms which plants have evolved to enable them to cope with naturally occurring environmental stresses. The movements of stomata in response to factors of the environment are extremely complex, and the mechanisms behind them are not fully understood. The purpose behind many of the changes in stomatal aperture which occur on a daily basis is the balancing of two

131

opposing priorities, the need for CO_2 for photosynthesis and the prevention of excessive water loss by transpiration. The ways in which stomata react to different evaporative conditions, to soil water status, and to water stress in the plant are well documented (Mansfield and Davies, 1981). An integral part of these responses is the effect of CO_2 on stomata, which seems to exert a significant regulatory role. Stomata close as the CO_2 concentration increases, but not by a constant amount. The reactivity of the stomata to CO_2 is regulated by two hormones, abscisic acid and indol-3-yl acetic acid (Snaith and Mansfield, 1982) and appears to vary according to the water status of the plant (Raschke, 1975). Guard cells are able to sense the CO_2 concentration in the substomatal cavity, not in the air external to the leaf (Heath, 1948). This means that when air movement breaks down the resistance to diffusion through the boundary layer adjacent to the leaf surface, more CO_2 will diffuse through the stomata and enter the substomatal cavity. Thus the effect of the response to CO_2 is to reduce stomatal aperture in wind and cut down the rate of transpiration (Mansfield and Davies, 1981). The regulation of the CO_2-sensing ability by two hormones means that this form of transpiration control can be modified according to the water status of the plant.

In the stomatal response to CO_2 we see, therefore, an example of an elegant gaseous control system which helps to defend the plant against excessive water loss. This constitutes an avoidance mechanism which plays a significant part in the adaptation of plants to the terrestrial environment.

We can make analogies with the control of water loss in considering how stomata could fulfil a role in providing resistance to the stress imposed by air pollutants. As the concentration of a gas in the atmosphere increases, the rate of diffusion into the leaf increases according to Fick's law (Meidner and Mansfield, 1968). Ideally, therefore, the stomata should close as the leaf begins to experience stress due to the increased entry of pollutant molecules. However, since stomatal closure also represents a stress because it reduces the availability of CO_2 for photosynthesis, it would be important for the closure to be incomplete. It should be sufficient to reduce the entry of pollutant molecules to a tolerable level, but insufficient to deprive the leaf of CO_2. Thus an ideal mechanism for avoiding pollution stress would involve a balancing of the leaf's priorities, just as is the case in the avoidance of water stress.

Stomatal responses to toxic agents

Mechanics of stomatal functioning

Opening and closing of stomata are achieved by changes in the turgor relationship between the guard cells and their neighbours. In normal functioning the turgor changes of the guard cells themselves play a dominant role: opening is achieved by a massive transport of potassium ions into the guard cells, and a consequent increase in their turgor. In some well defined situations, however, changes in stomatal aperture can occur without any change in the guard cells. This is because the opening movement results in a competition for space between the guard cells and the neighbouring cells (which are usually well-defined subsidiary cells). For the pore to open the guard cells have to press into space that would otherwise be occupied by the subsidiary cells. The subsidiary cells thus offer a resistance to opening and they do, consequently, have some control over the aperture that can be achieved by a given amount of guard cell turgor. Heath (1938) performed some very elegant experiments in which he punctured guard or subsidiary cells with microneedles. Stomatal closure to half-aperture was caused by

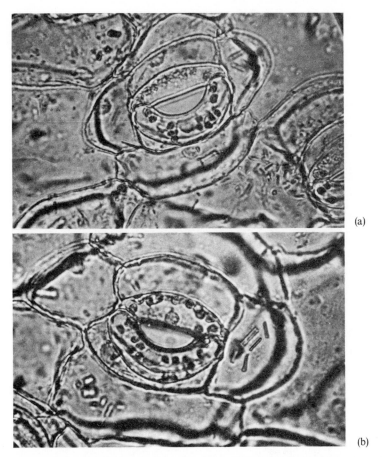

(a)

(b)

Figure 10.1. Asymmetrical stomatal pores in *Commelina communis* resulting from changes of turgor in guard cells or subsidiary cells. In (a), the guard cell at top has lost turgor, and in (b) both the guard cells are turgid but the subsidiary cell at the bottom is not

puncturing one guard cell, but a considerable increase in aperture resulted when the turgor of one subsidiary cell was eliminated. *Figure 10.1* illustrates how a similar appearance of the stomatal pore can result from two different situations. In *Figure 10.1a,* one guard cell has been damaged (note that the chloroplasts have disintegrated), its turgor has been lost, and the pore has closed on that side. In *Figure 10.1b* both guard cells are intact, but the subsidiary cell at the bottom has lost turgor, allowing the pore to expand more on that side. *Figure 10.2* provides a comparison of normally open and closed stomata (note the extent to which the guard cells push into the subsidiary cells when the pore opens) and also a distorted pore resulting from the collapse of one subsidiary cell.

These simple mechanical influences on stomatal aperture appear to be very important in determining the reaction to air pollutants. Let us consider two ways in which a toxic gas may act on cells of the epidermis:

(1) Molecules diffuse in through the stomata, and from the substomatal cavity enter the inner surfaces of guard and other cells of the epidermis. The guard cells are

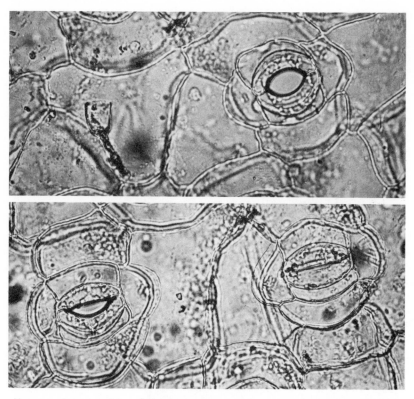

Figure 10.2. On the right, normal appearance of open and closed stomata of *Commelina communis*, for comparison with the abnormal opening (bottom left) resulting from the loss of turgor of one subsidiary cell

more tolerant of the strain imposed by the pollutant than are the subsidiary cells, and the latter lose turgor while the guard cells retain it. Stomatal opening occurs purely as a result of the release of surrounding pressure.

(2) Molecules enter the epidermis as in (1), but cause a loss of turgor only in the guard cells. The result is stomatal closure.

In both these cases the nature of the action of the pollutant at the cellular level could be the same, e.g. a structural effect on the tonoplast membrane leading to a loss of the solutes responsible for maintaining turgor. The relative sensitivity of subsidiary and guard cells to the pollutant is the difference between (1) and (2), but the measurable reaction to the pollutant in terms of stomatal aperture is an apparently quite opposite effect.

The complexity of the reactions of stomata to toxic chemicals is well known to plant physiologists. There has long been an interest in the possibility of imposing artificial controls on the behaviour of stomata in the field, in order to regulate the consumption of water by crops. Chemicals considered as possible candidates have included some known to inhibit particular metabolic steps, and others that are membrane active. One, phenylmercuric acetate, was once considered to have considerable potential for regulating water loss from forest trees, but subsequent work showed that the compound could both induce stomatal closure (the wanted response) or cause opening (unwanted). Differential destruction of guard cells and subsidiary cells was the likely cause of these contrary effects. The lesson learned was that even with known doses of a chemical sprayed in solution onto the leaf surface, the effect on stomatal aperture could not be controlled (Waisel, Borger and Kozlowski, 1969).

EFFECTS OF SO_2 ON EPIDERMAL MECHANICS

Observations of the effects of one pollutant, SO_2, on stomatal behaviour suggest that there are close parallels with the action of phenylmercuric acetate. Our own studies some years ago showed that SO_2 could induce wider stomatal opening than normal in the broad bean (*Vicia faba*) (Majernik and Mansfield, 1971). An interesting aspect of the effects observed was that the essential diurnal functioning of the stomata was not impaired. They retained the ability to close at night, and the effect of the pollutant was simply to cause them to open more widely during the day. Recent studies by Black and Black (1979) have provided an explanation for an effect of this kind. They confirmed that stomatal conductance increased by 20–25% when broad bean was exposed to SO_2, even with concentrations as low as 0.017 ppm. The enhanced stomatal opening was accompanied by extensive damage to the epidermal cells (there are no anatomically distinct subsidiary cells in broad bean), judging from the uptake of the vital stain neutral red (*Table 10.1*). Some damage does often occur to cells of the epidermis when it is stripped away from the mesophyll. The figures for cell survival in *Table 10.1* do not, therefore, necessarily reflect numbers alive and dead on the intact

TABLE 10.1. Cell survival estimated by neutral red uptake in strips of lower epidermis from control and polluted plants of *Vicia faba*

SO_2 concentration (ppm)	Cell survival (%)	
	Epidermal cells adjacent to guard cells	Guard cells
0	64.2 (2.5)	100.0 (0.0)
0.017	35.4 (1.0)	97.5 (2.5)
0	54.8 (2.5)	93.1 (3.1)
0.070	27.7 (2.8)	91.3 (2.9)

From Black and Black (1979).
The figures in parenthesis are the standard errors of the means.

leaf. They may show that ability to survive the process of stripping the epidermis from the leaf has been affected by prior fumigation with SO_2. Microscopic examination of the epidermis by Black and Black did, however, support the view that the epidermal cells had suffered structural injury as a result of exposure to the pollutant. They noted 'severe disorganization of epidermal cell protoplasts to form granular or stringy cytoplasmic material'.

Even if the epidermal cells *in situ* are not actually killed by exposure to SO_2, damage to their membranes would impair their ability to maintain full turgor. The guard cells, being less susceptible to injury as suggested by the figures in *Table 10.1*, would then be able to achieve wider stomatal apertures.

The inner walls of the epidermal cells which are adjacent to the stomatal pore are thought to represent major evaporation sites within leaves (*see Figure 2.1* of Meidner and Sheriff, 1976). This is because the diffusion paths of least resistance between the substomatal cavity and the atmosphere are those from cells closest to the pore. The converse will apply to SO_2 entering from the atmosphere: once a molecule has entered through a stoma the route to the surface of a nearby subsidiary or epidermal cell is very short. The cells of the epidermis are therefore in a vulnerable position. They may also be more susceptible to SO_2 injury because they lack chloroplasts (epidermal cells of higher plants only rarely contain chloroplasts). This could deprive them of the ability to convert SO_2 to H_2S using energy from photosynthetic electron transport. Sekiya *et al.* (1981) have reported that H_2S emission from cucurbit leaves is almost totally light dependent. Cells without chloroplasts may therefore lack an important detoxification mechanism for removing excess sulphur by the production of H_2S, and its diffusion back into the atmosphere.

The ability of SO_2 to damage epidermal and subsidiary cells preferentially can thus produce stomatal opening by a simple mechanical means, namely the reduction of the pressure normally imposed upon the guard cells by the turgor of their neighbours. This leads to wider stomatal apertures during the day, though does not necessarily impair their ability to respond to stimuli such as light and CO_2.

Consequences of enhanced stomatal opening

Biscoe, Unsworth and Pinckney (1973) estimated that the stomatal opening responses to SO_2 which they had observed in broad bean would increase transpiration of a field crop by 23%. This could obviously be of economic significance since the supply of soil water often imposes limitations on productivity. There are, however, other possible consequences of enhanced stomatal opening. Mansfield and Davies (1981) devised a simple experiment to determine the effect of excessive transpiration on the leaves of an ordinary mesophyte, *Commelina communis*. The whole plant was enclosed in a transparent container and illuminated for 2 h. The restricted gas exchange led to conditions of high humidity and low CO_2 concentrations which stimulated wide stomatal opening. When the container was suddenly removed and the leaves were exposed to a light wind, severe wilting occurred and parts of the leaves were permanently damaged (*Figure 10.3*). The brown necrotic areas that develop closely resemble those caused by SO_2 in many plants, and in future work it may be worth considering whether some forms of acute injury due to SO_2 are the result of excessive transpiration. It is important to emphasize that the plant in *Figure 10.3* was not deprived of water. There was an adequate supply in the soil, and the wilting occurred because the very wide stomatal apertures led to a high rate of transpiration which exceeded the rate at which water could be replaced by the vascular system. Stomatal apertures in the field have to be finely regulated to prevent damage of this kind occurring, and any agent which interferes with this regulation by stimulating opening is likely to cause acute injury at times when transpiration is high.

Apart from these possible effects on the water relations of the plant, stomatal opening in SO_2-polluted air is clearly not an appropriate reaction as far as resistance mechanisms are concerned. Resistance to SO_2 would be achieved by stomatal closure in response to the pollutant, not opening. Further studies of the responses of stomata of broad bean to SO_2 by Mansfield and Majernik (1970) and Black and Unsworth (1980) have shown that there is a further complication, with the nature of the response determined by water vapour content of the atmosphere. When the relative humidity is low (i.e. when the vapour pressure dificit is high) stomatal closure, rather than opening, occurs in SO_2-polluted air. Black and Unsworth (1980) found that this reversal of the response to SO_2 also occurred in sunflower and the tobacco plant, but not in pinto bean (*Figure 10.4*). A fundamental difference between pinto bean and the

Figure 10.3. (a) Severe wilting of a plant of *Commelina communis* after 15 min exposure to light wind when stomata had been induced to open to abnormally wide apertures. There was an adequate water supply in the soil. (b) A leaf from the same plant photographed four days later, showing permanent damage to the margins and tip in the form of light brown necrotic areas

others is that its stomata are not sensitive to atmospheric humidity. An increase in the vapour pressure deficit (vpd) causes stomata to close in many species, as will be seen to be the case for the control plants of broad bean, sunflower and tobacco in *Figure 10.4*. This vpd response is thought to be the result of evaporation from the guard cells themselves, probably through special unthickened areas in their walls (Appleby and Davies, 1982). The operation of a vpd-sensing mechanism using evaporation from the guard cells implies a resistance to water movement from elsewhere in the leaf to allow the guard cells to lose turgor. The epidermis is isolated hydraulically from the rest of the leaf, and the guard cells can only acquire water via the epidermal cells. Black and Unsworth (1980) suggested that exposure to SO_2 interacts with this mechanism, and that in vpd-sensitive species, loss of turgor by the guard cells in dry air will be greater in the presence of SO_2. At a critical point (where the curves cross over in *Figures 10.4a–c*) the evaporation will reach a point at which the vpd response becomes enhanced.

The fact that vpd determines the nature of the action of SO_2 on stomata means that no simple assessment of the importance of these effects will be possible in the field. *Figures 10.4a–c* show that in a climate where the air usually has a high vpd, the stomata may tend to have smaller apertures in the presence of SO_2, thus providing a mechanism for the avoidance of pollution stress.

Stomatal closure caused by SO_2

There have been several reports in the literature of stomatal closure induced by SO_2. The foregoing discussion indicates that such effects are not incompatible with the observations on broad bean and other vpd-sensitive species, but we must also recognize

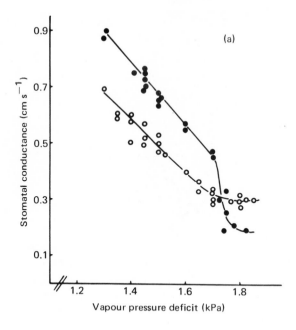

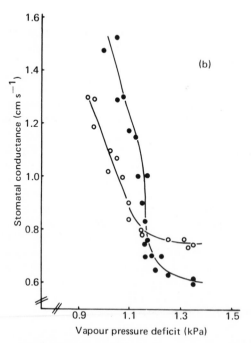

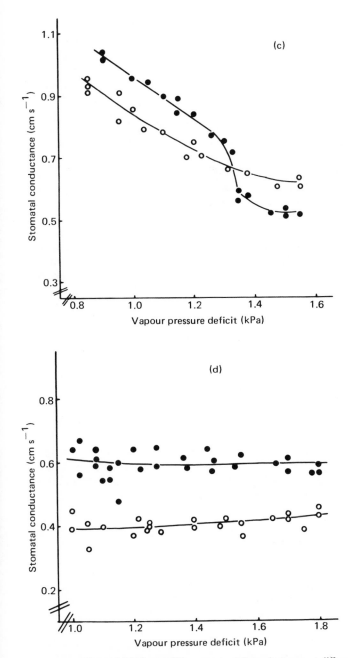

Figure 10.4. Effects of 0.035 ppm SO_2 on stomatal conductance at different vapour pressure deficits. (a) broad bean (*Vicia faba*), (b) sunflower (*Helianthus annuus*), (c) tobacco (*Nicotiana tabacum*), (d) pinto bean (*Phaseolus vulgaris*). O =control in charcoal-filtered air; ● =plants in 0.035 ppm SO_2. After Black and Unsworth (1980) with the permission of Oxford University Press

that not all reported effects are necessarily due to direct action on cells of the epidermis. Black (1982) has listed the many reports in the literature of SO_2-induced stomatal closure, or inhibition of transpiration. She does, however, draw attention to the fact that in many of the studies the concentrations of SO_2 used were relatively high. The effects of SO_2 on net photosynthesis vary considerably between different species (*see* Black, 1982, *Figure 4.4*). It is, however, generally found that concentrations above 0.5 ppm are strongly inhibitory. They must, therefore, exert an effect on the intercellular CO_2 concentration in the leaf, and this in turn will cause stomatal closure. This could occur without any direct effect of the pollutant on the guard cells themselves. Treatment of leaves with inhibitors of photosynthetic CO_2 fixation causes stomatal closure, but this can be reversed if the leaf is flushed with CO_2-free air (Allaway and Mansfield, 1967), and it is likely that the same would apply in the case of SO_2.

Stomatal closure which occurs as a result of the inhibitory action of a pollutant on photosynthesis in the mesophyll cannot be looked upon as a desirable way of avoiding stress due to the pollutant. As the stomata close there will, of course, be an increased diffusion resistance to SO_2, but a true avoidance mechanism would involve closure in advance of stress in the mesophyll, rather than as an event secondary to that stress.

We have recently completed some studies of the effects of SO_2 on silver birch (*Betula pendula*) which have suggested that in this species stomatal closure in the presence of the pollutant may operate as a useful avoidance mechanism. The stomata of silver birch respond to atmospheric humidity (Osonubi and Davies, 1980) and the experiments were conducted in fairly dry atmospheres (> 1.58 kPa water vapour pressure deficit). The closing reaction to SO_2 reported by Black and Unsworth (1980) (*Figure 10.4*) was thus expected to operate. *Figure 10.5* shows the light response curves for photosynthesis and transpiration in control plants and in plants exposed to 0.07 ppm SO_2. The pollutant caused a depression of net photosynthesis amounting to about 19%, but transpiration fell by 46%. Where stomatal closure is the cause of changes in photosynthesis and transpiration, a smaller proportional effect would be predicted for the former than for the latter, especially in C_3 plants where there is an appreciable 'internal resistance' to CO_2 intake (Mansfield, 1976). Calculations of the resistance to CO_2 uptake for silver birch suggested that the main effect of 0.07 ppm SO_2 was on the stomatal component with little influence on the internal resistance. Since the net sulphur flux is closely correlated with the stomatal conductance to water vapour (*see below*, and *Figure 10.7*), it appears that the response to SO_2 during the first day of exposure, as shown in *Figure 10.5*, may represent a mechanism for avoiding pollution stress without major interference with the supply of CO_2 for photosynthesis.

Responses of stomata to other pollutants

OZONE

There are several reports in the literature of stomatal closure in response to O_3, but it is still not clear whether the depressions in photosynthesis which occur at the same time are the cause or the consequence of changes in stomatal aperture (Unsworth and Black, 1981). Rich and Turner (1972) found that the stomata of an O_3-resistant variety of tobacco closed more in the presence of the pollutant than those of an O_3-sensitive variety. Other authors have similarly reported that stomatal responses may help to confer resistance to damage by O_3 (Unsworth and Black, 1981). Little work has, however, been done to identify sites of action in the stomatal complex or elsewhere, and priority should be given to these in future studies.

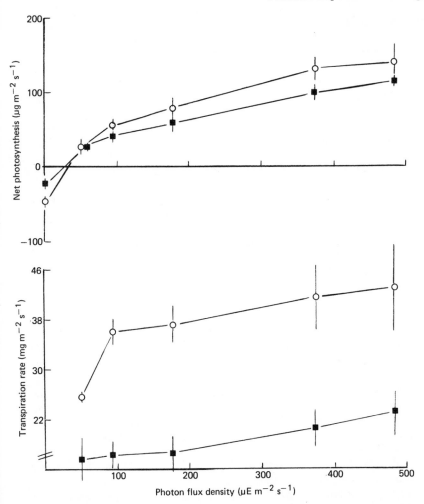

Figure 10.5. Effects of exposure to 0.07 ppm SO_2 on net photosynthesis and transpiration in silver birch. O, plants grown in clean air; ■, plants placed in SO_2 after growing in clean air. Measurements were made during the first day of exposure to the pollutant. Means and standard errors are shown

NITROGEN OXIDES

NO and NO_2 are the two oxides of nitrogen of interest as potentially toxic air pollutants, and both have been reported to inhibit photosynthesis (Hill and Bennett, 1970; Capron and Mansfield, 1976). Studies by Srivastava, Joliffe and Runeckles (1975a,b) showed that NO_2 increased the internal resistance to CO_2 uptake, and concluded that stomatal closure at high NO_2 concentrations was probably a consequence of raised intercellular CO_2 concentrations.

MIXTURES OF POLLUTANTS

A series of experiments by Beckerson and Hofstra (1979a,b) has drawn attention to the complex situation that can result when stomata are affected by two pollutants simultaneously. They confirmed other reports that the stomata of the pinto bean opened in response to SO_2 and closed in response to O_3. However, in a mixture of the two they closed even further than in O_3 alone. Similar effects were also found in three other species.

These results draw attention to the dangers of drawing too many conclusions about effects in the field from fumigations using single pollutants. Stomatal opening caused by SO_2 may be unimportant where episodes of SO_2 pollution are accompanied by O_3. More attention will have to be given to these combined effects in future studies.

We have now begun to study stomatal responses as part of our own investigations into the combined action of SO_2 and NO_2 on silver birch. Initially $SO_2 + NO_2$ caused a reduction in stomatal opening comparable to that in SO_2 alone (compare *Figure 10.5* with *Figures 10.6a* and *6b*). After 20 days' exposure, however, there was little difference between the polluted and control plants in the light, but stomatal closure in the dark (not shown in the figure) was inhibited. This suggests that there may be progressive injury to epidermal cells during prolonged exposures. After 20 days' treatment with 0.07 ppm $SO_2 + 0.07$ ppm NO_2, visible injury can be seen on the abaxial surfaces of the leaves, but it is not yet known whether this is a cause or a consequence of stomatal malfunctioning, or is independent of it.

Stomatal control of pollutant uptake

Unsworth, Biscoe and Black (1976) have summarized the principles of gaseous exchange in photosynthesis and transpiration, and have drawn useful parallels in relation to the transfer of pollutants from the atmosphere to leaves. They point out the difficulties of making simple assumptions about the similarities of pathways for diffusion for different gases, e.g. H_2O vapour and SO_2. For example, the resistance to movement through the cuticle may be less for SO_2 than for H_2O. Hällgren *et al.* (1982) have performed a detailed analysis of gaseous exchange in SO_2-polluted atmospheres for shoots of Scots pine. Fluxes of CO_2 and H_2O vapour were measured simultaneously with those of gaseous sulphur compounds. Although there was a marked diurnal variation in the uptake of SO_2, with values during the day about three times those at night, stomatal opening did not appear to be the primary controlling factor. Part of the discrepancy was probably due to the re-emission of H_2S from the needles, which was a light-dependent process. Such re-emissions would have to be taken into account in sulphur-balance calculations.

We have recently completed a study of net sulphur uptake into shoots of silver birch fumigated with 0.07 ppm SO_2 along with 0.07 ppm NO_2. These two pollutants normally occur together in urban air, and it is therefore realistic to include NO_2 in experimental fumigations with SO_2 (Mansfield and Freer-Smith, 1981). *Figure 10.7* shows the results obtained for plants fumigated for different periods of time, with net sulphur uptake plotted against stomatal conductance to H_2O vapour. The plant material consisted of birch seedlings approximately one month old and uptake was measured for the whole shoots with soil uptake excluded. Prior to the determinations the plants had been grown in 12-h days with a photon flux density of 240 $\mu E\ m^{-2}\ s^{-1}$. During the measurements changes in stomatal aperture were achieved by varying photon flux densities between 0 and 500 $\mu E\ m^{-2}\ s^{-1}$. Net sulphur uptake was determined by difference between a reference chamber and the plant chamber (both were identically constructed) using a Meloy SA285 flame photometric sulphur analyser (Meloy Labs. Inc., Springfield, VA, USA) and water vapour exchange was measured

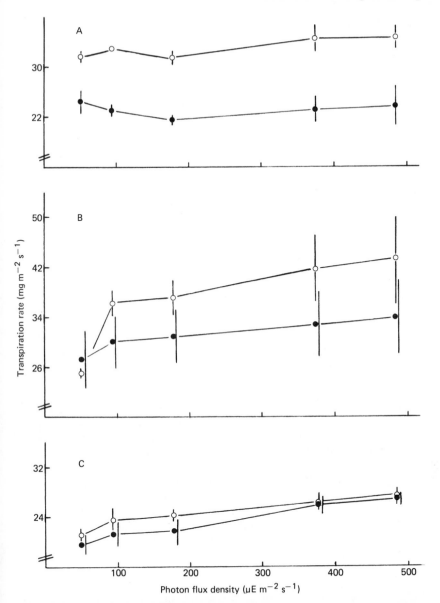

Figure 10.6. Transpiration rate of plants of silver birch grown in clean (○) and polluted (●) air. Measurements were made after the following periods of exposure to 0.07 ppm SO_2 + 0.07 ppm NO_2: A, 5 days; B, 12 days; C, 20 days. Means and standard errors are shown

using a model 911 DEW-ALL digital humidity analyzer (E.G.&G. Inc., Waltham, MA, USA). The stomatal component was calculated after taking into account aerodynamic resistance (estimated as nearly as possible using a model of the plant made of filter paper) and it is important to emphasize that various sources of error are drawn together so that there is always some uncertainty about absolute values of stomatal conductance. Too much emphasis should not, therefore, be placed on the

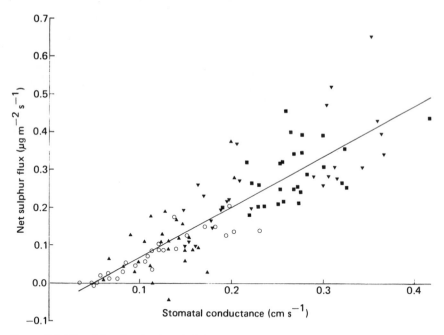

Figure 10.7. Net sulphur flux for shoots of silver birch plotted against stomatal conductance. Plants had been exposed to 0.07 ppm SO_2 + 0.07 ppm NO_2 for: $\bigcirc$, zero time; $\blacktriangle$, 5 days; $\blacktriangledown$, 12 days; $\blacksquare$, 20 days. Different photon flux densities were employed to produce changes in stomatal conductance, over the range 0–500 $\mu E\ m^{-2}\ s^{-1}$

slope of the line in *Figure 10.7*, nor on its points of intercept. The regression line is significant at $P < 0.001$ and 74% of the variance was accounted for. The control of net sulphur flux by factors other than stomata may also be important, and in particular we draw attention to the possibility of increased emissions of reduced sulphur compounds at higher light intensities (Hällgren *et al.*, 1982), and of the entry of SO_2 through the cuticle, especially when the leaf is beginning to show visible injury.

The data in *Figure 10.7* therefore demonstrate that stomatal control of SO_2 uptake is of major significance in silver birch. Responses of the stomata of this species to SO_2 may, consequently, be of importance in determining the pollutant dose.

Acknowledgements

We are grateful to the Natural Environment Research Council for a grant in support of one of the authors (Freer-Smith), and to members of staff of the Institute of Terrestrial Ecology at Edinburgh for supplying plant material and for advice on the physiology of woody plants.

References

APPLEBY, R. F. and DAVIES, W. J. A possible evaporation site in the guard cell wall and the influence of leaf structure on the humidity response by stomata of woody plants, *Oecologia* **56**, 30–40 (1982)

ALLAWAY, W. G. and MANSFIELD, T. A. Stomatal responses to changes in carbon dioxide concentration in leaves treated with 3-(4-chlorophenyl)-1, 1-dimethyl urea, *New Phytologist* **66**, 57–63 (1967)

BECKERSON, D. W. and HOFSTRA, G. Stomatal responses of white bean to O_3 and SO_2 singly and in combination, *Atmospheric Environment* **13**, 533–535 (1979a)

BECKERSON, D. W. and HOFSTRA, G. Response of leaf diffusive resistance of radish, cucumber and soybean to O_3 and SO_2 singly or in combination, *Atmospheric Environment* **13**, 1263–1268 (1979b)

BISCOE, P. V., UNSWORTH, M. H. and PINCKNEY, H. R. The effects of low concentrations of sulphur dioxide on stomatal behaviour in *Vicia faba*, *New Phytologist* **72**, 1299–1306 (1973)

BLACK, C. R. and BLACK, V. J. The effects of low concentrations of sulphur dioxide on stomatal conductance and epidermal cell survival in field bean (*Vicia faba* L.), *Journal of Experimental Botany* **30**, 291–298 (1979)

BLACK, V. J. and UNSWORTH, M. H. Stomatal responses to sulphur dioxide and vapour pressure deficit, *Journal of Experimental Botany* **31**, 667–677 (1980)

BLACK, V. J. Effects of sulphur dioxide on physiological processes in plants. In *Effects of Gaseous Air Pollution in Agriculture and Horticulture*, (Eds. M. H. UNSWORTH and D. P. ORMROD), pp. 67–91, Butterworths, London (1982)

CAPRON, T. M. and MANSFIELD, T. A. Inhibition of net photosynthesis in tomato in air polluted with NO and NO_2, *Journal of Experimental Botany* **27**, 1181–1186 (1976)

HÄLLGREN, J.-E., LINDER, S., RICHTER, A., TROENG, E. and GRANAT, L. Uptake of SO_2 in shoots of Scots pine: field measurements of net flux of sulphur in relation to stomatal conductance, *Plant, Cell & Environment* **5**, 75–83 (1982)

HEATH, O. V. S. An experimental investigation of the mechanism of stomatal movement, with some preliminary observations of the responses of guard cells to shock, *New Phytologist* **37**, 385–395 (1938)

HEATH, O. V. S. Control of stomatal movement by a reduction in the normal carbon dioxide content of the air, *Nature (London)* **161**, 179–181 (1948)

HILL, A. C. and BENNETT, J. H. Inhibition of apparent photosynthesis by nitrogen oxides, *Atmospheric Environment* **4**, 341–348 (1970)

LEVITT, J. *Response of Plants to Environmental Stress*, Academic Press, New York (1972)

MAJERNIK, O. and MANSFIELD, T. A. Effects of SO_2 pollution on stomatal movements in *Vicia faba*, *Phytopathologische Zeitschrift* **71**, 123–128 (1971)

MANSFIELD, T. A. Chemical control of stomatal movements, *Philosophical Transactions of the Royal Society of London, Series B* **273**, 541–550 (1976)

MANSFIELD, T. A. and MAJERNIK, O. Can stomata play a part in protecting plants against air pollutants? *Environmental Pollution* **1**, 149–154 (1970)

MANSFIELD, T. A. and DAVIES, W. J. Stomata and stomatal mechanisms. In *The Physiology and Biochemistry of Drought Resistance in Plants*, (Eds. L. G. PALE and D. ASPINALL), pp. 315–346, Academic Press, Sydney (1981)

MANSFIELD, T. A. and FREER-SMITH, P. H. Effects of urban air pollution on plant growth, *Biological Reviews* **56**, 343–368 (1981)

MEIDNER, H. and MANSFIELD, T. A. *Physiology of Stomata*, McGraw-Hill, London (1968)

MEIDNER, H. and SHERIFF, D. W. *Water and Plants*, Blackie, Glasgow and London (1976)

OSONUBI, O. and DAVIES, W. J. The influence of plant water stress on stomatal control of gas exchange at different levels of atmospheric humidity, *Oecologia (Berlin)* **46**, 1–6 (1980)

RASCHKE, K. Simultaneous requirement for carbon dioxide and abscisic acid for stomatal closing in *Xanthium strumarium* L., *Planta (Berlin)* **125**, 243–259 (1975)

RICH, S. and TURNER, N. C. Importance of moisture on stomatal behaviour of plants subjected to ozone, *Journal of the Air Pollution Control Association* **22**, 718–721 (1972)

SEKIYA, J., SCHMIDT, A., RENNENBERG, H., WILSON, L. and FILNER, P. Role of light in H_2S emission in response to sulfate, *16th Annual Report of the MSU-DOE Plant Research Laboratory*, Michigan State University, p. 57 (1981)

SNAITH, P. J. and MANSFIELD, T. A. Control of the CO_2 responses of stomata by indol-3yl-acetic acid and abscisic acid, *Journal of Experimental Botany* **33**, 360–365 (1982)

SRIVASTAVA, H. S., JOLLIFFE, P. A. and RUNECKLES, V. C. Inhibitions of gas exchange in bean leaves by NO_2, *Canadian Journal of Botany* **53**, 466–474 (1975a)

SRIVASTAVA, H. S., JOLLIFFE, P. A. and RUNECKLES, V. C. The effects of environmental conditions on the inhibition of leaf gas exchange by NO_2, *Canadian Journal of Botany* **53**, 475–482 (1975b)

TAYLOR, G. E. Plant and leaf resistance to gaseous air pollution stress, *New Phytologist* **80**, 523–534 (1978)

UNSWORTH, M. H., BISCOE, P. V. and BLACK, V. J. Analysis of gas exchange between plants and polluted atmospheres. In *Effects of Air Pollutants on Plants*, (Ed. T. A. MANSFIELD), pp. 5–16, Cambridge University Press, Cambridge (1976)

UNSWORTH, M. H. and BLACK, V. J. Stomatal responses to pollutants. In *Stomatal Physiology*, (Eds. P. G. JARVIS and T. A. MANSFIELD), pp. 187–263, Cambridge University Press (1981)

WAISEL, Y., BORGER, G. A. and KOZLOWSKI, T. T. Effects of phenylmercuric acetate on stomatal movement and transpiration of excised *Betula papyrifera* March leaves, *Plant Physiology* **44**, 685–690 (1969)

Chapter 11

Photosynthetic gas exchange in leaves affected by air pollutants

J.-E. Hällgren*

SWEDISH UNIVERSITY OF AGRICULTURAL SCIENCES, UMEÅ, SWEDEN

Introduction

Vegetation represents a very important sink for most airborne pollutants originating from anthropogenic and other sources. Fluxes of gases and plant assimilatory processes are of central interest for micrometeorologists concerned with the removal of pollutants from the atmosphere. Plant assimilation of gases is, of course, always of central interest to plant physiologists and researchers concerned with the environmental aspects of photosynthesis and plant productivity. All these researchers are faced with a number of problems, requiring the application of a whole range of methods for the measurements of micrometeorological factors such as energy, wind, water and carbon dioxide fluxes as well as of the concentrations and fluxes of air pollutants and nutrients in the field. Furthermore, a number of interacting environmental variables has to be taken into account as well as the effects of plant structure, developmental stage and water relations, just to mention a few.

In spite of all these problems much excellent research has been conducted under field conditions. Drawing theory from micrometeorological work on the exchange of heat, mass and momentum between the atmosphere and vegetation, a representation of gas transfer to the plant surface has been used to describe the process of gas deposition. The most appropriate method makes use of the resistance analogy described by Thom (1975) and is used in the field also for fluxes of pollutant gases (Fowler and Unsworth, 1979; Galbally, Garland and Wilson, 1979; Hill and Chamberlain, 1975). The analysis allows a distinction to be made between resistances which are functions of the aerodynamic, physiological or surface properties of plants, and it has been comprehensively described earlier (Unsworth, Biscoe and Black, 1976; Unsworth, 1981). *Figure 11.1* gives a representation of the resistances of the leaf.

Most of the work by micrometeorologists and plant physiologists in analysing the dry deposition of gases has concentrated on SO_2. The theoretical and methodological aspects have been comprehensively described by Fowler (1980); the literature in this field has also been recently reviewed (Hosker and Lindberg, 1982).

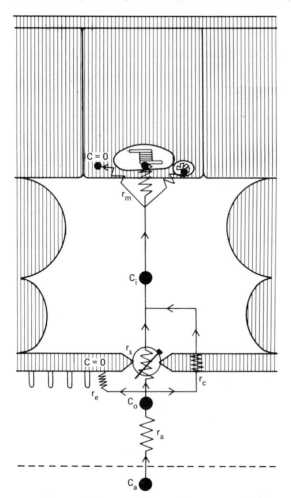

Figure 11.1. A resistance analogue for SO_2 uptake by a model leaf, showing possible paths. C_a—SO_2 concentration in the air; C_o—SO_2 concentration at the leaf surface; C_i—SO_2 concentration in the substomatal cavity; r_a—boundary layer resistance, boundary layer is indicated by a dotted line; r_e—epidermis or surface resistance; r_c—cuticular resistance; r_s—stomatal resistance; r_m—internal resistance, a chloroplast and a mitochondrion being indicated as a possible sink

Interaction with plant surfaces

The range of plant species in the field for which dry deposition measurements are available is limited. However, most measurements make it clear that stomata generally represent the major sink for uptake of SO_2, NO_2, HNO_3, O_3 and other gaseous pollutants (Fowler, 1980; Hosker and Lindberg, 1982; Bennett, Hill and Gates, 1973). It is at present not possible to make a satisfactory integrated and unified model for pollutant (dry deposition) fluxes to plant canopies. The reason is that a complex interaction of pollutant gases with foliar surfaces occurs. Surfaces of different species vary tremendously and the amount of epicuticular wax, the degree of pubescence, the number of trichomes, glandular hair exudates, epiphytes and dead tissue provide numerous microsites for the sorption of reactive gases like SO_2, O_3, NO_2 and HNO_3. Leaves also emit volatile compounds such as terpenes, which can react with pollutants. The surface may also mediate the thermal decomposition of labile gases. For reactive gases such as O_3, a limited number of reactive sites on the leaf surface have been reported (Bennett, Hill and Gates, 1973; Bennett and Hill, 1973a,b), and this might provide one explanation for the often-reported decrease in pollutant uptake with time during experiments (Elkiey and Ormrod, 1981).

A very large number of important variables are easily tabulated; however we do not yet know their relative contribution to the uptake of air pollutants. For the dry deposition of many important air pollutants such as NO_2 and HNO_3 surprisingly few measurements have been made (Yoneyama *et al.*, 1980; Bengtsson *et al.*, 1980); more data are clearly needed to discuss the detail of surface processes (Fowler, 1980). The reactivity of HNO_3 implies larger deposition rates than SO_2.

Boundary layer effects

As air pollutants approach plant leaves, they will pass through the leaf boundary layer, where transfer is no longer determined by turbulent but rather by molecular diffusion. Here, in the leaf–air micro-environment, concentrations of transpired water vapour and other organic/inorganic compounds emitted by the leaf are higher in the inner boundary layer, closest to the leaf surface. To what extent they affect the transfer of gases to and/or uptake on the leaf surface is not known. Other micrometeorological factors that affect the surface environment are temperature, humidity and absorbed radiation. These can vary considerably along or across the leaf surface. The boundary layer will vary during the leaf development since it is dependent upon leaf size.

The relative humidity (RH) of the air has recently been shown to affect significantly the uptake rate of SO_2 (McLaughlin and Taylor, 1981). An increase of the RH from 35 to 75% in the air led to almost three- and fourfold increases in the uptake of SO_2 and O_3, respectively, in contrast to earlier findings (Fowler and Unsworth, 1974). The authors surprisingly claimed that the increased uptake rate of O_3 and SO_2 could not be explained by an effect of RH on the stomata (McLaughlin and Taylor, 1981). Although it might be plausible that one effect of RH on absorption of gases could include their adsorption by a film of water on the surface of the leaf (McLaughlin and Taylor, 1981), the reported increase in the uptake of O_3 and SO_2 does not fit the solubility data for these gases.

Sorption on plant surfaces

Foliar wetness is expected to increase considerably the uptake rate of SO_2 and other water-soluble pollutants (Hill and Chamberlain, 1975; Fowler, 1980; Hosker and Lindberg, 1982; Bennett and Hill, 1973b). Morphological features which trap water may also enhance SO_2 adsorption rates (Fowler and Unsworth, 1974, 1979). Adsorption on the leaf surface depends on the composition and the distribution of water on the surface; for example, if the surface water is already in equilibrium with the ambient SO_2 concentration then rates of uptake will be small and dependent on the removal of S(IV) from solution (Fowler and Unsworth, 1979). If the pH of the water film on the surface becomes low (equal to the pH of the throughfall) this may eventually limit SO_2 uptake and CO_2 uptake. In practice, water will occlude the stomatal antechambers, and if the water is acid, both CO_2 and SO_2 flux would be restricted. This may then eventually influence the photosynthetic process.

Some air pollutants are readily absorbed or adsorbed on the surface of leaves. Fowler and Unsworth (1979) estimated that during the day, 70% of the SO_2 was absorbed through stomata and most of the remaining 30% was adsorbed by the wheat (*Triticum aestivum*) cuticular surface. Similar results have been obtained with other techniques and plants (Garsed and Read, 1977a, 1977b) and other pollutants (Elkiey and Ormrod, 1981; Jacobson *et al.*, 1966). The sorption rate increases as pollutant solubility increases (Bennett, Hill and Gates, 1973). Some gases such as fluoride are reported to be adsorbed onto the leaf surface in percentages as large as 80% (Garsed and Read, 1977b): such high values are often reported in areas with high amounts of particulate fallout. It is also plausible that the washing procedure employed to determine the surface deposition removes more pollutant than is actually sorbed on the leaf surfaces. It would be very useful if a standardized procedure could be worked out. Although the uptake on the leaf surface is complex, it is possible to analyse accurately the resistances limiting the rate of surface deposition, as shown by Black and Unsworth (1979a). They have described useful equations for analysing the resistances from measurements of SO_2 and water vapour fluxes (Unsworth, 1981; Black and Unsworth, 1979a). At a typical value of stomatal resistance of 300 s m^{-1} about 10% of the SO_2 was deposited on the surface and/or diffused through the cuticle. It has been shown by Fowler (1976) that the relatively small cuticular uptake should not be ignored, as this process is continuous. For a detailed discussion about the permeability of plant cuticles to gaseous air pollutants *see* Lendzian (this volume).

It must be concluded that the stomata play a very important role in determining the rate of uptake of gaseous pollutants and in controlling the deposition rate of several gaseous air pollutants. Since the role of the stomata in pollutant uptake and the effects of SO_2 on stomatal movements have been described comprehensively elsewhere in this volume (*see* Mansfield and Freer-Smith), it will not be discussed here.

The average rates of dry deposition of SO_2 may be predicted from a knowledge of the average components of the resistance pathway (Wesely and Hicks, 1977); it is often considered to be around 1 cm s^{-1} for SO_2 and similar values are given for O_3 and HF. However, these average values should not be accepted and used uncritically. For example, the field measurements over forests indicate a lower rate of deposition in dry conditions (Galbally, Garland and Wilson, 1979).

Deposition velocities for the uptake of H_2S by needles of Monterey pine (*Pinus radiata*) and leaves of perennial ryegrass (*Lolium perenne*) have been measured in the field, using $H_2{}^{35}S$ and an unexpectedly high deposition velocity (max. 0.17 cm s^{-1}) was observed (Cope and Spedding, 1982). On the other hand, NO is reported to be taken up very slowly (Bennett, Hill and Gates, 1973; Bengtson *et al.*, 1980) and this has been explained by its low water solubility (Bennett, Hill and Gates, 1973). However, solubility data alone cannot explain the high rate of H_2S uptake in comparison with that of SO_2 and NO, and it is evident that an improved understanding of the chemical reactions between air pollutants and plant leaves is needed. Differences between plants to absorb air pollutants due to differences in external features may certainly account for some of the difference in pollutant resistance (Cope and Spedding, 1982). Since species have differences in their capacity to adsorb SO_2 on the surface, the SO_2 concentration around the stomata may differ, the amount that reaches the photosynthetic cells may differ considerably.

Internal resistance

The internal resistances for the diffusion of SO_2 to cells are generally considered small or non-existent. This is the general assumption made when calculating stomatal resistances for SO_2; it is also assumed that the cuticles are impermeable to gases. Hällgren *et al.* (1982) tested this hypothesis and surprisingly found that the internal resistance for SO_2 could not be zero in needles of Scots pine (*Pinus sylvestris*) subjected to 4–250 μg SO_2 m^{-3} air. The theoretical value for SO_2 flux, derived from measurements of transpiration and calculation of the stomatal conductance to water vapour, was compared with the conductance values obtained from actual measurements of sulphur flux, calculated using the difference in sulphur concentration between inlet and outlet of the assimilation chamber and the airflow rate. If the internal resistance for SO_2 was zero then these two conductances should have been identical (Hällgren *et al.*, 1982). This was not the case most of the time. At least some of the difference between the theoretical and experimental values observed could be explained by a light-dependent H_2S evolution from the SO_2 fumigated needles (Hällgren and Fredriksson, 1982).

General effects on photosynthesis

The inhibitory effects of SO_2 and NO_x pollutants on the photosynthetic CO_2 exchange of plants are documented in several reviews (Ziegler, 1975; Mudd and Kozlowski, 1975; Hällgren, 1978; Heath, 1980; Black, 1982). In spite of all this information, more or less generalized, we cannot yet confidently predict the effect of a given air pollutant on the photosynthetic CO_2 assimilation by the leaves of a given plant. There are fortunately a few exceptions from this statement (Black and Unsworth, 1979a, 1979b; Winner and Mooney, 1982). I will therefore concentrate the discussion on a few central physiological aspects of photosynthetic CO_2 assimilation. Photosynthetic CO_2 assimilation is affected by the amount of light quanta absorbed, the light regimen, temperature, leaf age, plant nutritional status, water stress and a number of biochemical and physiological parameters. Changes in the net rate of CO_2 assimilation may reflect changes in both stomatal conductance and 'mesophyll' capacity. Stomatal resistances often correlate well with net gas exchange of CO_2 and mesophyll conductances for unpolluted plants (Farquhar, von Caemmerer and Berry, 1980).

Measurements of CO_2 exchange may also serve different purposes; they can be one approach for developing a capacity to predict the SO_2 sensitivity of plants from their physiological responses (Taylor and Tingey, 1981), and they may also serve to give information about enzymatic processes, e.g. RuBP-carboxylase activity (Farquhar, von Caemmerer and Berry, 1980). Some relationships between the biochemistry of photosynthesis and the gas excahnge of leaves have been presented elsewhere (Farquhar, von Caemmerer and Berry, 1980; von Caemmerer and Farquhar, 1981). The next step will then be to define how pollutant induced changes in photosynthesis become translated into changes in biomass productivity: such a discussion of alterations in biomass productivity is beyond the scope of this chapter.

In plants subjected to air pollutants it can be difficult to separate stomatal from non-stomatal effects on gas exchange. Winner and Mooney (1980a, 1980b) were the first to successfully employ a type of diagnostic gas exchange technique to explain the nature of the impact of SO_2 on ecologically diverse plant species. By analysing the resistances for water vapour, carbon dioxide and sulphur dioxide simultaneously, the component resistances of the plant leaves could be determined, and the effects of SO_2 on the CO_2 flux could be analysed. Hence this series of papers described a method for partitioning changes in photosynthesis between stomatal and non-stomatal components.

The effect on CO_2 dependent assimilation rate

The non-stomatal component depends, *inter alia*, on the activity of ribulose-1,5-bisphosphate carboxylase-oxygenase (RuBPCase) and on the capacity of the C_3 cycle regenerate ribulose-1,5-bisphosphate (Farquhar, von Caemmerer and Berry, 1980; von Caemmerer and Farquhar, 1981). When the CO_2 assimilation rate of a Scots pine intercellular partial pressure of CO_2 (C_i), at the ordinary partial pressure of oxygen and 'light saturating' conditions, the assimilation rate increases and reaches its steady state value at a higher C_i than normally is experienced (C_3-plants). The slope of CO_2 assimilation rate versus internal partial pressure of CO_2 (C_i) is sometimes called the 'mesophyll conductance'. It is also possible to correlate the initial slope with *in vitro* measurements of RuBPCase activity (Farquhar, von Caemmerer and Berry, 1980; von Caemmerer and Farquhar, 1981). When the CO_2 assimilation rate of a Scots pine shoot is plotted versus C_i, a decrease is observed as a response to SO_2 fumigation (*Figure 11.2*). Earlier studies also indicated that the slope of the curve decreases with increasing SO_2 concentration (Winner and Mooney, 1980b). The inhibition observed at saturating internal partial pressures of CO_2 is obviously independent of stomatal closure. Hence, the plot of percentage inhibition of CO_2 assimilation rate versus rate of uptake of the pollutant gives the direct response between the pollutant and the CO_2 assimilation process. This plot may also give the biochemically-oriented researcher indications of the type of interaction occurring with CO_2 fixation. It is also obvious that several other environmental parameters could be altered together with the pollutant concentration to gain more information on their effects on rates of CO_2 assimilation. The response of 'mesophyll conductance' could then be compared for plant leaves of different ages, species or varieties.

Effects at high CO_2 concentrations

The introduction of higher than ambient concentrations of CO_2 during SO_2 fumigation had been employed in earlier studies to test the assumed competitive inhibition of RuBPCase. The general result in these studies was that the effects of SO_2 was decreased at higher CO_2 concentrations (Furukawa *et al.*, 1979; Carlson, 1983). However, the results are in most cases simply explained by stomatal closure, and therefore do not give evidence to prove competitive inhibition of the enzyme. However, a high CO_2 concentration in air might, exclusive of the stomatal effect, give rise to other alterations that might influence the SO_2 effect. The data presented by Black (1982) are the first, and so far the only published analysis, that gives the inhibition of photosynthesis at different CO_2 concentrations as a function of SO_2 flux through the stomata. Her data clearly show that the inhibition is lower at higher CO_2 concentrations; these data are supported by others (Furukawa *et al.*, 1979). It can thus be concluded from these results that the protection against SO_2 conferred by an increase in the CO_2 concentration in air may both be due to stomatal closure and by a unknown 'mesophyll' component.

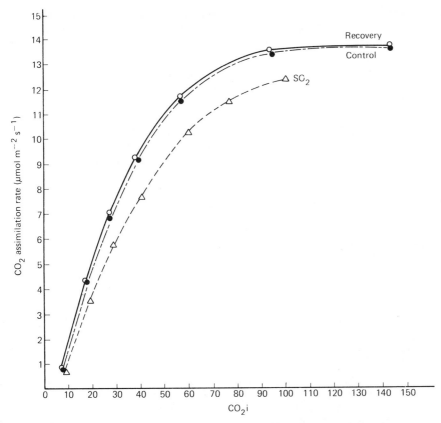

Figure 11.2. The CO_2 assimilation rate of a current year pine shoot before, during and after fumigation with SO_2 plotted against internal partial pressure of CO_2. The pine shoot was fumigated with 190 μg m^{-3} for ten days, and the recovery was measured about 12 h after termination of fumigation. The quantum flux density was 1000 μmol m^{-2} s^{-1}

Effects on the CO_2 compensation point and on photorespiration

The plot of CO_2 assimilation rate versus C_i also gives the CO_2 compensation point, which significantly differs between C_3 and C_4 plants. The CO_2 compensation point in C_3 plants is light and temperature dependent, and dependent on the partial pressure of oxygen. Changes in the CO_2 compensation point are difficult to interpret since photosynthesis, photorespiration and dark respiration all may affect the balance point. Furukawa, Natori and Totsuka (1980) measured the CO_2 compensation point in a sunflower (*Helianthus annuus*) leaf and found that the CO_2 compensation point increased as a result of SO_2 fumigation, while the dark respiration rate was unaffected. Other data (Winner and Mooney, 1980b) indicate either a very small or negligible effect on the CO_2 compensation point or a variable effect (Lorenc-Plucinska, 1978a).

Changes in the photorespiratory processes have been suggested as an alternative to explain the effects of SO_2 on photosynthesis (Koziol and Cowling, 1978). The observed temporary stimulation of photosynthesis by SO_2 has sometimes been suggested to reflect a reduction of photorespiration. There are only a very limited number of estimations on the effects of air pollutants on photorespiratory gas exchange. Lorenc-Plucinska (1978b) estimated photorespiration from CO_2 compensation point and net photosynthetic rates in provenances of Scots pine and larch (*Larix leptolepis*). The results were dependent on the clones of the trees, but generally, photorespiration was reduced by the high concentration of SO_2 used. Hällgren and Gezelius (1982) found no specific effect on photorespiration when estimated by the difference between the CO_2 flux rate at the normal partial pressure of oxygen and at 2 kPa of oxygen. Furukawa, Natori and Totsuka (1980) used the same technique, as well as that of measuring the rate of CO_2 evolution into CO_2-free air in the light. The authors concluded that SO_2 resulted in an inhibition in the rate of photorespiration. Firstly, both these methods for estimating photorespiration can be quondly, much more data are needed before any specific effect of SO_2 on photorespiratory CO_2 production can be discussed. Until then, I will take the opinion that no specific effect of SO_2 on photorespiration has been demonstrated *in vivo* from gas exchange measurements.

Effects on the quantum yield

Generally, when effects of air pollutants are discussed, a more severe inhibition of photosynthesis is observed at higher light intensities (Black, 1982; Keller, 1978; Sisson, Booth and Throneberry, 1981; Taniyama *et al.*, 1972; Matsuoka, 1978). However, the opposite has been reported for plant growth processes (Davies, 1980; Bell, 1982). Information on the effect of irradiance on CO_2 assimilation can be obtained by measuring the quantum yield of CO_2 uptake (Hällgren and Gezelius, 1982).

The quantum yield of photosynthesis — the slope of the amount of quanta absorbed $(mol\ s^{-1}\ m^{-2})$ versus the amount of CO_2 absorbed $(mol\ s^{-1}\ m^{-2})$ gives the efficiency of photosynthesis in normal air when the quantum flux density is the rate limiting factor for photosynthesis. No effect of SO_2 on the quantum yield of Scots pine seedlings was observed by Hällgren and Gezelius (1982). When photosynthesis was light-saturated, the inhibition was dependent on the SO_2 concentration. Similar results were reported for the apparent quantum yield by Black and Unsworth (1979b) for broad bean (*Vicia faba*) plants and by Winner and Mooney (1980c) for the 'salt bush' *Atriplex sabulose*, a C_4 plant. However, they observed a lowered apparent quantum efficiency for the C_3 'salt bush' plant, *Atriplex triangularis*. The reason for the observed discrepancies are not known, but different SO_2 concentrations, different experimental conditions, different species and plant ages as well as different techniques were used. It is possible that a higher concentration of SO_2 causes one type of inhibition that is different from the one caused by lower pollutant concentrations. It is also obvious that other electron transport dependent processes, such as sulphur or nitrogen reduction, might alter the observed quantum yield of photosynthetic CO_2 fixation.

The observed response at low quantum flux densities need not necessarily imply that the electron transport is completely unaffected by SO_2 (Hällgren and Gezelius, 1982). Simultaneous measurements of CO_2 flux and fluorescence measurements during SO_2 fumigation of leaves might give information of value for a mechanistic approach.

Photosynthetic rate

Winner and Mooney (1981) noted that a greater relative photosynthetic decline could be noted for the deciduous species *Diplacus aurantius* (monkey flower) than for the evergreen species *Heteromeles arbutifolia* (Christmas berry). They concluded that the capacity to sustain photosynthesis in relation to SO_2 absorption appeared inversely related to the photosynthetic rate of the plant in SO_2 free air. This would be an interesting observation and a fairly simple hypothesis to test.

Photosynthetic induction

Photosynthetic induction phenomena can be studied in intact leaves, although a slightly different technique needs to be applied, since non-steady state gas exchange measurements are required. Several aspects such as the effects on the length of the induction period can be studied. This phenomenon has been studied extensively in England by Delieu and Walker (1981) using oxygen exchange measurements. A sound biochemical/physiological understanding therefore exists of what is happening inside the leaf in the light. A computer simulation of metabolite fluxes would make it possible to compare the results obtained by simulation of inhibition with those actually measured in terms of gas exchange (CO_2 and O_2).

Recovery and rate of recovery

Inhibition of photosynthesis in the absence of visible injury has often been found to be reversible. The recovery is often complete within 24 h after the end of fumigation. The photosynthetic recovery in several species is more rapid, and full recovery can be obtained within 30 min to 2 h (White, Hill and Bennett, 1974). Both stomatal and mesophyll components are involved in this observed recovery of photosynthesis and it is not surprising that small differences in the recovery rate has been reported. However, the mechanisms responsible for recovery from photosynthetic inhibition are not known. The rapid rate of recovery from a slight inhibition may give the biochemically-oriented researcher some ideas to test about the mechanism of inhibition. The ability to recover from inhibition is of ultimate importance for plants in the field where they can experience transient concentrations of pollutants sufficient to inhibit photosynthesis. However, very few data indicating the recovering capacity in the field are available.

It is desirable for laboratory experiments to be performed to allow extrapolation to the field situation. Müller, Miller and Sprugel (1979) investigated the effect of fluctuating ambient concentrations of SO_2 on soybeans. This type of approach towards experimentation in the field is needed in the future, as plants in the field experience all kinds of combined stress factors (Keller, 1981).

'Dose'-response relationships

The lowest concentration of SO_2 reported to result in inhibition of photosynthesis seems dependent rather more on the precision and sensitivity of the equipment (and approach) than on the sensitivity of the plants. However, there are documented differences in relative sensitivity of resistance between plants, and examples of this are given by Black (1983). Another example is that C_3 plants are more sensitive than C_4 plants to similar exposure concentrations (Winner and Mooney, 1980c). The response to a pollutant is likely to depend either on the rate of absorption or on the amount absorbed over a period of time. We have earlier assumed that only the SO_2 absorbed through stomata could contribute to photosynthetic inhibition. As was indicated above, attempts to correlate the degree of photosynthetic inhibition with the concentration in the air gives only qualitative assessments of SO_2 resistance. Surprisingly few comparable data are available, and any firm conclusions can therefore not be drawn.

Generally, all forms of sulphur compounds can be involved in the normal sulphur metabolism of plants. Consequently, it can be argued that it is the amount of SO_2 ions in excess of 'normal' quantities that might be injurious to photosynthesis. The usual 'dose-response' analysis will then become more difficult to perform. Sulphur and nitrogen compounds require special attention since both the reduction of sulphur and of nitrogen are true photosynthetic processes. These processes might also differ considerably in their 'normal' rate during a season. It is therefore extremely difficult to determine the limit between normal and excess amounts during different stages of development and during different environmental conditions. Other approaches, which involve an easily identified 'abnormal' signal have therefore been emphasized. The production of ethylene or ethane has been suggested as an appropriate indicator of SO_2 injury (Bressan et al., 1979). However, these signals are not specific for an air pollution effect. Therefore, I would rather suggest H_2S emission when SO_2 is concerned (Hällgren and Fredriksson, 1982). The emission of H_2S may indicate an avoidance reaction, and be specific for excess SO_2 or excess amounts of SO_4^{-2}. However, it is not yet known whether this 'photosynthetic' H_2S emission can be observed in all kinds of plants. The situation is more complex for NO_x, where no similar process has been described so far.

A few laboratory experiments with a single pollutant have been discussed in this chapter. However, I am aware that plants in the real world outside the laboratory experience all possible interactions with the atmospheric environment. It is also demonstrated that combinations of air pollutants can cause rapid inhibition of CO_2 exchange (Bull and Mansfield, 1974; Ormrod, Black and Unsworth, 1981). However, we still know so little about the relation between air pollutants and the effects on photosynthesis that laboratory and field experimentation with plants can add much valuable information. Hopefully, the data collected will make predictions of the interactions between plant photosynthetic processes and atmospheric environmental factors possible. A challenge for the future will be to describe the relation between deposition velocities of air pollutants, the influence on plant photosynthesis and productivity in the field.

Acknowledgements

I want to thank the organizing committee for the invitation to contribute this book, and especially Mike Koziol for his efforts. The Swedish Natural Science Research Council is acknowledged for financial support.

References

BELL, J. N. B. Sulphur dioxide and the growth of grasses. In *Effects of Gaseous Air Pollution in Agriculture and Horticulture,* (Eds. M. H. UNSWORTH and D. P. ORMROD), pp. 225–246, Butterworths, London (1982)
BENGTSON, C., BOSTRÖM, C. Å., GRENNFELT, P., SKARBY, L. and TROENG, E. Deposition of nitrogen oxides to Scots pine (*Pinus sylvestris* L.). In *Ecological Impact of Acid Precipitation.* Proceedings of an International Conference, Sandefjord, Norway, 1980, (Eds. D. DRABLØS and A. TOLLAN), pp. 154–155, Oslo-Ås (1980)

BENNETT, J. H. and HILL, A. C. Inhibition of apparent photosynthesis by air pollutants, *Journal of Environmental Quality* **2**, 526–530 (1973a)

BENNETT, J. H. and HILL, A. C. Absorption of gaseous air pollutants by a standardized plant canopy, *Journal of the Air Pollution Control Association* **23**, 203–206 (1973b)

BENNETT, J. H., HILL, A. C. and GATES, D. M. A model for gaseous pollutant sorption by leaves, *Journal of the Air Pollution Control Association* **23**, 957–962 (1973)

BLACK, V. J. Effects of sulphur dioxide on physiological processes in plants. In *Effects of Gaseous Air Pollution in Agriculture and Horticulture*, (Eds. M. H. UNSWORTH and D. P. ORMROD), pp. 67–91, Butterworths, London (1982)

BLACK, V. J. and UNSWORTH, M. H. Resistance analysis of sulphur dioxide fluxes to *Vicia faba*, *Nature* **282**, 68–69 (1979a)

BLACK, V. J. and UNSWORTH, M. H. Effects of low concentration of sulphur dioxide on net photosynthesis and dark respiration of *Vicia faba*, *Journal of Experimental Botany* **30**, 473–483 (1979b)

BRESSAN, R. A., LECUREUX, L., WILSON, L. G. and FILNER, P. Emission of ethylene by leaf tissue exposed to injurious concentrations of sulphur dioxide or bisulphite ion, *Plant Physiology* **63**, 924–930 (1979)

BULL, J. N. and MANSFIELD, T. A. Photosynthesis in leaves exposed to SO_2 and NO_2, *Nature* **250**, 443–444 (1974)

CARLSON, R. W. The effect of SO_2 on photosynthesis and leaf resistance at varying concentrations of CO_2, *Environmental Pollution* (in press) (1983)

COPE, D. M. and SPEDDING, D. J. Hydrogen sulphide uptake by vegetation, *Atmospheric Environment* **16**, 349–353 (1982)

DAVIES, T. Grasses more sensitive to SO_2 pollution in conditions of low irradiance and short days, *Nature* **284**, 483–485 (1980)

DELIEU, T. and WALKER, D. A. Polarographic measurement of photosynthetic O_2 evolution by leaf discs, *New Phytologist* **89**, 165–178 (1981)

ELKIEY, T. and ORMROD, D. P. Absorption of ozone, sulphur dioxide, and nitrogen dioxide by petunia plants, *Environmental and Experimental Botany* **21**, 63–70 (1981)

FARQUHAR, G. D., VON CAEMMERER, S. and BERRY, J. A. A biochemical model of photosynthetic CO_2 assimilation in leaves of C_3 species, *Planta* **149**, 78–90 (1980)

FOWLER, D. *Uptake of SO_2 by Crops and Soil*, PhD thesis, University of Nottingham (1976)

FOWLER, D. Removal of sulphur and nitrogen compounds from the atmosphere in rain and by dry deposition. In *Ecological Impact of Acid Precipitation*. Proceedings of an International Conference, Sandefjord, Norway, 1980, (Eds. D. DRABLØS and A. TOLLAN), pp. 22–32, Oslo-Ås (1980)

FOWLER, D. and UNSWORTH, M. H. Dry deposition of sulphur dioxide on wheat, *Nature* **249**, 389–390 (1974)

FOWLER, D. and UNSWORTH, M. H. Turbulent transfer of sulphur dioxide to a wheat crop, *Quarterly Journal of the Royal Meteorological Society* **105**, 767–783 (1979)

FURAKAWA, A., KOIKE, A., HOZUMI, K. and TOTSUKA, T. The effect of SO_2 on photosynthesis in poplar leaves at various CO_2 concentrations, *Journal of the Japanese Forest Society* **61**, 351–356 (1979)

FURAKAWA, A., NATORI, T. and TOTSUKA, T. The effect of SO_2 on net photosynthesis in sunflower leaf. In *Studies on the Effects of Air Pollutants on Plants and Mechanisms of Phototoxicity*. Research Report from The National Institute for Environmental Studies, No. 11, pp. 1–8 (1980)

GALBALLY, I. E., GARLAND, J. A. and WILSON, M. J. G. Sulfur uptake from the atmosphere by forest and farmland, *Nature* **280**, 49–50 (1979)

GARSED, S. G. and READ, D. J. SO_2 metabolism in soybean, *Glycine max* var. Biloxi. I. The effects of light and dark on the uptake and translocation of $^{35}SO_2$. *New Phytologist* **78**, 111–119 (1977a)

GARSED, S. G. and READ, D. J. The uptake and metabolism of $^{35}SO_2$ in plants of differing sensitivity to SO_2, *Environmental Pollution* **13**, 173–186 (1977b)

HÄLLGREN, J.-E. Physiological and biochemical effects of sulfur dioxide on plants. In *Sulfur in the Environment: Part II Ecological Impacts*, (Ed. J. O. NRIAGU), pp. 163–209, J. Wiley, New York (1978)

HÄLLGREN, J.-E. and FREDRIKSSON, S.-Å. Emission of hydrogen sulfide from sulfur dioxide-fumigated pine tress, *Plant Physiology* **70**, 456–459 (1982)

HÄLLGREN, J.-E. and GEZELIUS, K. Effects of SO_2 on photosynthesis and ribulose bisphosphate carboxylase in pine tree seedlings, *Physiologia Plantarum* **54**, 153–161 (1982)

HÄLLGREN, J.-E., LINDER, S., RICHTER, A., TROENG, E. and GRANAT, L. Uptake of SO_2 in shoots of Scots pine: field measurements of net flux of sulphur in relation to stomatal conductance, *Plant, Cell and Environment* **5**, 75–83 (1982)

HEATH, R. L. Initial events in injury to plants by air pollutants, *Annual Review of Plant Physiology* **31**, 395–431 (1980)

HILL, A. C. and CHAMBERLAIN, E. M. The removal of water soluble gases from the atmosphere by vegetation, *ERDA Symposium No. 38, Atmosphere-Surface Exchange Particles and Gaseous Pollutants,* Richland, Washington 1974, pp. 153–170 (1975)

HOSKER, R. P., Jr. and LINDBERG, S. E. Review: Atmospheric deposition and plant assimilation of gases and particles, *Atmospheric Environment* **16**, 889–910 (1982)

JACOBSOEN, J. S., WEINSTEIN, L. H., McCUNE, D. C. and HITCHCOCK, A. E. The accumulation of fluorine by plants, *Journal of the Air Pollution Control Association* **16**, 412–417 (1966)

KELLER, T. H. Einfluss niedriger SO_2-konzentrationen auf die CO_2-Aufnahme von Fichte und Tanne, *Photosynthetica* **12**, 316–322 (1978)

KELLER, Th. Folgen einer winterlichen SO_2-Belastung für die Fichte, *Gartenbauwissenschaft* **46**, 170–178 (1981)

KOZIOL, M. J. and COWLING, D. W. Growth of ryegrass (*Lolium perenne* L.) exposed to SO_2. II. Changes in the distribution of photoassimilated ^{14}C, *Journal of Experimental Botany* **29**, 1431–1439 (1978)

LORENC-PLUCINSKA, G. Effect of sulphur dioxide on photosynthesis, photorespiration and dark respiration of Scots pines differing in resistance to this gas, *Arboretum Kornickie* **23**, 133–144 (1978a)

LORENC-PLUCINSKA, G. The effect of SO_2 on the photosynthesis and dark respiration of larch and pine differing in resistance to this gas, *Arboretum Kornickie* **23**, 121–132 (1978b)

McLAUGHLIN, S. B. and TAYLOR, G. E. Relative humidity: Important modifier of pollutant uptake by plants, *Science* **211**, 167–169 (1981)

MATSUOKA, Y. Experimental studies of sulphur dioxide injury to rice plants and its mechanism, *Special Bulletin of the Chiba-ken Agricultural Experimental Station*, No. 7, pp. 1–63 (1978)

MONSON, R. K., STIDHAM, M. A., WILLIAMS, G. J., EDWARDS, G. E. and URIBE, E. G. Temperature dependence of photosynthesis in *Agropyron smithii* Rydb, *Plant Physiology* **69**, 921–928 (1982)

MUDD, J. B. and KOZLOWSKI, T. T. (Eds.) *Response of Plants to Air Pollutants*, Physiol. Ecol. Ser. Academic Press, NY (1975)

MÜLLER, R. N., MILLER, J. E. and SPRUGEL, D. G. Photosynthetic response of field grown soybeans to fumigation with sulphur dioxide, *Journal of Applied Ecology* **16**, 567–576 (1979)

ORMROD, D. P., BLACK, V. J. and UNSWORTH, M. H. Depression of net photosynthesis in *Vicia faba* L. exposed to sulphur dioxide and ozone, *Nature* **291**, 585–586 (1981)

SISSON, W. B., BOOTH, J. A. and THRONEBERRY, G. O. Absorbtion of SO_2 by pecan (*Carya illinoensis* (Wang.) K. Koch) and alfalfa (*Medicago sativa* L.) and its effect on net photosynthesis, *Journal of Experimental Botany* **32**, 523–534 (1981)

TANIYAMA, T., ARIKADO, H., IWATA, T. and SAWANAKA, K. Studies on the mechanism of injurious effects of toxic gases on crop plants. On photosynthesis and dark respiration of the rice plant fumigated with sulphur dioxide for long periods, *Proc. Crop Sci. Soc. Japan* **41**, 120–125 (1972)

TAYLOR, G. E. Jr. and TINGEY, D. T. Physiology and ecotypic plant response to sulphur dioxide in *Geranium carolinianum* L., *Oecologia* **49**, 76–82 (1981)

THOM, A. S. Momentum mass and heat exchange of plant communities. In *Vegetation and the atmosphere I* (Ed. J. L. MONTEITH), Academic Press, London (1975)

UNSWORTH, M. H. The exchange of carbon dioxide and air pollutants between vegetation and the atmosphere. In *Plants and their Atmospheric Environment*, 21st Symposium of the British Ecological Society, (Eds. J. GRACE, E. D. FORD and P. G. JARVIS), pp. 111–138, Blackwell Scientific Publications, Oxford (1981)

UNSWORTH, M. H., BISCOE, P. V. and BLACK, V. Analysis of gas exchange between plants and polluted atmospheres. In *Effects of Air Pollutants on Plants*, (Ed. T. A. MANSFIELD), pp. 5–16, Cambridge University Press, Cambridge (1976)

VON CAEMMERER, S. and FARQUHAR, G. D. Some relationships between the biochemistry of photosynthesis and the gas exchange of leaves, *Planta* **153**, 376–387 (1981)

WESELY, M. L. and HICKS, B. B. Some factors that affect the deposition rates of SO_2 and similar gases on vegetation, *Journal of the Air Pollution Control Association* **27**, 1110–1116 (1977)

WHITE, K. L., HILL, A. C. and BENNETT, J. H. Synergistic inhibition of apparent photosynthesis rate of alfalfa by combinations of sulphur dioxide and nitrogen dioxide, *Environmental Science and Technology* **8**, 574–476 (1974)

WINNER, W. E. and MOONEY, H. E. Ecology of SO_2 resistance. I. Effects of fumigations on gas exchange of deciduous and evergreen shrubs, *Oecologia* **44**, 290–295 (1980a)

WINNER, W. E. and MOONEY, H. E. Ecology of SO_2 resistance. II. Photosynthetic changes of shrubs in relation to SO_2 absorption and stomatal behavior, *Oecologia* **44**, 296–300 (1980b)

WINNER, W. E. and MOONEY, H. E. Ecology of SO_2 resistance. III. Metabolic changes of C_3 and C_4 *Atriplex* species due to SO_2 fumigations, *Oecologia* **46**, 49–54 (1980c)

WINNER, W. E. and MOONEY, H. E. Ecology of SO_2 resistance. IV. Predicting metabolic responses of fumigated shrubs and trees, *Oecologia* **52**, 16–21 (1982)

YONEYAMA, T., TOTSUKA, T., HAYAKAWA, N. and YAZAKI, J. Absorption of atmospheric NO_2 by plants and soils. V. Day and night NO_2-fumigation effect on the plant growth and estimation of the amount of NO_2-nitrogen absorbed by plants. In *Studies on Effects of Air Pollutants on Plants and Mechanisms of Phytotoxicity*, Research Report from the National Institute for Environmental Studies, No. 11, pp. 31–50 (1980)

ZIEGLER, I. The effect of SO_2 pollution on plant metabolism, *Residue Reviews* **56**, 79–105 (1975)

Chapter 12

Effects of gaseous air pollutants on stromal reactions

M. A. J. Parry and C. P. Whittingham

ROTHAMSTED EXPERIMENTAL STATION, HARPENDEN, UK

Introduction

There is considerable evidence that the presence of air pollutants under certain conditions decreases plant productivity and this may be, in large part, related to a decrease in photosynthetic activity. There are a number of contributing factors which determine the overall photosynthetic activity, such as the influence of pollutants on stomatal movement. In this chapter, however, we are concerned with the influence of pollutants on the reactions taking place in the stroma of the chloroplast.

In addition to the direct effect of the pollutants on stomatal movement, the opening and closing of the stomata will regulate the concentration of pollutant within the leaf tissue and thus influence the amount that has access to and can accumulate in the chloroplast. Moreover, the pollutant may solvate and dissociate within the water phase of the cell and its cytoplasm so that the chemical form of the pollutant arriving at the chloroplast may not be identical with that which is present in the atmosphere outside the plant. For example, sulphur dioxide will presumably enter the chloroplast as an uncharged molecule ($SO_2 \cdot H_2O$) and then dissociate to bisulphite and sulphite. Thus it is important to identify the chemical form a pollutant has when entering the chloroplast and to determine the reactivity of its chemical species in relation to the metabolic reactions taking place within the chloroplast.

Physical effects

It has been long established that exposure to a number of pollutants results in a change in the physical structure of the stroma of the chloroplast. The first effect of ozone on bean (*Phaseolus vulgaris*) leaves observed is an increase in the granulation and electron density of the chloroplast stroma, which precedes alterations in the ultrastructure of the protoplasts, and occurs long before the appearance of any external visual symptoms of damage to the leaves (Thompson, Dugger and Palmer, 1966). The final stages of damage occur only when most of the organelles within the leaf have been ruptured. There is evidence that crystalline arrays form within the stroma of fumigated leaves of tobacco (*Nicotiana*), cotton (*Gossypium*) and bean (*Phaseolus*) but it has not yet been established to what extent these changes are reversible and at what stage they become irreversible. Comparable crystalline arrays have been observed in leaves exposed to water stress, infection and herbicides. In the case of water stress there is evidence that the change is reversible when the leaf is returned to full water content. Crystalline arrays in the stroma of the chloroplast have also been observed as a result of fumigating bean leaves with NO_2 (Dolzmann and Ullrich, 1966). This change preceded a change in grana structure, swelling of the grana compartments and the eventual breakdown of all internal structure. In this case the structural alterations were reversible. Granulation of the stroma has also been observed when bean leaves are exposed to sulphur dioxide (Fischer, Kramer and Ziegler, 1973). Again the effect was reversible when the plants, after exposure to the pollutant for 1 h, were returned to unpolluted air. Prolonged exposure resulted in the degradation of the chloroplast envelope so that the loss of selective permeability caused the chloroplast to swell. The relationship between these changes in ultrastructure and the effect of pollutants on specific enzymes has not yet been established. However, since the stroma is the site of most of the enzymes involved in CO_2 fixation and these account for some 50% of the total protein content of the chloroplast it would be surprising if significant changes in the stromal structure did not reflect a change in the physical state of the enzymes present, resulting in some change in their catalytic activity. Steer et al. (1968) have suggested that the crystalloids observed in several C_3 species under stress are a crystalline state of RuBP carboxylase.

Biochemical effects

The main reactions within the stroma involving photosynthetic carbon metabolism are the well known reactions of the pentose phosphate reaction sequence. It is common to all autotrophic organisms and is the exclusive mechanism whereby acids produced from the fixation of carbon dioxide are converted to sugars either in the form of starch synthesized in the chloroplast or of free sugars synthesized in the cytoplasm. The main reactions constituting the carbon cycle can be separated into four categories:

(1) the carboxylation reaction in which RuBP is converted to phosphoglycerate by reaction with carbon dioxide or to phosphoglycolate by reaction with oxygen;
(2) the reduction of phosphoglycerate to the aldehyde, a process requiring reduced pyridine nucleotide and ATP, both of which are generated by the light reactions of photosynthesis;
(3) a series of epimerization, isomerization and transfer reactions which convert some of the triosephosphate formed back to ribulose-5-phosphate; and finally, to complete the cycle;
(4) the phosphorylation of ribulose-5-phosphate to form the carbon dioxide acceptor molecule ribulose-1,5-bisphosphate.

It is clearly not possible to state that any single reaction exclusively regulates such a complex sequence of reactions without specifying the conditions under which the effects are to be observed. The influence of light on the supplies both of phosphorylative capacity (ATP) and of reductive capacity ($NADPH_2$) will regulate the reaction in category (2) in relation to the other three. Again, under optimal conditions of illumination the overall balance of the cycle may depend on the reaction in category (1), i.e. the rate of carboxylation regulated by the supply of carbon dioxide. Enzymes which have been considered the most critical with regard to the reactions in category (3) include   fructose-1,6-bisphosphatase,   sedoheptulose-1,7-bisphosphatase   and phosphoribulose kinase. However, there have been relatively few studies on the effects of air pollutants on these enzymes. Most studies of the effects of pollutants have concentrated on the major enzyme concerned with carboxylation, namely RuBP carboxylase.

RuBP carboxylase is a complex enzyme having two functional activities. The same protein can either catalyse a carboxylation of the substrate RuBP to form two molecules of 3-phosphoglycerate or it can catalyse reaction with oxygen to give one molecule of 2-phosphoglycolate and one molecule of 3-phosphoglycerate. Both these catalyses are thought to involve a single active site, and in these circumstances the simple view would be that air pollutants should influence both types of catalysis equally. However, so far there have been few definitive experiments which have attempted to determine whether the ratio of these two activities is influenced by exposure to pollutants. Since carboxylation essentially results in net photosynthesis, and oxygenation essentially in photorespiration, the balance between the two processes of photosynthesis and photorespiration should not be influenced by the presence of air pollutants as this balance is mainly regulated by the ratio of the concentration of oxygen and carbon dioxide available to the plant.

One experimental approach is to examine the effect of air pollutants on the balance of the sequence of reactions resulting in photosynthesis through a study of the influence of air pollutants on the accumulation of specific intermediates. Libera, Ziegler and Ziegler (1975) demonstrated that, with isolated chloroplasts and concentrations of sulphite greater than 1 mM, fixation of $^{14}CO_2$ declined rapidly, and at 5mM was reduced to 20%. The relative amounts of radioactivity in phosphoglycerate and sugar phosphate were decreased whereas those in aspartate and malate were increased. This indicated a possible shift towards the C_4 dicarboxylic-type of fixation and may indicate a higher sensitivity of RuBP carboxylase than of PEP carboxylase towards sulphite. Some stimulation of glycolate accumulation was observed. In whole leaves, the accumulation of glycolate due to the inhibition of glycolate oxidase is greater than that observed in isolated chloroplasts, as might be expected since glycolate oxidase is located in the peroxisomes. An increase in glycolate concentrations in chloroplasts may indicate an increased synthesis of this compound. In *Pinus* species which had been exposed to ozone Wilkinson and Barnes (1973) observed a reduction in the radioactivity of free sugars but an increase in the radioactivity of sugar phosphates and free amino acids. Koziol and Cowling (1978) using perennial ryegrass cultivar S23 which had been grown with exposure concentrations of SO_2 up to 400 μg m^{-3} (0.15 ppm) found an increased rate of photosynthesis and a small but significant decrease in radioactivity in glycine and serine and an increase in sucrose which was not significant. They interpreted this as indicating that the rate of photorespiration had been decreased relative to the rate of photosynthesis. However, in all these types of observations it is necessary to appreciate that when a constant time of fixation with $^{14}CO_2$ is used changes in the rates of photosynthesis will result in changes in the distribution of radioactivity. For example, the higher the rate of photosynthesis for a given time of fixation, the greater the percentage of radioactivity in sucrose and the smaller than in glycine and serine; this, however, results not from any change in metabolic balance but merely reflects the different pool sizes of the various intermediates.

A second approach is to observe the effect of pollutants on the properties of certain enzymes isolated from treated plants. Thus, Horsham and Wellburn (1975) exposed *Pisum sativum* var. Feltham in air polluted with known amounts of SO_2 and/or NO_2 for six days under constant conditions of temperature, light and relative humidity. At the end of this period RuBP carboxylase was extracted and assayed. Whilst little change was observed with the lower concentrations, at concentrations in excess of 1.5

to 2.0 ppm SO_2, RuBP carboxylase activity was reduced; with NO_2 stimulation was observed with concentrations greater than 1.0 ppm. Stimulation has also been observed with tomato plants exposed to 0.4 to 0.5 ppm NO.

In experiments using NO_2 or NO as fumigants there is a possibility that these substances can act as sources of nitrogen and result in increased protein (including RuBP carboxylase) synthesis. In these circumstances an increase in enzyme activity in the leaf may result from an increase in the total amount of enzyme protein present rather than in an increase in the catalytic activity of pre-existing protein. When activities are expressed on the basis of unit soluble protein extracted, this complication is minimized.

Miszalski and Ziegler (1980) have investigated the use of RuBP carboxylase in plant material as a measure of toxicity of previous exposure to SO_2. These experiments had the advantage of using an assay system for RuBP carboxylase which ensured full activation and optimum catalytic rates. As we shall see later this was not always the case in earlier experiments. Miszalski and Ziegler (1980) confirmed RuBP carboxylase activity was decreased after exposure to higher concentrations of SO_2. Additional proof is provided by Hällgren and Gezelius (1982), who showed that fumigations with 'low' SO_2 concentrations (≈ 400 μg SO_2 m^{-3}, 0.15 ppm) decreased RuBP carboxylase activity when expressed on a dry weight basis. However, no significant differences were observed between fumigated and control plants when the enzyme activity was calculated on a protein basis. This indicates a decrease in the amount of active enzyme present rather than in its specific activity, but the reason for the decrease is not known. Any inhibition of the enzyme activity by SO_2 would probably be lost during extraction procedures.

Direct effects may be best observed in the third type of investigation, in which the effect of pollutants on the catalytic activity of isolated enzymes *in vitro* has been studied. In view of the similar size and structure of the molecules SO_2 and CO_2 the influence of SO_2 on RuBP carboxylase has been extensively studied. Since SO_2 is largely thought to be active in the form of SO_3^{2-} at physiological pH, the effect on the enzyme of dissolved SO_3^{2-} has been investigated. Ziegler (1972) found SO_3^{2-} inhibited RuBP carboxylase competitively with respect to bicarbonate; presumably SO_3^{2-} replaces HCO_3^- by reacting at the same enzyme site. SO_3^{2-} showed a non-competitive inhibition with respect to RuBP and Mg^{2+}. The non-competitive type of inhibition suggests that SO_3^{2-} does not react with the keto group of RuBP. Since SO_2 binds to the enzyme in the same way as CO_2, the degree of inhibition by SO_3^{2-} will be independent of the RuBP and Mg^{2+} concentrations but highly dependent on the concentration of CO_2 at the reaction site. If this is the case it follows that in plants with the C_4 type of photosynthesis and an increased concentration of CO_2 in the bundle sheath cells, SO_2 should be a less powerful inhibitor.

Gezelius and Hällgren (1980) demonstrated an inhibitory effect of 10 mM SO_4^{2-} of approximately the same order as for 10 mM SO_3^{2-}. Paulsen and Lane (1966) found that ammonium sulphate inhibited RuBP carboxylase competitively with respect to RuBP and suggested that there was competition between the phosphate groups of the RuBP and the SO_4^{2-} ion. This result has recently been confirmed by Parry and Gutteridge (1983) who found a K_i of 1 mM SO_4^{2-} suggesting that SO_4^{2-} is an inhibitor of RuBP binding. A mixed pattern of inhibition with respect to HCO_3^- was observed.

Gezelius and Hällgren (1980) also examined crude extracts of spinach (*Spinacia oleracea*) using the same assay conditions as Ziegler (1972) and found that SO_3^{2-} was a less potent inhibitor than claimed previously. They observed K_i values with respect to HCO_3^- between 9 and 13 mM compared with 3 mM found by Ziegler, and found in addition that the pattern of inhibition was non-competitive. Part of the confusion between these results may well be due to the conditions under which the enzyme is assayed. It has only recently been shown by Lorimer, Badger and Andrews (1976) that the enzyme must be preincubated with Mg^{2+} and CO_2 to be in a fully activated state and before the true affinity with respect to bicarbonate can be observed. In recent experiments (Parry and Gutteridge, 1983) special attention has been given to ensure that the enzyme was fully activated. They found using enzyme purified from wheat and spinach that the inhibition of catalytic activity was complex. In the presence of SO_3^{2-} the time course for the reaction was biphasic so that over the first 30 s carboxylation occurred rapidly with little inhibition, but this rate declined over the next 2 min to a much lower constant value. With higher concentrations of SO_3^{2-} the inactivation was more marked. The biphasic curves showed changes both in the apparent patterns of inhibition by SO_3^{2-} and in the kinetic constants with time. Thus the inhibition pattern for SO_3^{2-} versus RuBP was mixed over the first period of the assay but became non-competitive over longer periods. The K_i increased from 2.5 mM SO_3^{2-} at 15 s to 9 mM at 4 min. The inhibition pattern for SO_3^{2-} versus HCO_3^{-} was mixed throughout the assay period but the K_i decreased from 8 to 1.2 mM during the assay. It is clearly important to follow the progress of the enzyme reaction in the presence of the inhibitor as a function of time, rather than attempt to deduce rates after a set reaction period. Further studies to explain the nature of SO_3^{2-} inhibition indicated that preincubation of the substrates or enzymes with SO_3^{2-} prior to initiating the reaction did not alter the biphasic form of the reaction curves; moreover no potent inhibitor was accumulated during the course of the reaction, since a further addition of the enzyme produced a two-phase curve almost identical to the first. The results of Parry and Gutteridge (1983) suggest that the chemistry of the catalytic reactions are so affected that some form of the enzyme common to both the carboxylase and the oxygenase reactions becomes modified in such a way that further substrate turnover proceeds at a much reduced rate. The effect of the progressive inactivation of the enzyme even by low concentrations (1 mM SO_3^{2-}) suggests that the potential effects of SO_2 on this enzyme may have been under-estimated. Certainly the complex changes noted here provide the basis for a plausible explanation for the discrepancies between the types of inhibition patterns reported previously on SO_3^{2-} effects.

In addition Parry and Gutteridge (1983) have demonstrated that if activation is performed in the presence of either SO_3^{2-} or SO_4^{2-} significantly less CO_2 and Mg^{2+} is required for full activation. This effect on activation is separate from effects on catalysis, whether the same anion molecule causes both phenomena has yet to be established.

Ziegler (1973, 1974) showed in her earlier studies that sulphite inhibits PEP carboxylase non-competitively with regard to the substrate PEP; it also inhibits malic enzyme non-competitively with respect to pyruvate. PEP carboxylase from corn (*Zea mays*), a C_4 plant, has a greater affinity for HCO_3^{-} than RuBP carboxylase and shows a smaller inhibition by SO_2. Mukerji and Yang (1974) showed that PEP carboxylase from spinach chloroplasts was more sensitive to SO_2 than PEP carboxylase from corn but less sensitive than RuBP carboxylase from spinach. Winner and Mooney (1980) compared the influence of SO_2 fumigations on two species of 'salt bush' (*Atriplex*), one of which was C_3 and the other C_4. The C_3 species was more sensitive than the C_4 but this difference in response must also reflect the difference in the morphological arrangement of the mesophyll tissues and the different response of stomatal opening following exposure to SO_2. Since the C_4 species has a lower conductance than the C_3 it absorbed less SO_2 during fumigation. Furthermore, the C_3 plant was more vulnerable to stimulation of stomatal opening by SO_2. Thus rates of SO_2 flux into the C_3 leaf and water flux out of the leaf were both increased. C_4 plants therefore seem better adapted to polluted environments. Although malate dehydrogenase can be inhibited severely by SO_3^{2-} this effect is relatively small on the overall process because it is generally believed that PEP carboxylase is the rate-limiting step. Thus CO_2 fixation should be less inhibited with the C_4 type of fixation than with C_3 which depends solely on RuBP carboxylase.

There is only a limited amount of information on the effects of pollutants on fructose-1,6-bisphosphatase and sedoheptulose-1,7-bisphosphatase. Purczeld et al. (1978) have investigated the effect of nitrite on the chloroplast stroma. Intact chloroplasts were extracted from spinach and exposed to different levels of nitrite. Inhibition of CO_2 fixation by nitrite was dependent on pH, e.g. at pH 7.3, 1 mM $NaNO_2$ and at pH 7.9, 5 mM $NaNO_2$ were required for 50% inhibition. Purczeld et al. (1978) suggested that the addition of $NaNO_2$ leads to an acidification of the stroma. When the levels of the intermediates of the CO_2 fixation cycle are measured in chloroplasts exposed to NO_2, a large increase in the concentrations of fructose and sedoheptulose bisphosphates and a parallel decrease in the corresponding monophosphates was observed, suggesting that the fructose and sedoheptulose bisphosphatases were inhibited. At low concentrations NO_2 can be reduced to glutamine by the normal cell machinery and it is only at higher concentrations that NO_2 can affect membrane permeability, stromal pH and the activities of the fructose and sedoheptulose bisphosphatases. Although there is no direct evidence, it is likely that many other pollutants which cause acidification of the stroma will also inhibit these two enzymes.

Both bound and free forms of fructose-1,6-bisphosphatase undergo light activation in vitro. It has been suggested that SO_3^{2-} can activate the enzyme, but Alscher-Herman (1982) found light activation of the enzyme was inhibited by 1 mM SO_3^{2-} in SO_2-susceptible and by 5 mM SO_3 in SO_2-resistant cultivars of soybean (Glycine max). She proposed that there was competition between the SO_3^{2-} and the stromal enzyme for binding sites on the thylakoid membrane and that the differences in susceptibilities between cultivars might be related to the nature or number of binding sites which exist on membranes.

Another biochemical pathway of major importance in the stroma is that of nitrate reduction and amino acid synthesis. Until recently the enzyme thought to control this pathway was nitrate reductase, which is either bound to the outside of the chloroplast or is located in the cytoplasm, i.e. outside the scope of this review. NO_3, NO_2, NO and NH_3 are all extremely reactive and hence toxic, but sufficient enzyme is normally present to prevent their accumulation by incorporation into glutamine. Unless there is severe inhibition of the enzymes involved in the pathway, the effect of these pollutants should be small. Assuming that all the oxides of nitrogen are natural intermediates of this metabolic pathway, they might be expected to have a beneficial effect as nitrogen sources. Yoneyama and Sasakawa (1979) have shown that ^{15}N labelled NO_2 may be converted into nitrate and nitrite and that the reduced ^{15}N from this source can be assimilated into amino acids. However, NO might not be a natural substrate and might be inhibitory. In experiments in which tomato plants (Lycopersicon esculentum) were fumigated with 0.4–2.5 ppm NO for 3 h, nitrite reductase activity was increased by 50 to 500% (Wellburn, Wilson and Aldridge, 1980). Large increases in the activity of nitrite reductase have also been recorded in response to NO_2 (Wellburn et al., 1981). It has been established that the reduced nitrogen is assimilated into amino acids through the glutamine synthetase (GS), glutamate synthase (GOGAT) pathway and not by glutamate dehydrogenase (GDH) activity. However, the apparently redundant GDH has been shown to be stimulated by oxides of nitrogen and ammonia and has been referred to as the best enzymic indicator of pollution stress (Wellburn *et al.*, 1981). Under the conditions of excess NH_4^+ (2–10 mM), external GDH may become more important as GS activity is reduced, but the level of GS activity remaining may still be sufficient to remain the primary route. In fumigation of plants with up to 0.80 ppm NO no significant change in GS activity was measured, so under similar conditions it is unlikely that there would be any shift to the GDH-mediated reductive amination. Why plants respond to an increase in NH_4^+ by increased GDH activity is unclear.

Experiments with other pollutants have demonstrated inhibition of nitrite reductase activity by O_3 and SO_2. In plants exposed to SO_2 and NO_2, the presence of SO_2 appeared to destroy the plant's ability to respond to NO_2 by blocking a pathway for detoxification. For the reasons outlined above, low levels of these pollutants are unlikely to affect nitrate reduction or amino acid synthesis; with higher concentrations the effects will probably be secondary due to changes in pH or disrupted electron transport and reduced photophosphorylation. The effects of pollutants on GS require further investigation.

Conclusion

Some fumigation experiments have clearly demonstrated effects on plant metabolism. To analyse these effects it is necessary to understand the extent of penetration and distribution of pollutant between different tissues and cells. In addition knowledge is required concerning the effects of pollutants on enzyme systems *in vitro*. Most of this information is still lacking. Attention has been concentrated on certain well-known enzymes, e.g. those of the C_3 cycle. Even with regard to RuBP carboxylase the position is unclear since inhibition by sulphite is complex and time dependent. Effects on enzymes concerned with nitrogen metabolism are also established, but *in vivo* it is likely that the effects on nitrogen metabolism will be smaller than those on carbon metabolism.

Acknowledgements

We wish to thank Dr S. Gutteridge, Dr A. J. Keys and Dr P. J. Lea for helpful discussion during the preparation of this manuscript.

References

ALSCHER-HERMAN, R. The effect of sulphite on light activation of chloroplast fructose-1,6-bisphosphatase in two cultivars of soybean, *Environmental Pollution (Series A)* **27**, 83–96 (1982)

DOLZMANN, P. and ULLRICH, H. Einige Beobachtungen über Beziehungen zwishen Chloroplasten und Mitochondrien in Palisadenparenchyma von *Phaseolus vulgaris*, *Zeitschrift für Pflanzenphysiologie* **55**, 165–180 (1966)

FISCHER, K., KRAMER, D. and ZIELGER, H. Electronenmikroskopische untersuchungen SO_2-begaster Blätter von *Vicia faba*. I. Beobachtungen an Chloroplasten mit akuter Schadingung, *Protoplasma* **76**, 83–96 (1973)

GEZELIUS, K. and HALLGREN, J.-E. Effects of SO_3^{2-} on the activity of ribulose bisphosphate carboxylase from seedlings of *Pinus sylvestris*, *Physiologia Plantarum* **49**, 354–358 (1980)

HALLGREN, J.-E. and GEZELIUS, K. Effect of sulphur dioxide on photosynthesis and ribulose bisphosphate carboxylase in pine tree seedlings, *Physiologia Plantarum* **54**, 153–161 (1982)

HORSMAN, D. C. and WELLBURN, A. R. Synergistic effect of sulphur dioxide and nitrogen dioxide polluted air upon enzyme activity in pea seedlings, *Environmental Pollution* **8**, 123–133 (1975)

KOZIOL, M. J. and COWLING, D. W. Growth of ryegrass (*Lolium perenne* L.) exposed to sulphur dioxide. II. Changes in the distribution of assimilated ^{14}C, *Journal of Experimental Botany* **29**, 1431–1439 (1978)

LIBERA, W., ZIEGLER, I. and ZIEGLER, H. The action of sulfite on the HCO_3^- fixation and the fixation pattern of isolated chloroplasts and leaf tissue slices, *Zeitschrift für Pflanzenphysiologie* **74**, 420–433 (1975)

LORIMER, G. H., BADGER, M. R. and ANDREWS, T. J. The activation of ribulose-1,5-bisphosphate carboxylase by carbon dioxide and magnesium ions. Equilibria, kinetics, a suggested mechanism, and physiological implications, *Biochemistry* **15**, 529–536 (1976)

MISZALSKI, Z. and ZIEGLER, H. 'Available SO_2' — a parameter for SO_2 toxicity, *Phytopathologische Zeitschrift* **97**, 144–147 (1980)

MUKERJI, S. K. and YANG, S. F. Phosphoenolpyruvate carboxylase from spinach leaf tissue, *Plant Physiology* **53**, 829–834 (1974)

PARRY, M. A. J. and GUTTERIDGE, S. Effects of SO_3^{2-} and SO_4^{2-} on ribulose bisphosphate carboxylase activation and activities (in preparation) (1983)

PAULSEN, J. M. and LANE, M. D. Spinach ribulose diphosphate carboxylase. I. Purification and properties of the enzyme, *Biochemistry* **5**, 2350–2357 (1966)

PURCZELD, P., CHON, C. J., PORTIS, A. R., HELDT, H. W. and HEBER, U. The mechanism of the control of carbon fixation by the pH in the chloroplast stroma, *Biochemica Biophysica Acta* **501**, 488–498 (1978)

STEER, M. W., GUNNING, B. E. S., GRAHAM, T. A. and CARR, D. J. Isolation, properties and structure of fraction I protein from *Avena sativa* L., *Planta* **79**, 256–267 (1968)

THOMPSON, W. W., DUGGER, W. M. Jr. and PALMER, R. L. Effects of ozone on the fine structure of the palisade parenchyma cells of bean leaves, *Canadian Journal of Botany* **44**, 1677–1682 (1966)

WELLBURN, A. R., WILSON, J. and ALDRIDGE, P. H. Biochemical responses of plants to nitric oxide polluted atmospheres, *Environmental Pollution (Series A)* **22**, 219–228 (1980)

WELLBURN, A. R., HIGGINSON, C., ROBINSON, D. and WALMSLEY, C. Biochemical explanations of more than additive inhibitory effects of low atmospheric levels of sulphur dioxide plus nitrogen dioxide upon plants, *New Phytologist* **88**, 223–237 (1981)

WILKINSON, T. G. and BARNES, R. L. Effects of ozone on $^{14}CO_2$ fixation patterns in pine, *Canadian Journal of Botany* **51**, 1573–1578 (1973)

WINNER, W. E. and MOONEY, H. A. Ecology of sulphur dioxide resistance. III. Metabolic changes of C_3 and C_4 *Atriplex* species due to sulphur dioxide fumigation, *Oecologia (Berl.)* **46**, 49–54 (1980)

YONEYAMA, T. and SASAKAWA, H. Transformation of atmospheric NO_2 absorbed in spinach leaves, *Plant and Cell Physiology* **20**, 263–266 (1979)

ZIEGLER, I. The effects of SO_3^{2-} on the activity of ribulose-1,5-diphosphate carboxylase in isolated spinach chloroplasts, *Planta* **103**, 155–163 (1972)

ZIEGLER, I. Effects of sulphite on phosphoenolpyruvate carboxylase and malate formation in extracts of *Zea mays*, *Phytochemistry* **12**, 1027–1030 (1973)

ZIEGLER, I. Action of sulphite on plant malate dehydrogenase, *Phytochemistry* **13**, 2411–2416 (1974)

Chapter 13

Effects of air pollutants on light reactions in chloroplasts

K. Sugahara

DIVISION OF ENVIRONMENTAL BIOLOGY, NATIONAL INSTITUTE FOR
ENVIRONMENTAL STUDIES, IBARAKI, JAPAN

Introduction

Air pollutants can inhibit both the light and dark reactions of photosynthesis. Within the chloroplasts, the light reactions which produce ATP and NADPH are mainly associated with the lamellae or thylakoids, while the dark reactions which enzymatically fix CO_2 into acid-stable compounds occur within the stroma. To determine the effects of air pollutants on chloroplast metabolism two different approaches have generally been used:

(1) the effects of air pollutants on isolated chloroplasts have been studied by bubbling the gases through chloroplast suspensions, and
(2) analyses have been conducted on chloroplasts isolated from pollutant-exposed leaves.

There are several problems with both approaches. In the first system, the mechanical agitation caused by bubbling pollutants through chloroplast suspensions can inactivate certain reactions, confounding interpretation of the pollutant-induced effects. Further, this system cannot be used to study the long-term effects of pollutants because isolated chloroplasts do not retain their photosynthetic activity over long periods of time. Moreover, it is difficult to interpret results when pollutant mixtures are used because the gases have different solubilities in the reaction medium. In the second system, certain reactions of chloroplasts can be inactivated during their isolation from pollutant-exposed leaves. Further, there is also the possibility that pollutant-injured chloroplasts are more easily ruptured during isolation so that fewer intact chloroplasts are obtained than usual. This system, however, can be used for plants exposed for prolonged periods to gaseous pollutants. For the reasons cited above, it is very important to include the appropriate controls in order to obtain reproducible results when using either system.

In the present study, the second system was adopted because the plants were exposed to SO_2, NO_2 and O_3, either singly or in combination, for fumigation periods longer than those which could be used with chloroplast suspensions. The light reactions associated with chloroplast lamellae were analysed for effects of these pollutants on photosystems I and II (PSI and PSII) individually and working in tandem. The results obtained are compared with those of other workers.

Materials and methods

Plant material

Spinach (*Spinacia oleracea* L. cv. New Asia) and lettuce (*Lactuca sativa* L. cv. Romaine) plants were grown in pots (115 mm diameter) containing vermiculite, peat moss, perlite and fine gravel (2:2:1:1, by volume) at 20 °C day/15 °C night temperatures with a relative humidity of 70% in a glasshouse under sunlight. As nutrients, 4 g Magamp K (NPK = 6:40:6, W. R. Grace Co., USA) and 8 g magnesia of lime were applied in dry form to each litre of soil mixture and 200 ml of a solution of 1 g 1^{-1} Hyponex (NPK = 6.5:6:19) was supplied to each pot every five days thereafter. Plants were used for experimentation when four to six weeks old.

Fumigation conditions

Plants were fumigated with air pollutants in a growth cabinet (230 × 190 × 170 cm) at 20 °C with a relative humidity of 75%; wind velocity in each cabinet was 0.22 m s^{-1}. Illumination was provided with heat-filtered stannous halide vapour lamps (Toshiba Yoko Lamp, 400 W, Toshiba Co. Ltd, Tokyo, Japan) giving a light intensity of 25 000–35 000 lux at the leaf level. Plants were preconditioned for 2 h under illumination in the growth cabinet for clean air controls, after which half of the plants were transferred quickly into another growth cabinet receiving the appropriate concentration of pollutant gas or gas mixture. The lengths of the fumigation periods were varied and are

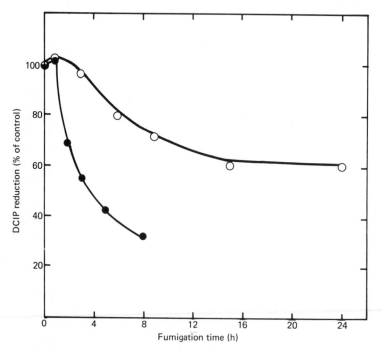

Figure 13.1. Inhibition of DCIP photoreduction in chloroplasts isolated from SO_2-fumigated spinach leaves. Fumigation was performed at 1.0 (○) and 2.0 (●) ppm; other conditions as described in the text

given with the experimental results. Chloroplasts were isolated from plants from each growth cabinet as described below.

Preparation of chloroplasts

After pollutant fumigation, leaves were homogenized at 0°C in 0.05M Tricine-NaOH buffer (pH 7.5) containing 0.02 NaCl and 0.4M sucrose. After the homogenate had been filtered through four layers of gauze, the filtrate was centrifuged at $200 \times g$ for 5 min and the chloroplasts were isolated from the supernatant by centrifugation at $1500 \times g$ for 7 min.

Measurement of photosynthetic electron transport

The rates of dichloroindophenol (DCIP) and NADP photoreduction were determined according to the method of Shimazaki and Sugahara (1979). The reaction mixture for DCIP photoreduction contained in 4 ml final volume, 12.5 mM Tricine-NaOH buffer (pH 7.5), 100 mM sucrose, 5 mM NaCl, 50 μM DCIP and chloroplasts containing 20 μg chlorophyll. The reaction mixture (4 ml) for NADP reduction contained 12.5 mM Tricine-NaOH buffer (pH 7.5), 100 mM sucrose, 5 mM NaCl, 5 μM NADP, a saturating amount of spinach ferredoxin and chloroplasts containing 40 μg chlorophyll. The DCIPH$_2$–NADP system contained in addition 50 μM DCMU, 50 μM DCIP, 2.5 mM sodium ascorbate and 25 mM NH$_4$Cl (the last to act as an uncoupler).

The rate of O$_2$ exchange was determined with a Clark-type oxygen electrode according to the method of Shimazaki and Sugahara (1980).

Results

Effects of SO$_2$

Effects of SO$_2$ on the electron transport system have been reported elsewhere (Shimazaki and Sugahara, 1979, 1980).

Figure 13.1 shows the effect of SO$_2$ on DCIP photoreduction (PS II) in chloroplasts isolated from fumigated spinach leaves. No significant effect was observed after 1 h, but thereafter the inhibition proceeded rapidly during fumigation. The rate of DCIP photoreduction was reduced to 40% of the control by fumigation at 2.0 ppm for 5 h, while 1.0 ppm SO$_2$ fumigation for 6 h decreased the rate to 80% of the control. The inactivation of the photoreduction was not reversed by washing in 15 mM Tris-HCl buffer, pH 7.4 (data not shown).

To determine whether this inhibition might be due to certain toxic substances formed in fumigated leaves, chloroplasts isolated from control leaves not treated with SO$_2$ were incubated in the supernatant obtained during isolation of chloroplasts from leaves fumigated with SO$_2$ at 2.0 ppm for 5 h. After 10 min incubation in an ice bath the chloroplasts were collected by centrifugation and resuspended in the isolation medium. No inhibitory action on the DCIP photoreduction was observed (*Table 13.1*). These results indicate that the inhibition did not occur as a result of the isolation process, but was induced as an irreversible damage by SO$_2$ fumigation.

The inhibitory action of SO$_2$ on the activities of photosystems I and II in chloroplasts are shown in *Table 13.2*. Electron flow from H$_2$O to DCIP was inhibited,

TABLE 13.1. Effect of supernatants obtained from SO$_2$-fumigated leaves on DCIP photoreduction in chloroplasts isolated from non-fumigated leaves [a]

	DCIP photoreduction (chlorophyl h^{-1} μmol mg^{-1})
Non-fumigated [b]	129
Fumigated [c]	128

[a] Incubation was performed for 10 min in an ice bath; other conditions as described in the text.
[b] Residual supernatant after isolation of chloroplasts from non-fumigated leaves.
[c] Residual supernatant after isolation of chloroplasts from SO$_2$-fumigated leaves. Fumigation was performed at 2.0 ppm SO$_2$ for 5 h.

while that from reduced DCIP to NADP (DCIPH$_2$–NADP) was not affected when electron transport was uncoupled by 2 mM NH$_4$Cl. SO$_2$ inhibited the overall electron flow from H$_2$O to NADP to the same degree as the electron flow from H$_2$O to DCIP. From these results, we concluded that SO$_2$ inhibited the electron flow driven by photosystem II but not that by photosystem I.

Next, we investigated the site of SO$_2$ inhibition in electron transport chain using chloroplasts isolated from SO$_2$ fumigated leaves of lettuce. Electron transfer from H$_2$O to DCIP was inhibited, whereas electron-flow from reduced DCIP to methyl viologen (photosystem I reaction) was not affected (*Table 13.3*). The results obtained were the same as those for spinach leaves. The rate of DCIP photoreduction inhibited by SO$_2$ could not be recovered by the addition of diphenylcarbazide (DPC), an artificial electron donor for photosystem II (*Table 13.3*). This indicates that the site of the SO$_2$ action is either closer to the reaction centre of photosystem II than the site where DPC acts as donor, or that it is on the reducing side of photosystem II.

As shown in *Figure 13.2*, O$_2$ evolution in chloroplasts with ferricyanide as oxidant was inhibited by SO$_2$ fumigation. When 7 μM DCMU was added to the SO$_2$-inactivated chloroplasts, O$_2$ evolution was suppressed completely. However, the subsequent addition of silicomolybdic acid, a lipophilic electron acceptor of photosystem II, restored the rate of O$_2$ evolution to the original values of SO$_2$-inactivated chloroplasts.

Thus, the site inactivated by SO$_2$ fumigation must be located on the electron transfer path from the water-splitting system to Q (the primary electron acceptor of photosystem II), since silicomolybdic acid is thought to accept electrons directly from Q. From analyses of carotenoid photobleaching and fluorescence transients in SO$_2$-inactivated chloroplasts, we further concluded that SO$_2$ inactivated the primary electron donor in photosystem II or possibly the reaction centre itself.

TABLE 13.2. Effects of SO$_2$ on electron transport activities

Reaction measured	SO$_2$ fumigation (h)		
	0	2	4
	(μmol acceptor reduced mg^{-1} chlorophyll h^{-1})		
H$_2$O–NADP	170	107	66
DCIPH$_2$–NADP (+DCMU)	95	97	108
H$_2$O–DCIP	217	124	70

SO$_2$ fumigation was performed at 2.0 ppm; other conditions as described in the text.

TABLE 13.3. Effects of SO$_2$ on electron transport activities and the effect of DPC^a on the SO$_2$-inhibited DCIP photoreduction

Reaction measured	SO$_2$ fumigation (h)			
	0	2	3	5
H$_2$O–DCIP (μmol DCIP reduced mg^{-1} chlorophyl h^{-1})	148	86	45	13
H$_2$O–DCIP (+DPC) (μmol DCIP reduced mg^{-1} chlorophyl h^{-1})	166	93	49	15
DCIPH$_2$–MV^b (+DCMU) (μmol O$_2$ uptake mg^{-1} chlorophyl h^{-1})	216	206	193	190

SO$_2$ fumigation was performed at 2.0 ppm.
Other conditions are described in the text.
^{a} Diphenylcarbazide.
^{b} Methyl viologen.

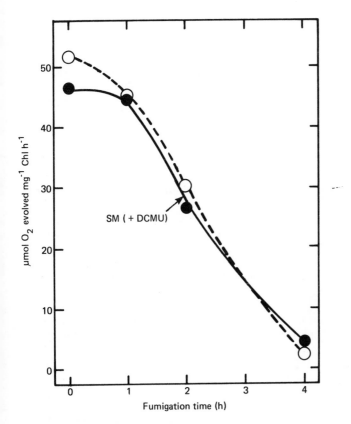

Figure 13.2. Effects of SO$_2$ on DCMU-sensitive and -insensitive O$_2$ evolution. The rate of O$_2$ evolution in the absence of DCMU (○) and in the presence of DCMU and silicomolybdic acid (SM) (●). Fumigation was performed at 2.0 ppm SO$_2$. SM was added to the chloroplasts suspension after the addition of DCMU during the measurement in the light

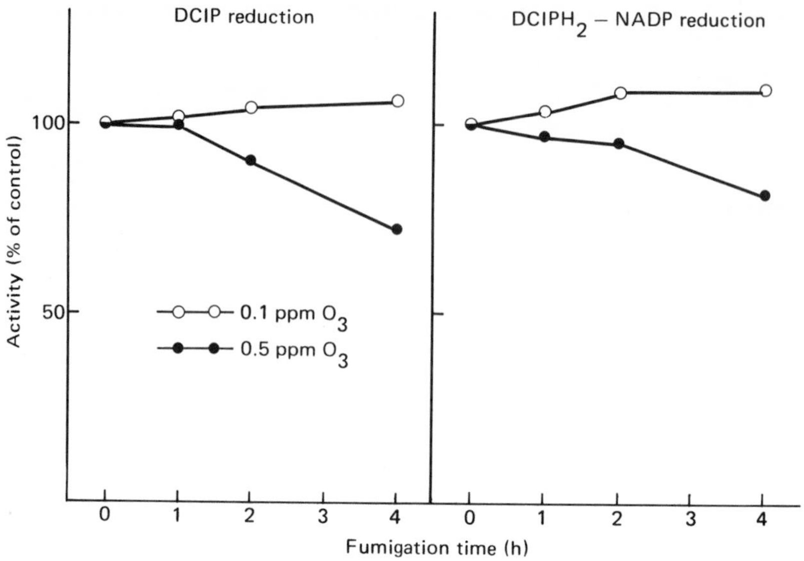

Figure 13.3. Effects of O_3 on DCIP and $DCIPH_2$–NADP photoreduction. Experimental conditions are described in the text

Experiments using low concentrations of SO_2 and/or NO_2 were conducted with perennial ryegrass (*Lolium perenne*) by Wellburn *et al.* (1981). Exposure to 0.25 ppm SO_2 for 11 days did not have any effect on either of the reactions of photosystem I and II. We found that in spinach exposed to 0.5 ppm SO_2 for 20–30 h photosystem II was inhibited but photosystem I was not (*Figures 13.4* and *13.6*).

Effects of ozone

Figure 13.3 shows the effect of O_3 on the photoreduction of DCIP by H_2O and of NADP by $DCIPH_2$; 0.1 ppm O_3 did not suppress the electron transport in either photosystem. On the other hand, 0.5 ppm O_3 inhibited the photoreactions in both photosystems after 4 h exposure. Unlike the effects of SO_2, O_3 did not preferentially inhibit photosystem II, but affected both photosystem reactions at the same time. Coulson and Heath (1974) also reported that O_3 bubbled into a suspension of isolated spinach chloroplasts inhibited electron transport in both photosystems. Murabayashi *et al.* (1981) and Suzuki, Murabayashi and Matsuno (1982) investigated the effect of O_3 on electron transport in spinach chloroplasts more closely. They performed experiments using both chloroplasts isolated from O_3-fumigated leaves and suspensions of normal isolated chloroplasts through which O_3 had been bubbbled and found that electron transport in both photosystems was inhibited by O_3.

The maintenance of the normal permeability characteristics and integrity of the membranes of the chloroplast lamellae is necessary for the production of the proton gradient that is the driving force for ATP formation and O_3 may perhaps affect these.

Effects of nitrogen dioxide

In general, NO_2 fumigation does not affect plants severely, even at relatively high concentrations. As shown in *Figures 13.4* and *13.6*, 4 ppm NO_2 caused little inhibition of electron transport in either photosystem I or II after 10 h fumigation, and only a slight inhibition was found after 20 h fumigation.

Wellburn *et al.* (1981) reported that long-term fumigation at a low concentration of NO_2 (0.25 ppm for 11 days) had no effect on electron transport in either photosystem, but did enhance the production of ATP.

Effects of pollutant mixtures

In a series of experiments using pollutant mixtures, the following concentrations of pollutants were selected on the basis of the results obtained by exposure of plants to these pollutants singly: namely SO_2, 0.5 ppm; NO_2 4.0 ppm; O_3, 0.1 ppm. The fumigation of plants with any one of the pollutants at these concentrations showed either no effect or a slight and gradual effect on electron transport with time. The results of fumigation by mixtures at these concentrations are summarized in *Figures 13.4–13.9*.

Sulphur dioxide and ozone

DCIP photoreduction was not inhibited by 0.1 ppm O_3 alone but was inhibited slightly by 0.5 ppm SO_2 alone after 10 h of fumigation (*Figure 13.4*). The inhibition of photoreduction of DCIP by H_2O in response to SO_2 and O_3 given together was not significantly different from that observed with SO_2 alone (*Figure 13.5*). Photoreduction of NADP by $DCIPH_2$ was not affected by 0.5 ppm SO_2 even after 30 h fumigation (*Figure 13.6*). Fumigation with 0.1 ppm O_3 or a mixture of O_3 and SO_2 also had no effect on photosystem I (*Figure 13.7*). Total photosystem activity (NADP photoreduction by H_2O) was enhanced by 0.1 ppm O_3 for the first 10 h of fumigation

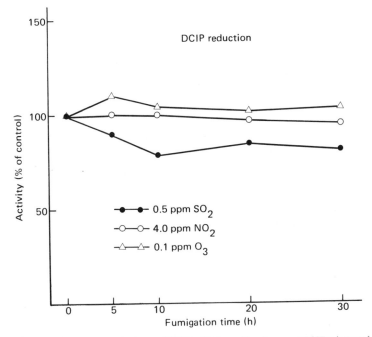

Figure 13.4. Effects of exposure with SO_2, NO_2 or O_3 alone on DCIP photoreduction (PSII). Experimental conditions are described in the text

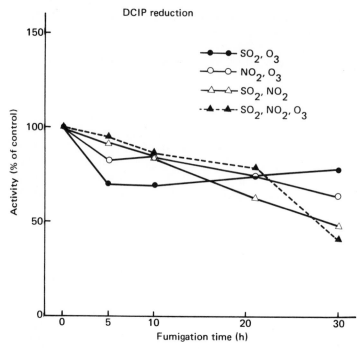

Figure 13.5. Effects of exposure with SO_2, NO_2 and O_3 in combination on DCIP photoreduction (PSII). Experimental conditions are described in the text

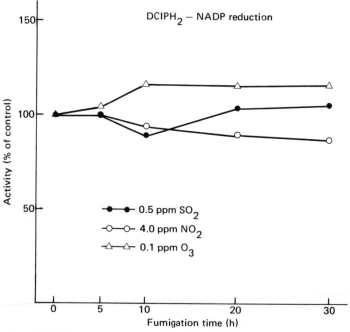

Figure 13.6. Effects of exposure with SO_2, NO_2 or O_3 alone on $DCIPH_2$–NADP system (PSI). NH_4Cl (25 mM) was added as an uncoupler; other conditions as described in the text

(Figure 13.8). The gradual decrease in the rate of photoreduction of NADP by H_2O caused by exposure to 0.5 ppm SO_2 was not significantly increased by the additional presence of O_3 *(Figure 13.9).*

Nitrogen dioxide and ozone

The photoreduction of DCIP was not inhibited by fumigation either with 4 ppm NO_2 or 0.1 ppm O_3 singly *(Figure 13.4)*, but in combination an inhibition was observed *(Figure 13.5).* The activity of photosystem I decreased only slightly after a 30 h exposure to 4 ppm NO_2 *(Figure 13.6).* However, the inhibition of DCIP photoreduction by H_2O obtained by the mixture of NO_2 and O_3 was not reflected in

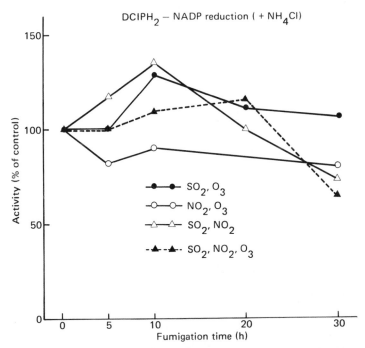

DCIPH$_2$ – NADP reduction (+ NH$_4$Cl)

Figure 13.7. Effects of exposure with SO_2, NO_2 and O_3 in combination on DCIPH$_2$–NADP system (PSI). NH$_4$Cl (25 mM) was added as an uncoupler; other conditions are described in the text

any inhibition in photosystem I activity *(Figure 13.7).* The total photosystem activity was inhibited gradually with the time in response to fumigation with 4 ppm NO_2; this inhibition was enhanced synergistically by the combination of NO_2 and O_3 *(Figures 13.8 and 13.9).*

Sulphur dioxide and nitrogen dioxide

In combination, 0.5 ppm SO_2 and 4 ppm NO_2 inhibited the photoreduction of DCIP by H_2O after 20 h. On the other hand, this gas mixture increased the DCIPH$_2$–NADP photoreduction during the first 10 h of exposure, after which the photoreduction of DCIPH$_2$–NADP decreased with time to a level representing a significant inhibition

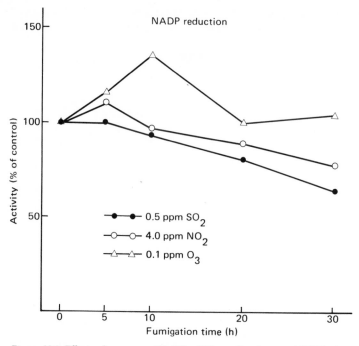

Figure 13.8. Effects of exposure with SO$_2$, NO$_2$ or O$_3$ alone on NADP photoreduction (PSI +II). Experimental conditions are described in the text

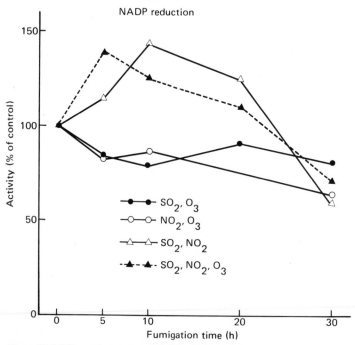

Figure 13.9. Effects of exposure with SO$_2$, NO$_2$ and O$_3$ in combination of NADP photoreduction (PSI +II). Experimental conditions are described in the text

after 30 h (*Figures 13.5* and *13.7*). Exposure to the mixture of SO_2 and NO_2 also enhanced the total photosystem activity during the first 20 h of exposure although an inhibition appeared finally after 30 h fumigation (*Figure 13.9*).

Sulphur dioxide, nitrogen dioxide and ozone

The inhibition by a mixture of SO_2, NO_2 and O_3 of photoreduction of DCIP by H_2O followed a similar pattern to the inhibition caused by a mixture of SO_2 and NO_2 (*Figure 13.5*). Ozone did not enhance the inhibitions caused by SO_2 or NO_2, singly or in combination. However, in the photoreduction of NADP by $DCIPH_2$ it appears that the enhancement of the activity observed by the fumigation with SO_2 and NO_2 in combination for 10 h was suppressed by O_3 treatment. When the overall photosystem was measured (NADP reduction by H_2O) the enhancement observed during 20 h fumigation with a mixture of SO_2 and NO_2 still remained.

Discussion

The inhibition of photosynthetic processes by air pollutants has been reported by many workers. Their results have shown that following exposure to low concentrations of these pollutants for long periods, no inhibition of photosynthetic electron transport occurred but there was a suppression of growth. This may indicate that membrane-associated light reactions were more resistant to the pollutants than dark reactions concerned with CO_2 fixation. Further, the inhibition of the light reaction was mostly irreversible and it took a long time to restore its activity. In the present study, we therefore used relatively high concentrations of air pollutants in order to get clear inhibitory effects on the activity of the light reaction.

SO_2 preferentially inhibited photosystem II. It was shown that the site of inhibition was at the primary electron donor site or at the reaction centre itself. However, O_3 and NO_2 inactivated electron transport in both photosystems I and II. This may suggest that inactivation of the reaction by O_3 or NO_2 was probably the result of denaturation or destruction of constituents contained in the membrane structure of both photosystems.

The effects of mixed pollutants on the light reactions were very complex. On fumigation with SO_2 and O_3 in combination, the inhibition of photosystem II, and of the total photosystem, did not increase beyond the level of 10 h inhibition even after 30 h fumigation. The result may indicate that from 10 to 30 h injury by SO_2 was prevented by the presence of O_3. A speculative mechanism of this 'protection' could be that SO_3^{2-} ion is accumulated in the cytoplasm at a relatively high concentration following SO_2 fumigation and that O_3 introduced into cytoplasm may react with SO_3^{2-} directly to produce the SO_4^{2-} ion, which is less toxic than SO_3^{2-}.

On exposure to a combination of O_3 and NO_2, both photosystem I and II reactions were inhibited significantly, although they were both slightly inhibited by 30 h exposure to NO_2 alone. It is possible that when O_3 is present with NO_2 the nitrite reductase system is inhibited and nitrite is accumulated to a toxic concentration. It is also possible that the combined effect of NO_2 and O_3 results in the formation of free radicals, which damage the chloroplast membrane.

On exposure to a mixture of SO_2 and NO_2, photosystem I reaction was enhanced at 10 h but had become inhibited at 30 h. The temporary enhancement was also observed when the total photosystem (PSI + PSII) was measured. Such an enhancement has never been observed in the effect of an air pollutant on photosynthetic processes in intact systems. The enhancement cannot be explained at present. We can only suggest that a new biochemical product was produced by the combined effect of sulphite and nitrite and believe that the transient phenomenon obtained by the fumigation with relatively high concentrations of SO_2 and NO_2 may occur even at low concentrations if the new product were to accumulate in sufficient amount.

On exposure to a mixture of SO_2, NO_2 and O_3 an enhancement of photosystem I reaction was observed as well as a total photosystem reaction (PSI + PSII). The inhibition pattern of photosystem II reaction resembled that caused by the mixture of SO_2 and NO_2. No synergistic inhibition was observed.

All of the fumigations that included NO_2 caused injury in both photosystems I and II after 30 h. This implies that fumigation with 4 ppm NO_2 was more damaging to plants than exposure to the other pollutants in our experiments, and it is suggested that 4 ppm NO_2 exposure could have gone beyond the threshold of tolerance. If O_3 had been given at 0.2 ppm (double that in the present study) the effect of O_3 on photosynthetic reaction might have been more clearly observed.

The effects of the mixed pollutants on plant metabolism are very complex. It is necessary to perform the fumigation with several pollutants singly or in combination, to get more clear information on the mechanism of their effects.

References

COULSON, C. and HEATH, R. L. Inhibition of photosynthetic capacity of isolated chloroplasts by ozone, *Plant Physiology* **53**, 32–38 (1974)

MURABAYASHI, M., AWAYA, M., TSUJI, H. and MATSUNO, T. Effects of ozone on photosynthetic electron transport in spinach (II), *Bulletin of the Institute of Environmental Science and Technology, Yokohama National University* **7**, 43–49 (1981)

SHIMAZAKI, K. and SUGAHARA, K. Specific inhibition of photosystem II activity in chloroplasts by fumigation of spinach leaves with SO_2, *Plant and Cell Physiology* **20**, 947–955 (1979)

SHIMAZAKI, K. and SUGAHARA, K. Inhibition site of the electron transport system in lettuce chloroplasts by fumigation of leaves with SO_2, *Plant and Cell Physiology* **21**, 125–135 (1980)

SUZUKI, S., MURABAYASHI, M. and MATSUNO, T. Effects of ozone on photosynthetic electron transport in spinach (III), *Bulletin of the Institute of Environmental Science and Technology, Yokohama National University* **8**, 81–87 (1982)

WELLBURN, A. R., HIGGINSON, C., ROBINSON, D. and WALMSLEY, C. Biochemical explanations of more than additive inhibitory effects of low atmospheric levels of sulphur dioxide plus nitrogen dioxide upon plants, *New Phytologist* **88**, 223–237 (1981)

Chapter 14

Effects of SO$_2$ on light-modulated enzyme reactions

R. Alscher

BOYCE THOMPSON INSTITUTE AT CORNELL UNIVERSITY, ITHACA, NY, USA

Introduction

Sulphur dioxide has been long established as an inhibitor of net photosynthesis (Thomas and Hill, 1937; Katz, 1949). Its action is thought to derive from effects both on stomatal behavior and also on the photosynthetic process itself (Hällgren, 1978; Müller, Miller and Sprugel, 1979). Exposures to concentrations of SO$_2$ that do not produce necrosis sometimes cause a temporary inhibition of net photosynthesis, with recovery to control rates after removal of the plants from the SO$_2$ atmosphere (McLaughlin *et al.*, 1979). Thus, SO$_2$ can, under certain circumstances, reversibly disrupt the operation of the photosynthetic machinery.

A considerable body of evidence has accumulated over the past 15 years which demonstrates the existence of a light-controlled regulation system for carbon fixation called light modulation. Sulfur dioxide, by virtue of its chemical properties, is well-suited to interact with the light modulation system, and some evidence has accumulated which suggests that this is indeed the case. This review will provide:

(1) an outline of what is known about the light modulation system and its role in photosynthesis,
(2) an account of the properties of SO$_2$ which would enable it to interact with the light modulation system,
(3) evidence that a SO$_2$/sulfite–light modulation interaction does take place,
(4) a proposed scheme for the mode of action of SO$_2$ on light modulation, and
(5) proposed bases for differences in SO$_2$ tolerance among cultivars and species arising from the scheme proposed in (4).

Light and the 'dark reaction' of photosynthesis

The classic studies of Bassham and his colleagues at the University of California at Berkeley (Calvin and Bassham, 1962) established the sequence of interconversions within the chloroplast which lead to carbon fixation (*Figure 14.1*). This phase of photosynthesis was initially thought to be light-independent, although it used ATP and NADPH produced during the 'photochemical' phase. The Zieglers, in their pioneering work (Ziegler, Ziegler and Schmidt-Clausen, 1965; Ziegler and Ziegler, 1965, 1967)

181

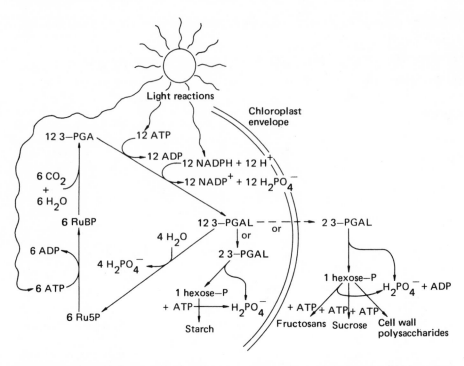

Figure 14.1. Photosynthetic carbon fixation — a summary. Points are indicated at which products of the light reactions are utilized (from Salisbury and Ross, 1978; courtesy of Wadsworth Publ. Co.)

TABLE 14.1. Light modulation in different types of plants. Values are for activation (x-fold)

	C_3 plants		C_4 plants		CAM
	Pea (Pisum)	Spinach (Spinacia)	Maize (Zea)	Tidestromia	Kalanchoe
Light-activated enzymes			Light stimulation x-fold		
NADP-linked malic dehydrogenase	14	∞	50	3.3	1.7
NADP-linked glyceraldehyde phosphate dehydrogenase	2.4	5[a]	2	3.1	2
Ribulose-5-phosphate kinase	7.7	3.2	1.6	4	4.4
FBPase	1.7	2.2	Nil		Nil
SBPase	1.8	1.7			1.7
Pyruvate orthophosphate dikinase			~12		
Dark activated enzymes			Dark stimulation x-fold		
Glucose-6-phosphate dehydrogenase	1.3	3			Nil
Phosphofructokinase	11				3

[a] Glyceraldehyde-3-P-dehydrogenase is activated in *Vicia faba, Nicotiana tabacum, Brassica napus, Valerianella olitoria, Beta vulgaris, Myrothamnus flabellifolia, Triticum aestivum, Avena sativa, Lemna gibba, Saccharum officinale.*
(After Anderson, 1979; courtesy of Springer-Verlag.)
For individual references *see* Anderson (1979).

demonstrated a light-stimulated increase in the activity of NADP–glyceraldehyde-3-phosphate dehydrogenase (NADP–GPD). This dehydrogenase is the enzyme which catalyzes the reductive step of the pentose phosphate or C_3 cycle. Further results established that this effect of light was a reversible one (Müller, Ziegler and Ziegler, 1969). The work of Buchanan, Kalberer and Arnon (1967) during the same period demonstrated that alkaline fructose-1,6-bisphosphatase (FBPase) was also light-activated. The possible significance of light activation of these various enzymes for photosynthesis was suggested by results from Bassham's group (Pedersen, Kirk and Bassham, 1966), showing light-induced changes in levels of photosynthetic intermediates in *Chlorella* cells.

In subsequent years more enzymes were added to the list of those known to be light-modulated (*see Table 14.1* and *Figure 14.2* for a current inventory) and some of the requirements for the process began to be understood. It was found that dithiothreitol (DTT) could substitute for light (Anderson and Lim, 1972; Anderson, 1974; Anderson, Ng and Park, 1974; Baier and Latzko, 1975) and also that the presence of sulfhydryl reagents inhibited activation (Wolosiuk and Buchanan, 1976). These findings suggested that activation involves a light-dependent reduction of disulfide bonds, a reaction which is blocked by sulfhydryl reagents and stimulated by thiols. Anderson and Avron (1976) showed that glucose-6-phosphate dehydrogenase, the first enzyme of the oxidative pentose phosphate pathway is inactivated by light in a mechanism also involving sulfhydryl groups. (The oxidative pentose phosphate pathway can supply NADPH and pentose sugars starting with glucose-6-phosphate, which is produced as a result of starch breakdown.)

$$\text{Glucose-6-phosphate} + \text{NADP}^+ \xrightleftharpoons[\text{dehydrogenase}]{\text{glucose-6-phosphate}} \rightarrow 6 \text{ phosphoglucono-}$$

$$\delta\text{-lactone} \dashrightarrow 6\text{-phosphogluconate}.$$

$$6\text{-Phosphogluconate} + \text{NADP}^+ \xrightarrow[\text{dehydrogenase}]{6\text{-phosphogluconate}}$$

$$\longrightarrow \text{D-ribulose-5-phosphate} + CO_2 + \text{NADPH} + H^+.$$

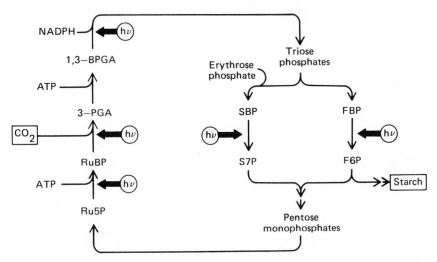

Figure 14.2. Role of light in the activation of enzymes of the reductive pentose phosphate cycle (from Buchanan, 1980; courtesy of Annual Reviews)

The work of several laboratories demonstrated that photosynthetic electron transport was necessary for light activation (Anderson and Avron, 1976; Champigny and Bismuth, 1976; Kelly *et al.*, 1976). Data obtained with photosynthetic inhibitors implicated at least two sites on the electron transport chain as sources of reductant for the process.

Other light-induced changes were also found to control these chloroplast enzymes. Werdan, Heldt and Milovancev (1975), Portis *et al.* (1977), Purczeld *et al.* (1978) demonstrated that, in the light, stromal pH increases from pH 7.0 to 8.0. Carbon fixation, which is optimal at pH 8.1, cannot proceed at the lower pH. Light-induced changes in magnesium concentrations were also detected. It was found that if pH and Mg^{2+} concentrations were artificially altered, carbon fixation did not proceed in illuminated chloroplasts. Under the altered conditions, the concentrations of fructose-1,6-bisphosphate (FBP) and sedoheptulose-1,7-bisphosphate (SBP) increased in the stroma, indicating an inhibition at the bisphosphatase step(s) in carbon fixation.

Charles and Halliwell (1980a, 1981a) showed that the oxidized or 'dark' inactive form of fructose-1,6-bisphosphatase has a K_m* (a measure of the affinity of the enzyme for its substrate) for FBP of 800 μM. The reduced or active form of the enzyme has a K_m for FBP of 33 μM. The values obtained by Heldt *et al.* (1980) for stromal concentrations of FBP are in the range of 100 μM. Thus, the oxidized form of the enzyme would not be active within the chloroplast under *in vivo* conditions whereas the reduced form would be fully active.

The mechanism by which light activates these various chloroplast enzymes is still not fully understood. There is a general agreement that photochemically generated reductant causes either a reduction of disulfide bonds or a thiol-disulfide exchange on the enzyme molecule which leads to a configurational change allowing the enzyme to

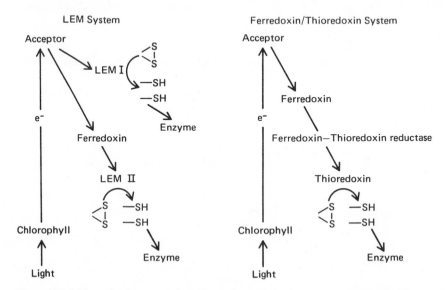

Figure 14.3. Light activation — two hypotheses (from Buchanan, 1980). LEM (light effect mediator) system has been proposed by Anderson (Anderson, 1979), ferredoxin/thioredoxin system by Buchanan (Buchanan, 1980). (Courtesy of Annual Reviews)

* K_m = substrate concentration required to yield half the maximum velocity of the reaction. Thus, the lower the K_m, the greater the affinity of the enzyme for its substrate.

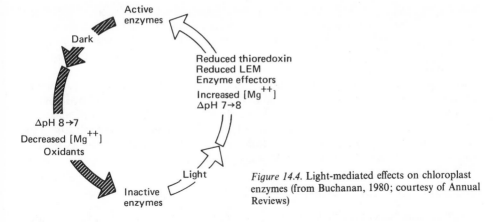

Figure 14.4. Light-mediated effects on chloroplast enzymes (from Buchanan, 1980; courtesy of Annual Reviews)

become more active. Two alternative mechanisms have been proposed (*Figure 14.3*). In the Anderson scheme ('LEM' system) light and electron transport causes the appearance of new thiol groups on the chloroplast membrane, with a subsequent interaction of the enzyme molecule with the membrane dithiol groups. In the Buchanan scheme, soluble thiol-containing proteins (thioredoxins) are involved. The Anderson group has now discovered a soluble factor they call 'protein modulase' (Ashton and Anderson, 1981), and have modified their scheme accordingly. Mohamed and Anderson (1981) have succeeded in extracting a protein essential for light activation from chloroplast membranes and in reconstituting light activation from its parts (depleted thylakoids, released protein, enzyme and protein modulase).

What then is the function of the light activation mechanism(s) in photosynthesis? The various known effects of light on chloroplast enzyme activity are summarized in *Figure 14.4*. The plant cell, to function efficiently, must possess mechanisms that enable it to switch from carbon fixation (via the reductive pentose phosphate pathway) in the light to starch breakdown (via the oxidative pentose phosphate pathway and glycolysis) in the dark. Enzymes for each of these pathways exist within the chloroplast and the various effects of light on them are shown in *Figure 14.5*. In the light, inactivation of glucose-6-phosphate dehydrogenase would lead to a shutdown of starch degradation (glycolytic pathway) while the light activated enzymes, such as fructose-1,6-bisphosphatase (FBPase), would allow carbon fixation to proceed. These changes would be clearcut only if the inactivation of the various enzymes was complete, e.g. if glucose-6-phosphate dehydrogenase was entirely inactive in the light. This does not appear to be the case. Bassham (1973) demonstrated that light entirely switched off the oxidative pathway in chloroplasts, leaves or algal cells, even though glucose-6-phosphate dehydrogenase activity is still about 50% of its dark level. Activity of the enzyme was then found to be also regulated by $NADPH/NADP^+$ ratios, with the 'light' ratio switching off the oxidative pathway altogether (Lendzian, 1980).

Fructose-1,6-bisphosphatase is a light-activated enzyme whose activity is also controlled by pH and magnesium. It is one of the rate-limiting enzymes of the C_3 cycle (*see above*). Leegood and Walker (1980), Robinson and Walker (1980) and Charles and Halliwell (1981b) demonstrated that this enzyme does not, however, directly control the rate of carbon fixation under *optimal* physiological conditions. Enzyme activity measured under the Mg^{2+} and substrate conditions known to exist within the chloroplast *in vivo* (Heldt *et al.*, 1980) decreased by only 18–39% after 15 min in the dark. Rates of carbon fixation occurring in chloroplasts which have been in the dark are much lower than they would be if they were only controlled by light-modulated enzymes such as FBPase. Leegood and Walker (1981) found that the rates of light activation of five chloroplast enzymes were much faster than were rates of photosynthetic induction.

It seems, therefore, that the light modulation system does not control directly the rate of carbon fixation under *optimal* conditions. However, it is important not to forget that if enzymes such as FBPase are completely inactivated, then carbon fixation will be totally inhibited. It follows, therefore, that any factor which drastically influences the activation status of the light-modulated enzymes has the potential to influence the rate at which carbon fixation, or other relevant pathways, proceed.

The influence of hydrogen peroxide on carbon fixation — *in vitro* studies

Low concentrations of hydrogen peroxide strongly inhibit carbon fixation in isolated chloroplasts (Kaiser, 1976). Under these conditions, inhibition occurs at the bisphosphatase steps, with accumulation of FBP and SBP (Kaiser, 1979). Charles and Halliwell (1980a, 1980b) showed that isolated FBPase was inhibited 70% by 300 μM H$_2$O$_2$ and that activity could be restored by incubating with dithiothreitol. This is consistent with the proposal that the inhibitory effect of peroxide on carbon fixation is due to an oxidation of essential thiol groups on the enzyme,

$$R—SH + 3H_2O_2 \rightarrow R—SO_3H + 3H_2O$$

with a resultant conformational rearrangement and inactivation of the enzyme. The reactivity of hydrogen peroxide with the —SH groups of cysteine is well established (Means and Feeney, 1971).

Brennan and Anderson (1980) demonstrated that the presence of hydrogen peroxide can stimulate the dark inactivation of glucose-6-phosphate dehydrogenase. This, presumably, would occur via an oxidation of sulfhydryl groups (—SH) on the enzyme.

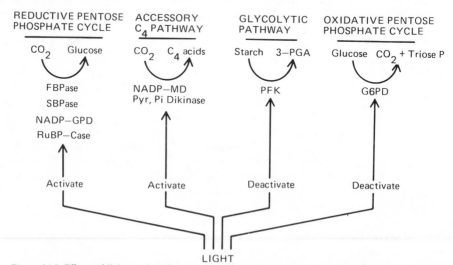

Figure 14.5. Effects of light modulation on regulatory enzymes of four metabolic pathways (from Buchanan, 1980; courtesy of Annual Reviews.) RuBP Case = RuBP carboxylase, Pyr, Pi Dikinase = pyruvate ortho-phosphate dikinase, PFK = phosphofructokinase

TABLE 14.2. ATP imbalance created during electron transport from water to NADP[a]

Photosynthesis	ATP/NADPH requirement	Assumed ATP/2e ratio of NADP reduction (ATP imbalance in μmol mg^{-1} chlorophyll h^{-1})		
		1.0	1.33	2.0
Via C_3 cycle	1.5	-150	-50	$+150$
Via C_4 pathway	2.5	-450	-350	-150

[a] The rate of CO_2 reduction is assumed to be 150 μmol mg^{-1} chlorophyll h^{-1} and coupling ratios are as indicated. Minus signs show ATP deficiency, plus signs ATP surplus.
(From Huber, 1976; courtesy of Plenum Press.)

The *in vivo* production of hydrogen peroxide

The studies described above were all carried out with added or exogenous hydrogen peroxide. It has become clear now, however, that hydrogen peroxide is produced 'normally' in photosynthesizing leaf cells (Halliwell, 1981).

The photoreduction of molecular oxygen by isolated chloroplasts was discovered by Mehler (1951). The work of several laboratories (Egneus, Heber and Kirk, 1975; Radmer and Kok, 1976; Heber *et al.*, 1978; Ziem-Hanck and Heber, 1980; Steigner and Beck, 1981) has established that oxygen is reduced in the light in whole leaves and in algal cells as well as in isolated chloroplasts. ATP synthesis is thought to occur as a result of this photoreduction; the process is called pseudocyclic photophosphorylation. The linear electron transport pathway may not provide an adequate ATP to NADPH ratio for CO_2 fixation (*see Table 14.2* for sample values) and the pseudocyclic process may boost ATP levels to the ATP/NADPH ratio of 1.5 required for the operation of the C_3 cycle. The photoreduction of molecular oxygen is a one-electron transfer, however, and the product of the reaction is therefore the superoxide free radical of oxygen, $\cdot O_2^-$. The superoxide radical has been shown to be 'dismutated' to oxygen and hydrogen peroxide by a cyanide insensitive superoxide dismutase located within the chloroplast (Jackson *et al.*, 1978). These events are outlined in *Figure 14.6*. Oxygen and hydrogen peroxide can also react with each other to form the hydroxyl radical, $\cdot OH$, which is a highly toxic species to both lipids and nucleic acids (Halliwell, 1981). Hydrogen peroxide can also inactivate the light-modulated bisphosphatases as was described above. Nakano and Asada (1981) have calculated that if even 10% of the electrons from photosystem I were diverted to molecular oxygen, hydrogen peroxide production would occur at 10 μM s^{-1}. This amount of H_2O_2 is sufficient to inhibit CO_2 fixation by 50% (Kaiser, 1976, 1979). Thus, it is essential that the chloroplast possess an efficient removal system for hydrogen peroxide. Nakano and Asada (1980, 1981) have demonstrated that such a system does exist. It is shown in *Figure 14.7* and involves glutathione and ascorbate, both of which are known to exist at high concentrations within the chloroplast.

Halliwell (1981), in his discussion of oxygen–chloroplast interactions, makes the suggestion that the levels of superoxide dismutase and ascorbate in the chloroplast are sufficient to 'neutralize' the normal amount of $\cdot O_2^-$ produced *in vivo*. Any increased production of $\cdot O_2^-$ would lead to a persistence of hydrogen peroxide at concentrations inhibitory to carbon fixation. It is my proposal that one indirect consequence of SO_2 exposure for the plant is just such an overproduction of $\cdot O_2^-$, and hence H_2O_2.

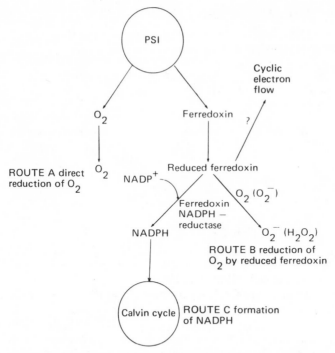

Figure 14.6. Electron flow from reduced Photosystem I. The univalent reduction of molecular oxygen by reduced ferredoxin is indicated (Route B) (from Halliwell, 1981; courtesy of Clarendon Press)

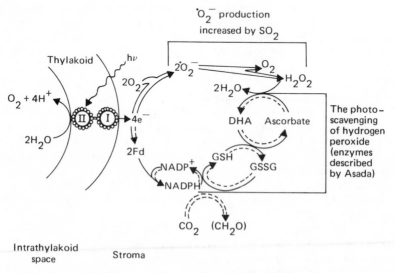

Figure 14.7. $\cdot O_2^-$ production and the photoscavenging of hydrogen peroxide within the chloroplast. Fd = ferredoxin, GSH = reduced glutathione, GSSG = oxidized glutathione, DHA = dehydroascorbate (after Nakano and Asada, 1981; courtesy of Japanese Society of Plant Physiologists)

The reactivity of sulfite with proteins

Sulfur dioxide in solution at the pH of the chloroplast exists mainly as sulfite. Disulfide bonds such as those which exist in proteins are known to react with sulfite to form S-sulfonates

$$R\text{---}S\text{---}S\text{---}R + SO_3^{2-} \rightarrow R\text{---}S\text{---}SO_3^{2-} + R\text{---}S\text{---}$$

(Cecil and McPhee, 1955). A reaction such as this would result in a disruption of the protein's tertiary structure and in the case of an enzyme molecule, inactivation would most probably occur. The reaction can be reversed by thiols and native polypeptide chains can be regenerated in this manner (Dixon and Wardlaw, 1960). It should be noted also that this reaction results in the production of an R—S— group which may react again with another SO_3^{2-}, if available. If not, a new —SH group has been generated.

Sulfur dioxide and light modulation

Direct interactions

Inhibition

Conformational changes in protein structure can be blocked by reaction with sulfite. Changes such as these are thought to occur during light modulation of the chloroplast enzymes (discussed above). Membrane-bound and possibly stromal dithiol groups generated in the light are known to participate in the modulation mechanism. Ziegler, Marewa and Schoepe (1976) showed that the presence of sulfite blocked the DTT-induced transformation of NADP-dependent glyceraldehyde-3-phosphate dehydrogenase (NADP-GPD) to its activated form. Anderson and Avron (1976) demonstrated the sulfite sensitivity of a light-generated chloroplast membrane-bound reductant necessary for the modulation of glucose-6-phosphate dehydrogenase and NADP-linked malic dehydrogenase. (Recall that glucose-6-phosphate dehydrogenase is the first enzyme in the oxidative pentose phosphate pathway and is inoperative in the light.) Ziegler (1977) performed the crucial experiment to determine whether atmospheric SO_2 and/or sulfite in solution binds to chloroplast membranes. She found that exposure of spinach (*Spinacia oleracea*) leaves to $^{35}SO_2$ or $^{35}SO_3^{2-}$ for 1 h resulted in rapid and preferential association with chloroplast membranes. *This did not occur when* $^{35}SO_4^{2-}$ *was applied to the leaves.* Her results are shown in *Table 14.3.* Ziegler and Hampp (1977) presented evidence that the light-induced generation of chloroplast membrane-bound —SH groups is a prerequisitie for this preferential association of sulfite with the membranes. They place this binding in the context of the incorporation

TABLE 14.3. Percentage distribution of specific radioactivity in osmotically ruptured chloroplasts after discontinuous sucrose density gradient centrifugation

	Chloroplast lamellae	Chloroplast envelopes	Supernatant (stroma-bound ^{35}S +free $^{35}SO_3^{2-} + {}^{35}SO_4^{2-}$)
Feeding of $^{35}SO_3^{2-}$	57.2	6.4	33.1
Feeding of $^{35}SO_4^{2-}$	21.0	11.5	61.8

(From Ziegler, 1977; courtesy of Springer-Verlag.)

of exogenous sulfite into the sulfate assimilation pathway of chloroplasts through binding to a carrier —SH protein known to occur in the membranes (Schiff and Hodson, 1973). Anderson and Duggan (1977) extended the study of the effects of SO$_2$ and sulfite on light modulation to several other enzymes. They found that the activation of NADP-GPD was inhibited by 200 μM sulfite, but the activation of sedoheptulose bisphosphatase was stimulated by 5 mM sulfite. The activation of NADP-GPD was also inhibited in pea (*Pisum sativum*) seedlings which had been exposed to 5 ppm SO$_2$ for 1 h. Anderson and Duggan were unable, however, to detect any effect of sulfite (5 mM) on light modulation of four other chloroplast enzymes.

Mohamed and Anderson (1981) have succeeded in removing a protein from pea chloroplast membranes which is both partially sulfite-sensitive and essential for light modulation. They can reconstitute light activation by adding the sulfite-sensitive membrane protein back to depleted membranes (*see above*).

Thus, the light modulation system of chloroplasts is sulfite-sensitive and one protein which is responsible for this sensitivity can be removed from the membranes. Atmospheric SO$_2$ preferentially binds to chloroplast membranes (Ziegler, 1977). Interaction of sulfite with either the membrane site of the generation of dithiol groups or the regulatory enzymes themselves could result in an inhibition of the light

TABLE 14.4. Effect of sulphite on light activation of chloroplast FBPase. Chloroplast samples from Hark and Beeson cultivars, were prepared, light-treated and FBPase activity determined

Sample	FBPase activity[a]		
	Light-treated	Dark	L/D
Reconstituted Hark[b] chloroplasts	3.9	2.8	1.39
Reconstituted Hark chloroplasts +5 mM sulphite	2.7	5.9	0.46
Reconstituted Hark chloroplasts +1 mM sulphite	2.7	3.9	0.69
Reconstituted Beeson[c] chloroplasts	6.7	4.0	1.68
Reconstituted Beeson chloroplasts +1 mM sulphite	8.0	5.4	1.49

Stromal protein concentrations: Hark, 5.6 μg ml^{-1}; Beeson, 7.0 μg ml^{-1}.
[a] nmol P$_i$ produced min^{-1}.
[b] SO$_2$ susceptible.
[c] SO$_2$ resistant.
(From Alscher-Herman, 1982a; courtesy of Applied Science Publishers.)

TABLE 14.5. Effect of 1 mM sulfite on *in vitro* light activation of FBPase in broken chloroplasts reconstituted from Hark and Beeson

Treatment	Sample		Condition		
	Stroma	Membranes	Light	Dark	L/D
			FBPase activity[a]		
Control	Hark[b]	Beeson[c]	2.0	0.9	2.2
1 mM sulfite	Hark	Beeson	1.6	1.2	1.3
Control	Beeson	Hark	0.8	0.5	1.6
1 mM sulfite	Beeson	Hark	0.7	1.6	0.4

[a] nmol F6P produced min^{-1}.
Chlorophyll concentration in reaction mixtrues: 5 μg chl ml^{-1}.
Stromal protein concentrations: Beeson, 8.5 μg ml^{-1}; Hark, 9.5 μg ml^{-1}.
[b] SO$_2$-susceptible.
[c] SO$_2$-resistant.

TABLE 14.6. Effect of the presence of sulfite on light activation of membrane-bound FBPase from soybean chloroplasts

Sample	Treatment	FBPase activity[a]	
		Dark	Light
Hark	Control	0.25	1.7
Hark	5 mM sulfite	1.08	1.6
Beeson	Control	0.2	1.8
Beeson	5 mM sulfite	1.6	1.5

[a] nmol P_i produced μg^{-1} chl.

modulation process. Mohamed and Anderson (1981) have demonstrated that the chloroplast membrane contains such a site. Ziegler's work (Ziegler, Marewa and Schoepe, 1976) with DTT-induced activation of NADP-GPD demonstrates that sulfite can act directly on a regulatory enzyme also.

I have carried out a study of in vitro sulfite effects (Alscher-Herman, 1982a) and of SO_2 in vivo (Alscher-Herman and Jeske, in preparation) on light activation of alkaline FBPase in two cultivars, 'Beeson' and 'Hark', of soybean (Glycine max) whose SO_2 susceptibilities in the field have been established (Amundson, in preparation). The effects of sulfite on light activation of chloroplast FBPase are shown in Table 14.4. FBPase was activated by light in each case. The presence of sulfite had little effect on the light activation process in Beeson, whereas light activation was completely inhibited in Hark. In fact, dark values for Hark FBPase were higher than those obtained for illuminated samples. Sulfite was found to have no effect on the activity of FBPase present in the stromal (soluble) phase of the soybean chloroplasts in the absence of chloroplast membranes. Thus, the differential sulfite-susceptibility appears to reside within the light modulation process and hence perhaps with the membranes, rather than with the enzyme itself in the soluble or stromal phase of the chloroplast. To test this hypothesis, a 'cross-over' experiment was performed, in which stroma from one cultivar was mixed with chloroplast membranes from the other. The effect of sulfite on light activation of FBPase in this mixed system was then tested. The results are shown in Table 14.5. Sulfite (1 mM) completely inhibited light activation of FBPase when Beeson stroma was incubated with Hark membranes. Some activation was detected in the presence of 1 mM sulfite when Hark stroma was incubated with Beeson membranes. This result should be contrasted with those of the previous table (Table 14.4) where it is shown that 1 mM sulfite completely inhibited light activation of FBPase in Hark chloroplasts. Thus, differential sulfite susceptibility is associated with the soybean chloroplast membranes and not with the soluble stromal phase of that organelle. The sulfite-sensitive membrane protein described by Mohamed and Anderson (1981) and Heuer, Hansen and Anderson (1982) is a likely candidate for this site.

Another factor must also be taken into consideration at this point. The results of Ben-Bassat and Anderson (1981), Fischer and Latsko (1979), Scheibe and Beck (1979) and Alscher-Herman (1982b) have demonstrated the existence of membrane-bound enzymes in the chloroplasts of soybean, pea and spinach. Neither the binding process nor its functional significance is yet fully understood. It may represent an essential step in the light modulation process, i.e. the enzyme is reductively activated or inactivated while bound and subsequently released to the stroma. Membrane-bound FBPase of soybean chloroplasts was stimulated by sulfite (Table 14.6) in the light and in the dark. Thus, an association of the enzyme with the chloroplast membrane alters its properties

TABLE 14.7. The effect of illumination of FBPase activity associated with soybean thylakoids

Sample	Condition		L/D
	Light	Dark	
		(Membrane-associated FBPase activity nM FBP produced min^{-1} mg^{-1} chl)	
Washed thylakoids from illuminated intact chloroplasts	90	61	1.48
Washed thylakoids illuminated directly	128	79	1.62

The activity of FBPase associated with illuminated thylakoids was compared with that of thylakoids prepared from illuminated chloroplasts. Initially, intact chloroplasts or washed thylakoids were isolated in darkness from mature soybean leaves, then illuminated with 900 μE m^{-2} s^{-1} PAR for 5 min at 25 °C. A second set of washed thylakoids was then prepared from illuminated intact chloroplasts.
(From Alscher-Herman, 1982b; courtesy of American Society of Plant Physiologists.)

with respect to sulfite sensitivity. It is possible to activate this bound soybean enzyme without any added soluble factors as shown in *Table 14.7*. Activation of the stromal soybean FBPase is sulfite-sensitive (*see* the results shown in *Table 14.4*). I propose that one basis for the sulfite sensitivity of light activation is due to the binding of sulfite to the chloroplast membrane at a site which is crucial for activation.

Stimulation of carbon fixation by sulfur dioxide and sulfite — consequences for plant growth

Low SO$_2$ concentrations (<0.2 ppm) were long ago shown to cause increases in yield (Thomas *et al.*, 1943) and in net photosynthetic rates (Katz, 1949). Continuous exposure to low concentrations of SO$_2$ are also known to bring about premature senescence (Guderian, 1977). Libera, Ziegler and Ziegler (1973) demonstrated that exposure of isolated spinach chloroplasts to low concentrations of sulfite (below 1 mM) produced a stimulation of carbon fixation. Higher levels of sulfite (up to 3 mM) stimulated photosynthetic electron transport but inhibited carbon fixation. They were able to show that the stimulation by low concentrations of sulfite occurred at the bisphosphatase step. Ziegler's group extended this approach to the alga *Chlorella vulgaris* where they were able to show that the presence of low concentrations of sulfite (<1 mM) increased growth rates (expressed as cell number, protein and chlorophyll yield) even under conditions of sulfate sufficiency (*Figure 14.8*). This increased yield was accompanied by an increase in the rate of carbon fixation. At higher sulfite concentrations rates of carbon fixation were still higher than those of the control. However, yield had dropped below control levels (*Figure 14.9*). Miszalski and Ziegler (1979) showed that exposure of whole spinach plants to 0.67 ppm (1.8 mg m^{-3}) SO$_2$ for 1 h produced increases in chloroplast membrane thiol groups and an increase in the light activation of NADP-GPD. Paul and Bassham (1978) demonstrated a stimulation of carbon fixation by sulfite in isolated cells of the opium poppy (*Papaver somniferum*). Pierre (1977) and Pierre and Queiroz (1981, 1982) showed that exposing whole bean plants over a long term to low concentrations of SO$_2$ (0.1 ppm) increased rates of activity of several enzymes present in the soluble phase of leaf extracts. An increase in serine levels was also observed. These plants also became prematurely senescent as a result of the SO$_2$ treatment. Pierre did not measure leaf growth rates nor rates of protein production in her experiments. However, Ashenden and Mansfield (1978) demonstrated significant reductions in dry weight in three species of grass after prolonged exposure to SO$_2$ (0.068 ppm). Leaf area was not significantly affected by the

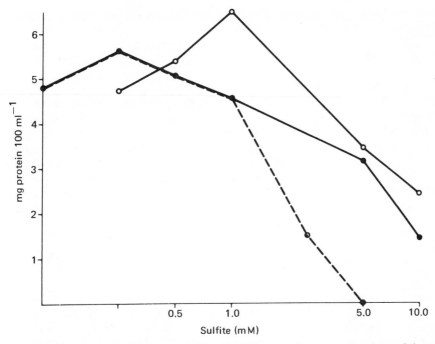

Figure 14.8. Effect of the presence of sulfite on protein production in *Chlorella vulgaris*. Cultures were grown for four days in ●—● sulfate-sufficient or ○—○ sulfate-deficient medium (from Soldatini *et al.*, 1978; courtesy of Springer-Verlag)

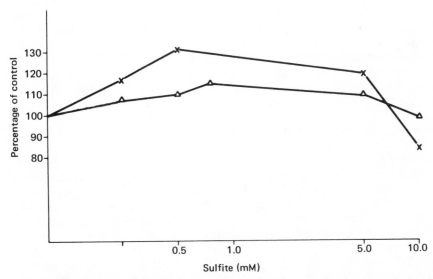

Figure 14.9. Effect of the presence of sulfite on $^{14}CO_2$ fixation and O_2 evolution per aliquot of protein during steady state photosynthesis. Sulfite was present during growth of the culture but was absent during the assay. ×—× $^{14}CO_2$ fixation, △—△ O_2 evolution (from Soldatini *et al.*, 1978; courtesy of Springer-Verlag)

prolonged exposure. This suggests that the grass leaves had undergone premature senescence as a result of exposure to SO_2.

Results obtained in our laboratory show a stimulation under certain conditions by 100 μM sulfite of carbon fixation in isolated cucumber cells (*Cucumis sativa*) (Rothermel and Alscher-Herman, in preparation). Results from INRA-Montardon showed that exposure of cucumber plants to 0.05 ppm SO_2 for six weeks produced premature senescence of the leaves (Bonte and Bonte, personal communication).

Taken together, these data suggest that sulfite at low concentrations can stimulate light activation through increasing the concentrations of membrane-bound thiol groups. This in turn can bring about higher carbon fixation rates, the consequence of which for algal cells can be either increased or decreased productivity. In higher plants, it appears that a consequence of this increased rate of metabolism can be premature senescence. Serine has been implicated in metabolic changes accompanying senescence (Noodén, 1980). Perhaps the increases in serine concentrations which were detected in fumigated bean leaves (*Phaseolus vulgaris*) (Pierre and Queiroz, 1982) are an expression of the shift towards senescence. Clearly, this phenomenon is not understood as yet.

Indirect effects

It is well established that some of the excess sulfur which is introduced into the leaf cell from atmospheric SO_2 is oxidized and converted to sulfate (Weigl and Ziegler, 1962). In this form it is relatively harmless. Miller and Xerikos (1979) demonstrated a correlation between 'residence time' of sulfite in various cultivars of soybean and their respective resistance to SO_2, estimated on the basis of foliar injury. However, the work of the INRA group at Montardon (de Cormis, 1968; de Cormis and Bonte, 1970) has established that some of the excess sulfur may also be reduced and be eliminated from the plant as H_2S gas. The reports of Wilson, Bressan and Filner (1978), Spaleny (1977) and Winner *et al.* (1981) have confirmed these findings. Asada (1967) has partially characterized the enzyme which catalyzes this reduction and Tamura and Itoh (1974) showed that photosynthetically generated reductant (as reduced ferredoxin) is the physiological electron donor for the process. Sawhney and Nicholas (1975) and Silvius *et al.* (1976) demonstrated that the process takes place within the chloroplast.

Thus, sulfite originating from atmospheric SO_2 may be either oxidized or reduced. Reduction takes place within the chloroplast. Asada and Kiso (1973) demonstrated that light-dependent sulfite oxidation also can take place within the chloroplast. Light-independent sulfite oxidation is known to take place within the mitochondrion (Tager and Rautanen, 1955; Ballantyne, 1977). Sulfite oxidation by chloroplasts is a chain reaction which is initiated by the superoxide radical $^{\cdot}O_2^-$ (Asada and Kiso, 1973; Asada, 1980), which is produced as a result of the univalent reduction of O_2 from ferredoxin as described above. This occurs when CO_2 is limited or when carbon fixation cannot take place due to an imbalance of ATP/NADPH. This sequence involves the production of $^{\cdot}OH$, a toxic radical species. The chain can be terminated by the action of superoxide dismutase, producing hydrogen peroxide. Superoxide dismutase was shown to inhibit the photo-oxidation of sulfite by spinach chloroplasts, thus providing evidence for this proposed sequence (Asada, 1980).

The production of hydrogen peroxide through the action of superoxide dismutase has potentially deleterious consequences for the light modulation system. The increased levels of H_2O_2 produced as superoxide dismutase terminates the chain reaction of sulfite photo-oxidation may oxidize light activated enzymes such as FBPase. A point may be reached at which carbon fixation cannot proceed because of the inhibition at rate-limiting steps, such as the one catalyzed by FBPase, that lead to the ultimate regeneration of RuBP. The degree to which this would occur would be partially predicated on the rate at which H$_2$O$_2$ was removed by the H$_2$O$_2$-photoscavenging system described by Nakano and Asada (1980, 1981) (*Figure 14.7*) and also by the efficiency with which the oxidized form of the enzyme could be re-reduced via the light modulation system. Halliwell (1981) has proposed that the light modulation system is not in itself a regulatory system but 'exists to generate and protect the reduced forms of certain Calvin cycle enzymes, which are then controlled by other means'. If this were so, light modulation exists as a type of safety valve for the regulation of carbon metabolism within the green leaf cell. The photo-oxidation of sulfite puts an extra burden on the system which 'normally' functions to re-reduce enzyme oxidized by levels of H$_2$O$_2$ 'normally' produced as a consequence of $\cdot$O$_2^-$ production. The removal of sulfite by the oxidative route, in other words, is in itself a metabolic stress for the leaf cell. Whether the light modulation system functions as a safety valve, or whether it has the role of an 'on-off switch' (summaries by Anderson, 1979; Buchanan, 1980), the photo-oxidation of sulfite has potentially harmful consequences for photosynthetic carbon metabolism.

The mode of action of SO$_2$ on light modulation — a summary (*Figure 14.10*)

Sulfite can stimulate light modulation by increasing the availability of free thiol groups on the chloroplast membrane. This is caused by sulfitolysis of a disulfide protein by the reaction shown earlier where a S-sulfonate and a thiol group are produced.

Formation of sulfonate groups from disulfide bridges on the regulatory enzymes themselves, on the membrane protein and/or soluble factors involved in light modulation would block the process.

Sulfite detoxification via the oxidative pathway can result in the production of hydrogen peroxide which de-activates light-modulated enzymes such as FBPase and

A. Stimulation of regulatory enzymes through increase of free —SH groups.

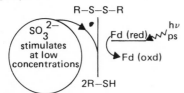

B. Inhibition of regulatory enzymes.
   (1) Sulfonate formation from disulfide bridges on regulatory enzymes.
   (2) Oxidation of —SH groups on regulatory enzymes by H$_2$O$_2$. H$_2$O$_2$ formed as a result of sulfite oxidation and SOD activity.

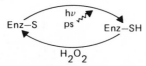

Figure 14.10. Proposed effects of sulfite and hydrogen peroxide on light modulation

activates enzymes which are normally active in the dark such as glucose-6-phosphate dehydrogenase. The degree to which this occurs would depend on the relative efficiencies of both the H$_2$O$_2$ scavenging system and the light modulation system itself.

Possible sources of sulfur dioxide tolerance in plants

Miller and Xerikos (1979) were able to correlate the rapidity with which sulfite was detoxified with relative resistances of several soybean cultivars to SO$_2$. Results obtained in our laboratory (Alscher-Herman and Jeske, in preparation) show that sulfite is accumulated to a greater extent in the leaves of a less SO$_2$-tolerant cultivar after a brief SO$_2$ exposure than it is in the leaves of a more tolerant cultivar. Yet sulfite detoxification via the oxidative pathway is potentially harmful to the cell (*see above*). The relative degree of tolerance to SO$_2$ which any particular cultivar or species possesses may therefore be due either to a more efficient reductive detoxification pathway for SO$_2$ ('Expulsion Route' of *Figure 14.11*) or to a more efficient photoscavenging system for hydrogen peroxide, an end-product of the oxidative detoxification pathway ('Storage Route' of *Figure 14.11*). The rate of regeneration of membrane dithiol groups may also contribute (*Table 14.8*). The longer the residence time of sulfite within the cell in general, and within the chloroplast in particular, the

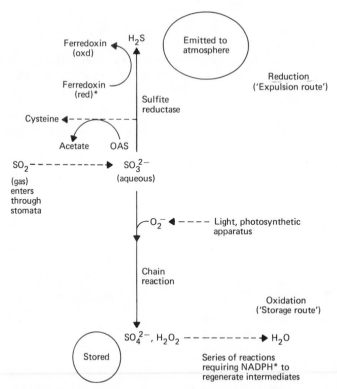

Figure 14.11. Possible fates of SO$_2$ within the chloroplast. The two steps requiring the furnishing of reductant generated as a result of photosynthesis are indicated by *. All of the conversions indicated here have been shown to occur either *in vivo* or *in vitro*. Their relative importance with regard to SO$_2$ tolerance remains to be investigated

TABLE 14.8. Chloroplast metabolism: bases for SO$_2$ tolerance

(1) Rate of regeneration of membrane dithiol groups.
(2) Rate of photoscavenging of hydrogen perioxide — SOD, glutathione, ascorbate.
(3) Activity (or inducibility) of sulfite reductase.

more likely it is that disturbances in systems such as that of light modulation will occur. High concentrations of sulfite may cause an inhibition of carbon fixation. Lower concentrations may bring about a stimulation which can bring premature senescence of the whole plant in its wake. The mechanism by which the shift to senescence is brought about is not yet understood. Further research investigating the role of serine and of rates of carbon fixation in the metabolic changes underlying senescence is warranted.

References

ALSCHER-HERMAN, R. Effect of sulphite on light activation of fructose-1,6-bisphosphatase in two cultivars of soybean, *Environmental Pollution Series A: Ecological and Biological* 27, 83–96 (1982a)

ALSCHER-HERMAN, R. Chloroplast alkaline fructose-1,6-bisphosphatase exists in a membrane-bound form, *Plant Physiology* 70, 728–734 (1982b)

ANDERSON, L. E. Interaction between photochemistry and activity of enzymes. In *Encyclopedia of Plant Physiology: New Series Vol. 5, Photosynthesis I*, (Eds. A. TREBST and M. AVRON), pp. 271–281, Springer-Verlag, Heidelberg (1979)

ANDERSON, L. E. Activation of pea leaf chloroplast sedoheptulose-1,7-diphosphate phosphatase by light and dithiothreitol, *Biochemical and Biophysical Research Communications* 59, 907–913 (1974)

ANDERSON, L. E. and DUGGAN, J. X. Inhibition of light modulation of chloroplast enzyme activity by sulfite, *Oecologia* 28, 147–151 (1977)

ANDERSON, L. E. and AVRON, M. Light modulation of enzyme activity in chloroplasts. Generation of membrane-bound vicinal dithiol groups by photosynthetic electron transport, *Plant Physiology* 57, 209–213 (1976)

ANDERSON, L. E. and DUGGAN, J. X. Light modulation of glucose-6-phosphate dehydrogenase. Partial characterization of the light inactivation system and its effects on the chloroplastic and cytoplasmic forms of the enzyme, *Plant Physiology* 58, 135–139 (1976)

ANDERSON, L. E., NG, T. C. L. and PARK, K. E. Y. Inactivation of pea leaf chloroplastic and cytoplasmic glucose-6-phosphate dehydrogenase by light and dithiothreitol, *Plant Physiology* 53, 835–839 (1974)

ANDERSON, L. E. and LIM, T.-C. Chloroplast glyceraldehyde-3-P-dehydrogenase: light-dependent changes in the enzyme, *FEBS Letters* 27, 189–191 (1972)

ASADA, K. Purification and properties of a sulfite reductase from leaf tissue, *Journal of Biological Chemistry* 242, 3646–3654 (1967)

ASADA, K. Formation and scavenging of superoxides in chloroplasts, with relation to injury by sulfur dioxide, *National Institute for Environmental Studies, Research Report No. 11*, 165–169 (1980)

ASADA, K. and KISO, K. Initiation of aerobic oxidation of sulfite by illuminated spinach chloroplasts, *European Journal of Biochemistry* 33, 253–257 (1973)

ASHENDEN, T. W. and MANSFIELD, T. A. Extreme pollution sensitivity of grasses when SO$_2$ and NO$_2$ are present in the atmosphere together, *Nature (London)* 273, 142–143 (1978)

ASHTON, A. R. and ANDERSON, L. E. Resolution of the light-dependent modulation system of pea chloroplasts, *Biochimica et Biophysica Acta* 638, 242–249 (1981)

BAIER, D. and LATZKO, E. Properties and regulation of C-1 fructose-1,6-diphosphatase from spinach chloroplasts, *Biochemica et Biophysica Acta* 396, 141–147 (1975)

BALLANTYNE, D. J. Sulphite oxidation by mitochondria from green and etiolated peas, *Phytochemistry* 16, 49–50 (1977)

BASSHAM, J. A. Control of photosynthetic carbon metabolism, *Symposia of the Society for Experimental Biology* 27, 461–483, Cambridge University Press, Cambridge (1973)

BEN-BASSAT, D. and ANDERSON, L. E. Light-induced release of bound glucose-6-phosphate dehydrogenase to the stroma in pea chloroplasts, *Plant Physiology* 68, 279–283 (1981)

BRENNAN, T. and ANDERSON, L. E. Inhibition by catalase of dark-mediated glucose-6-phosphate dehydrogenase activation in pea chloroplasts, *Plant Physiology* **66**, 815–817 (1980)

BUCHANAN, B. B. Role of light in the regulation of chloroplast enzymes, *Annual Review of Plant Physiology* **31**, 341–433 (1980)

BUCHANAN, B. B., KALBERER, P. P. and ARNON, D. I. Ferredoxin-activated fructose diphosphatase in isolated chloroplasts, *Biochemical and Biophysical Research Communications* **29**, 74–79 (1967)

CALVIN, M. and BASSHAM, J. A. *The Photosynthesis of Carbon Compounds*, Benjamin Company, New York (1962)

CECIL, R. and MCPHEE, J. R. A kinetic study of the reactions on some disulphides with sodium sulphite, *Biochemical Journal* **60**, 496 (1955)

CHAMPIGNY, M. L. and BISMUTH, E. Role of photosynthetic electron transfer in light activation of Calvin cycle enzymes, *Physiologia Plantarum* **36**, 95–100 (1976)

CHARLES, S. A. and HALLIWELL, B. Properties of freshly purified and thiol-treated spinach chloroplast fructose bisphosphatase, *Biochemical Journal* **185**, 689–693 (1980a)

CHARLES, S. A. and HALLIWELL, B. Effect of hydrogen peroxide on spinach (*Spinacia oleracea*) chloroplast fructose bisphosphatase, *Biochemical Journal* **189**, 373–376 (1980b)

CHARLES, S. A. and HALLIWELL, B. Light activation of fructose bisphosphatase in isolated spinach chloroplasts and deactivation by hydrogen peroxide, *Planta* **151**, 242–246 (1981a)

CHARLES, S. A. and HALLIWELL, B. Light activation of fructose bisphosphatase in photosynthetically competent pea chloroplasts, *Biochemical Journal* **200**, 357–363 (1981b)

DE CORMIS, L. Dégagement d'hydrogène sulfuré par des plantes soumises à une atmosphère contenant de l'anhydride sulfureux, *Comptes Rendus Hebdomadaires des Seances de l'Academie des Sciences, Série D: Sciences Naturelles* **266**, 683–685 (1968)

DE CORMIS, L. and BONTE, J. Étude du dégagement d'hydrogène sulfuré par des feuilles de plantes ayant recue du dioxyde de soufre, *Comptes Rendus Hebdomadaires des Seances de l'Academie des Sciences, Série D: Sciences Naturelles* **270**, 2078–2080 (1970)

DIXON, G. H. and WARDLAW, A. C. Regeneration of insulin activity from the separated and inactive A and B chains, *Nature (London)* **188**, 721 (1960)

EGNEUS, H., HEBER, U. and KIRK, M. Reduction of oxygen by the electron transport chain of chloroplasts during assimilation of carbon dioxide, *Biochimica et Biophysica Acta* **408**, 252–268 (1975)

FISCHER, K. H. and LATZKO, E. Chloroplast ribulose-5-phosphate kinase: light-mediated activation and detection of both soluble and membrane-associated activity, *Biochemical and Biophysical Research Communications* **89**, 300–306 (1979)

GUDERIAN, R. *Air Pollution*, Springer-Verlag, Berlin (1977)

HÄLLGREN, J.-E. Physiological and biochemical effects of sulfur dioxide on plants. In *Sulfur in the Environment Part II*, (Ed. J. O. NRIAGU), pp. 164–209, John Wiley & Sons, New York (1978)

HALLIWELL, B. *Chloroplast Metabolism*, Clarendon Press, Oxford (1981)

HEBER, U. Energy coupling in chloroplasts, *Journal of Bioenergetics and Biomembranes* **8**, 157–172 (1976)

HEBER, U., EGNEUS, H., HANCK, U., JENSEN, J. and KÖSTER, S. Regulation of photosynthetic electron transport and photophosphorylation in intact chloroplasts and leaves of *Spinacia oleracea* L., *Planta* **143**, 41–49 (1978)

HELDT, H. W., PORTIS, A. R., McCLILLEY, R., MOSBACH, A. and CHON, C. J. Assay of nucleotides and other phosphate-containing compounds in isolated chloroplasts by ion exchange chromatography, *Analytical Biochemistry* **101**, 278–287 (1980)

HEUER, B., HANSEN, M. and ANDERSON, L. E. Light modulation of phosphofructokinase in pea leaf chloroplasts, *Plant Physiology* **69**, 1404–1406 (1982)

JACKSON, C., DENCH, J., MOORE, A. L., HALLIWELL, B., FOYER, C. H. and HALL, D. O. Subcellular localization and identification of superoxide dismutase in the leaves of higher plants, *European Journal of Biochemistry* **91**, 339–344 (1978)

KAISER, W. The effect of hydrogen peroxide on CO₂ fixation of isolated intact chloroplasts, *Biochimica et Biophysica Acta* **440**, 476–482 (1976)

KAISER, W. Reversible inhibition of the Calvin cycle and activation of oxidative pentose phosphate cycle in isolated intact chloroplasts by hydrogen peroxide, *Planta* **145**, 377–382 (1979)

KATZ, M. Sulfur dioxide in the atmosphere and its relation to plant life, *Industrial and Engineering Chemistry* **41**, 2450–2465 (1949)

KELLY, G. J., ZIMMERMAN, G. and LATZKO, E. Light induced activation of fructose-1,6-bisphosphate in isolated intact chloroplasts, *Biochemical and Biophysical Research Communications* **70**, 193–199 (1976)

LEEGOOD, R. C. and WALKER, D. A. Modulation of fructose bisphosphatase activity in intact chloroplasts, *FEBS Letters* **116**, 21–24 (1980)

LEEGOOD, R. C. and WALKER, D. A. Photosynthetic induction in wheat protoplasts and chloroplasts. Autocatalysis and light activation of enzymes, *Plant, Cell and Environment* **4**, 59–66 (1981)

LENDZIAN, K. J. Modulation of glucose-6-phosphate dehydrogenase by NADPH, NADP$^+$ and dithiothreitol at variable NADPH/NADP$^+$ ratios in an illuminated reconstituted spinach (*Spinacia oleracea* L.) chloroplast system, *Planta* **148**, 1–6 (1980)

LIBERA, W., ZIEGLER, H. and ZIEGLER, I. Förderung der Hillreaktion und der CO_2-fixierung in isolierten Spinatchloroplasten durch niedere sulfit konzentrationen, *Planta* **109**, 269–279 (1973)

McLAUGHLIN, S. B., SHRINER, D. S., MCCONATHY, R. K. and MANN, L. K. The effects of SO_2 dosage kinetics and exposure frequency on photosynthesis and transpiration of kidney beans (*Phaseolus vulgaris* L.), *Environmental and Experimental Botany* **19**, 174–191 (1979)

MEANS, G. E. and FEENEY, R. E. *Chemical Modification of Proteins*, Holden-Day, Inc., San Francisco (1971)

MEHLER, A. H. Studies on reactions of illuminated chloroplasts. II. Stimulation and inhibition of the reaction with molecular oxygen, *Archives of Biochemistry and Biophysics* **34**, 339–351 (1951)

MILLER, J. E. and XERIKOS, P. B. Residence time of sulphite in SO_2 'sensitive' and 'tolerant' soybean cultivars, *Environmental Pollution* **18**, 259–264 (1979)

MISZALSKI, Z. and ZIEGLER, I. Increase in chloroplastic thiol groups by SO_2 and its effect on light modulation of NADP-dependent glyceraldehyde 3-phosphate dehydrogenase, *Planta* **145**, 383–387 (1979)

MOHAMED, A. H. and ANDERSON, L. E. Extraction of chloroplast light effect mediator(s) and reconstitution of light activation of NADP-linked malate dehydrogenase, *Archives of Biochemistry and Biophysics* **209**, 606–612 (1981)

MÜLLER, R. N., MILLER, J. E. and SPRUGEL, D. G. Photosynthetic response of field-grown soybeans to fumigations with sulphur dioxide, *Journal of Applied Ecology* **16**, 567–576 (1979)

MÜLLER, B., ZIEGLER, I. and ZIEGLER, H. Licht induzierte, reversible aktivitätssteigurung der NADP-abthängigen glycerinaldehyd-3-phosphat-dehydrogenase in chloroplasten, *European Journal of Biochemistry* **9**, 101–106 (1969)

NAKANO, Y. and ASADA, K. Spinach chloroplasts scavenge hydrogen peroxide on illumination, *Plant and Cell Physiology* **21**, 1295–1307 (1980)

NAKANO, Y. and ASADA, K. Hydrogen peroxide is scavenged by ascorbate-specific peroxidase in spinach chloroplasts, *Plant and Cell Physiology* **22**, 867–880 (1981)

NOODÉN, L. Senescence in the whole plant. In *CRC Series in Aging — Senescence in Plants* (Ed. K. V. THIMANN), pp. 219–259, CRC Press, Boca Raton, FL (1980)

PAUL, J. S. and BASSHAM, J. A. Effect of sulfite on metabolism in isolated mesophyll cells from *Papaver somniferum*, *Plant Physiology* **62**, 210–214 (1978)

PEDERSON, T. A., KIRK, M. and BASSHAM, J. Light-dark transients in levels of intermediate compounds during photosynthesis in air adapted Chlorella, *Physiologia Plantarum* **19**, 219–231 (1966)

PIERRE, M. Action du SO_2 sur le métabolisme intermédiare. II. Effet de doses subnecrotiques de SO_2 sur des enzymes de feuilles de Haricot, *Physiologie Végétale* **15**, 195–205 (1977)

PIERRE, M. and QUEIROZ, O. Enzymic and metabolic changes in bean leaves during continuous pollution by subnecrotic levels of SO_2, *Environmental Pollution: Series A: Ecological and Biological* **25**, 41–51 (1981)

PIERRE, M. and QUEIROZ, O. Modulation by leaf age and SO_2 concentration of the enzymic response to subnecrotic SO_2 pollution, *Environmental Pollution: Series A: Ecological and Biological* **28**, 209–217 (1981)

PORTIS, A. R., CHON, C. J., MOSBACH, A. and HELDT, H. W. Fructose and sedoheptulose bisphosphatase. The sites of a possible control of CO_2 fixation by light-dependent changes of the stromal Mg^{2+} concentration, *Biochimica et Biophysica Acta* **461**, 313–325 (1977)

PURCZELD, P., CHON, C. J., PORTIS, A. R., Jr., HELDT, H. W. and HEBER, U. The mechanism of the control of carbon fixation by the pH in the chloroplast stroma, *Biochimica et Biophysica Acta* **501**, 488–498 (1978)

RADMER, R. J. and KOK, B. Photoreduction of O_2 primes and replaces CO_2 assimilation, *Plant Physiology* **58**, 336–340 (1976)

ROBINSON, S. P. and WALKER, D. A. The significance of light activation of enzymes during the induction phase of photosynthesis in isolated chloroplasts, *Archives of Biochemistry and Biophysics* **202**, 617–623 (1980)

SALISBURY, F. B. and ROSS, C. W. *Plant Physiology*, 2nd edition. Wadsworth Publishing Company, Belmont, California (1978)

SAWHNEY, S. K. and NICHOLAS, D. J. D. Nitrite hydroxylamine and sulphite reductases in wheat leaves, *Phytochemistry* **14**, 1499–1503 (1975)

SCHEIBE, R. and BECK, E. On the mechanism of light activation of the NADP-dependent malate dehydrogenase in spinach chloroplasts, *Plant Physiology* **64**, 744–748 (1979)

SCHIFF, J. A. and HODSON, R. C. The metabolism of sulfate, *Annual Review of Plant Physiology* **24**, 381–414 (1973)

SILVIUS, J. E., BAER, C. H., DODRILL, S. and PATRICK, H. Photoreduction of sulfur dioxide by spinach leaves and isolated spinach chloroplasts, *Plant Physiology* **57**, 799–801 (1976)

SOLDATINI, C. F., ZIEGLER, I. and ZIEGLER, H. Sulfite: preferential sulfur source and modifier of CO$_2$ fixation in *Chlorella vulgaris*, *Planta* **143**, 225–231 (1978)

SPALENY, J. Sulphate transformation to hydrogen sulphide in spruce seedlings, *Plant and Soil* **48**, 557–563 (1977)

STEIGNER, H. M. and BECK, E. Formation of hydrogen peroxide and oxygen dependence of photosynthetic CO$_2$ assimilation by isolated chloroplasts, *Plant and Cell Physiology* **22**, 561–576 (1981)

TAGER, J. M. and RAUTANEN, N. Sulphite oxidation by a plant mitochondrial system, *Biochimica et Biophysica Acta* **18**, 111–121 (1955)

TAMURA, G. and ITOH, S. Photoreduction of sulfite by spinach leaf preparations in the presence of grana system, *Agricultural and Biological Chemistry* **38**, 225–226 (1974)

THOMAS, M. D., HENDRICKS, R. H., COLLIER, T. R. and HILL, G. R. The utilization of sulfate and SO$_2$ for the S nutrition of alfalfa, *Plant Physiology* **18**, 345–371 (1943)

THOMAS, M. D. and HILL, G. R. Relation of sulphur dioxide in the atmosphere to photosynthesis and respiration of alfalfa, *Plant Physiology* **12**, 309–383 (1937)

WEIGL, J. and ZIEGLER, H. R. Die räumliche Verteilung von ^{35}S und die Art der markierten Verbindurgen in Spinatblättern nach Begasung mit ^{35}SO$_2$, *Planta* **58**, 435–437 (1962)

WERDAN, K., HELDT, H. W. and MILOVANCEV, M. The role of pH in the regulation of carbon fixation in the chloroplast stroma, *Biochimica et Biophysica Acta* **396**, 276–292 (1975)

WILSON, L. G., BRESSAN, R. A. and FILNER, P. Light dependent emission of hydrogen sulfide from plants, *Plant Physiology* **61**, 184–189 (1978)

WINNER, W. E., SMITH, C. L., KOCH, G. W., MOONEY, H. A., BEWLEY, J. D. and KROUSE, H. R. Rates of emission of H$_2$S from plants and patterns of stable sulphur isotope fractionation, *Nature (London* **289**, 672–673 (1981)

WOLOSIUK, R. A. and BUCHANAN, B. B. Studies on the regulation of chloroplast NADP-linked glyceraldehyde-3-phosphate-dehydrogenase, *Journal of Biological Chemistry* **251**, 6456–6461 (1976)

ZIEGLER, I. Subcellular distribution of ^{35}S-sulfur in spinach leaves after application of ^{35}SO$_4^{2-}$, ^{35}SO$_3^{2-}$, ^{35}SO$_2$, *Planta* **135**, 25–32 (1977)

ZIEGLER, I. and HAMPP, R. Control of ^{35}SO$_4^{2-}$ and ^{35}SO$_3^{2-}$ incorporation into spinach chloroplasts during photosynthetic CO$_2$ fixation, *Planta* **137**, 303–307 (1977)

ZIEGLER, I., MAREWA, A. and SCHOEPE, E. Action of sulfite on the substrate kinetics of chloroplastic NADP-dependent glyceraldehyde-3-phosphate dehydrogenase, *Phytochemistry* **15**, 1627–1632 (1976)

ZIEGLER, H. and ZIEGLER, I. The influence of light on the NADP$^+$-dependent glyceraldehyde-3-phosphate dehydrogenase, *Planta* **65**, 369–380 (1965)

ZIEGLER, H. and ZIEGLER, I. The light-induced increase in the activity of NADP$^+$-dependent glyceraldehyde-3-phosphate dehydrogenase, *Planta* **72**, 162–169 (1967)

ZIEGLER, H., ZIEGLER, I. and SCHMIDT-CLAUSEN, H. J. The influence of light intensity and light quality on the increase in activity of the NADP$^+$-dependent glyceraldehyde-3-phosphate-dehydrogenase, *Planta* **67**, 344–356 (1965)

ZIEM-HANCK, K. U. and HEBER, U. Oxygen requirement of photosynthetic CO$_2$ assimilation, *Biochemica et Biophysica Acta* **591**, 266–274 (1980)

Part IV

Effects on photophosphorylation and respiration

Chapter 15

The influence of atmospheric pollutants and their cellular products upon photophosphorylation and related events

A. R. Wellburn

DEPARTMENT OF BIOLOGICAL SCIENCES, UNIVERSITY OF LANCASTER, UK

Introduction

In any discussion of the effect of atmospheric pollutants such as O_3, PAN, SO_2 and NO_x (or their derivatives sulphite and nitrite at $pH > 7$) upon light-induced ATP formation, there are many central and peripheral aspects which must be taken into account and defined.

In field and fumigation studies it is relatively easy with modern bioluminescence assays to determine and compare the concentrations of total ATP and adenylates, but very much more difficult to discriminate the amounts in different cellular compartments or to compare rates of ATP formation in isolates of chloroplasts from control and polluted tissue because of the difficulties and variations imposed by the procedures of plastid isolation. Consequently, *in vitro* treatment of portions of chloroplast suspensions from unpolluted sources with different pollutants or their derivatives is the more normal practice.

In a chemiosmotic explanation of ATP formation, electron flow is responsible for the creation of a proton gradient which, in turn, is harnessed by the coupling factor complex $(CF_1–CF_0)$ to form ATP. At the same time electrons reduce nitrite or $NADP^+$ to form ammonia or NADPH which are then rapidly consumed by the glutamine synthetase/glutamate synthase (GOGAT) and C_3 cycles respectively. Clearly, pollutants which penetrate beyond the plastid envelopes may affect any or all of these different processes at one or more critical points in each of these events, depending upon the concentration(s) and duration(s) of the prevailing pollutants or their products within the chloroplasts. In some of the tests available (e.g. bicarbonate or 3-PGA-dependent O_2 evolution) virtually all these functions are encompassed together so that only the net effect of the pollutants may be assessed, but the use of more specific probes such as the quenching of 9-amino-acridine fluorescence or electron spin resonance (ESR) spectroscopy allows a more detailed examination of the mechanisms involved to be made.

Changes in the levels of adenine nucleotides due to atmospheric pollutants

One of the most popular parameters to define the bioenergetic state of living systems has been the energy charge ratio, but assessments of the phosphorylation potential may actually be more valuable. Endogenous concentrations of ATP, ADP and orthophosphate change considerably during light to dark transitions (and *vice versa*), which may temporarily change the cellular energy charge ratios or phosphorylation potentials. These levels are usually restored quite rapidly. There appears to be quite a variation in pool sizes of different adenine nucleotides across the different cellular compartments (Hampp, 1980). Amounts and energy charge ratios appear surprisingly to be highest in the 'cytosol' (i.e. cytoplasm plus vacuole plus broken nuclei), lower in plastids and lowest of all in mitochondria.

A number of field and fumigation studies have been undertaken which have compared the amounts of adenine nucleotides in polluted and unpolluted plant tissues. In the case of O_3 the observations have been rather mixed. There has been a report of a 30% reduction in ATP content in both O_3-resistant and O_3-susceptible beans (*Phaseolus vulgaris*) (Tomlinson and Rich, 1968) but Pell and Brennan (1973) reported a significant increase in the levels of ATP and total adenylates in beans after a 3 h exposure to O_3. Perennial ryegrass (*Lolium perenne*) laminae exposed to NO_2 (250 ppb) also have significantly higher levels of ATP but these are much reduced in the presence of SO_2 as well (Wellburn *et al.*, 1981). It is also known that there is a close inter-regulating relationship between ATP, NO_3^-, NO_2^- and light (Sawhney, Naik and Nicholas, 1978; Larsson, 1980) which affect and closely modulate the rates of nitrate and nitrite reduction and also electron transport in both plastids and mitochondria.

Fumigation with SO_2 (0.3 ppm) reduces significantly the levels of ATP and other adenylates (Hoffman, Pahlich and Steubing, 1976) but there is no significant effect upon the energy charge ratio. Even at lower levels of SO_2 (50–200 ppb) there is an inverse linear relationship between ATP content and SO_2 concentration (Harvey and Legge, 1979) in field and also in field-preadapted and then laboratory-fumigated foliage. Interestingly, this preadaption requirement may be linked to a slight deficiency in phosphate nutrition due to soil acidification by SO_2. Bisulphite derivatives are also known to reduce the cellular levels of ATP (Lüttge *et al.*, 1972).

Events associated with the chloroplast envelopes

After using the technique of silicone oil-filtering centrifugation to compare rates of uptake of orthophosphate, sulphite and sulphate, Hampp and Ziegler (1977) and Mourioux and Douce (1978) suggested that the transport of sulphite and sulphate is associated, at least in part, with the phosphate translocator system of plastid envelopes. However the rates of exchange for both sulphur anions are much lower than those for orthophosphate and it would appear that there is no competition between orthophosphate and sulphite or sulphate. On the contrary it would appear from the results of Hampp and Ziegler (1977) that orthophosphate enhances the influx of sulphur anions. This process is further accentuated by exchange for photosynthetically-synthesized dihydroxyacetone phosphate and 3-PGA thereby providing a mechanism for regulation for optimum assimilatory sulphate reduction through the phosphorylating steps (Ziegler and Hampp, 1977). Un-ionized sulphur species such as $SO_2 \cdot H_2O$, however, appear to diffuse readily across the plastid envelopes and are not under metabolic control (Spedding *et al.*, 1980). Such uptake rates are greatly enhanced by reduced pH and emphasize the importance of pH as a major factor influencing the toxicity of SO_2 within the plant cell.

Plastid envelopes are also permeable to nitrite, ammonia and un-ionized nitrous acid but not to ammonium ions (Heber and Purczeld, 1978). As ammonia may move across the envelope freely in either direction the ionization relationship with ammonium, in the stroma for example, will be disturbed as the protons are left behind causing an acidification of the space vacated. However, when both an anion and its neutral protonation product (e.g. NO_2^- and HNO_2) can permeate a membrane barrier, shuttle transfer of protons will abolish the pH gradient across the membranes. In the light, alkalization of the stroma normally activates ribulose-1,5-bisphosphate carboxylase/oxygenase. Indirect proton uptake via a shuttle involving both nitrite and HNO_2 may consequently interfere with such events by causing a breakdown of the trans-envelope pH gradient (Heber and Purczeld, 1978) and this may be a partial explanation for some of the known inhibitory effects of nitrite upon CO_2 fixation (Hiller and Bassham, 1965).

The penetration of O_3 further than the plasma membrane has been the subject of some discussion but the fact that O_3 can rapidly effect chlorophyll fluorescence (Schreiber et al., 1978) means that O_3 probably enters the plastids and gains access to the thylakoids. How much damage it does as it crosses the plastid envelopes and stroma is not known.

In studies using isolated chloroplasts there are certain measurable functions which are totally dependent upon the presence of the envelopes to retain the stromal contents around the thylakoid membranes. Measurement of rates of oxygen evolution dependent upon bicarbonate, ribulose-1,5-bisphosphate or 3-PGA are quite different to those polarographic determinations of electron flow from water or artificial donors of electrons using preparations of thylakoids. These latter functions are outside the scope of this chapter and are covered elsewhere in this book. In general the effect of pollutants upon them is extremely variable and highly dependent upon the degree of coupling (or lack of it) to the phosphorylation of ATP. By contrast, effective measurement of bicarbonate or 3-PGA-dependent O_2 evolution relies upon a variety of functions encompassing all aspects of photosynthesis within the intact plastids from water splitting to CO_2 fixation, not the least of which are ATP formation by the thylakoids and its consumption by various enzymic events within the C_3 cycle.

When the literature concerning the effects of SO_2, sulphite and sulphate upon 3-PGA-dependent O_2 evolution are compared concentrations of at 1 mM (Table 15.1) a pleasing harmony, by comparison with determinations of the effects of pollutants upon electron flow, emerges. Sulphite and SO_2 have a greater detrimental effect than sulphate but the magnitude of inhibition is determined in an inverse manner by the concentrations of orthophosphate present. In other words, the lower the orthophosphate concentration the greater the reduction of 3-PGA-dependent O_2 evolution. Silvius, Ingle and Baer (1975) in their experiments unfortunately employed pyrophosphate rather than orthophosphate, but it is presumed from the similar relationships they obtained between their results that proportional amounts of orthophosphate were present after endogenous partial hydrolysis of the pyrophosphate supplied. Included in Table 15.1 are some unpublished results of our own on pea (Pisum sativum) chloroplasts isolated and assayed in very similar manner to and confirming those of Plesničar and Kalezić (1980). These Yugoslav workers have gone further by studying the kinetics of the inhibition of 3-PGA-dependent O_2 evolution by varying the amounts of sulphite in relation to different orthophosphate concentrations. By means of a Dixon plot (Figure 15.1) they have established that there is a competitive inhibition of sulphite upon O_2 evolution associated with photosynthetic carbon assimilation, and determined a K_i of 0.8 mM for sulphite competition with orthophosphate which they ascribed to events associated with photophosphorylation.

Unfortunately no comparable studies of the effects of NO_x, nitrite, nitrate, O_3 or

TABLE 15.1. Summary of the literature concerning the effect of 1 mM sulphite, sulphate or SO$_2$ upon the rates of 3-phosphoglycerate dependent oxygen evolution by isolated intact chloroplast preparations. (In some cases the quoted depressions were obtained by extrapolation)

Additive	Plant	Phosphate conc. (mM)	Difference from controls	Reference
1 mM sulphate	spinach	0	−60%	Baldry, Cockburn and Walker (1968)
	spinach	0.2	−40%	Baldry, Cockburn and Walker (1968)
	spinach	0[a]	−35%	Silvius, Ingle and Baer (1975)
	spinach	0[b]	−3%	Silvius, Ingle and Baer (1975)
1 mM sulphite	spinach	0[a]	−35%	Silvius, Ingle and Baer (1975)
	spinach	0[b]	−10%	Silvius, Ingle and Baer (1975)
	pea	0.05	−70%	Plesničar and Kalezić (1980)
	pea	0.4	−54%	Plesničar and Kalezić (1980)
	pea	0.6	−33%	Plesničar and Kalezić (1980)
	pea	1	−23%	Plesničar and Kalezić (1980)
	pea	0.5	−43%**	(previously unpublished)
	pea	2	−20%*	(previously unpublished)
1 mM SO$_2$	spinach	0[a]	−43%	Silvius, Ingle and Baer (1975)
	spinach	0[b]	−18%	Silvius, Ingle and Baer (1975)

[a] No orthophosphate but 4 mM pyrophosphate.
[b] No orthophosphate but 40 mM pyrophosphate.
Differences from controls * $P < 0.05$; ** $P < 0.01$.

PAN have been reported for 3-PGA-dependent O$_2$ evolution by isolated intact plastids but Grant, Labelle and Mangat (1972) showed that spinach (*Spinacia oleracea*) plastids fixing HCO$_3^-$ in the presence of NO$_2^-$ have a greater proportion of their NADP present in reduced form than in the absence of nitrite. This effect is not alleviated by 3-PGA instead of bicarbonate but the observation that ribose-5-phosphate is partially effective is consistent with the original proposal by Hiller and Bassham (1965) for sites of nitrite inhibition in the C$_3$ cycle around fructose-6-phosphate and sedoheptulose-7-phosphate.

Effect of pollutants and their products upon rates of ATP formation

There is a considerable literature concerning the effects of different sulphur compounds upon photophosphorylation, some of it done not with atmospheric pollution or even sulphur metabolism in mind but to further our understanding of the basic mechanisms involved in the formation of ATP. Previously there was a tendency to use the phrases 'cyclic or non-cyclic photophosphorylation' to describe electron flow determined polarographically or spectrophotometrically rather than the actual formation of ATP or, at least, the measurable disappearance of orthophosphate. *Table 15.2* shows a summary of the literature concerned only with the latter forms of assay, choosing 1 mM as the comparative level in each case. For 1 mM sulphate, sulphite or SO$_2$ there is a consistency in that each causes a depression of light-induced ATP formation; but even when levels of nitrite are raised to 10 mM there appears to be no direct effect upon photophosphorylation. The levels of orthophosphate in the assay media have again been included but, taken as a whole, there is a concealment of any orthophosphate interaction because a wide variety of different electron donor/acceptor systems have been used. Phenazine methosulphate (PMS), for example, is now known

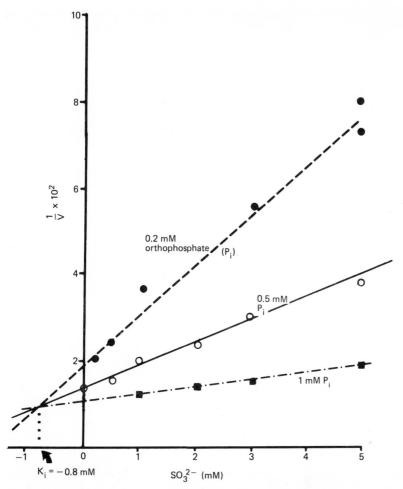

Figure 15.1. Reciprocal values of the reaction velocity of 3-PGA-dependent O_2 evolution by isolated intact pea chloroplasts plotted as a function of sulphite concentration for different orthophosphate concentrations (this Dixon plot is redrawn from Plesničar and Kalezić (1980) by kind permission of Dr M. Plesničar)

to react directly with P-700$^+$ to support cyclic electron flow at high rates with almost no light saturation (Witt, Rumberg and Junge, 1968), whereas ascorbate/diamino-durene (DAD) uses some of the natural electron transport intermediates whilst generating proton gradients. Electron flow from water to ferricyanide uses other intermediates and supports the lowest rates of ATP formation of all three models. Also included in *Table 15.2* are some previously unpublished determinations of ascorbate/DAD to methylviologen (MV)-supported photophosphorylation by both oat (*Avena sativa*) and pepper (*Capsicum frutesceus*) preparations using the procedures described in Wellburn *et al.* (1981). In the case of sulphite, the concentration of orthophosphate was found to have an inverse effect on the degree of depression of the rates of ATP formation; similar to the results of Plesničar and Kalezić (1980) concerning 3-PGA-dependent O_2 evolution by pea chloroplasts.

TABLE 15.2. Summary of the literature concerning the effect of 1 mM sulphate, sulphite, SO_2 or 10 mM nitrite upon the rates of light-induced ATP formation by isolated thylakoid preparations. (In some cases the quoted depressions were obtained by extrapolation. Where not stated the plant source for the preparations was spinach)

Donor/acceptor system	Phosphate conc. (mM)	Difference from controls	Reference
1 mM sulphate			
Pyocyanine	5	-10%	Ryrie and Jagendorf (1971)
PMS	5	-10%	Ryrie and Jagendorf (1971)
PMS	2.5	-24%	Hall and Telfer (1969)
PMS	1.3	-21%	Asada, Deura and Kasai (1968)
PMS	2	-20%	Silvius, Ingle and Baer (1975)
H_2O/ferricyanide	2	-9%	Silvius, Ingle and Baer (1975)
H_2O/ferricyanide	1.3	-11%	Asada, Deura and Kasai (1968)
H_2O/ferricyanide	2.5	-31%	Hall and Telfer (1969)
H_2O/ferricyanide	5	-10%	Ryrie and Jagendorf (1971)
Ascorbate/DAD/MV	5	-10%	Ryrie and Jagendorf (1971)
1 mM sulphite			
PMS	2	-27%	Silvius, Ingle and Baer (1975)
H_2O/ferricyanide	2	-8%	Silvius, Ingle and Baer (1975)
H_2O/ferricyanide	2.5	-33%	Hall and Telfer (1969)
Ascorbate/DAD/MV (peppers)	2.5	-13%**	(previously unpublished)
Ascorbate/DAD/MV (oats)	0.2	-48%**	(previously unpublished)
Ascorbate/DAD/MV (oats)	1.2	-22%**	(previously unpublished)
1 mM sulphur dioxide			
PMS	2	-25%	Silvius, Ingle and Baer (1975)
H_2O/ferricyanide	2	-14%	Silvius, Ingle and Baer (1975)
H_2O/ferricyanide (peas)	1.7	-19%	Cerović, Kalezić and Plesničar (1983)
10 mM nitrite			
PMS	1.3	No effect	Asada, Deura and Kasai (1968)
H_2O/ferricyanide	1.3	No effect	Asada, Deura and Kasai (1968)
Ascorbate/DAD/MV (peppers)	2.5	NS	(previously unpublished)

Differences from control ** $P < 0.01$; NS = no significant differences.

Recently Cerović, Kalezić and Plesničar (1983) have investigated in some detail the effect of SO_2 upon the rates of ATP formation by pea envelope-free chloroplasts. The rates of photophosphorylation were depressed by increasing amounts of SO_2, especially at low orthophosphate concentrations (*Figure 15.2*). When these results are replotted using a Dixon plot (*Figure 15.3*) a competitive inhibition of SO_2 upon photophosphorylation is revealed. Moreover the intersect reveals an inhibition constant (K_i) of 0.8 mM, precisely the same as for 3-PGA-dependent O_2 evolution (Plesničar and Kalezić, 1980). This is strong evidence that the two are actually the same phenomenon and that SO_2 or sulphite competes with orthophosphate for a binding site probably upon the CF_1 particles. This inhibition of photophosphorylation by SO_2 has also been shown by Cerović, Kalezić and Plesničar (1983) to be fully reversible and that by using an ATP-regenerating system (phosphocreatine) they have demonstrated that their previously reported depression of 3-PGA-dependent O_2 evolution by sulphite was solely a consequence of inhibited photophosphorylation.

Other pollutants are also capable of inhibiting light-induced ATP formation. Ammonia in the form of ammonium ions is a classic example of an uncoupler (Krogmann, Jagendorf and Avron, 1959) even at relatively low concentration. This effect is freely reversible. However, O_3 and PAN also inhibit photophosphorylation

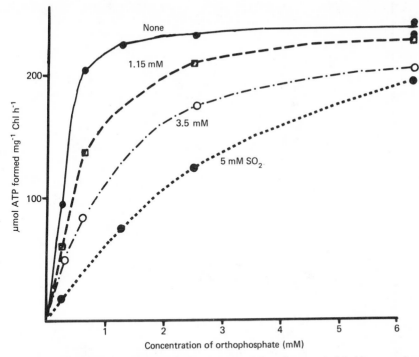

Figure 15.2. Inhibition of photophosphorylation by envelope-free pea thylakoid preparations by different concentrations of SO_2 as a function of phosphate concentration. (Hitherto unpublished data kindly supplied by Dr M. Plesničar)

irreversibly (Koukol, Dugger and Palmer, 1963, 1967; Coulson and Heath, 1974). At the time they were thought not to act as uncouplers, but these interpretations of the mode of action of O_3 and PAN were made before chemiosmotic principles were fully appreciated. In the light of the known effects of oxidants (*see* Mudd, 1982) and the effect of O_3 upon the ΔpH described below it would appear that extensive and permanent damage may be induced by such oxidants to the integrity of the thylakoid membranes so as to cause a discharge or non-reversible uncoupling of the proton gradient which then gives rise to decreased rates of ATP formation.

The effect of pollutants and their products on the ΔpH across thylakoid membranes

Hitherto, with the possible exception of the studies of Coulson and Heath (1974) on the effects of O_3 on photoinduced amine-supported swelling of plastids, there have been no reports of experiments to determine the effect of pollutants and their products directly upon proton gradients across thylakoids, although O_3 is already known to have a dramatic effect on the efflux of potassium ions from the alga *Chlorella* (Chimiklis and Heath, 1975; Heath and Frederick, 1980).

Various methods for the determination of the ΔpH generated across photosynthetic membranes in the light and the dark exist but one of the most convenient, based upon the uptake of fluorescent amines, was suggested by Schuldiner, Rottenberg and Avron

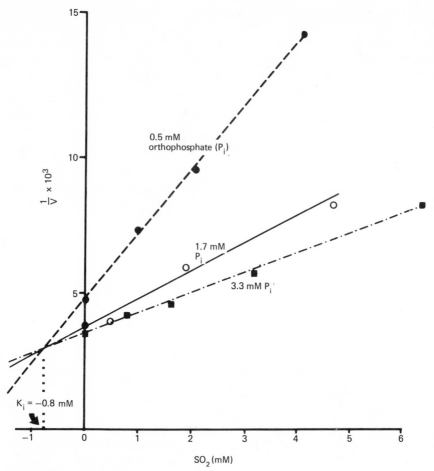

Figure 15.3. Reciprocal values of the rates of formation of ATP at different orthophosphate concentrations similar to that shown in *Figure 15.2* replotted as a function of SO_2 concentration. (Redrawn from data kindly supplied by Dr M. Plesničar)

(1972). This uptake is dependent upon the dissociation constant of the amine and the number of ionizable amines in the probe. 9-Amino-acridine (9-AA), was found to have highly suitable properties for these measurements and has recently been widely adopted. The quenching of 9-AA fluorescence by illuminated thylakoid suspensions is partly due to the redistribution of the probe in response to the proton gradient and also to increased binding of the probe to thylakoid membranes (Haraux and Kouchkovsky, 1979). However corrections can be made for binding of the probe by measuring light-induced quenching as a function of chlorophyll concentration, which then allows meaningful assessments of ΔpH to be made (Slovacek and Hind, 1981). The technique has the advantage that for comparative purposes the relative percentage change in light-induced quenching quickly provides an indication of differential change across a membrane with respect to protons due to a different treatment.

Using a fluorescence apparatus very similar to that described by Mills, Slovacek and Hind (1978) with identical filters and light intensities, we have followed the procedures and calculations of Slovacek and Hind (1981) in order to employ the technique of 9-AA light-induced fluorescence quenching to follow the effects of O_3, sulphite, sulphate, nitrite and nitrate singly and in combination upon thylakoid preparations from oats (*Avena sativa*) (Robinson and Wellburn, in preparation). The following data show only some of the single treatments upon 9-AA quenching but serve to demonstrate the potential of the technique for the study of changes in ΔpH induced by pollutants or their derivatives.

Figure 15.4 illustrates examples of the traces from sulphite, nitrite and sulphate-treated and unpolluted controls. Typically the level of 9-AA fluorescence of 1.5 ml samples of thylakoid preparations (10–20 µg chlorophyll) in TRIS-Tricine buffer (pH 8.1) in the dark is considerably above instrumental zero (left-hand side of the traces) and taken to be 100% relative fluorescence. Upon illumination with red light (>630 nm) the 9-AA fluorescence of the samples declines in proportion to the formation of ΔpH across the thylakoids. At concentrations of 1 mM, sulphite reduces the light-induced fluorescence quenching although in some (e.g. 1 mM sulphate) there may be an enhancement rather than a reduction of the quenched signal in the light. The percentage change of light-induced 9-AA fluorescence quench after additions of either sulphite, sulphate, nitrite or nitrate relative to the untreated controls (measured immediately prior to the addition in each case) are summarized in *Figure 15.5*. The buffering capacity of the medium was checked over all ranges of additions. Only in the range of 1–2 mM of the added ions did the pH have to be recorrected back to pH 8.1 before turning on the red light. The degree of reproducibility was extremely high. Although not shown each of the treatments was carried out four times using

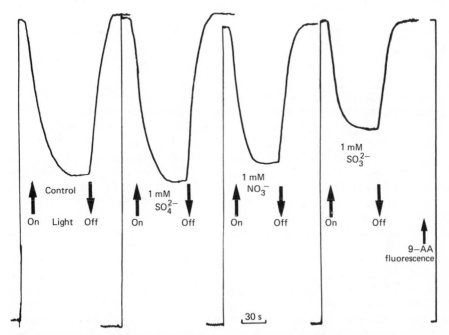

Figure 15.4. Examples of traces recorded of the 9-amino-acridine (9-AA) fluorescence from oat thylakoid preparations given different treatments in the dark and then quenched as a result of illumination by red light signifying redistribution of the probe and the generation of different ΔpH gradients across the thylakoid membranes in the presence or absence of sulphate, nitrite or sulphite

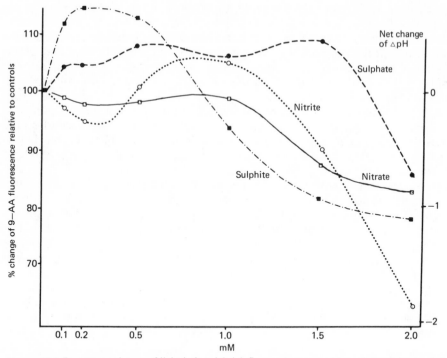

Figure 15.5. Percentage change of light-induced 9-AA fluorescence quenching relative to untreated oat thylakoid preparations as the result of the single addition of different amounts of sulphate, sulphite, nitrate or nitrite and an approximate estimation of the net ΔpH change as a consequence of the addition. The measurements were made at equivalent chlorophyll concentrations (15 μg/1.5 ml)

different plastid preparations and the standard deviation in any case never rose above ±5% for these ions. Greater problems were encountered during the gassing with O_3 (*see later*). Only at very high concentrations did both sulphite and nitrite appear to have a detrimental effect upon the proton gradients established across the thylakoid membranes, whereas sulphate and nitrate had little effect. Surprisingly sulphate, nitrate and sulphite show enhanced quenching of 9-AA fluorescence over different concentration ranges. These are especially difficult to interpret and must mean that multiple events such as changes in the nature of the light-induced binding of the 9-AA to the membranes may also be involved. It is difficult therefore to ascribe a 'beneficial' effect to the lower ranges of the sulphite treatment for example. Preliminary results with mixtures of nitrite and sulphite (not given) clearly indicate the abolition of this enhanced quenching and, instead, significant ('more than additive') reductions of ΔpH.

When the technique was applied to O_3 using gassing procedures very similar to those of Coulson and Heath (1974), a 50 ml syringe controlled by a Perfusor control unit (B. Braun, Melsungen, FRG) was used to measure accurately and dilute the O_3 samples. There are considerable problems inherent to the use of O_3 and for physiological comparisons of doses between those in solution and those in air. These have been discussed at length elsewhere (Coulson and Heath, 1974; Heath, 1980). The exposure to O_3 in solution is of a pulse nature rather than a continuous concentration-related event as with sulphite or nitrite. This feature also makes direct comparisons difficult.

However the use of pulses of O_3 into dilute samples of thylakoid membranes and measuring the degree of 9-AA light-induced fluorescence quenching repeatedly before and after gassing reveals different effects of O_3 even at very low dosages as compared to those of sulphite and nitrite.

Figure 15.6 shows an example of one such experiment. Normal and repeatable light-induced 9-AA fluorescence quenching was rapidly and progressively destroyed following a pulse of O_3. Furthermore, this destruction is clearly light-induced. Instead of attaining a steady mimimum of light-induced fluorescence quenching as normally observed, even with sulphite or nitrite, the signal increases in the light (i.e. the quenching, and hence the ΔpH, declines) in a progressive manner. Over the first light on/off/on cycle this light-dependency is clearly shown but afterwards an additional feature appears. During the intervening and subsequent dark periods a partial 'repair' occurs which causes a temporary increase in quenching during the onset of the following light-pulse. The nature of the 'repair' mechanism is not known at present but may be related to the activities of superoxide dismutase or reactions involved in the formation of antioxidants. A fuller representation of the overall decline in light-induced 9-AA fluorescence quenching after pulses of different amounts of O_3 is shown in *Figure 15.7*. Pulses of ozone greater than 50 nmol have a significant effect upon ΔpH. Very large pulses (2.4 μmol O_3) completely abolish the effective proton gradient within seconds. Consequently, even allowing for the difficulties of comparison, it would appear that in terms of the action of single pollutants upon the integrity of the thylakoid membrane to build and effectively support a proton gradient, O_3 has the capability of causing the greatest damage and this may be a partial explanation for the decline in orthophosphate esterification due to ozone observed by Coulson and Heath (1974), the original interpretation of which was later corrected by Heath (1980).

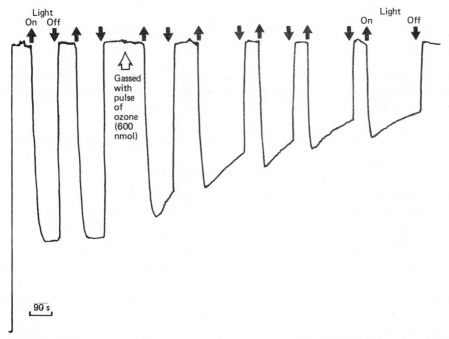

Figure 15.6. A trace of the 9-AA fluorescence emitted by an oat thylakoid preparation in the dark or illuminated with red light before and after gassing with a single pulse (600 nmol) of O_3

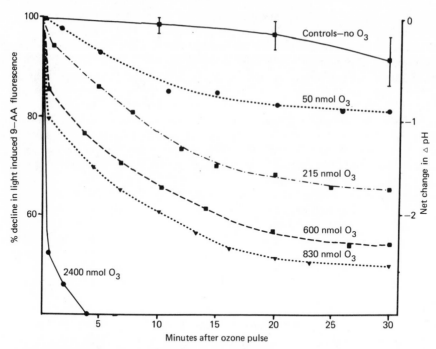

Figure 15.7. Summary of the percentage decline in light-induced 9-AA fluorescence quenching with time by oat thylakoid preparations given different pulses of O_3

The use of electron spin resonance to study the effect of pollutants upon thylakoid structure

The oxidation of water to molecular O_2 in photosynthesis involves manganese in some manner (Radmer and Cheniae, 1977) and there appear to be at least three pools of manganese in thylakoids:

(1) a loosely-bound pool associated with the photolysis mechanism;
(2) tightly-bound manganese associated with light-harvesting chlorophyll-protein complexes; and
(3) a very loosely-bound pool unrelated to O_2 evolution (Khanna *et al.*, 1981).

Manganese has a characteristic electron spin resonance (ESR) spectrum at both room and cryogenic temperatures. Unfortunately functional manganese in chloroplast membranes is ESR silent and can only be detected when it has been liberated from its functional site. This release is dependent upon the immediate environmental conditions around the membrane which suggests that ESR may well be a suitable probe for determining the effect of atmospheric pollutants upon photosynthetic membranes and give an indication as to their influence upon bioenergetic functions. A considerable number of studies involving the use of ESR to detect changes in the membranes of animals due to pollutants have already been undertaken (Rowlands, Allen-Rowlands and Gause, 1977).

Figure 15.8 shows the room temperature ESR spectra (first derivative) around the manganese signal of pepper (*Capsicum annuum*) chloroplast thylakoid membrane preparations after different treatments recorded by a Varian E3 spectrometer at the following instrumental settings: microwave power, 75 mW; modulation amplitude, 40G; time constant, 0.3 s; frequency, 9.29 GHz and receiver gain the same for all six spectra. Spectrum A was of untreated thylakoids (4 mg chlorophyll ml^{-1}) and spectrum B was of the same sample treated for 2 min with 2 mM sulphite with a dilution factor of 0.75. Spectrum C was likewise a treatment of B with an additional 8 mM sulphite. Spectrum D was taken of a similar preparation isolated in parallel to A but gently washed with dilute TRIS buffer (80 mM, pH 8). Although treatment with 0.8 M TRIS is known to release all the manganese from photosynthetic membranes and hence to give the full six-line manganese signal (Blankenship, Babcock and Sauer, 1975), this less drastic treatment gave an enhanced but single line manganese signal (4 mg chlorophyll

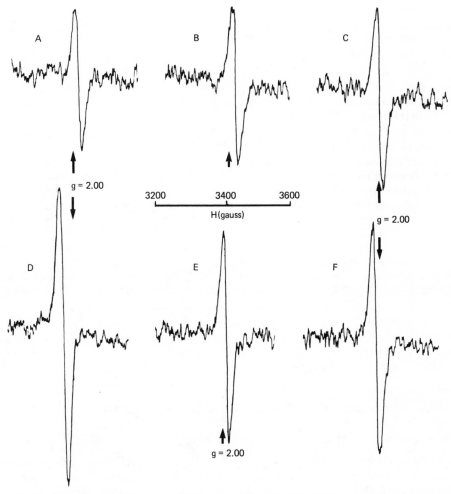

Figure 15.8. Room temperature electron resonance spectra (ESR) around the manganese region (first derivative) of pepper chloroplasts. The full description of the treatments and instrumental conditions are given in the text and summarized in *Table 15.3*. (A = untreated thylakoids; B = A + 2 mM sulphite; C = B + 8 mM sulphite; D = 80 mM TRIS-washed thylakoids; E = D + 8 mM sulphite; F = A + 2 mM nitrite)

TABLE 15.3. Summary of ESR data from treatment of pepper thylakoid preparations with sulphite and nitrite as shown in spectra A–F of *Figure 15.8*

Spectrum	Preparation	Addition	Chlorophyll content (mg ml^{-1})	Effective percentage increase in signal after allowing for dilutions
A	Thylakoid preparation	none	4	—
B	(as A above)	+2 mM sulphite	3	+50% on A
C	(as B above)	+8 mM sulphite	2.25	+104% on A
F	(as A above)	+2 mM nitrite	3	+115% on A
D	TRIS^a-washed thylaioids	none	4	—
E	(as D above)	+8 mM sulphite	3	+9.5% on D

^{a} 0.08 M, pH 8.

ml^{-1}), indicating that only a portion of the manganese had been dislodged from the membrane. Treatment E with 8 mM sulphite upon D (dilution factor 0.75) indicates that little further disturbance of the manganese occurred with sulphite. However treatment F is analogous to treatment B upon A but with 2 mM nitrite instead of 2 mM sulphite. Clearly nitrite, as the summary *Table 15.3* shows, is a much more effective releasing agent for bound manganese and must be capable of creating greater disturbance in certain functions of photosynthetic membranes than sulphite. This may be part of the basic mechanism of nitrite toxicity and undoubtedly involves various free-radical events similar to those predicted by Mudd (1982). In view of the fact that nitrite is also a natural electron acceptor of photosynthetic membranes, mechanisms that regulate nitrite levels must be arranged so as to ensure that nitrite concentrations within plastids are normally kept at low and within tightly controlled concentration ranges to avoid these additional reactions.

This preliminary use of ESR as a probe for effects of atmospheric pollution upon photosynthetic membranes demonstrates that the technique has further potential and may be especially useful in studies of NO$_x$ and nitrite toxicity. It has also shown that there are a range of different sites across the range of bioenergetic functions within the plastid that respond differently to the various toxic products of atmospheric pollution. As has been shown here, functions associated with electron transport and oxygen evolution may be more sensitive to nitrite and its derivatives whereas ΔpH is affected most by O$_3$ and coupling factor activities by sulphite. The effect of pollutant mixtures will be to compound these various lesions.

Effects of atmospheric pollutants on reductive capacity

Photosynthetic electron transport is responsible for:

(1) the formation of proton gradients and the formation of ATP and/or the reduction of NADP$^+$;
(2) providing electrons either for thiosulphonate reductase associated with the assimilatory sulphate pathway involving APS-sulphotransferase and the direct reduction of sulphite by sulphite reductase (both processes yielding thiol groups of various compounds); and
(3) the reduction of nitrite to ammonia by ferredoxin-linked nitrite reductase activity.

Photoreduction of $NADP^+$ has already been covered in the preceding discussion and elsewhere in this book as an intimate part of the mechanisms leading to the formation of reductant or ATP by thylakoids. Likewise sulphur metabolism, especially in relation to the effects of SO_2, is adequately covered in this volume and also in a comprehensive review by Hällgren (1978). Consequently it is proposed to devote this section to a discussion of the known effects of various pollutants upon the reduction of nitrite, a process which is known to be intimately associated with plastids, and to extend it to include the mechanisms by which nitrite is formed outside the plastid by means of nitrate reductase activity.

Carbon monoxide is an effective inhibitor of photosynthesis and chloroplast development. Sensitivity to CO, however, is strongly affected by the nitrogen status. The CO inhibition of photosynthesis is much larger than can be accounted for by reduced oxygen evolution (Vennesland and Jetschmann, 1971). It would appear that CO affects nitrite reductase but not nitrate reductase activities, causing an increase in toxic nitrite levels, and at the same time inhibits a CO-sensitive oxidase as well as several photosynthetic functions. By contrast, O_3 inhibits both nitrate and nitrite reductase activities (Leffler and Cherry, 1974; Tingey, Fites and Wicklife, 1973) but, at low concentrations of O_3, nitrite reductase activities appear to be the most sensitive.

Studies with $^{15}NO_2$ have revealed that nitrogen from NO_2 may be utilized and converted to nitrate, nitrite and ammonia and assimilated into amino acids through the glutamine synthetase/glutamate synthase (GOGAT) pathway (Yoneyama and Sasakawa, 1979). Accumulations of nitrite due to NO_2 fumigation have been measured, but the responses in terms of increases in nitrite reductase activities differ according to the plant species and the conditions under which the exposure takes place (Yoneyama et al., 1979). Exposure to nitric oxide has also indicated stimulations of nitrite reductase activities (Wellburn, Wilson and Aldridge, 1980), implying that the nitrogen of NO may also enter nitrogen pathways despite the low solubility of NO. Recent results (Murray, Mansfield and Wellburn, unpublished) have indicated that different species adopt different strategies to combat NO_x pollution. Some of these observations are included in *Table 15.4*. In the case of NO, tomatoes (*Lycopersicon esculentum*) either reduce their levels of nitrate reductase activities or increase those of nitrite reductase, but with NO_2 treatment a sensitive cultivar (Ailsa Craig) increases the levels of both enzymes. A more NO_x^- tolerant tomato cultivar (Eurocross BB), by contrast, appears to be unchanged.

TABLE 15.4. Effect of NO_2 and NO (1.5 ppm) upon the levels of nitrate and nitrite reductase activities in the leaves of tomatoes and peppers grown under glasshouse conditions

Treatment of plant species (cultivar)	Nitrate reductase activities[a]		Nitrite reductase activities[b]	
	Clean air	Polluted	Clean air	Polluted
Nitrogen dioxide fumigation				
Tomato (Eurocross BB)	1.23 ± 0.07	1.33 ± 0.08	1.95 ± 0.1	2.15 ± 0.48
Tomato (Ailsa Craig)	0.45 ± 0.02	1.85 ± 0.29***(↑)	1.37 ± 0.33	2.36 ± 0.07*(↑)
Pepper (Bell Boy)	0.98 ± 0.09	0.81 ± 0.04	1.78 ± 0.23	0.66 ± 0.17**(↓)
Nitric oxide fumigation				
Tomato (Eurocross BB)	1.29 ± 0.04	0.7 ± 0.09***(↓)	2.4 ± 0.18	2.8 ± 0.3
Tomato (Ailsa Craig)	0.52 ± 0.04	0.59 ± 0.15	1.3 ± 0.2	2.1 ± 0.28*(↑)
Pepper (Bell Boy)	0.7 ± 0.1	0.6 ± 0.2	0.61 ± 0.07	0.26 ± 0.07*(↓)

These preliminary results are quoted from Murray, Mansfield and Wellburn (in preparation).
Significant differences * $P < 0.05$; ** $P < 0.01$; *** $P < 0.001$; rest not significant.
[a] (μmol NO_2^- produced h^{-1} mg^{-1} protein).
[b] (μmol NO_2^- reduced min^{-1} mg^{-1} protein).

Peppers adopt a different strategy. Levels of nitrite reductase activities in pepper leaves are severely reduced by either NO or NO_2, drawing in fewer oxidized nitrogen-containing compounds into their leaf metabolism. Preliminary results indicate that this response is controlled by the levels of nitrogen-containing compounds available to the roots.

Cereals and grasses appear to respond like tomato to low concentrations of NO_2 by increases in the levels of nitrite reductase activities. SO_2 by itself has no effect. However, if the two pollutants (NO_2 plus SO_2) are present together, the SO_2 appears to remove the ability of the NO_2 to stimulate fresh increases in the levels of nitrite reductase activities (Wellburn et al., 1981). This inhibition of a potential detoxification mechanism of nitrite is believed to be one of the reasons why the $SO_2 + NO_2$ combination exhibits such drastic 'more than additive' effects upon the growth of grasses.

Concluding remarks

A great deal is known about exchange of pollutants, such as SO_2, between the plant and the atmosphere from field and fumigation studies. The importance of the boundary layer and stomatal resistances over cuticular resistance has been stressed (see Hällgren, 1978) and estimates have been made of the various internal resistances (see Unsworth, 1982). The fates of the S and N atoms are also known from studies using $^{35}SO_2$ and $^{15}NO_2$ (Ziegler, 1975; Yoneyama and Sasakawa, 1979; Garsed and Mochrie, 1980) and the importance of the nutrient supply in the case of either SO_2 or NO_2 pollution damage has been realized (Leone and Brennan, 1972; Klein and Jäger, 1976; Matsumaru et al., 1979). Physiological, ultrastructural and biochemical studies of fumigated tissue have revealed that the principal effect of pollutants is upon various mechanisms associated with photosynthesis to which must be added photophosphorylation, as the sections above reveal.

As a consequence of these observations, and largely in parallel to them, a large number of in vitro experiments have been undertaken which have studied the effect of the likely products of atmospheric pollutants (for SO_2 see Petering, 1977) upon organelles (principally chloroplasts), detached membrane systems (usually thylakoids), simpler organisms (algae), model systems (artificial micelles) or enzymes. In vitro studies like some of those described in the preceding sections have revealed a host of likely sites of action of the toxic products. Most of them are concentration-dependent and may enhance as well as inhibit, e.g. $^{14}CO_2$ fixation is increased below 1 mM sulphite and decreased above (Ziegler and Libera, 1975).

It has been evident for some time, however, that between the two approaches to the study of the effects of atmospheric pollutants upon plants (i.e. field and fumigation v. in vitro) a gap exists which hampers and threatens further progress. The major difficulty in establishing which event or function has significance rests upon the concentration of SO_2 or sulphite or NO_2 or nitrite chosen, the period of exposure (in vitro it is necessarily short), the events which remove or add to the toxic levels and other critical factors such as pH and buffering capacity. The problem is best put by quoting directly from two of the best reviews.

Ziegler (1975) states that 'the actual concentration of sulphite within a cell or a cell compartment after fumigation is still completely unknown; however, this is a question that merits central attention and demands resolution' whilst Hällgren (1978) asks 'what are the actual concentrations inside the mesophyll cells and in the chloroplasts, after fumigation with SO_2, at different dosages? There are several good reasons that warrant an answer to this question'.

A similar situation prevails for NO_x. Examination of the nitrogen metabolism and nitrite toxicity literature has failed to reveal assessments of concentrations of nitrate and nitrite within the various cellular compartments. Even the assessments within the plastids, cytoplasm or vacuole of orthophosphate (with which sulphite competes) have been obtained in an indirect manner using ^{32}P and are far from definitive (Bieleski, 1973). Clearly in the future the range of internal concentrations at sites of action must be known if we are to progress in our understanding not just of effects of pollutants upon events like photophosphorylation but to appreciate the normal mechanisms involved in plant metabolism as a whole.

Acknowledgements

I am grateful to Dr Marijana Plesničar (Department of Pesticides, University of Belgrade, Yugoslavia) for making available to me the recent results of her group, to Dr Brian Tabner (Department of Chemistry, University of Lancaster) for undertaking and advising upon the ESR spectroscopy and to all the other members of the air pollution research group at Lancaster.

References

ASADA, K., DEURA, R. and KASAI, Z. Effect of sulphate ions on photophosphorylation by spinach chloroplasts, *Plant & Cell Physiology* 9, 143–146 (1968)

BALDRY, C. W., COCKBURN, W. and WALKER, D. A. Inhibition, by sulphate, of the oxygen evolution associated with photosynthetic carbon assimilation, *Biochimica et Biophysica Acta* 153, 476–483 (1968)

BIELESKI, R. L. Phosphate pools, phosphate transport, and phosphate availability, *Annual Review of Plant Physiology* 29, 225–252 (1973)

BLANKENSHIP, R. E., BABCOCK, G. T. and SAUER, K. Kinetic study of oxygen evolution parameters in Tris-washed, reactivated chloroplasts, *Biochemica et Biophysica Acta* 387, 165–175 (1975)

CEROVIĆ, Z. G., KALEZIĆ, R. and PLESNIČAR, M. The role of photophosphorylation in SO_2 and SO_3^{2-} inhibition of photosynthesis in isolated chloroplasts, *Planta* 156, 249–254 (1983)

CHIMIKLIS, P. E. and HEATH, R. L. Ozone-induced loss of intracellular potassium ion from *Chlorella sorokiniana*, *Plant Physiology* 56, 723–727 (1975)

COULSON, C. L. and HEATH, R. L. The interaction of peroxyacetyl nitrate (PAN) with the electron flow of isolated chloroplasts, *Atmospheric Environment* 9, 231–238 (1975)

GARSED, S. G. and MOCHRIE, A. Translocation of sulphite in *Vicia faba* L., *New Phytologist* 84, 421–428 (1980)

GRANT, B. R., LABELLE, R. and MANGAT, B. S. The action of nitrite on NADP reduction by intact spinach chloroplasts, *Planta* 106, 181–184 (1972)

HALL, D. O. and TELFER, A. The effect of sulphate and sulphite on photophosphorylation by spinach chloroplasts, *Progress in Photosynthesis Research* 3, 1281–2387 (1969)

HÄLLGREN, J.-E. Physiological and biochemical effects of sulphur dioxide on plants. In *Sulphur in the Environment*, (Ed. J. O. NRIAGU), pp. 163–209, John Wiley & Sons, New York (1978)

HAMPP, R. Rapid separation of the plastid, mitochondrial and cytoplasmic fractions from intact leaf protoplasts of *Avena*. Determination of *in vivo* ATP pool sizes during greening, *Planta* 150, 291–298 (1980)

HAMPP, R. and ZIEGLER, I. Sulfate and sulfite translocation via the phosphate translocation of the inner envelope membrane of chloroplasts, *Planta* 137, 309–312 (1977)

HARAUX, F. and KOUCHKOVSKY, Y. de. Quantitative estimation of the photosynthetic proton binding inside the thylakoids by correlating internal acids-fixation to external alkalinisation and to oxygen evolution in chloroplasts, *Biochimica et Biophysica Acta* 546, 455–471 (1979)

HARVEY, G. W. and LEGGE, A. H. The effect of sulphur dioxide upon the metabolic level of adenosine triphosphate, *Canadian Journal of Botany* 57, 759–764 (1979)

HEATH, R. L. Initial events in injury to plants by air pollutants, *Annual Review of Plant Physiology* **31**, 395–431 (1980)

HEATH, R. L. and FREDERICK, P. E. Ozone alteration of membrane permeability in *Chlorella*. I. Permeability of potassium ion as measured by ^{86}Rb tracer, *Plant Physiology* **64**, 455–459 (1980)

HEBER, U. and PURCZELD, P. Substrate and product fluxes across the chloroplast envelope during bicarbonate and nitrite reduction. In *Proceedings of the IV International Congress on Photosynthesis*, (Eds. D. O. HALL, J. COOMBS and T. W. GOODWIN), pp. 107–118, The Biochemical Society, London (1978)

HILLER, R. G. and BASSHAM, J. A. Inhibition of CO_2 fixation by nitrous acid, *Biochimica et Biophysica Acta* **109**, 607–610 (1965)

HOFFMANN, J., PAHLICH, E. and STEUBING, L. Enzymatischanalytische Untersuchungen zum Adenosinphosphatgehalt SO_2-begaster Erbsen, *International Journal of Environmental Analytical Chemistry* **4**, 183–196 (1976)

KHANNA, R., RAJAN, S., GOVINDJEE, and GUTOWSKY, H. S. NMR and EST studies of thylakoid membranes. In *Proceedings of the Vth International Congress on Photosynthesis*, (Ed. G. AKOYUNOGLOU), Vol. II, pp. 307–316, Balaban International Science Services, Philadelphia (1981)

KLEIN, H. and JÄGER, M.-J. Einfluss der Nährstoffversorgung auf die SO_2-Empfindlichkeit von Erbsenpflanzen, *Zeitschrift für Pflanzenkrankheiten und Pflanzenschutz* **83**, 555–568 (1976)

KOUKOL, J., DUGGER, W. M. and BELSER, N. O. The inhibition of cyclic photophosphorylation by peroxyacetyl nitrate, *Plant Physiology* **38** (Supplement) xii (1963)

KOUKOL, J., DUGGER, W. M. and PALMER, R. L. Inhibitory effect of peroxyacetyl nitrate on cyclic photophosphorylation by chloroplasts from black valentine bean leaves, *Plant Physiology* **42**, 1419–1422 (1967)

KROGMANN, D. W., JAGENDORF, A. T. and AVRON, M. Uncouplers of spinach chloroplast photosynthetic phosphorylation, *Plant Physiology* **34**, 272–277 (1959)

LARSSON, C.-M. Photophosphorylation in *Scenedesmus in vivo*: O_2 evolution, ATP pools and transients, and phosphate binding during photoreduction of NO_3^-, NO_2^- and CO_2, *Physiologia Plantarum* **48**, 326–332 (1980)

LEFFLER, H. R. and CHERRY, J. H. Destruction of enzymatic activities of corn and soybean leaves exposed to ozone, *Canadian Journal of Botany* **52**, 1233–1238 (1974)

LEONE, I. A. and BRENNAN, E. Sulphur nutrition as it contributes to the susceptibility of tobacco and tomato to SO_2 injury, *Atmospheric Environment* **6**, 259–266 (1972)

LÜTTGE, U., OSMOND, C. B., BALL, E., BRINCKMANN, E. and KINZE, G. Bisulphite compounds as metabolic inhibitors: non specific effects on membranes, *Plant and Cell Physiology* **13**, 505–514 (1972)

MATSUMARU, T., YONEYAMA, T., TOTSUKA, T. and SHIRATORI, K. Absorption of atmospheric NO_2 by plants and soils, *Soil Science and Plant Nutrition* **25**, 255–265 (1979)

MILLS, J. D., SLOVACEK, R. E. and HIND, G. Cyclic electron transport in isolated intact chloroplasts, *Biochimica et Biophysica Acta* **504**, 298–309 (1978)

MOURIOUX, G. and DOUCE, R. Transport spécifique du sulphate à travers l'enveloppe des chloroplastes d'Épinard, *Compte rendu de l'Académie des Sciences, Paris* **286**, 277–280 (1978)

MUDD, J. B. Effects of oxidants on metabolic function. In *Effects of Gaseous Air Pollution in Agriculture and Horticulture*, (Eds. M. H. UNSWORTH and D. P. ORMROD), pp. 189–203, Butterworths, London (1982)

PELL, E. J. and BRENNAN, E. Changes in respiration, photosynthesis, adenosine 5'-triphosphate, and total adenylate content of ozonated pinto bean foliage as they relate to symptom expression, *Plant Physiology* **51**, 378–381 (1973)

PETERING, D. H. Sulphur dioxide: A view of its reactions with biomolecules. In *Biochemical Effects of Environmental Pollutants*, (Ed. S. D. LEE), pp. 293–306, Ann Arbor Science Publishers Inc., Ann Arbor (1977)

PLESNIČAR, M. and KALEZIĆ, R. Sulphite inhibition of oxygen evolution associated with photosynthetic carbon assimilation, *Periodicum Biologorum* **82**, 297–301 (1980)

RADMER, R. and CHENIAE, G. Mechanisms of oxygen evolution. In *Primary Processes of Photosynthesis*, (Ed. J. BARBER), pp. 303–351, Elsevier/North Holland, Amsterdam (1977)

ROWLANDS, J. R., ALLEN-ROWLANDS, C. J. and GAUSE, E. M. Effects of environmental agents on membrane dynamics. In *Biochemical Effect of Environmental Pollutants*, (Ed. S. D. LEE), pp. 203–246, Ann Arbor Science Publishers Inc., Ann Arbor (1977)

RYRIE, I. J. and JAGENDORF, A. T. Inhibition of photophosphorylation in spinach chloroplasts by inorganic sulphate, *Journal of Biological Chemistry* **246**, 582–588 (1971)

SAWHNEY, S. K., NAIK, M. S. and NICHOLAS, D. J. D. Regulation of nitrate reduction by light, ATP and mitochondrial respiration in wheat leaves, *Nature* **272**, 647–648 (1978)

SCHREIBER, U., VIDAVER, W., RUNECKLES, V. C. and ROSEN, P. Chlorophyll fluorescence assay for ozone injury in intact plants, *Plant Physiology* **61**, 80–84 (1978)

SCHULDINER, S., ROTTENBERG, H. and AVRON, M. Determination of ΔpH in chloroplasts, *European Journal of Biochemistry* **25**, 64–70 (1972)

SILVIUS, J. E., INGLE, M. and BAER, C. H. Sulphur dioxide inhibition of photosynthesis in isolated spinach chloroplasts, *Plant Physiology* **56**, 434–437 (1975)

SLOVACEK, R. E. and HIND, G. Correlation between photosynthesis and the trans-thylakoid proton gradient, *Biochemica et Biophysica Acta* **635**, 393–404 (1981)

SPEDDING, D. J., ZIEGLER, I., HAMPP, R. and ZIEGLER, H. Effect of pH on the uptake of ^{35}S-sulphur from sulphate, sulphite and sulphide by isolated spinach chloroplasts, *Zeitschrift für Pflanzenphysiologie* **96**, 351–364 (1980)

TINGEY, D. T., FITES, R. C. and WICKLIFE, C. Ozone alteration of nitrate reduction in soybean, *Physiologia Plantarum* **29**, 33–38 (1973)

TOMLINSON, H. and RICH, S. The ozone resistance of leaves as related to their sulfhydryl and adenosine triphosphate content, *Phytopathology* **58**, 808–810 (1968)

UNSWORTH, M. H. Exposure to gaseous pollutants and uptake by plants. In *Effects of Gaseous Air Pollution in Agriculture and Horticulture*, (Eds. M. H. UNSWORTH and D. P. ORMROD), pp. 43–63, Butterworths, London (1982)

VENNESLAND, B. and JETSCHMANN, C. The nitrate dependence of the inhibition of photosynthesis by carbon monoxide in *Chlorella*, *Archives of Biochemistry and Biophysics* **144**, 428–437 (1971)

WELLBURN, A. R., HIGGINSON, C., ROBINSON, D. and WALMSLEY, C. Biochemical explanations of more than additive inhibitory effects of low atmospheric levels of sulphur dioxide plus nitrogen dioxide upon plants, *New Phytologist* **88**, 223–237 (1981)

WELLBURN, A. R., WILSON, J. and ALDRIDGE, P. H. Biochemical responses of plants to nitric oxide polluted atmospheres, *Environmental Pollution (Series A)* **22**, 219–228 (1980)

WITT, H. T., RUMBERG, B. and JUNGE, W. Electron transfer, field changes, proton translocation and phosphorylation in photosynthesis. Coupling in the thylakoid membrane. In *Biochemie des Sauerstoff*, (Eds. B. HESS and H. J. STAUDINGER), pp. 262–306, Springer-Verlag, Berlin (1968)

YONEYAMA, T. and SASAKAWA, H. Transformation of atmospheric NO_2 absorbed in spinach leaves, *Plant and Cell Physiology* **20**(1), 263–266 (1979)

YONEYAMA, T., SASAKAWA, H., ISHIZUKA, S. and TOTSUKA, T. Absorption of atmospheric NO_2 by plants and soils, *Soil Science and Plant Nutrition* **25**(2), 267–275 (1979)

ZIEGLER, I. The effect of SO_2 pollution on plant metabolism, *Residue Reviews* **56**, 79–105 (1975)

ZIEGLER, I. and HAMPP, R. Control of $^{35}SO_4^{2-}$ and $^{35}SO_3^{2-}$ incorporation into spinach chloroplasts during photosynthetic CO_2 fixation, *Planta* **137**, 303–307 (1977)

ZIEGLER, I. and LIBERA, W. The enhancement of CO_2 fixation in isolated chloroplasts by low sulphite concentrations and by ascorbate, *Zeitschrift für Naturforschung* **30c**, 634–637 (1975)

Chapter 16

Phytotoxic air pollutants and oxidative phosphorylation

D. J. Ballantyne

DEPARTMENT OF BIOLOGY, UNIVERSITY OF VICTORIA, CANADA

Introduction

While there has not been a great deal of work carried out on the effects of phytotoxic air pollutants on oxidative phosphorylation in plants, such effects could help to explain the many adverse effects of air pollutants on the agricultural productivity and appearance of plants as outlined by such authors as Treshow (1970), Jacobson and Hill (1970) and Ormrod (1978). Most authors of basic texts of plant physiology including Bidwell (1979), Salisbury and Ross (1978) and Ting (1982) have outlined the role of ATP produced by oxidative phosphorylation in such processes as growth, ion transport and membrane permeability, phloem translocation and the synthesis of macromolecules such as nucleic acids, proteins and lipids. Perhaps one of the principal reasons for a lack of work in this area has been the relative difficulty of preparing phosphorylating mitochondria from green leaves, although such procedures have long been available (Smillie, 1955; Pierpoint, 1960; Macdowall, 1965; Lee, 1967; Douce, Moore and Neuberger, 1977). Because of the manipulations involved, such techniques have generally been limited to a few plants such as spinach (*Spinacea oleracea*) and tobacco (*Nicotiana tabacum*) with relatively large leaves, which is a further limitation on the widespread use of such studies in investigating the effects of phytotoxic air pollutants. The effects of ozone (O_3), F^- and Cd^{2+} have all been studied on oxidative phosphorylation in preparations of plant mitochondria although only the *in vivo* effects of O_3 and F^- have been investigated with mitochondria isolated from treated plants (Macdowall, 1965; Lee, 1967; Miller and Miller, 1974). Studies with Cd^{2+} have employed aetiolated shoots of maize (*Zea mays*) (Miller, Bittell and Koeppe, 1973). Similarly, in studies on the effect of SO_3^{2-} on mitochondrial ATP formation I have used hypocotyls of aetiolated bean (*Phaseolus vulgaris*) seedlings (Ballantyne, 1973).

In this chapter I will discuss the inhibiting effect of phytotoxic air pollutants on oxidative phosphorylation of plant mitochondria together with possible reasons suggested for the observed inhibitions.

Ozone

The initial report on inhibition of oxidative phosphorylation in plant mitochondria by O_3 was made by Macdowall (1965). He demonstrated a considerable reduction in oxidative phosphorylation of tobacco (*Nicotiana tabacum*) leaf mitochondria following gassing of leaf tissue (discs) with O_3. The oxygen uptake of the tobacco leaf mitochondria was increased following O_3 treatments of leaf tissue, suggesting that mitochondrial uncoupling was taking place. Macdowall considered that O_3 might have induced a bypass 'of cytochrome oxidase mediated electron transport'.

Lee (1967, 1968) carried out a detailed series of investigations on the inhibition of tobacco leaf mitochondrial oxidative phosphorylation by O_3. He demonstrated a dramatic decrease in oxidative phosphorylation in O_3 treated leaves, accompanied by a decrease in O_2 uptake of the mitochondria. The O_3-induced decrease in O_2 uptake was not as great as the decrease in phosphorylation. Treatment of mitochondria with O_3 *in vitro* also resulted in decreases in mitochondrial O_2 uptake and phosphorylation. Again mitochondrial phosphorylation was more sensitive to O_3 than mitochondrial O_2 uptake. Lee noted that although exposure to O_3 induced mitochondrial swelling, oxidative phosphorylation seemed to be more sensitive to O_3. Lee (1968) studied the O_3-induced swelling of tobacco leaf mitochondria *in vitro*. This swelling was accompanied by a loss of protein and other substances (presumably nucleotides) from mitochondria, indicating an O_3-induced increase in membrane permeability.

Evans and Ting (1973) and Perchorowicz and Ting (1974) have described the increase in plant cell permeability (both of efflux and influx) induced by O_3, and have considered that cell membranes are a primary target site for O_3. Heath (1980) has recently discussed the reaction of O_3 with double bonds of fatty acids — a factor which could explain O_3-induced membrane responses, although he pointed out a number of other possible modes of action. Pauls and Thompson (1981) have recently described O_3-induced reductions in both unsaturated fatty acids and in sterol/phosphate ratios in microsomal membranes from bean cotyledons. Certainly it is tempting to consider that O_3-induced reductions in oxidative phosphorylation are due to O_3 effects on mitochondrial membranes.

Correlations of ATP levels in O_3-treated tissues could be an indication of the relative importance of oxidative phosphorylation responses. Tomlinson and Rich (1968) have reported that O_3-treatments considerably reduced the ATP levels in tobacco leaves. However, Pell and Brennan (1973) have found increases in ATP levels in leaves from O_3-treated bean plants. They have suggested that the increase in ATP could be due to an increase in nucleic acid breakdown, a decrease in nucleic acid synthesis or to an increase in respiration that could be 'a reflection of cellular injury and not its cause'.

Sulphite (sulphur dioxide)

While phosphorylative mitochondrial preparations have not been made from plants fumigated with SO_2, *in vitro* experiments have been carried out with SO_3^{2-} solutions. After SO_2 enters the cell it is considered to be present as SO_3^{2-} (Heath, 1980). ATP formation in aetiolated bean or maize mitochondria has been inhibited by 3–10 mM SO_3^{2-}, and in aetiolated bean shoot mitochondria by SO_3^{2-} concentrations up to 100 mM (Ballantyne, 1973). Within SO_3^{2-} concentrations that inhibited ATP formation in plant mitochondria a SO_3^{2-}-stimulated mitochondrial oxygen uptake (CN-sensitive) has been detected as has a stimulation of mitochondrial ATPase activity (Ballantyne, 1977; Ballantyne and Black, 1980). SO_3^{2-} stimulation of plant mitochondrial oxygen uptake had been previously reported by Tager and Rautanen (1955, 1956), Arrigoni (1959) and Stickland (1961). The oxidation of SO_3^{2-} by intact plants has been reported by Miller and Xerikos (1979), and by Ayazloo, Garsed and

Bell (1982), in each case taking place most rapidly in SO_2-tolerant cultivars or selections.

Tominaga (1978) has reported a SO_3^{2-}-stimulation of ATPases prepared from bacteria, fungi and leaves of higher plants. Also, SO_3^{2-} has been found to stimulate mitochondrial ATPases of animal tissues (Lambeth and Lardy, 1971) and of yeast (Takeshige et al., 1976). Recktenwald and Hess (1977), in studying the stimulation of yeast mitochondrial ATPase by anions including HSO_3^-, have suggested that anions 'control the affinity of ATP to the site of its hydrolysis in yeast mitochondrial ATPase'. Schmiz (1980), in studying the inhibitory effects of SO_2 or SO_3^{2-} on colony formation in yeast (Saccharomyces cerevisiae) cells, has concluded that oxidative phosphorylation may not be necessary for the inhibition, but that the SO_3^{2-} stimulation of ATPase itself is responsible by inducing a depletion of ATP.

ATP levels have been found to be depressed by fumigation of higher plants with SO_2 or by SO_3^{2-} treatments of algae (Hoffman, Pahlich and Steubing, 1976; Harvey and Legge, 1979; Keck and Schlee, 1981), although in the case of peas (Pisum sativum) there was an initial rise in ATP levels following SO_2 fumigation (Hoffman, Pahlich and Steubing, 1976). Glyoxal bisulfite, which was found in the leaves of SO_2-fumigated wheat (Triticum aestivum) plants (Tanaka, Takanashi and Yatazawa, 1972), can also induce a drop in the levels of ATP in maize leaf slices when applied exogenously (Lüttge et al., 1972). These observations are consistent with the suggestion that SO_2 can inhibit oxidative phosphorylation, although other SO_2-induced phenomena could also induce decreases in ATP levels.

Some investigators have found that SO_2 can influence membrane constituents, although not studying mitochondrial membranes per se. Khan and Malhotra (1977) have found that SO_2 fumigation of pine needles (Pinus contorta) can result in a drop in glycolipids, and in the linolenic content of glycolipids. Also, in SO_2-fumigated pine (P. contorta and P. banksiana) needles there was a reduced biosynthesis of lipids (Malhotra and Khan, 1978). Lipid peroxidation in SO_2-fumigated leaves from singlet oxygen produced from superoxide radicals has been considered (Shimazaki et al., 1980). If leaves contained more superoxide dismutase activity they were less likely to suffer damage from SO_2 (Tanaka and Sugahara, 1980). Malhotra and Hocking (1976) have reviewed earlier work on SO_2 influences on both membranes and proteins (SO_2 may break disulphide bonds in proteins).

Observations on the inhibition of dehydrogenases by SO_3^{2-} are consistent with SO_3^{2-} inhibition of oxidative phosphorylation. SO_3^{2-} has been found to inhibit the activity of malate dehydrogenase (Ziegler, 1974a, 1974b). Sulphur dioxide fumigations have reduced malate dehydrogenase activity in pine (Pinus banksiana) needles (Sarkar and Malhotra, 1979) and isocitrate dehydrogenase activity in leaves of alfalfa (Medicago sativa) and pansy (Viola tricolor) (Rabe and Kreeb, 1980). Lederer (1978) has suggested that SO_3^{2-} may bind to the active site of a dehydrogenase.

Fluoride

Initial studies on F^- inhibition of tricarboxylic acid cycle dehydrogenases and oxidative phosphorylation were carried out by Lovelace and Miller (1967). Miller and Miller (1974) carried out detailed studies on the inhibition of soybean (Glycine max) mitochondrial oxidative phosphorylation by F^-. They noted a decrease in mitochondrial oxidative phosphorylation of SO_2-fumigated leaves as well as an increase in mitochondrial ATPase activity. In vitro studies with mitochondria prepared from aetiolated soybean hypocotyls indicated a F^--induced reduction in oxidation of succinate, malate and NADH and a decrease in the rate of phosphorylation. Also, F^- induced an increased rate of swelling in mitochondria prepared from aetiolated corn shoots, and a loss of protein from such mitochondria.

More recently, work has focused on the influence of F^- on membranes. Psenak *et al.* (1977) reported that while F^- could inhibit mitochondrial malic dehydrogenase it had little effect on isolated malate dehydrogenase isozymes. They interpreted these data as an indication that F^- was having more influence on mitochondrial membranes than on enzymes *per se*. Simola and Koskimies (1980) found that F^- inhibited lengthening of fatty acid chains in leaves of moss, *Sphagnum fimbriata*, and that with F^- treatment there was more linolenate and less oleate and linoleate in leaves. Perhaps such changes could occur in mitochondrial membranes.

F^- has induced increased levels of phosphorylated nucleotides in plants (McNulty and Lords, 1960) which, however, is not consistent with reduced rates of oxidative phosphorylation or increased rates of ATPase activity. Treshow and Harner (1968) have discussed work by the same authors demonstrating that F^- has induced increased ATP concentrations in bean leaf discs.

Other pollutants (Cd^{2+}, Pb^{2+} and NH_4^+)

Ormrod (1978) has considered cadmium to be an air pollutant. The effects of Cd^{2+} on plant mitochondrial oxidative phosphorylation have been studied extensively. *In vitro* experiments have demonstrated that Cd^{2+} inhibited oxidative phosphorylation in mitochondria prepared from aetiolated maize hypocotyls (Miller, Bittell and Koeppe, 1973). However, Cd^{2+} was even more effective in inhibiting State 4 or minus phosphate respiration, indicating that the site of Cd^{2+} action may be early in electron transport. Cd^{2+} induced mitochondrial swelling, indicating a change in membrane properties. Because dithiothreitol prevented Cd^{2+} induced mitochondrial swelling it was thought that sulphydryl groups could be involved in a Cd^{2+} membrane interaction. Further studies demonstrated that other heavy metals (Zn^{2+}, Ni^{2+}, Co^{2+}) could also uncouple mitochondrial metabolism and that Zn^{2+} and Pb^{2+} could induce mitochondrial swelling (Bittell, Koeppe and Miller, 1974). Keck (1978) and De Filippis, Hampp and Ziegler (1981) have reported decreases in plant ATP levels induced by Cd^{2+}. Lue-Kim, Wozniak and Fletcher (1980) found that Cd^{2+} applied to the alga, *Chlorella*, induced a disruption of mitochondria. The mitochondrial membrane would appear to be one of the sites of Cd^{2+} action.

Pb^{2+} particles have also been described as air pollutants (Lerman and Darley, 1975). In *in vitro* experiments, Pb^{2+} has been found to inhibit succinate oxidation in mitochondria prepared from aetiolated corn shoots (Koeppe and Miller, 1970) and to induce mitochondrial swelling (Bittell, Koeppe and Miller, 1974). However, because Pb^{2+} reacts with phosphate, the effects of Pb^{2+} on oxidative phosphorylation have not been investigated.

NH_4^+ has been applied to cucumber (*Cucumis sativus*) plants and the leaves harvested and mitochondria prepared from the leaves (Matsumoto, Wakiuchi and Takahashi, 1971). The activity of all mitochondrial enzymes except ATPase increased due to NH_4^+ treatments. Thus, there did not seem to be uncoupling of mitochondrial metabolism. Levels of ATP in treated cucumber leaves were investigated later and rose considerably due to NH_4^+ treatments (Matsumoto and Wakiuchi, 1974). NH_4^+ did not appear to inhibit oxidative phosphorylation in cucumber leaf mitochondria.

Table 16.1 Effects of phytotoxic air pollutants on plant mitochondria

Pollutant	Type of experiment	ATP formation	Effect on				References
			Mito-chondrial ATPase	Mito-chondrial swelling	Mito-chondrial protein loss		
O_3	In vivo	Decrease	—	—	—		Macdowall (1965)
O_3	In vivo	Decrease	—	—	—		Lee (1967)
	In vitro	Decrease	—	Increase	Increase		Lee (1968)
SO_3^{2-}	In vitro	Decrease	Increase	—	—		Balantyne (1973) Ballantyne and Black (1980)
F^-	In vivo	Decrease	Increase	—	—		Miller and Miller (1974)
	In vitro	Decrease	—	Increase	Increase		Miller and Miller (1974)
Cd^{2+}	In vitro	Decrease	—	Increase	—		Miller, Bittell and Koeppe (1973) Bittell, Koeppe and Miller (1974)

Discussion

Phytotoxic air pollutants appear to be inhibitors of mitochondrial phosphorylation, at least in the case of O_3, SO_3^{2-}, F^- and Cd^{2+}. These pollutants either stimulate mitochondrial ATPase activity or increase mitochondrial membrane permeability (*Table 16.1*). Other stress factors may influence mitochondrial phosphorylation in an adverse manner. Flowers and Hanson (1969) have demonstrated the inhibiting effects of drought on the phosphorylation of aetiolated soybean hypocotyl mitochondria. Heber and Santarius (1964) found that freezing could eliminate phosphorylation in spinach leaf mitochondria. Lyons and Raison (1970) reported on the inhibiting influence of chilling temperatures (0–10 °C) on the rates of oxidative phosphorylation of mitochondria prepared from sweet potato roots, and cucumber and tomato fruits. Certainly the mitochondria are a definite target for a number of stress factors, including air pollutants. On the other hand, other phytoactive chemicals like phytohormones rarely have a direct effect on the phosphorylative activity of mitochondria.

Schemes on the mode of action of phytotoxic air pollutants have been presented by Heck (1973) and Heath (1980). From these schemes, which have been prepared from available research, it is easy to conclude that pollutant induced *in vivo* mitochondrial inhibitions could be due to pollutant induced changes to the genetic code, enzyme effects, osmotic imbalances, or pollutant poisoning of other cytoplasmic organelles or membranes. In any event mitochondrial damage may account for a considerable amount of pollutant induced necrosis and decreased agricultural productivity.

References

ARRIGONI, O. The enzymic oxidation of sulphite in mitochondrial preparations of pea internodes, *Italian Journal of Biochemistry* **8**, 181–186 (1959)

AYAZLOO, M., GARSED, S. G. and BELL, J. N. B. Studies on the tolerance to sulphur dioxide of grass populations in polluted areas. II. Morphological and physiological investigations, *New Phytologist* **90**, 109–126 (1982)

BALLANTYNE, D. J. Sulphite inhibition of ATP formation in plant mitochondria, *Phytochemistry* **12**, 1207–1209 (1973)

BALLANTYNE, D. J. Sulphite oxidation by mitochondria from green and etiolated peas, *Phytochemistry* **16**, 49–50 (1977)

BALLANTYNE, D. J. and BLACK, M. W. Sulphite stimulation of bean mitochondrial adenosine triphosphatase, *Phytochemistry* **19**, 2021 (1980)

BIDWELL, R. G. S. *Plant Physiology*, 2nd edn, pp. 192–233; 293–302; 562, Macmillan, New York (1979)

BITTELL, J. E., KOEPPE, D. E. and MILLER, R. J. Sorption of heavy metal cations by corn mitochondria and the effects on electron and energy transfer reactions, *Physiologia Plantarum* **30**, 226–230 (1974)

DOUCE, R., MOORE, A. L. and NEUBERGER, M. Isolation and oxidative properties of intact mitochondria isolated from spinach leaves, *Plant Physiology* **60**, 625–628 (1977)

DE FILIPPIS, L. F., HAMPP, R. and ZIEGLER, H. The effect of sublethal concentrations of zinc, cadmium and mercury on *Euglena gracilis*: Adenylates and energy charge, *Zeitschrift für Pflanzenphysiologie* **103**, 1–8 (1981)

EVANS, L. S. and TING, I. P. Ozone-induced membrane permeability changes, *American Journal of Botany* **60**, 155–162 (1973)

FLOWERS, T. J. and HANSON, J. B. The effect of reduced water potential on soybean mitochondria, *Plant Physiology* **44**, 939–945 (1969)

HARVEY, G. and LEGGE, A. H. The effect of sulfur dioxide upon the metabolic level of adenosine triphosphate, *Candian Journal of Botany* **57**, 759–764 (1979)

HEATH, R. L. Initial events in injury to plants by air pollutants, *Annual Review of Plant Physiology* **31**, 395–431 (1980)

HEBER, U. W. and SANTARIUS, K. A. Loss of adenosine triphosphate synthesis caused by freezing and its relationship to frost hardiness problems, *Plant Physiology* **39**, 712–719 (1964)

HECK, W. W. Air pollution and the future of agricultural production. In *Air Pollution Damage to Vegetation*, (Ed. J. A. NAEGELE), pp. 118–129, *Advances in Chemistry Series 122*, American Chemical Society, Washington, DC (1973)

HOFFMAN, J., PAHLICH, E. and STEUBING, L. Enzyme analytic studies on the adenosine phosphate content of peas exposed to sulphur dioxide gas, *International Journal of Environmental Analytical Chemistry* **4**, 183–196 (In German) (original not seen; 1976. *Biological Abstracts* **62**, 563) (1976)

JACOBSON, S. and HILL, A. C. *Recognition of Air Pollution Injury to Vegetation: a Pictorial Atlas*, Air Pollution Control Association, Pittsburgh (1970)

KECK, R. W. Cadmium alteration of root physiology and potassium ion fluxes, *Plant Physiology* **62**, 94–96 (1978)

KECK, M. and SCHLEE, D. Effect of sulphite on adenine nucleotides of the green alga *Trebouxia*, *Phytochemistry* **20**, 2089–2092 (1981)

KHAN, A. A. and MALHOTRA, S. S. Effects of aqueous sulphur dioxide on pine needle glycolipids, *Phytochemistry* **16**, 539–543 (1977)

KOEPPE, D. E. and MILLER, R. J. Lead effects on corn mitochondrial respiration, *Science* **167**, 1376–1378 (1970)

LAMBETH, D. O. and LARDY, H. A. Purification and properties of rat-liver-mitochondrial adenosine triphosphatase, *European Journal of Biochemistry* **22**, 355–363 (1971)

LEDERER, F. Sulfite binding to a flavodehydrogenase cytochrome β from baker's yeast, *European Journal of Biochemistry* **88**, 425–432 (1978)

LEE, T. T. Inhibition of oxidative phosphorylation and respiration by ozone in tobacco mitochondria, *Plant Physiology* **42**, 692–696 (1967)

LEE, T. T. Effect of ozone on swelling of tobacco mitochondria, *Plant Physiology* **43**, 133–139 (1968)

LERMAN, S. L. and DARLEY, E. F. Particulates. In *Responses of Plants to Air Pollution*, (Eds. J. B. MUDD and T. T. KOZLOWSKI), p. 153, Academic Press, New York (1975)

LOVELACE, C. J. and MILKLER, G. W. *In vitro* effects of fluoride on tricarboxylic acid cycle dehydrogenases and oxidative phosphorylation, *Journal of Histochemistry and Cytochemistry* **15**, 195–201 (1967)

LUE-KIM, H., WOZNIAK, P. C. and FLETCHER, R. A. Cadmium toxicity on synchronous populations of *Chlorella ellipsoida*, *Canadian Journal of Botany* **58**, 1780–1788 (1980)

LÜTTGE, U., OSMOND, C. B., BALL, E., BRINCKMANN, E. and KINZE, G. Bisulfite compounds as metabolic inhibitors: nonspecific effects on membranes, *Plant and Cell Physiology* **13**, 505–514 (1972)

LYONS, J. M. and RAISON, J. K. Oxidative activity of mitochondria isolated from plant tissues sensitive and resistant to chilling injury, *Plant Physiology* **45**, 386–389 (1970)

MACDOWALL, F. D. H. Stages of ozone damage to respiration of tobacco leaves, *Canadian Journal of Botany* **43**, 419–427 (1965)

MALHOTRA, S. S. and HOCKING, D. Biochemical and cytological effects of sulphur dioxide on plant metabolism, *New Phytologist* **16**, 227–237 (1976)

MALHOTRA, S. S. and KHAN, A. A. Effects of sulfur dioxide fumigation on lipid biosynthesis in pine needles, *Phytochemistry* **17**, 241–244 (1978)

MATSUMOTO, H. and WAKIUCHI, N. Changes in ATP in cucumber leaves during ammonium toxicity, *Zeitschrift für Pflanzenphysiologie* **73**, 82–85 (1974)

MATSUMOTO, H., WAKIUCHI, N. and TAKAHASHI, E. Changes of some mitochondrial enzyme activities of cucumber leaves during ammonium toxicity, *Physiologia Plantarum* **25**, 353–357 (1971)

McNULTY, I. B. and LORDS, J. L. Possible explanation of fluoride-induced respiration in *Chlorella pyrenoidosa*, *Science* **132**, 1553–1554 (1960)

MILLER, J. E. and MILLER, G. W. Effects of fluoride on mitochondrial activity in higher plants, *Physiologia Plantarum* **32**, 115–131 (1974)

MILLER, J. E. and XERIKOS, P. B. Residence time of sulphite in SO_2 sensitive and tolerant soybean cultivars, *Environmental Pollution* **18**, 259–264 (1979)

MILLER, R. J., BITTELL, J. E. and KOEPPE, D. E. The effect of cadmium on electron and energy transfer reactions in corn mitochondria, *Physiologia Plantarum* **28**, 166–171 (1973)

ORMROD, D. P. *Pollution in Horticulture*, pp. 3–64; 114–195, Elsevier, Amsterdam (1978)

PAULS, K. P. and THOMSPON, J. E. Effects of *in vitro* treatment with ozone on the physical and chemical properties of membranes, *Physiologia Plantarum* **53**, 255–262 (1981)

PELL, E. J. and BRENNAN, E. Changes in respiration, photosynthesis, adenosine-5'-triphosphate, and total adenylate content of ozonated pinto bean foliage as they relate to symptom expression, *Plant Physiology* **51**, 378–381 (1973)

PIERPOINT, W. S. Mitochondrial preparations from the leaves of tobacco (*Nicotiana tabacum*), *Biochemical Journal* **75**, 504–511 (1960)

PERCHOROWICZ, J. T. and TING, I. P. Ozone effects on plant cell permeability, *American Journal of Botany* **61**, 787–793 (1974)

PSENAK, M., MILLER, G. W., YU, M. H. and LOVELACE, C. J. Separation of malic dehydrogenase tissue in relation to fluoride treatment, *Fluoride* **10**, 63–72 (1977)

RABE, R. and KREEB, K. H. Effects of sulfur dioxide upon enzyme activity in plant leaves, *Zeitschrift für Pflanzenphysiologie* **97**, 215–226 (1980)

RECKTENWALD, D. and HESS, B. Allosteric influence of anions on mitochondrial ATPase of yeast, *FEBS Letters* **76**, 25–28 (1977)

SALISBURY, F. B. and ROSS, C. W. *Plant Physiology*, pp. 76–78; 107; 197–205; 208–212; 247, Wadsworth, Belmont, California (1978)

SARKAR, S. K. and MALHOTRA, S. S. Effects of sulfur dioxide on organic acid content and malate dehydrogenase activity in jack pine (*Pinus banksiana*) needles, *Biochemie und Physiologie der Pflanzen* **174**, 438–445 (1979)

SCHMIZ, K.-L. The effect of sulfite on the yeast, *Saccharomyces cerevisiae*, *Archives of Microbiology* **125**, 89–96 (1980)

SHIMAZAKI, K.-I., SAKAKI, T., KONDO, D. and SAGAHARA, K. Active oxygen participation in chlorophyll destruction and lipid peroxidation in sulfur dioxide-fumigated leaves of spinach (*Spinacea oleracea* cultivar New Asia), *Plant and Cell Physiology* **21**, 1193–1204 (1980)

SIMOLA, L. K. and KOSKIMIES, K. The effect of fluoride on the growth and fatty acid composition of *Sphagnum fimbriatum* at two temperatures, *Physiologia Plantarum* **50**, 74–77 (1980)

SMILLIE, R. M. Enzyme activity of particles isolated from various tissues of the pea plant, *Australian Journal of Biological Sciences* **8**, 186–195 (1955)

STICKLAND, R. G. Oxidation of reduced pyridine nucleotides and of sulphite by pea root mitochondria, *Nature* **190**, 648–649 (1961)

TAGER, J. M. and RAUTANEN, N. Sulphite oxidation by a plant mitochondrial system, *Biochimica et Biophysica Acta* **18**, 111–121 (1955)

TAGER, J. M. and RAUTANEN, N. Sulphite oxidation by a plant mitochondrial system. Enzymic and non-enzymic oxidation, *Physiologia Plantarum* **9**, 665–673 (1956)

TAKESHIGE, K., HESS, B., BOHM, M. and ZIMMERMAN-TELSCHOW, R. Mitochondrial adenosine triphosphatase from yeast, *Saccharomyces cerevisiae*: purification, subunit structure and kinetics, *Hoppe-Seyler's Zeitschrift der Physiologische Chemie* **357**, 1605–1622 (1976)

TANAKA, H. and SUGAHARA, K. Role of superoxide dismutase activity with sulfur dioxide fumigation, *Plant and Cell Physiology* **21**, 601–612 (1980)

TANAKA, H., TAKANASHI, T. and YATAZAWA, M. Experimental studies on sulfur dioxide injuries in higher plants. I. Formation of glyoxal-bisulfite in plant leaves exposed to sulfur dioxide, *Water, Air, and Soil Pollution* **1**, 205–211 (1972)

TING, I. P. *Plant Physiology*, pp. 161–164; 200–205; 228–231; 341–342; 365, Addison-Wesley, Reading, Mass. (1982)

TOMINAGA, N. Comparative studies on the effects of anions on ATPase and acid phosphatase activities from different sources, *Plant and Cell Physiology* **19**, 627–636 (1978)

TOMLINSON, H. and RICH, S. The ozone resistance of leaves as related to their sulphydryl and adenosine triphosphate content, *Phytopathology* **58**, 808–810 (1968)

TRESHOW, M. *Environment and Plant Response*, pp. 245–376, McGraw-Hill, New York (1970)

TRESHOW, M. and HARNER, F. M. Growth responses of pinto bean and alfalfa to sublethal fluoride concentrations, *Canadian Journal of Botany* **46**, 1207–1210 (1968)

ZIEGLER, I. Action of sulphite on plant malate dehydrogenase, *Phytochemistry* **13**, 2411–2416 (1974a)

ZIEGLER, I. Malate dehydrogenase in *Zea mays*: properties and inhibition by sulfite, *Biochimica et Biophysica Acta* **364**, 28–37 (1974b)

Chapter 17

The effect of air pollutants on apparent respiration

V. J. Black

DEPARTMENT OF HUMAN SCIENCES, UNIVERSITY OF TECHNOLOGY,
LOUGHBOROUGH, UK

Introduction

Exposure of plants to air pollutants may result not only in leaf damage and reduction in growth and yield of crops, but also in interference with physiological processes, in particular photosynthesis and transpiration (*see* Mudd and Kozlowski, 1975; Weinstein, 1977; Unsworth and Ormrod, 1982, for reviews). The available evidence for effects of air pollutants permits a relatively confident prediction of plant responses to pollutants. Exposure of plants to pollutants at high concentrations, or for long periods, generally results in the development of symptoms of visible injury and associated physiological disturbances. These responses are largely irreversible and may lead ultimately to reductions in plant growth, development and yield. In contrast, visible injury does not result when exposures are short or pollutant concentrations low, but physiological disturbances and growth reductions are frequently observed. When plants are exposed to very low pollutant concentrations, they may show no physiological responses or alternatively, may exhibit small physiological disturbances. These are usually reversible and do not necessarily lead to reduced growth or dry matter production. Indeed, a stimulation of growth may be observed occasionally (Bennett, Resh and Runeckles, 1974; Cowling and Koziol, 1982). Within this general scheme of plant responses, it is generally recognized that many factors, plant, pollutant and environmental, will influence relative sensitivity to a range of air pollutants. These include the toxicity of the pollutant itself, the concentration, frequency and duration of exposure to pollutant, stomatal behaviour and pollutant uptake by the plant, and prevailing environmental conditions such as light, humidity and temperature. In addition, responses may be modified by the tissue examined, leaf age, nutrient status, previous plant history and the species or variety of the plant itself.

Although the respiratory processes, i.e. dark respiration and photorespiration, are important components of the carbon budget, evidence for pollutant-induced modifications of respiration are less well documented than for photosynthesis. Respiration is vital to supply energy and metabolites, not only for normal metabolism and growth, but also for repair and detoxification of pollutants. These processes are likely to be vulnerable to pollutant attack, since they occur in several sites in the cell, including the mitochondria, peroxisomes and cytoplasm. Indeed, there is evidence to show that respiratory pathways and organelles may be affected by air pollutants (Ballantyne, this volume). However, whether these metabolic effects result in changes in

231

the respiratory gas exchange of oxygen uptake and carbon dioxide, has not been well defined. Therefore most of the evidence for pollutant induced effects on respiration in plants is limited to responses to exposures to high pollutant concentrations which have often yielded contradictory results. In addition, few investigations have considered the effects on photorespiration.

Fluoride

Fluoride is taken up by plants from the atmosphere, water and soil, and once accumulated by leaves, there is little evidence of subsequent translocation. It is therefore one of the most phytotoxic of air pollutants; exposure to very low concentrations leads to a variety of floral and foliar symptoms of damage, reductions in yield, and biochemical and physiological disturbances (Thomas and Alther, 1966; Chang, 1975; Weinstein and Alscher-Herman, 1982). Thus of all the air pollutants, the effects of fluoride on respiration have been the most extensively studied, although responses have, as yet, not been elucidated fully, nor the mechanisms whereby fluorides affect metabolism understood. Weinstein (1977) has proposed that this may be due to the complexity of fluoride action, the pollutant interfering with processes that occur at different levels of organization and at different times.

Both stimulatory and inhibitory effects on respiration have been recorded. Plant responses seem to be related to three major factors:

(1) the concentration of gaseous hydrogen fluoride, or fluoride in solution, to which plants are exposed;
(2) the concentration of fluoride within the tissues; and
(3) the presence of visibly damaged tissue.

This development of visible injury usually results from the accumulation of toxic levels of pollutant in the tissue, either from exposure to high fluoride concentrations or to low concentrations for a long period. Such damage is usually irreversible and is associated with a permanent impairment of photosynthesis. However, visible injury induced by the pollutant is accompanied either by a stimulation or inhibition of respiration. Initially a stimulation of respiration accompanies the early stages of injury. However, if damage is severe, this stimulation is superseded by an inhibition of respiration. This has been observed by Applegate, Adams and Carriker (1960) in bush beans (*Phaseolus vulgaris* var. *humilis*) and by Yu and Miller (1967) when tissue injury was pronounced. Associated with this degree of damage were conformational changes in the mitochondria (Pilet and Roland, 1972).

However, there are conflicting reports as to whether respiratory processes are disturbed if plants are exposed to fluoride concentrations that do not lead to visible damage. McNulty and Newman (1957) and Ross, Wiebe and Miller (1962) reported that when leaves of bush bean, gladiolus, knotweed (*Polygonum orientale*) and the nettle-leaved goosefoot (*Chenopodium murale*) were exposed to hydrogen fluoride, both the necrotic areas and the green portions surrounding these damaged areas exhibited enhanced rates of respiration. This stimulation in undamaged portions of leaves may not have been due to the HF *per se*, but could have occurred in response to the adjacent necrotic damage. For example Hill *et al.* (1958) demonstrated that necrotic tissue will cause a stimulation of oxygen uptake in adjacent undamaged tissue.

There is evidence that fluoride can act directly on respiratory processes, and that effects are not solely as a response to tissue injury. Indeed McCune and Weinstein

(1971) stated that the pathways of respiratory activity were perhaps the most likely sites of fluoride toxicity. These conclusions are drawn from the observations that respiratory processes may be altered by exposure to fluoride in the absence of visible injury. Verkroost (1974) for example, reported that not only was respiration of the alga (*Euglena gracilis*) affected by fluoride before the appearance of visible injury, but respiratory effects were observed before those on photosynthesis.

However, even in tissue that was not visibly damaged, fluoride has been shown to induce both increases and decreases in respiration. This difference in response may be related to fluoride concentration. Enhanced rates of respiration have been observed in wheat (*Triticum aestivum*) seedlings (Luštinec, Krekule and Pokorná, 1960) and bush bean (Applegate, Adams and Carriker, 1960) exposed to low fluoride concentrations, while exposure to higher concentrations had an inhibitory effect in the absence of visible damage. Stimulation of respiration by fluoride has been reported also for many other species, e.g. pea (*Pisum sativum*) (Christiansen and Thiman, 1950), tomato (*Lycopersicon esculentum*) (Weinstein, 1961) and grape (*Vitis vinifera*) (Pilet, 1963). More recent work by McLaughlin and Barnes (1975) showed that the presence of low concentrations of fluoride in the tissue of pine trees (*Pinus strobus* and *Pinus taeda*) resulted in an inhibition of photosynthesis but a stimulation of respiration (*Figure 17.1*). The pine trees were found to be more sensitive than hard woods and young pine needles more sensitive than old. Tissue containing high fluoride concentrations exhibited an inhibition of respiration. These authors suggested that even in the absence of foliar necrosis, these respiratory responses could lead to a reduction in growth rate.

Several workers do propose, however, that exposure to fluoride results in an inhibition of respiration (McNulty and Newmann, 1956; Hill *et al.*, 1958; Givan and Torrey, 1968). Difficulties in defining respiratory responses to fluoride are exacerbated by the modifying influence of other factors in determining plant responses. Respiratory responses to the pollutant seem to be influenced not only by the concentration of fluoride in the tissue (Applegate, Adams and Carriker, 1960) but by many other plant and environmental factors (Weinstein, 1977). For example, the response is determined by the age of the tissue (Coutrez-Geerinck, 1973; McLaughlin and Barnes, 1975; Béjaoui and Pilet, 1975), young and expanding leaves in general being more sensitive than old leaves. Duration of exposure and nutrient status of plants are also important. Applegate and Adams (1960a,b), Hill *et al.* (1958) and Givan and Torrey (1968) have observed that some tissues are relatively insensitive to fluoride, whereas Hitchcock, Zimmerman and Coe (1963) reported that plants may be sensitive only at particular stages of development. They found that the yield of seeds, and the above ground portions of the plant, was reduced when *Sorghum* was exposed to fluoride during the periods of anthesis and tassel shooting, whereas exposures before or after these short periods of development were without effect.

This variation in response led Weinstein (1977) to suggest that plant responses to fluoride were complex and depended not only on the various pool sizes of respiratory and photosynthetic intermediates, but also on the exchange between metabolite pools and the relative activity of various pathways. Indeed, Béjaoui and Pilet (1975) suggested that the greater sensitivity observed in the respiration of young as compared with older tissue, could be due to shift from the glycolytic to the pentose phosphate pathway in leaves exposed to fluoride. A change in relative proportions of these two respiratory pathways has been observed by other workers (Lee, Miller and Welkie, 1966; Ross, Wiebe and Miller, 1962). Weinstein (1977) states that such a shift would result in changes in the energy produced and in the production of intermediates more suitable to the metabolism of an older leaf. Thus, although an increase in the pentose

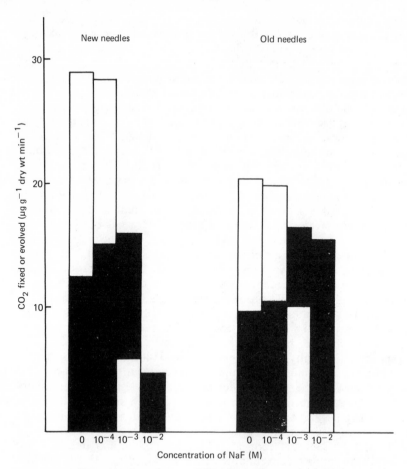

Figure 17.1. Effects of NaF on the rate of CO_2 fixed ($\square$) or evolved ($\blacksquare$) (expressed as μg CO_2 g^{-1} dry weight min^{-1}) in new and young needles of white pine (after McLaughlin and Barnes, 1975)

phosphate pathway may facilitate repair processes, this action could lead ultimately to cellular disruption.

However, it has been proposed that a change in the relative activity of these two pathways could result from an inhibition of the glycolytic pathway as a result of the metabolism of hydrogen fluoride to fluoro-organic acids such as fluorocitrate, a compound toxic to intermediates of the Krebs cycle (Verkroost, 1974). Similarly, a change in relative proportion of the glycolytic and pentose phosphate pathways could result from changes in the activity of particular enzymes of the two pathways. For example, Miller (1958) showed that enolase was fluoride sensitive, while Lee, Miller and Welkie (1966) observed substantial increases in glucose-6-phosphate dehydrogenase, the first enzyme of the pentose phosphate pathway, in soybeans (*Glycine max*) exposed to hydrogen fluoride. In contrast, Lords and McNulty (1965) have proposed that a stimulation of respiration may result from a blockage of $ADP \rightarrow ATP$ interconversion. Yu and Miller (1967) suggest that a stimulation may be due to an increased activity of the entire respiratory cycle or in response to an energy requirement for the alleviation of cell injury. These conflicting hypothesis agree with Weinstein's (1977) assessment that although the pathways of respiratory activity are likely to be one of the primary sites of fluoride toxicity, the specific metabolic steps affected *in vivo* have not been elucidated.

It has been proposed that there is a threshold below which fluoride does not affect respiratory gas exchange, but this threshold varies from species to species (Thomas, 1958; Hill *et al.*, 1958). However above this threshold, the magnitude of response may change during a long-term exposure, or may be reversible after a period of exposure (Applegate and Adams, 1960a). The mechanisms for these changes in sensitivity have not been elucidated, but McCune and Weinstein (1971) suggest that the recovery may involve the development of fluoride insensitive pathways in glycolysis.

Ozone

Since ozone is a very reactive pollutant, it is not surprising that effects on respiration have been observed. However it is often difficult to separate the direct effects of this pollutant on the processes of respiration themselves from changes in respiration due to injury, or as a result of the action of O_3 on other processes such as photosynthesis. One piece of evidence that respiratory responses were not due solely to changes in photosynthetic rates or the size of metabolite pools, was demonstrated by Anderson and Taylor (1973). They reported increased carbon dioxide evolution in tobacco callus exposed to O_3. These cultures were non-photosynthetic and contained no chlorophyll. The threshold for this stimulation was $0.10\ \mu l\ l^{-1}$ for 2 h. However effects were enhanced as O_3 dosage was increased. They reported, also, that varieties of tobacco varied in their response to O_3; the sensitive variety Bel-W3 exhibited a greater stimulation of respiration on exposure to low O_3 concentrations, than the more resistant variety Bel-B (*Table 17.1*).

TABLE 17.1. Per cent increase in the respiration rates of callus tissue of two varieties of tobacco, Bel-W3 and Bel-B, exposed to ozone

Ozone concentration ($\mu l\ l^{-1}$)	Increase in respiration (%)	
	Bel-W3	*Bel-B*
0.1	4	4
0.2	58	36
0.3	67	58
0.4	68	68

After Anderson and Taylor (1973).

It has also been proposed that O_3 did not influence respiration in photosynthetic tissue unless plants were exposed to pollutant concentrations which led to the appearance of visible injury. However other workers have observed that low concentrations of O_3 can induce respiratory disturbances in plants in the absence of, or prior to, the appearance of visible injury. Todd (1958) reported that rates of respiration in pinto beans were stimulated by O_3 only in the presence of visible injury, but found that the greater the stimulation, the more complete was recovery after fumigation. Similar responses were observed in citrus leaves exposed to O_3, but respiration in Valencia orange leaves could be stimulated by O_3 in the absence of visible injury

(Todd and Garber, 1958). The observation that respiration in leaf discs of bean (*Phaseolus vulgaris*) was stimulated after 24 h exposure to O_3 led Pell and Brennan (1973) to propose that enhanced respiration was a consequence, rather than a cause of cellular injury. Stimulated respiration rates were reported by Macdowall and Ludwig (1962), but only after the appearance of visible damage, and as a consequence of the uncoupling of oxidative phosphorylation.

Macdowall (1965) did report, however, that although larger dosages of O_3 resulted in increased respiration and visible damage in tobacco (*Nicotiana tabacum*), exposure to low concentrations of O_3 which did not lead to visible damage, resulted in an inhibition of respiration and a reduction in mitochondrial phosphorylation. Freebairn (1957) also reported an inhibition of O_2 uptake and a reduction in the activity of the citric acid cycle. A reduction in respiration rates of the alga *Scenedesmus obtusiusculus* exposed to O_3 has been observed (Verkroost, 1974). He found that this inhibition increased with the duration of exposure and the temperature during exposure. A maximum inhibition of respiration of $\sim 50\%$ was observed when *Scenedesmus* was polluted with O_3 for 3 h at 35 °C, while exposure for a similar period at 15 °C only had a small effect.

In contrast, there is considerable evidence that exposure to low concentrations of O_3 can result in a stimulation of respiration in the absence of visible damage. For example, Dugger and Ting (1970) observed that one of the first noticeable changes after the start of an oxidant exposure period was an increase in respiration rates. This substantiated the work of Todd and Propst (1963) who found that ozone caused a two or three fold stimulation in the respiration of *Coleus* and tomato plants in the absence of visible injury. Similarly, Dugger, Koukol and Palmer (1966) and Dugger and Palmer (1969) reported enhanced respiration rates in lemon leaves exposed to 0.15–0.25 ppm O_3 for several weeks. Barnes (1972) demonstrated a stimulation of respiration when pine seedlings were exposed to 0.05–0.15 ppm O_3. The magnitude of these increases was as high as 90% during exposure to 0.15 ppm O_3. This led Barnes to propose that a stimulation in respiration, in association with the observed inhibition of photosynthesis, would be likely to be a drain on carbohydrate supply and would result in a reduction in growth and vigour, even in the absence of foliar symptoms of damage. Barnes also found that there was a great variation in response depending on the age of the tissue; younger needles appeared sensitive to O_3 whereas older needles responded unpredictably on exposure.

Not only photosynthetic tissue responds to ozone. Hofstra *et al.* (1981) reported that the metabolic activity of roots was very sensitive to the changes induced in the leaves of the bean exposed to O_3 and $O_3 + SO_2$ (*Table 17.2*). Within 24 h of the beginning of exposure to 0.15 ppm O_3, not only was root growth inhibited but there was also a

TABLE 17.2. Per cent reduction in CO_2 evolution of whole roots and root tips of bean plants exposed to 0.15 ppm O_3

Duration of exposure (h)	Reduction in CO_2 evolution (%)	
	Whole roots	Root tips
6	0	16
24	13	33
48	32	21
72	16	24

After Hofstra *et al.* (1981).

reduction in the CO_2 evolution of the whole root system and root tips. However when plants were exposed to 0.15 ppm O_3 in the presence of a similar concentration of SO_2, the effects on the roots were reduced. Exposure to SO_2 alone had no effect on CO_2 evolution by roots. Inhibition of root activity occurred before the appearance of visible injury and a reduction in photosynthetic area. Changes in root respiratory activity have also been found in the lentil (*Lens culinaris*) treated with fluoride, but the response observed was an increase in the respiratory rates of the roots (Pilet, 1964).

Associated with physiological changes are ultrastructural alterations in mitochondria. Injury to mitochondrial membranes by O_3 was observed by Pell and Weissberger (1976), and increased permeability of mitochondrial membranes reported by Lee (1968). Exposure to low concentrations of O_3 can also lead to increases in mitochondrial volume (Swanson, Thomson and Mudd, 1973). Such ultrastructural changes are likely to alter the components of the electron transport system. Changes in ATP levels have been observed (Tomlinson and Rich, 1968) while Lee (1967) reported that an uncoupling of oxidative phosphorylation in mitochondria occurred when tobacco leaves were exposed to O_3. These changes in phosphorylation were observed before changes in respiration, which became evident only after 1-h exposure to O_3 (*Table 17.3*).

TABLE 17.3. Per cent inhibition of oxidative phosphorylation and oxygen uptake of mitochondria isolated from tobacco leaves exposed to 1 ppm O_3

Duration of exposure (h)	Per cent inhibition of	
	Oxidative phosphorylation	Oxygen uptake
1	17	—
3	59	34
5	64	39

After Lee (1967).

Sulphur dioxide

Over the last decade, there have been several reviews which have documented the photosynthetic responses to SO_2 (Nash, 1973; Hällgren, 1978; Heath, 1980; Black, 1982). In contrast, respiratory responses to this pollutant are less well understood, although a number of workers have studied the effects of a wide range of concentrations of SO_2 on a variety of organisms including algae, lichens, mosses and higher plants. Most of the work has concentrated on defining effects on dark respiration and Baddeley, Ferry and Finegan (1973) have proposed that respiratory processes may be three to five times less sensitive than photosynthetic processes of lichens.

Both stimulation and inhibition of respiration has been reported in plants exposed to SO_2. For example several workers have reported that an inhibition of respiration resulted when mosses (Gilbert, 1968) and lichens (Pearson and Skye, 1965; Klee, 1970; Baddeley, Ferry and Finegan, 1971, 1972, 1973) were exposed to high concentrations of this pollutant. Nieboer *et al.* (1976) have hypothesized that this inhibition of respiration observed in lichens was a measure of the response of the fungal partner. In contrast, when Pearson and Skye (1965) and Baddeley, Ferry and Finegan (1971, 1972) exposed lichens to a lower concentration of SO_2, a stimulation of respiration was observed.

Many other workers have reported only enhanced rates of respiration in lower plants exposed to SO_2. Dekoning and Jegier (1968) reported a 14% increase in the respiration of the alga, *Euglena gracilis*, while Showman (1972) has observed a stimulation of respiration in lichens. Türk, Wirth and Lange (1974) found significant increases when lichens were exposed to the relatively low concentration of 0.2 ppm SO_2 for 14 h. They suggested that this indicated that for SO_2 detoxification mechanisms were operating, a view expressed also by Le Blanc and Rao (1975) who postulated that the energy produced in enhanced respiration is used by the plants for the quick oxidation of sulphite to sulphate. Türk, Wirth and Lange (1974) did find, however, that after 25 days' exposure, rates of respiration in polluted plants fell to control values. A similar stimulation was also observed in bryophytes exposed to 5 ppm SO_2, this stimulation increasing as exposure time was extended (Syratt and Wanstall, 1969).

Thus it seems that SO_2 or sulphite does cause changes in the respiration of lower plants. However the plant response depends on the concentration of pollutant to which plants are exposed and factors such as the wetness of the thallus. For example, Türk, Wirth and Lange (1974), have shown that dry lichens are very resistant to SO_2.

A variety of respiratory responses to SO_2 have been reported also in higher plants. Thomas and Hill (1937) found no effects on dark respiration in plants exposed to 1 ppm SO_2 for 1 h. Similar responses to high concentrations have also been observed by Katz (1949), Sij and Swanson (1974) and Furukawa, Natori and Totsuka (1980). In addition, Shimazaki and Sugahara (1979) report that the changes in dark respiration that they observed in plants exposed to 2 ppm SO_2 were too small to have an appreciable effect on rates of net photosynthesis. However many of the exposure systems were constructed to measure photosynthetic responses to pollutants, and may have lacked the capacity to measure small changes in carbon dioxide release or oxygen uptake.

Effects of SO_2 on dark respiration that have been reported include both an inhibition (Taniyama, 1972; Lüttge *et al.*, 1972) and a stimulation (Keller, 1957; Börtitz, 1964; Vogl, Börtitz and Polster, 1974; Vogl and Börtitz, 1965; Taniyama *et al.*, 1972; Black and Unsworth, 1979 *(see Figure 17.2)*. Enhanced respiratory rates have been observed in a number of species (pine (*Pinus*), larch (*Larix*), spruce (*Picea*), rice (*Oryza*) and bean (*Vicia*) exposed to a wide range of concentrations (0.04–2 ppm). These changes in respiratory rates may reflect a number of responses to the pollutants; for example the operation of detoxification processes as suggested for lichens, repair processes, or the direct interference with specific respiratory pathways or organelles. Indeed ultrastructural changes in the mitochondria of lodgepole pine (*Pinus contorta*) have been observed (Malhotra, 1976) and an inhibition of ATP formation and phosphorylating activity of mitchondria observed in plants exposed to SO_2 (Ballantyne, 1973; Malhotra and Hocking, 1976; Harvey and Legge, 1979). Nikolaevskii (1966, 1968, cited in Horsman and Wellburn, 1976) reported that exposure of *Betula* and *Acer* to very high concentrations (125 ppm) resulted in alterations in the activity of the glycolytic and pentose phosphate pathways and the citric acid cycle.

Reports in the literature indicate, therefore, that SO_2 may stimulate or inhibit respiration in the absence of visible injury. Respiratory rates may return to control values following exposure, as long as concentrations are relatively low and exposure short. This behaviour may reflect the capacity of plants to detoxify sulphite or repair damage incurred by exposure to toxic concentrations of SO_2.

Attempts have been made to define the effects of SO_2 on photorespiration, but these are limited in number. Assessments of pollutant induced changes on this process are of particular importance, if rates of photorespiration are significantly higher than those of dark respiration, as was suggested by Zelitch (1971). These are also necessary to permit

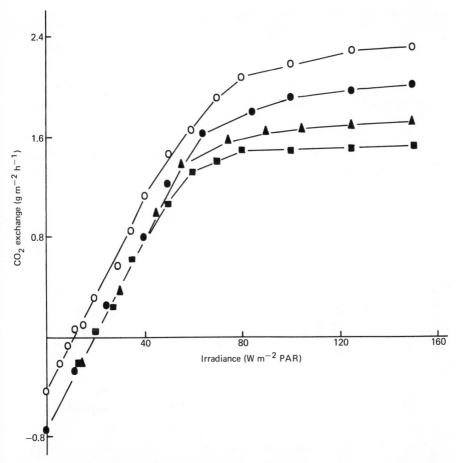

Figure 17.2. Effects of zero (○), 0.035 (●), 0.088 (▲) and 0.175 (■), ppm SO_2 on CO_2 exchange in *Vicia faba* over a range of irradiances (after Black and Unsworth, 1979)

the identification of the primary site for reductions in rates of net photosynthesis observed in plants exposed to SO_2. However there are several major difficulties in studying effects on photorespiration. Quantitative and unequivocal measurements of photorespiratory rates are difficult to make (Ludlow and Jarvis, 1971). Estimates may be made, but these involve experiments with radioactive carbon dioxide or a manipulation of the CO_2 or O_2 concentrations within the exposure system. Few investigators have had the facilities to carry out these procedures, and thus photorespiratory responses to pollutants are rarely studied especially using intact plants. Koziol and Jordan (1978) (*Figure 17.3*) however, estimated photorespiration from the rate of CO_2 release in the dark period immediately following a light period in which bean (*Phaseolus vulgaris*) plants had been exposed in high concentrations of SO_2. They reported an exponential increase in photorespiration with increasing SO_2 concentration, which they attribute to a greater use of energy in repair and replacement processes. Rates of respiration did diminish, however, to rates comparable to control plants towards the end of the dark period. Pearson (1973) has also reported that

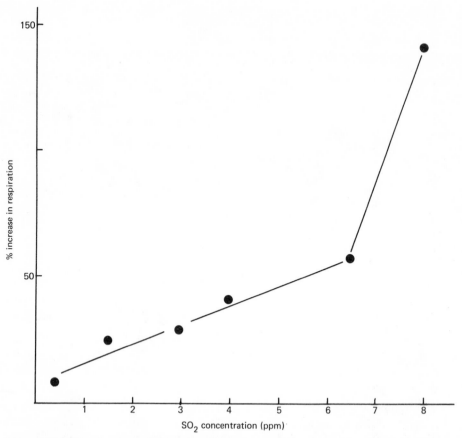

Figure 17.3. Effects of a range of SO_2 concentrations on respiration in *Phaseolus vulgaris* (after Koziol and Jordan, 1978)

photorespiration is enhanced in the lichen *Parmelia sulcata*, collected from a polluted city centre.

In contrast, Sij and Swanson (1974) proposed that SO_2 does not affect photorespiration, a hypothesis based on the evidence that C_3 and C_4 plants acted similarly in their photosynthetic responses to SO_2. However Ziegler (1975) has stated the view that SO_2 is likely to 'drastically reduce or completely abolish rates of photorespiration'. Furukawa, Natori and Totsuka (1980) examined photorespiratory rates of sunflower plants exposed to 1.5 ppm SO_2 for 30 min. By measuring CO_2 release into CO_2 free air within their chambers, they were able to observe that photorespiration was inhibited by SO_2 (*Figure 17.4*). They postulated that this response could have been due to either an inhibition of the metabolic pathways of photorespiration itself, or indirectly as a result of inhibition of photosynthetic intermediates. However, rates of photorespiration were too rapid to result from photosynthetic reductions; but stomatal closure may have contributed to the results observed. Koziol and Cowling (1978) reported, that when ryegrass was exposed to SO_2, an increase in $^{14}CO_2$ photoassimilation was observed, which they attributed to an inhibition of the photorespiratory cycle.

Many of the hypotheses which propose that SO_2 may influence photorespiration have been based on evidence obtained from plants exposed to high concentrations of

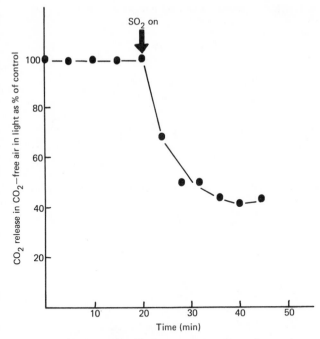

Figure 17.4. Rates of photorespiration as percentage of control, measured as CO_2 release into CO_2 free air, in sunflower plants exposed to 1.5 ppm SO_2 (after Furukawa, Natori and Totsuka, 1980)

this pollutant, or more often tissue exposed to SO_2 or sulphonates in solution. For example Bethge (1958) and Libera, Ziegler and Ziegler (1974) demonstrated that one of the photorespiratory intermediates, glycolic acid, accumulated in leaves of barley (*Hordeum vulgare*) and spinach (*Spinacia oleracea*) plants exposed to a high concentration of SO_2. They attributed this increase to a blocking of glycolate oxidation, and thus of photorespiration. Indeed Spedding and Thomas (1973) and Soldatini and Ziegler (1979) have shown that sulphite may inhibit glycolate oxidase. However, Soldatini and Thomas found that the specific activity of this enzyme increased in proportion to the SO_2 concentration in tobacco exposed to SO_2 for 18 h. Zelitch (1966) has also reported that sulphonates may induce glycolate accumulation and the inhibition of glycolate oxidase.

From the available evidence therefore, it is difficult to assess confidently the effects of SO_2 on photorespiration, especially in whole plants exposed to low concentrations of pollutant. Considerable careful investigation is necessary before this can be done.

Oxides of nitrogen, ammonia, PAN

Literature which documents the effects of these pollutants on plant metabolism are very rare. For example, Wellburn, Wilson and Aldridge (1980) stated that there was little solid evidence at the biochemical or physiological level for the mechanisms by which nitric oxide acts. However Srivastava, Jolliffe and Runeckles (1975a,b) have published the results of a comprehensive piece of research in which physiological

TABLE 17.4. Per cent reduction in rates of dark respiration in bean plants exposed to 3 ppm NO_2 in a range of temperatures

Temperature (°C)	Per cent reduction in rates of dark respiration
15	34
20	38
25	43
30	48
35	51

After Srivastava, Jolliffe and Runeckles (1975b).

TABLE 17.5. Per cent reduction in photorespiratory rates, measured as CO_2 evolution into CO_2 free air, in bean plants exposed to 3 and 7 ppm NO_2

Duration of exposure	Reduction in rates of photorespiration (%)	
(h)	3.0	7.0 ppm NO_2
0.5	9.4	18.8
1.0	15.5	28.1
1.5	25	43.8
2.0	28.1	51.6

After Srivastava, Jolliffe and Runeckles (1975b).

responses of bean (*Phaseolus vulgaris*), exposed to 1–7 ppm NO_2 under a range of environmental conditions, were investigated. By studying the evolution of CO_2 into CO_2 free air, they were able to show that both respiration and photorespiration were inhibited by exposure to NO_2 (*Tables 17.4* and *17.5*). The magnitude of the reduction in rates of respiration was larger than the inhibition of photosynthesis, and was enhanced with increasing NO_2 concentration, duration of exposure and temperature. These responses were observed in all ages of leaves studied. Although visible injury was only recorded after large dosages, inhibition of respiration was not readily reversible. This evidence suggests that if respiration and photorespiration are inhibited by low concentrations of pollutants, exposure to NO_2 may result in significant changes in the carbon budget of plants. Although Dolzmann and Ullrich (1966) reported that ultrastructural changes occurred when plants were exposed to 1% NO_2, and showed enhanced ATP formation in plants exposed to lower concentrations, respiratory effects were not observed by Bengston, Grennfelt and Skärby (1982) when they exposed one year old pine seedlings to 0.04–0.4 ppm NO_2 or NO for 6 h.

Exposure to ammonia can result in changes in respiration rates in plants. For example, work by Vines and Wedding (1960) has shown that ammonia inhibits respiration in excised roots, leaf discs and isolated mitochondria. They suggest that ammonia acts on the sites of the electron transport system, although Matsumoto, Wakiuchi and Takahashi (1971) and Wakiuchi, Matsumoto and Takahashi (1971) have reported that the activity of many of the respiratory enzymes in the glycolysis pathway and TCA cycle were stimulated during ammonium toxicity. Evidence for effects of PAN on respiratory processes is even less well documented, although little effect of PAN on respiration of *Chlamydomonas reinhardtii* was observed by Gross and Dugger (1969).

Conclusions

Although most workers have investigated photosynthetic responses to air pollutants, effects on respiratory processes have been observed. In general, exposure of plants to high concentrations of pollutants leads to the appearance of visible injury to the tissue. As a consequence of this visible damage, respiration may either be stimulated or inhibited, the type of response depending on the degree of injury. Stimulation of respiration has been observed also in non-damaged tissue adjacent to these necrotic areas, perhaps as a consequence of the utilization of energy in repair processes. However, there is evidence that respiration may be altered by exposure to pollutants in the absence of visible injury, indicating that respiratory processes themselves may be affected by pollutants; changes in respiration, then, do not result solely as a response to visible damage. Although very low concentrations of pollutants may not cause changes in respiration, both a stimulation and inhibition of dark respiration has been observed in plants exposed to low concentrations of pollutants. However, the magnitude, duration and degree of permanence of these effects has seldom been thoroughly defined, nor have any thresholds for these effects been identified. Since plant responses to pollutants are often complex, elucidation of respiratory effects is difficult. For example responses may be modified by the species or variety of plant investigated, the concentration, duration and frequency of exposure to the pollutant, the degree of pollutant accumulation within sensitive tissues, biotic variables such as plant age and nutrient status, and prevailing environmental conditions. This variation in response may reflect the rates and magnitude of pollutant uptake by the plant, which are usually determined by the degree of opening of stomata, the major pathway for pollutant uptake for many higher plants. Environmental and biotic factors will not only modify stomatal behaviour, but will also influence the solubility of pollutants and the speed of biochemical reactions of the respiratory processes themselves, as well as the activity of any detoxification, compartmentalization or repair processes. Modification of any of these factors will influence the sensitivity of plants to withstand toxic levels of pollutants in the environment.

Do pollutant induced effects on dark respiration and photorespiration result in changes in plant growth and yield? A significant enhancement of respiration in response to high concentrations of pollutants may result in wasteful loss of carbohydrate and energy normally used for growth. Even if this stimulation is due to the operation of detoxification and repair processes, these may not have the capacity to prevent physiological and visible damage. If exposures are not too extreme, physiological processes are altered but no visible damage results. Thus any observed change in respiration rates may reflect the balance between the inhibition of respiratory pathways and a stimulation of those processes concerned with repair and detoxification. These effects may lead to reductions in growth in the long term if stress periods are prolonged.

On exposure to low concentrations of pollutants, plants usually exhibit a stimulation of respiration, which may be due primarily to the operation of detoxification and repair mechanisms. Indeed a shift from the glycolytic pathway to the pentose phosphate pathway in response to pollutants is often observed. This enhanced use of energy by the plant is therefore likely to be of benefit, especially if the pollutant exposure periods are short. Such responses may prevent toxic concentrations of pollutants reaching the sensitive metabolic sites, for example, the photosynthetic pathways within the cell. Indeed respiration can be affected prior to any observed depression of photosynthesis.

Evidence is generally not available to allow an assessment of effect of pollutants on photorespiration. This is partly due to the fact that early investigators were unaware of the existence of this process, and partly because of the difficulties involved in measuring rates of respiration in the light. If this is a beneficial process, as was proposed by Tolbert and Ryan (1976), an inhibition of photorespiration by pollutants, may be ultimately deleterious to plants, although in the short-term the rates of net photosynthesis will increase. However, if indeed photorespiration is a wasteful process, pollutant induced effects may be beneficial to the growth of the plants.

Changes in respiration which are induced by pollutants, may be more meaningful if discussed in association with the photosynthetic performance of the plant. If photosynthetic rates are high, a small change in respiration will not have a significant effect on the carbon balance of the plant. However, if rates of photosynthesis are intrinsically low or are reduced, respiratory effects may result in significant reductions in growth and yield. For example, in the natural environment, photosynthetic rates may be severely reduced if pollutant concentrations or dosages are high, plants are exposed to several pollutants, or environmental conditions such as light or temperature are limiting for photosynthesis. Under these conditions, a change in respiration rates could alter the carbon balance of the plant significantly, so that negative rates of carbon fixation result. This may lead to premature senescence and leaf drop, especially in leaves where metabolite pools are low. Indeed Davies (1980) and Jones and Mansfield (1982) have reported that plants exposed to pollutants under light limited conditions showed greater reductions than plants exposed under higher irradiances. In addition, species with a low intrinsic capacity for photosynthesis, carboxylation and electron transport may be more vulnerable to pollutant attack. A large proportion of the biomass of these species, which are often perennial in habit, comprises non-photosynthetic respiring tissue. Thus a small change in rates of respiration, especially if associated with a pollutant induced reduction in photosynthesis, may have a significant effect on the dry matter production of the plant. Such changes in respiration and photosynthesis have been observed in pine seedlings by McLaughlin and Barnes (1975). There is also evidence that responses to pollutants may occur in non-photosynthetic portions of plants such as roots (Hofstra *et al.*, 1981). A reduction in root activity will have far reaching consequences, not only for root growth, but also for the whole plant, especially if it is growing in a stressful environment.

Therefore, before we can estimate the effect of pollutants, singly and in combination, on respiration, and the consequence of these on growth and yield, considerable research needs to be carried out. The responses of intact plants to low concentrations of pollutants, in exposure conditions typical of a range of environmental conditions, need to be studied. However it is vital to design exposure facilities where these respiratory responses can be examined in association with measurements of photosynthesis, the activity of biochemical respiratory pathways, and growth and yield of plants. In addition, observed changes in respiratory behaviour in plants exposed to pollutants in the laboratory and in the natural environment are urgently needed.

References

ANDERSON, W. C. and TAYLOR, O. C. Ozone induced carbon dioxide evolution in tobacco callus cultures, *Physiologia Plantarum* **28**, 419–423 (1973)
APPLEGATE, G. H. and ADAMS, D. F. Effect of atmospheric fluoride on respiration in bush beans, *Botanical Gazette* **121**, 223–227 (1960a)
APPLEGATE, G. H. and ADAMS, D. F. Invisible injury of bush beans by atmospheric and aqueous fluorides, *International Journal of Air Pollution* **3**, 231–248 (1960b)

APPLEGATE, H. G., ADAMS, D. F. and CARRIKER, R. C. Effect of aqueous fluoride solutions on respiration of intact bush bean seedlings, *American Journal of Botany* 47, 339–345 (1960)

BADDELEY, M. S., FERRY, B. W. and FINEGAN, E. J. A new method for measuring lichen respiration: Response of selected species to temperature, pH and sulphur dioxide, *Lichenologist* 5, 18–25 (1971)

BADDELEY, M. S., FERRY, B. W. and FINEGAN, E. J. The effects of sulphur dioxide on lichen respiration, *Lichenologist* 5, 283–291 (1972)

BADDELEY, M. S., FERRY, B. W. and FINEGAN, E. J. Sulphur dioxide and respiration in lichens. In *Air Pollution and the Lichens*, (Eds. B. W. FERRY, M. S. BADDELEY and D. L. HAWKSWORTH), pp. 299–313, Athlone Press, London (1973)

BALLANTYNE, D. J. Sulphite inhibition of ATP formation in plant mitochondria, *Phytochemistry* 12, 1207 (1973)

BARNES, R. L. Effects of chronic exposure to ozone on hotosynthesis and respiration of pines, *Environmental Pollution* 3, 133–138 (1972)

BÉJAOUI, M. and PILET, P. E. Effects du fluor sur l'absorption de l'oxygène de tissus de Ronce cultivés *in vitro*, *Comptes Rendu Hebdomadaires des Séances de l'Academie des Sciences Série D* 280, 1457–1460 (1975)

BENGSTON, C., GRENNFELT, P. and SKÄRBY, L. Deposition of nitrogen oxides to Scots pine (*Pinus sylvestris* L.). In *Effects of Gaseous Air Pollution in Agriculture and Horticulture*, (Eds. M. H. UNSWORTH and D. P. ORMROD), p. 461, Butterworths, London (1982)

BENNETT, J. P., RESH, H. M. and RUNECKLES, V. C. Apparent stimulations of plant growth by air pollutants, *Canadian Journal of Botany* 52, 35–41 (1974)

BETHGE, H. Spektralphotometric und Rauchschadendiagnostik. *Schriftenreihe des Vereins für Wasser Boden-, und Luft-hygiene* 13, 3–10 (1958)

BLACK, V. J. Effects of sulphur dioxide on physiological processes in plants. In *Effects of Gaseous Air Pollution in Agriculture and Horticulture*, (Eds. M. H. UNSWORTH and D. P. ORMROD), pp. 67–91, Butterworths, London (1982)

BLACK, V. J. and UNSWORTH, M. H. Effects of low concentrations of sulphur dioxide on net photosynthesis and dark respiration of *Vicia faba*, *Journal of Experimental Botany* 30, 473–483 (1979)

BÖRTITZ, S. Physiologische und biochemische Beiträge zur Rauchschadenforschung, *Biologisches Zentralblatt* 83, 501–513 (1964)

CHANG, C. W. Fluorides. In *Responses of Plants to Air Pollution*, (Eds. J. B. MUDD and T. T. KOZLOWSKI), pp. 57–95, Academic Press, New York (1975)

CHRISTIANSEN, G. S. and THIMAN, K. V. The metabolism of stem tissue during growth and its inhibition. II. Respiration and ether-soluble material, *Archives of Biochemistry* 26, 248–259 (1950)

COUTREZ-GEERINCK, D. Action de quelques effecteurs respiratoires sur des tissus de tubercules de *Solanum tuberosum* L. à plusieurs stades d'incubation, *Bulletin de la Société Royale de Botanique de Belgique* 106, 237–256 (1973)

COWLING, D. W. and KOZIOL, M. J. Mineral nutrition and plant response to air pollutants. In *Effects of Gaseous Air Pollution in Agriculture and Horticulture*, (Eds. M. H. UNSWORTH and D. P. ORMROD), pp. 349–375, Butterworths, London (1982)

DAVIES, T. Grasses more sensitive to SO_2 pollution in conditions of low irradiance and short days, *Nature* 284, 483–485 (1980)

DEKONING, H. W. and JEGIER, Z. A study of the effects of ozone and sulphur dioxide on the photosynthesis and respiration of *Euglena gracilis*, *Atmospheric Environment* 2, 321–326 (1968)

DOLZMANN, P. and ULLRICH, H. Einige Beobachtungen über Beziehungen zwischen Chloroplasten und Mitochondrien in Palisadeparenchym von *Phaseolus vulgaris*, *Zeitschrift für Pflanzenphysiologie* 55, 165–180 (1966)

DUGGER, W. M. and PALMER, R. L. Carbohydrate metabolism in leaves of rough lemon as influenced by ozone, *Proceedings of 1st International Citrus Symposium Riverside, California* 2, 711–715 (1969)

DUGGER, W. M. and TING, I. P. Air pollutant oxidants — their effects on metabolic processes in plants, *Annual Review of Plant Physiology* 21, 215–234 (1970)

DUGGER, W. M., KOUKOL, J. and PALMER, R. L. Physiological and biochemical effects of atmospheric oxidants on plants, *Journal of the Air Pollution Control Association* 16, 467–471 (1966)

FREEBAIRN, H. T. Reversal of inhibitory effects of ozone on oxygen uptake of mitochondria, *Science* 126, 303–304 (1957)

FURUKAWA, A., NATORI, T. and TOTSUKA, T. The effects of SO_2 on net photosynthesis in sunflower leaf. In *Studies on the Effects of Air Pollutants on Plants and Mechanisms of Phytotoxicity*, Research Report from the National Institute for Environmental Studies, Yatabe, Japan 11, 1–8 (1980)

GILBERT, O. L. *Biological indicators of air pollution*, PhD Thesis, University of Newcastle upon Tyne (1968)

GIVAN, C. V. and TORREY, J. G. Fluoride inhibition of respiration and fermentation in cultured cells of *Acer pseudoplatanus*, *Physiologia Plantarum* 21, 1010–1019 (1968)

GROSS, R. E. and DUGGER, W. M. Responses of *Chlamydomonas reinhardtii* to peroxyacetyl nitrate, *Environmental Research* **2**, 256–266 (1969)

HÄLLGREN, J.-E. Physiological and biochemical effects of sulphur dioxide on plants. In *Sulphur in the Environment: Part 2. Ecological Impact*, (Ed. J. O. NGRIAGU), pp. 164–209, John Wiley & Sons, New York (1978)

HARVEY, G. W. and LEGGE, A. H. The effect of sulphur dioxide upon the metabolic level of adenosine triphosphate, *Canadian Journal of Botany* **57**, 759–764 (1979)

HEATH, R. L. Initial events in injury to plants by air pollutants, *Annual Review of Plant Physiology* **31**, 395–431 (1980)

HILL, A. C., TRANSTRUM, L. G., PACK, M. R. and WINTERS, W. S. Air pollution with relation to agronomic crops. IV. An investigation of the 'hidden injury' theory of fluoride damage to plants, *Agronomy Journal* **50**, 562–565 (1958)

HITCHCOCK, A. E., ZIMMERMAN, P. W. and COE, R. R. The effects of fluoride on Milo maize (*Sorghum* sp), *Contributions from the Boyce Thompson Institute* **22**, 175–206 (1963)

HOFSTRA, G., ALI, A., WUKASCH, R. T. and FLETCHER, R. A. The rapid inhibition of root respiration after exposure of beans (*Phaseolus vulgaris* L.) plants to ozone, *Atmospheric Environment* **15**, 483–487 (1981)

HORSMAN, D. C. and WELLBURN, A. R. Guide to metabolic and biochemical effects of air pollutants on higher plants. In *Effects of Air Pollutants on Plants*, (Ed. T. A. MANSFIELD), pp. 185–199, Cambridge University Press (1976)

JONES, T. and MANSFIELD, T. A. The effects of growth and development of seedlings of *Phleum pratense* under different light and temperature environments, *Environmental Pollution* **28**, 199–207 (1982)

KATZ, M. Sulphur dioxide in the atmosphere and its relation to plant life, *Industrial and Engineering Chemistry, International Edition* **41**, 2450–2465 (1949)

KELLER, T. H. *Beiträge zur Erfassung der durch schweflige Säure hervorgrufenen Rauchschäden an Nadelhölzern*, PhD Dissertation, München (1957)

KLEE, R. Die Wirking von gas — und stabförmigen Immissionen auf Respiration und Inhaltstoffe von *Parmelia physodes*. *Angewandte Botanik* **44**, 253–261 (1970)

KOZIOL, M. J. and COWLING, D. W. Growth of ryegrass (*Lolium perenne* L.) exposed to SO_2. II. Changes in the distribution of photoassimilated ^{14}C, *Journal of Experimental Botany* **29**, 1431–1439 (1978)

KOZIOL, M. J. and JORDAN, C. F. Changes in carbohydrate levels in red kidney bean (*Phaseolus vulgaris* L.) exposed to sulphur dioxide, *Journal of Experimental Botany* **29**, 1037–1043 (1978)

LE BLANC, F. and RAO, D. N. Effects of air pollutants on lichens and Bryophytes. In *Responses of Plants to Air Pollution*, (Eds. J. B. MUDD and T. T. KOZLOWSKI), pp. 237–272, Academic Press, New York (1975)

LEE, C. J., MILLER, C. W. and WELKIE, G. W. The effects of hydrogen fluoride and wounding on respiratory enzymes in soybean leaves, *International Journal of Air and Water Pollution* **10**, 169–181 (1966)

LEE, T. T. Inhibition of oxidative phosphorylation and respiration by ozone in tobacco mitochondria, *Plant Physiology* **42**, 691–696 (1967)

LEE, T. T. Effect of ozone on swelling of tobacco mitochondria, *Plant Physiology* **43**, 133–139 (1968)

LIBERA, W., ZIEGLER, I. and ZIEGLER, H. The action of sulfite on the HCO_3^- fixation and the fixation pattern of isolated chloroplasts and leaf tissue slices, *Zeitschrift für Pflanzenphysiologie* **74**, 420–433 (1974)

LORDS, J. L. and McNULTY, I. B. Estimation of ATP in leaf tissue employing the firefly luminescent reactions, *Utah Academy of Science and Arts Letters* **42**, 163–164 (1965)

LUDLOW, M. M. and JARVIS, P. G. In *Plant Photosynthetic Production: A Manual of Methods* (Eds. Z. SESTAK, J. CATSKY and P. G. JARVIS), Junk, N. V., The Hague (1971)

LUSTINEC, J., KREKULE, J. and POKORNÁ, V. Respiratory pathways in gibberéllin treated wheat: The effect of fluoride on respiration rate, *Biologia Plantarum* **2**, 223–226 (1960)

LÜTTGE, U., OSMOND, C. B., BALL, E., BRINKMAN, E. and KINZE, G. Bisulfite compounds as metabolic inhibitors: non-specific effects on membranes, *Plant and Cell Physiology* **13**, 505–514 (1972)

McCUNE, D. C. and WEINSTEIN, L. H. Metabolic effects of atmospheric fluorides on plants, *Environmental Pollution* **1**, 169–174 (1971)

MACDOWALL, F. D. H. Stages of ozone damage to respiration of tobacco leaves, *Canadian Journal of Botany* **43**, 419–427 (1965)

MACDOWALL, F. D. H. and LUDWIG, R. A. Some effects of ozone on tobacco leaf metabolism, *Phytopathology* **52**, 740 (1962)

McLAUGHLIN, S. B. and BARNES, R. L. Effects of fluoride on photosynthesis and respiration of some South-East American forest trees, *Environmental Pollution* **8**, 91–96 (1975)

McNULTY, I. B. and NEWMAN, D. W. Effects of atmospheric fluoride on the respiration rate of bush bean and gladiolus leaves, *Plant Phytology* **32**, 121–124 (1957)

McNULTY, I. B. and NEWMAN, D. W. Effects of lime spray on the respiration rate and chlorophyll content of leaves exposed to atmospheric fluorides, *Utah Academy of Science and Arts Proceedings* **33**, 73–79 (1956)

MALHOTRA, S. S. Effects of sulphur dioxide on biochemical activity and ultrastructural organisation of pine needle chloroplasts, *New Phytologist* **76**, 239–245 (1976)

MALHOTRA, S. S. and HOCKING, D. Biochemical and cytological effects of SO_2 on plant metabolism, *New Phytologist* **76**, 227–237 (1976)

MATSUMOTO, H., WAKIUCHI, N. and TAKAHASHI, E. Changes in some mitochondrial enzyme activities of cucumber leaves during ammonium toxicity, *Physiologia Plantarum* **25**, 353–357 (1971)

MILLER, G. W. Properties of enolase in extracts from pea seed, *Plant Physiology* **33**, 199–206 (1958)

MUDD, J. B. and KOZLOWSKI, T. T. (Eds.). *Responses of Plants to Air Pollution*, Academic Press, New York (1975)

NASH, T. H. The effect of air pollution on other plants, particularly vascular plants. In *Air Pollution and the Lichens*, (Eds. B. W. FERRY, M. S. BADDELEY and D. L. HAWKSWORTH), pp. 192–223, Athlone Press, London (1973)

NIEBOER, E., RICHARDSON, D. H. S., PUCKETT, K. J. and TOMASSINI, F. O. The phytotoxicity of sulphur dioxide in relation to measurable responses in lichens. In *Effects of Air Pollutants on Plants* (Ed. T. A. MANSFIELD), pp. 61–85, Cambridge University Press, Cambridge (1976)

PEARSON, L. C. Air pollution and lichen physiology: progress and problems. In *Air Pollution and the Lichens*, (Eds. B. W. FERRY, M. S. BADDESLEY and D. L. HAWKSWORTH), pp. 224–237, Athlone Press, London (1973)

PEARSON, L. C. and SKYE, E. Air pollution affects pattern of photosynthesis in *Parmelia sulcata*, a corticolous lichen, *Science* **148**, 1600–1602 (1965)

PELL, E. J. and BRENNAN, E. Changes in respiration, photosynthesis, adenosine-5'-triphosphate, and total adenylate content of ozonated pinto bean foliage as they relate to symptom expression, *Plant Physiology* **51**, 378–381 (1973)

PELL, E. J. and WEISSBERGER, W. C. Histoplathological characterisations of ozone injury to soybean foliage, *Phytopathology* **66**, 787–793 (1976)

PILET, P. E. Action du fluor et de l'acide β-indolylacetique sur la respiration de disques de feuilles, *Bulletin de la Société Vaudoise des Sciences naturelles* **68**, 359–360 (1963)

PILET, P. E. Action du fluor et de l'acide β-indolylacetique sur la respiration des tissues radiculaires, *Révue générale de Botanique* **71**, 12–21 (1964)

PILET, P. E. and ROLAND, J. G. Effets physiologiques et ultrastructureaux du fluor sur des tissus de Ronce cultivés *in vitro*, *Berichte der Schweizerischen Botanischen Gesellschaft* **82**, 269–283 (1972)

ROSS, C. W., WIEBE, H. H. and MILLER, G. W. Effect of fluoride on glucose catabolism in plant leaves, *Plant Physiology* **37**, 305–309 (1962)

SHIMAZAKI, K. and SUGAHARA, K. Specific inhibition of photosystem II activity in chloroplasts by fumigation of spinach leaves with SO_2, *Plant and Cell Physiology* **20**, 947–955 (1979)

SHOWMAN, R. E. Residual effects of sulfur dioxide on the net photosynthetic and respiratory rates of lichen thalli and cultured lichen symbionts, *Bryologist* **75**, 335–341 (1972)

SIJ, J. W. and SWANSON, C. A. Short-term kinetic studies on the inhibition of photosynthesis by sulfur dioxide, *Journal of Environmental Quality* **3**, 103–107 (1974)

SOLDATINI, G. F. and ZIEGLER, I. Induction of glycolate oxidase by SO_2 in *Nicotiana tabacum*, *Phytochemistry* **18**, 21–22 (1979)

SPEDDING, D. J. and THOMAS, W. J. Effect of SO_2 on the metabolism of glycollic acid by barley (*Hordeum vulgare*) leaves, *Australian Journal of Biological Sciences* **26**, 281–286 (1973)

SRIVASTAVA, H. S., JOLLIFFE, R. A. and RUNECKLES, V. C. Inhibition of gas exchange in bean leaves by NO_2, *Canadian Journal of Botany* **53**, 466–474 (1975a)

SRIVASTAVA, H. S., JOLLIFFE, R. A. and RUNECKLES, V. C. The effects of environmental conditions on the inhibition of leaf gas exchange by NO_2, *Canadian Journal of Botany* **53**, 475–482 (1975b)

SWANSON, E. W., THOMSON, W. W. and MUDD, J. B. The effect of ozone on leaf cell membranes, *Canadian Journal of Botany* **51**, 1213–1219 (1973)

SYRATT, W. J. and WANSTALL, P. J. The effects of sulphur dioxide on epiphytic bryophytes. In *Air Pollution. Proceedings of the First European Congress on the Influence of Air Pollutants on Animals and Plants, 1968*, pp. 79–85, Centre for Agriculture Publishing and Research, Wageningen (1969)

TANIYAMA, T. *Bulletin of the Faculty of Agriculture, Mie University, Tsu Japan* **44**, 11–130 (1972)

TANIYAMA, T., ARIKADO, H., IWATA, Y. and SAWANKA, K. Studies on the mechanism of injurious effects of toxic gases on crop plants. On photosynthesis and dark respiration of rice plant fumigated with SO_2 for long period, *Proceedings of Crop Science Society of Japan* **41**, 120–125 (1972)

THOMAS, M. D. Air pollution with relation to agronomic crops. 1. General status of research on the effects of air pollutants on plants, *Agronomy Journal* **50**, 545–550 (1958)

THOMAS, M. D. and ALTHER, E. W. The effects of fluoride on plants. In *Handbook of Experimental Pharmacology. Part I Pharmacology of Fluorides*, pp. 251–306, Springer-Verlag, New York (1966)

THOMAS, M. D. and HILL, G. K. Relation of sulphur dioxide in atmosphere to photosynthesis and respiration in alfalfa, *Plant Physiology* **12**, 309–383 (1937)

TODD, G. W. The effect of ozone and ozonated 1-hexene on respiration and photosynthesis of leaves, *Plant Physiology* **33**, 416–420 (1958)

TODD, G. W. and GARBER, M. J. Some effects of air pollutants on the growth and productivity of plants, *Botanical Gazette* **120**, 75–80 (1958)

TODD, G. W. and PROPST, B. Changes in transpiration and photosynthetic rates of various leaves during treatments with ozonated hexene or ozone gas, *Physiologia Plantarum* **16**, 57–65 (1963)

TOLBERT, N. E. and RYAN, F. J. Glycolate biosynthesis and metabolism during photorespiration. In CO_2 *Metabolism and Plant Production*, (Eds. R. A. BURRIS and C. C. BLACK), pp. 141–159, University Park Press, Baltimore (1976)

TOMLINSON, H. and RICH, S. The ozone resistance of leaves as related to their sulfhydryl and ATP content, *Phytopathology* **58**, 808–810 (1968)

TÜRK, R., WIRTH, V. and LANGE, O. L. CO_2 — Gaswechsel-Untersuchungen zur SO_2 — Resistenz von Flechten, *Oecologia* **15**, 33–64 (1974)

UNSWORTH, M. H. and ORMROD, D. P. (Eds.). *Effects of Gaseous Air Pollution in Angriculture and Horticulture*, Butterworths, London (1982)

VERKROOST, M. The effect of ozone on photosynthesis and respiration of *Scenedesmus obtusiusculus* Chod. with a general discussion of effects of air pollutants in plants, *Mededelingen Landbouwhogeschool Te Wageningen* **74**, 1–78 (1974)

VINES, H. and WEDDING, R. T. Some effects of ammonia on plant metabolism and a possible mechanism for ammonia toxicity, *Plant Physiology* **35**, 820–825 (1960)

VOGL, M. and BÖRTIZ, S. Physiologische und biochemische Beiträge zur Rauchschädenforschung, *Flora* **155**, 347–352 (1965)

VOGL, M., BÖRTIZ, S. and POLSTER, H. Physiologische und biochemische Beiträge zur Rauchschädenforschung. 3. Mitteilung: Der einfluss stoszartiger, starker SO_2-begasung auf die CO_2-Absorption und einige Nadelinhaltsstoffe von Fichte (*Picea abies* L.) und Bergkiefer (*Pinus mugo*. TURRA) unter Laboratoriumsbedingungen, *Archiv für Forstwesen* **13**, 1031–1043 (1964)

WAKIUCHI, N., MATSUMOTO, H. and TAKAHASHI, E. Changes in some enzyme activities of cucumber during ammonium toxicity, *Physiologia Plantarum* **24**, 248–253 (1971)

WEINSTEIN, L. H. Effects of atmospheric fluoride on metabolic constituents of tomato and bean leaves, *Contributions from the Boyce Thompson Institute* **21**, 215–231 (1961)

WEINSTEIN, L. H. Fluoride and plant life, *Journal of Occupational Medicine* **19**, 49–78 (1977)

WEINSTEIN, L. H. and ALSCHER-HERMAN, R. Physiological responses of plants to fluorine. In *Effects of Gaseous Air Pollution in Agriculture and Horticulture*, (Eds. M. H. UNSWORTH and D. P. ORMROD), pp. 139–167, Butterworths, London (1982)

WELLBURN, A. R., WILSON, J. and ALDRIDGE, P. H. Biochemical responses of plants to nitric oxide polluted atmospheres, *Environmental Pollution* **22**, 219–228 (1980)

YU, M. H. and MILLER, G. W. Effect of fluoride on the respiration of leaves of higher plants, *Plant and Cell Physiology* **8**, 483–493 (1967)

ZELITCH, I. Increased rate of net photosynthetic CO_2 uptake caused by the inhibition of glycolate oxidase, *Plant Physiology* **41**, 1623–1631 (1966)

ZELITCH, I. *Photosynthesis, Photorespiration and Plant Productivity*, pp. 130–212, Academic Press, New York (1971)

ZIEGLER, I. The effect of SO_2 pollution on plant metabolism, *Residue Reviews* **56**, 79–105 (1975)

Part V

Effects on metabolite pools

Chapter 18

Interactions of gaseous pollutants with carbohydrate metabolism

M. J. Kozioł

BOTANY SCHOOL, UNIVERSITY OF OXFORD, UK

Introduction

Plant growth and development represent a complex series of coordinated events which are ultimately dependent upon the carbon economy of the plant; the early stages of seedling growth are dependent upon the organic reserves accumulated in the seed (cotyledons) while subsequent growth depends upon the translocation from mature leaves of carbohydrates in excess of their maintenance needs. The productivity of mature leaves, therefore, represents an asset to the total carbon economy of a plant while developing leaves, roots, flowers and fruit represent liabilities requiring an input of carbon. In most cases, the productivity of mature leaves is more than sufficient to meet these demands of carbon for growth and development. Exposure to gaseous pollutants can, however, alter the balance of a plant's carbon economy so that growth can be retarded and yields reduced; in the case of chronic exposures, reductions in growth and yield can occur without accompanying visible symptoms of injury.

The effects of gaseous pollutants on rates of photosynthesis and respiration are perhaps the most obvious of the induced changes in carbohydrate metabolism likely to affect the carbon economy of the plant. Of equal importance, however, are the numerous reactions of intermediate metabolism responsible for the interconversion of metabolites and synthesis of important molecules such as starch and fructosans (storage carbohydrates), cellulose, lipids and sterols. This chapter will first provide a review of the effects of the major gaseous air pollutants on photosynthesis and on the enzymes of intermediate carbohydrate metabolism; it will then review the data available on pollutant-induced changes in the standing concentrations of various carbohydrates and on changes in the translocation of carbohydrates from leaves, and finally it will review briefly the effects of pollutants on the development of root systems.

Photosynthesis

An abridged schema for carbohydrate metabolism in green plants is given in *Figure 18.1*; it incorporates pathways of C_3 and C_4 photosynthesis, photorespiration and some anabolic and catabolic reactions occurring in the cytosol. In C_3 plants the first step of carbohydrate metabolism is the photoassimilation of CO_2 into ribulose-1,5-

251

252

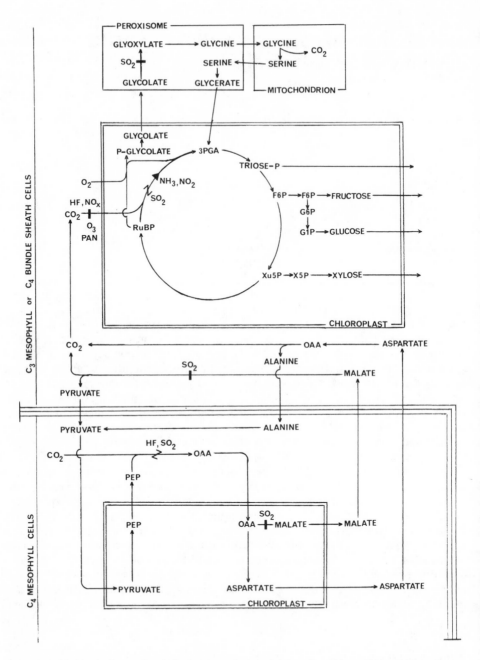

Figure 18.1. Abridged schema of carbohydrate metabolism showing sites of reported interactions by gaseous pollutants. For simplicity, no separate distinction is made between the gases, their ionic species or addition products, as these details are given in the text. Symbols: $\rightarrow\!\!\!\rightarrow$ = inhibition; $\longrightarrow\!\!\!\rightarrow$ = enhancement; $\rightarrow\!\!\!\!\!\!Z\!\!\!\rightarrow$ = mixed effect. Inhibition of photosynthetic CO_2 uptake by HF, NO_x, O_3 or PAN is shown by the vertical block intercepting arrow leading CO_2 into the C_3 chloroplast; the mixed response of glycolysis and pentose phosphate pathway to exposure to SO_2 is shown by the interception of the arrows leading to F6P and 6P-gluconate, respectively

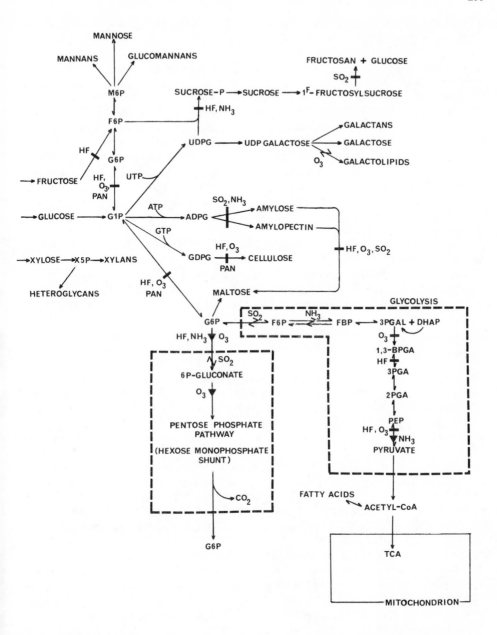

Metabolite abbreviations: RuBP, ribulose-1,5-bisphosphate; P-glycolate, phosphoglycolate; 3-PGA, 3-phosphoglycerate; triose-P, triose phosphates; F6P, fructose-6-phosphate; G6P, glucose-6-phosphate; G1P, glucose-1-phosphate; Xu5P, xylulose-5-phosphate; X5P, xylose-5-phosphate; M6P, mannose-6-phosphate; sucrose-P, sucrose phosphate; UTP, uridinetriphosphate; UDPG, uridinediphosphoglucose; UDPgalactose, uridinediphosphogalactose; ATP, adenosinetriphosphate; ADPG, adenosine-diphosphoglucose; GTP, guanosinetriphosphate, GDPD, guanosinediphosphoglucose; FBP, fructose-1,6-bisphosphate; 3-PGAL, 3-phosphoglyceraldehyde; DHAP, dihydroxyacetone phosphate; 1,3-BPGA, 1,3-bisphosphate glycerate; 2-PGA, 2-phosphoglycerate; PEP, phosphoenolpyruvate; 6P-gluconate, 6-phosphogluconate; OAA, oxaloacetic acid bisphosphate (RuBP) to produce two molecules of 3-phosphoglyceric acid (3-PGA), which via 1,3-bisphosphoglycerate (BPGA) is reduced to the triose phosphate, 3-phosphoglyceraldehyde (3-PGAL). An isomerase converts some of the 3-PGAL to dihydroxyacetone phosphate (DHAP). In the presence of fructose-1,6-bisphosphate aldolase, DHAP and 3-PGAL are condensed to yield fructose-1,6-bisphosphate (FBP), which is dephosphorylated to yield fructose-6-phosphate (F6P). The F6P thus formed represents a branching point in the cycle at which carbon in the form of free hexoses may be removed for other anabolic or catabolic reactions; triose phosphates, hexoses and pentoses can move out of the chloroplast into the cytosol.

RuBP carboxylase also has an affinity for O_2. Incorporation of O_2 into RuBP produces one molecule of 3-PGA and one molecule of P-glycolate. Glycolate is further metabolized to glyoxylate and then to glycine in the peroxisomes; in the mitochondria, the condensation of two molecules of glycine produces serine + CO_2. The net result is that CO_2 recently fixed by RuBP carboxylase in the C_3 cycle can be released again by photorespiration in as little as 15 s.

In C_4 plants, CO_2 entering the mesophyll cells is initially assimilated into phosphoenolpyruvate (PEP) by the enzyme PEP carboxylase to form oxaloacetic acid (OAA) which, depending upon the plant species, is converted in the chloroplasts into either malate or aspartate. The 4-carbon organic acid thus formed is then transported to the bundle sheath cells and decarboxylated; the CO_2 liberated is reassimilated by the C_3 pathway operative in the bundle sheath cells.

Hydrogen fluoride

Interaction of atmospheric fluorine with photosynthetic processes seems to occur at the light reaction stage which may be due, in part, to inhibition of the Hill reaction (Ballantyne, 1972; Spikes, Lumry and Rieske, 1955). Exposures to low concentrations of HF ($10-40\ \mu g\ m^{-3} = 12-48$ ppb) result in reversible reductions in apparent photosynthesis, but although recovery can be quite rapid it is frequently incomplete (Weinstein and Alscher-Herman, 1982). Yang and Miller (1963c) found that dark fixation of CO_2 and the activity of PEP carboxylase were enhanced both in vivo and in vitro in soyabean (*Glycine max*) treated either with 0.03 ppm HF (25 $\mu g\ m^{-3}$) for three to five days or with 1 or 10 μmol KF in the reaction mixture. Rather than being converted to sugars, most of the CO_2 fixed by PEP carboxylase was retained as organic or amino acids. It is interesting that the activity of PEP carboxylase increased in plants exposed to HF compared with control plants as RuBP carboxylase is the major carboxylating enzyme in soyabean and in all C_3 plants. An increase in the activity of PEP carboxylase was also found when chloroplasts of spinach (*Spinacia oleracea*) were treated with SO_3^{2-}, resulting in an increase within the chloroplasts of intermediates of C_4 photosynthesis (Libera, Ziegler and Ziegler, 1975). McCune et al. (1964) on the other hand, found no increase in the activity of PEP carboxylase in the bean (*Phaseolus vulgaris*) exposed to 1.7 or 7.6 μg HF m^{-3} (2 or 9 ppb) continuously for ten days.

Ozone and peroxyacetyl nitrate

Hill and Littlefield (1969) found that exposure to 0.40–0.90 ppm O_3 (788–1773 $\mu g\ m^{-3}$) for 30–120 min reduced the rates of apparent photosynthesis in 13 plant species that included C_3 and C_4 plants. The photosynthetic uptake of CO_2 in seedlings of *Citrus aurantiifolia* was reduced within 4–5 min of the plants being placed in an atmosphere containing 0.6 ppm O_3 (1178 $\mu g\ m^{-3}$) (Taylor et al., 1961). In chloroplasts isolated from plants exposed to O_3 both Hill reaction and photophosphorylation were found to be inhibited, and the degree of inhibition was greater if plants were given a dark period of 48–72 h before exposure to O_3 (Ziegler, 1973a). PAN can also inhibit photosynthesis but there is an absolute requirement for light during and after exposure if any effect is to be seen (Dugger and Taylor, 1961; Koukol, Dugger and Belser, 1963; Koukol, Dugger and Palmer, 1967). This requirement of light for the inhibition of photosynthesis by exposure to PAN is most likely related to the oxidation of sulphhydryl groups of a photoreducible protein (Ziegler, 1973a). Exposure of leaves of bean or chloroplasts of spinach to PAN reduced the amount of $^{14}CO_2$ that was assimilated but had no effect on fixation patterns, suggesting a quantitative and not qualitative reduction in the early products of photosynthesis (Dugger et al., 1963a).

Sulphur dioxide

The effects of SO_2 on photosynthesis are quite varied and have been extensively studied. In aqueous solution SO_2 becomes H_2SO_3 (sulphurous acid), an acid that has never been isolated and can be characterized only in terms of its dissociation species (Falk and Giguère, 1958); at about pH 5.4 HSO_3^- and SO_3^{2-} occur in equal proportions, with SO_3^{2-} being the dominant species at higher pH values (Vas and Ingram, 1949). The absorption of SO_2 can increase the acidity of aqueous solutions; providing the dose of SO_2 is not acute an increase in the buffering capacity of leaf tissue can usually be demonstrated (see Nieboer et al., this volume; Thomas, Hendricks and Hill, 1944; Weigl and Ziegler, 1962). In addition to being able to cause localized alterations in pH, both HSO_3^- and SO_3^{2-} are strongly nucleophilic and can reversibly form addition products with metabolites containing carbonyl functional groups. Addition products can be formed with a variety of carbonyl-containing metabolites including sugars (ribose, xylose, fructose, glucose, galactose, mannose, arabinose), glyconic acids (galacturonic and 2-ketogluconic), aldehydes (glyceraldehyde and acetaldehyde) and organic acids (pyruvic, α-ketoglutaric, succinic, malic, citric and lactic) (Burroughs and Sparks, 1964, 1973; Gehman and Osman, 1954; Jiráček, Macháčková and Kostir, 1972; Kielhöffer, 1958). Thus, the interaction of SO_2 with metabolism can result from disturbances of pH, from competition with substrate or from the alteration of the normal substrate by bisulphite addition. Such a complex interaction with metabolism is perhaps best illustrated by the reported inhibitions by HSO_3^- of glycolate oxidase, RuBP carboxylase and PEP carboxylase from spinach, summarized in Table 18.1. The bisulphite addition product of glyoxylate (glyoxylate bisulphite or sulphoglycolate) has been isolated from rice (Oryza sativa) leaves exposed to SO_2 (Tanaka, Takanashi and Yatazawa, 1972) and has been shown to be an effective inhibitor of glycolate oxidase (Lüttge et al., 1972; Paul and Bassham, 1978; Zelitch, 1957). Although it was once considered desirable to inhibit photorespiration as a means of improving the CO_2-fixation efficiency of C_3 plants, any prolonged inhibition of glycolate oxidase is likely to be detrimental to the plant in the long term. Under normal atmospheric conditions (21% O_2, 0.03% CO_2), 4 mol of RuBP are carboxylated and 1 mol is oxygenated; the stoichiometries of photosynthesis and photorespiration are such that the products of RuBP carboxylation and oxygenation require the same amount of energy to be recycled to RuBP (Ogren, 1978). The important difference is in the amount of carbon remaining, i.e. net photosynthesis. Inhibition of glycolate oxidase or of other enzymes of the photorespiratory pathway leading to the accumulation of glycolate can divert so much carbon from the C_3 cycle that insufficient RuBP is generated to maintain normal rates of photosynthesis.

TABLE 18.1. Aspects of bisulphite inhibition of enzymes isolated from spinach

Enzyme	Substrate/ cofactor	Nature of inhibition	K_i (mM)	Reference
Glycolate oxidase	glycolate	competitive (through glyoxylate bisulphite)	0.05–0.1	Zelitch (1957)
Ribulose-1,5-bisphosphate carboxylase	HCO_3^-	competitive	0.9	Libera, Ziegler and Ziegler (1975)
	HCO_3^-	competitive	1.6	Libera, Ziegler and Ziegler (1975)
	HCO_3^-	non-competitive	3.0	Ziegler (1972)
	Mg^{2+}	non-competitive	9.5	Ziegler (1972)
	$RuBP^a$	non-competitive	14	Ziegler (1972)
Phosphoenol-pyruvate carboxylase	Mn^{2+}	non-competitive	2.4	Mukerji and Yang (1974)
	Mg^{2+}	non-competitive	10	Mukerji and Yang (1974)
	PEP^b	non-competitive	11	Mukerji and Yang (1974)
	HCO_3^-	competitive/mixed	17	Mukerji and Yang (1974)

[a] Ribulose-1,5-bisphosphate.
[b] Phosphoenolpyruvate.

In vitro, glyoxylate bisulphite was also found to inhibit the activities of PEP carboxylase and malate dehydrogenase, two enzymes involved in the initial steps of C_4 photosynthesis (Osmond and Avadhani, 1970). Sulphite was also found to inhibit these enzymes (Ziegler, 1973b) as well as the malic enzyme (Ziegler, 1974) which decarboxylates malate to release CO_2 for fixation by RuBP carboxylase. Inhibition of photosynthesis by the action of SO_3^{2-} on various enzymes is, however, concentration-dependent; Libera, Ziegler and Ziegler (1975) found that at SO_3^{2-} concentrations less than 1 mM, the fixation of HCO_3^- by intact chloroplasts of spinach was enhanced. Recovery of normal photosynthetic rates after exposure to SO_2 is dependent upon the dose received; acute exposure can cause ultrastructural changes in chloroplasts, as has been discussed by Huttunen and Soikkeli, and by Parry and Whittingham in this volume.

Nitrogen oxides and ammonia

Very little information is available on the effects of NO_x and NH_3 on photosynthesis. Exposure to 1.0–7.0 ppm NO_2 (1.88–13.15 mg m^{-3}) for 2–5 h inhibited photosynthesis and respiration in bean (Srivastava, Jolliffe and Runeckles, 1975). In a 4×4 factorial experiment both NO_2 and NO at concentrations of 0.10, 0.25 or 0.50 ppm (191, 477 or 953 μg NO_2 m^{-3}; 124, 311 or 622 μg NO m^{-3}) inhibited photosynthesis in tomato (*Lycopersicon esculentum*) to nearly the same extent when supplied singly, and approximately additively when supplied in combination (Capron and Mansfield, 1976). Hill and Bennett (1970) investigated the effects of NO and NO_2 on the inhibition of photosynthesis in alfalfa (*Medicago sativa*) and oat (*Avena sativa*) by exposing plants for 2 h to concentrations of NO up to 10 ppm (12.3 mg m^{-3}) or to concentrations of NO_2 up to 8 ppm (15.1 mg m^{-3}). They found that there was a threshold of exposure defined by a pollutant concentration of 0.5–0.7 ppm for 45–90 min for both gases (616–862 μg NO m^{-3} or 944–1231 μg NO_2 m^{-3}) which, when exceeded, resulted in inhibition of photosynthesis in both plant species. Horsman and Wellburn (1975), on the other hand, found exposure to 0.1 or 1.0 ppm NO_2 (191 or 1097 μg m^{-3}) for six days to stimulate the activity of RuBP carboxylase in seedlings of pea (*Pisum sativum*).

Ammonia, supplied as NH_4^+ (0.34–2.0 mM), was found to increase the photosynthetic assimilation of $^{14}CO_2$ in cells isolated from opium poppy (*Papaver somniferum*) (Hammel, Cornwell and Bassham, 1979; Paul, Cornwell and Bassham, 1978) and in cells isolated from spinach treated with 0.2 mM NH_4^+ at pH 8.0 or 1.0–1.7 mM NH_4^+ at pH 7.2 (Woo and Canvin, 1980a,b). Possible mechanisms for the increase in CO_2 fixation in response to treatment with ammonia include the stimulation of RuBP carboxylase activity (De Benedetti *et al.*, 1976) and the enhancement of the light activation of enzymes regulated by the ferredoxin-thioredoxin system, i.e. NADP-malate dehydrogenase, fructose-1,6-bisphosphatase, ribulose-5-phosphate kinase and NADP-glyceraldehyde-3-phosphate dehydrogenase (Rosa, 1981).

Intermediate carbohydrate metabolism

The preceding section discussed the effects of gaseous pollutants on photosynthesis. *In vivo*, however, gaseous pollutants must first enter the leaves, pass the barriers presented by the cell walls and plasmalemma and then diffuse through the cytosol before accumulating to those concentrations in the chloroplasts at which metabolism has been sufficiently altered to be detectable on a gross scale as changes in rates of CO_2 uptake. As most of the anabolic and catabolic reactions of carbohydrate metabolism occur in the cytosol the potential for pollutant interactions is great.

Hydrogen fluoride

The fluoride ion (F^-) has long been used as a metabolic inhibitor to study the activity of various enzymes, and such studies have identified fluoride-sensitive metabolic reactions. The mechanism of F^- sensitivity in most cases involves, through binding with F^-, the removal of Mg^{2+} or other divalent heavy metal cations that serve as cofactors of the enzymes (Weinstein, 1977; Ziegler, 1973a). Various enzymes have been shown to be inhibited by F^- including enolase (Miller, 1958), phosphatases (Bonner and Wildman, 1946; Kielley and Meyerhoff, 1948; Massart and Dufait, 1942; Rapp and Sliwinski, 1956), phosphorylase (Rapp and Sliwinski, 1956), hexokinase (Melchior and Melchior, 1956) phosphoglycerate kinase (Axelrod and Bandurski, 1953), sucrose synthetase [= uridinediphosphoglucose-fructose transglucosylase] (Yang and Miller, 1963b), α-amylase (Rockwood, 1919) and, on the basis of the studies by Ordin and Propst (1962) and Ordin and Skoe (1963) possibly cellulose synthetase [GDPglucose-β-glucan glucosyltransferase]. There are conflicting reports on the effects of F^- or phosphoglucomutase; F^- inhibited the enzyme *in vivo* in the yeasts, *Zygosaccharomyces acidifaciens* and *Saccharomyces cerevisiae* (Chung and Nickerson, 1954; Nickerson and Chung, 1952) and *in vitro* when the enzyme was extracted either from jack-bean (*Canavalia ensiformis*) (Yang and Miller, 1963b) or from oat (*Avena*) coleoptiles (Ordin and Altman, 1965) but *in vitro* studies on phosphoglucomutase extracted from maize (*Zea mays*) (Chang, 1968) or potato (*Solanum tuberosum*) (De Moura, Letourneau and Weise, 1973) showed the enzyme to be relatively insensitive to F^-. The rate of removal of Mg^{2+} from solution by F^- is a slow process in which the precipitation of the salt is probably rate-limiting (i.e. $Mg^{2+} + 2F^- \rightarrow MgF_{2(aq)} \rightarrow MgF_{2(s)}$), so that differences in incubation times can markedly affect the degree of inhibition observed (Melchior and Melchior, 1956). With hexokinase, Melchior and Melchior (1956) found the results of 15-min incubations with F^- to be erratic and they obtained reproducible results only with incubation periods of 30 min. In the study of DeMoura, Letourneau and Weise (1973) the total incubation period was 25 min (no times are given in the study of Chang, 1968), while in the studies of Chung and Nickerson (1954) and of Ordin and Altman (1965) the incubation periods were 48 h and 75 min, respectively (no times given in the study of Yang and Miller, 1963b). Further, DeMoura, Letourneau and Weise (1973) used concentrations of F^- at least an order of magnitude lower than those used in the other studies, and supplied Mg^{2+} in the reaction medium at a concentration two to five times higher than the amount of inhibitor used. These concentration and time differences among the studies might partially explain the differences observed on the effects of F^- on the inhibition of phosphoglucomutase.

Inhibition of phosphoglucomutase may account for the increases in the concentrations of reducing sugars and decreases in the concentrations of non-reducing sugars in leaves of soyabean exposed to HF compared with leaves from control plants (Yang and Miller, 1963a). Such an inhibition might also account for the decrease in starch concentrations and increase in the concentrations of ethanol-soluble sugars in ponderosa pine (Pinus ponderosa) exposed to HF (Adams and Emerson, 1961). Ordin and Altman (1965) and Ordin and Skoe (1963) found that the incorporation of label from $U[^{14}C]$glucose-1-phosphate into cellulose was inhibited by treatment with F^- and used the known inhibition of phosphoglucomutase by F^- as an explanation of this observation, accounting for the observed reductions in the growth of oat coleoptiles in this way. They had assumed, however, that G6P was an intermediate in the synthesis of cellulose, whereas it is now believed that G1P is converted to GDPG which is the substrate for GDPglucose-β-glucan glucosyltransferase in cellulose biosynthesis (Teng and Whistler, 1973). The data of Ordin and co-workers do suggest the possibility of an effect of F^- on the activity either of guanosine diphosphoglucose pyrophosphorylase or of GDPglucose-β-glucan glucosyltransferase.

Enolase, with an absolute requirement for Mg^{2+} or Mn^{2+} for activity, is strongly inhibited by F^-, particularly in the presence of phosphate (Miller, 1958), the inhibitory species being the Mg^{2+}-fluorophosphate complex. Inhibition of enolase blocks the conversion of 2-PGA to PEP which in turn inhibits glycolysis, the main pathway for the catabolism of sugars in seeds (Tewfik and Stumpf, 1951a). Homogenates of leaves of pea (Pisum sativum), however, were found to be able to metabolize FBP by an alternative pathway that was insensitive to F^- treatment and did not form 3-PGA (Tewfik and Stumpf, 1951b). Nielands and Stumpf (1955) later suggested that the presence of the oxidative pentose phosphate pathway (hexose monophosphate shunt) would account for the F^--insensitive catabolism of glucose in some plant tissues. Indeed, Ross, Wiebe and Miller (1962) found the activity of the pentose phosphate pathway to increase and that of glycolysis to decrease in leaves of the nettle-leaved goosefoot (Chenopodium murale) and knotweed (Polygonum orientale) supplied with F^- either by uptake through petioles from solution (5 mM KF for two days) or by exposure to 6 ppb gaseous HF (5 μg m^{-3}) for five to six days. They suggested that the accumulation of 3-PGA caused by the inhibition of enolase might be converted to triosephosphate and resynthesized into hexose that would then be metabolized via the pentose phosphate pathway. The sensitivity of glycolysis to inhibition by F^- is dependent upon the age of the tissue (McCune and Weinstein, 1971; Tewfik and Stumpf, 1951b). As leaves mature, glycolytic and Krebs cycle activity decrease while the activity of the pentose phosphate pathway increases; hence any inhibition of glycolysis by F^- is likely to cause a greater change in the relative activities of the two pathways in younger rather than in older leaves (McCune and Weinstein, 1971; Weinstein, 1977). It must be noted, however, that both Lee, Miller and Welkie (1966), working with soyabean, and McCune *et al.* (1964), working with bean (*Phaseolus vulgaris*) and *Sorghum vulgare*, reported increases in enolase activity in response to treatment with HF; McCune *et al.* (1964) also found the activity of pyruvate kinase to increase in *Sorghum*. In both cases, the enzymes were extracted from the HF-treated leaf tissue and partially purified before assay; Weinstein (1977) has suggested that such extraction and partial purification might remove the inhibitor, so that when assayed under optimal conditions the activity may appear to be stimulated. Arrigoni and Marré (1955) reported an enhancement of the activity of glucose-6-phosphate dehydrogenase (G6PD) in tips of germinating pea treated with F^-; its activity was also enhanced in soyabean exposed to gaseous HF (Lee, Miller and Welkie, 1966). Such an enhancement would be expected if F^-, by inhibiting enolase, caused a shift from glycolysis to the pentose phosphate pathway.

Investigating the effect of exposure to HF on the isoenzymes in leaves of F^--resistant and sensitive clones of *Solanum pseudocapsicum*, Weinstein (1977) found one or more of the isoenzymes of G6PD, acid phosphatase, phosphoglucomutase and peroxidase to increase only in the sensitive clone; treatment with HF had little effect on malate dehydrogenase, 6-phosphogluconate dehydrogenase or hexokinase. Weinstein (1977) suggested that F^- inhibition *in vivo* might induce the production of additional enzyme. Such an induction of the transcription and synthesis of isoenzymes, coupled with the removal of F^- inhibition of the constitutive enzyme during extraction and purification might also result in an apparent increase in activity when assayed under optimal conditions. Ziegler (1973a) has also raised the caveat that in studies of F^- inhibition conducted on crude enzyme preparations, homogenizing the tissue destroys the cellular compartmentalization of F^- so that the results of such *in vitro* studies may not represent true *in vivo* conditions.

Ozone and peroxyacetyl nitrate

Both O_3 and PAN are strong oxidants with redox potentials of $+2.07$ and about $+0.4$ V, respectively (Dugger and Ting, 1970). It is now well established that photochemical oxidants can oxidize sulphhydryl groups, and such oxidation seems sufficient to account for the loss of enzymic activity. In this respect, the configuration of the protein becomes important as the oxidants must have access to the sulphhydryl groups (Ziegler, 1973a). Thus Todd (1958) found that papain, an enzyme dependent upon its sulphhydryl groups for activity was much more severely inhibited by O_3 than peroxidase, catalase or urease, and King (1961) found that exposure to O_3 decreased the thiol groups in crystalline glyceraldehyde-3-phosphate dehydrogenase to 5.75 thiol groups mol^{-1} compared with 8 groups mol^{-1} in the untreated enzyme.

Ordin and Altman (1965) found phosphoglucomutase extracted from oat coleoptiles to be inhibited by exposure to O_3 or PAN *in vitro*, but only by PAN *in vivo*. Ray and Koshland (1962) reported that oxidation by PAN of a methionine group in phosphoglucomutase resulted in a loss of activity. Although UDPG pyrophosphorylase was found to be insensitive to treatment with O_3 or PAN *in vivo* (Hall and Ordin, 1967), GDPG- and UDPG-β-glucan glucosyltransferases were found to be inhibited by PAN and to a lesser extent by O_3, suggesting a possible site of interference by these photochemical oxidants on cellulose biosynthesis (Ordin and Hall, 1967; Ordin, Hall and Kindinger, 1969). Treatment of UDPG-β-glucan glucosyltransferase with iodoacetamide or *p*-chloromercuribenzoate also inactivated the enzyme, further suggesting that the mechanism of inactivation by photochemical oxidants might involve reaction with sulphhydryl groups (Ordin and Hall, 1967). In a later study on the utilization of U[^{14}C]glucose by oat coleoptiles, exposure to PAN significantly inhibited the incorporation of ^{14}C into UDPG, sucrose and cellulose but not into G1P or G6P, implying an inhibition of nucleotide synthesis (Gordon and Ordin, 1972). Regarding the earlier report that UDPG pyrophosphorylase was insensitive to PAN treatment (Hall and Ordin, 1967), Gordon and Ordin (1972) suggested that a decrease in the amount of ^{14}C incorporated into UDPG could occur if the pool size of UDPG were regulated by a feed-back inhibition; the metabolic regulator, and not PAN, would therefore be responsible for the inhibition of UDPG pyrophosphorylase. Ordin and Hall (1967) have suggested that the greater effect of PAN on the inhibition of cellulose biosynthesis, compared with O_3, may be due to its inhibition of two enzymes in the synthetic pathway, namely phosphoglucomutase and UDPG-β-glucan glucosyltransferase. In addition, the growth of oat coleoptiles may also be inhibited by the oxidation products of indole-3-acetic acid resulting from treatment with PAN (Hall, Brown and Ordin, 1971).

The hydrolysis of reserve starch was inhibited by exposure to 0.05 ppm O_3 (99 μg m^{-3}) for 2–6 h in cucumber (*Cucumis sativus*), bean (*Phaseolus vulgaris*), senna (*Cassia occidentalis*) and monkey flower (*Mimulus cardinalis*) (Hanson and Stewart, 1970), suggesting an inhibition of amylase or phosphorylase. Carbon flow through various pathways in soyabean exposed for 2 h to 0.5 ppm O_3 (972 μg m^{-3}) was altered as evidenced by a decrease in the activity of glyceraldehyde-3-phosphate dehydrogenase and an increase in the activity of G6PD, implying a reduction of glycolysis and increased activity of the pentose phosphate pathway (Tingey, Fites and Wickliff, 1975). A second enzyme of the pentose phosphate pathway, 6-phosphogluconate dehydrogenase, was also found to increase in activity (Tingey, 1974).

PAN and O_3 can alter cellular concentrations of NADH and NADPH through their oxidizing effect, but whereas the oxidation of these nucleotides by PAN does not alter their biological activity (Mudd, 1965; Mudd and Dugger, 1963), the products of oxidation with O_3 inhibit the enzymatic activity in the metabolic reactions requiring these reduced nucleotides as coenzymes (Dugger and Ting, 1970). The reversible oxidation of these coenzymes may provide some protection to the enzymes with which they are associated. *In vitro*, exposure of G6PD on its own to PAN almost completely inactivated the enzyme, whereas the presence of NADPH was found to offer some protection (Mudd, 1963).

Glycolipid biosynthesis was inhibited in chloroplasts of spinach treated with O_3: the formation of di- and trigalactosyl diglycerides was more strongly inhibited than the formation of monogalactosyl diglyceride, sterol glycoside and acylated sterol glycoside (Mudd *et al.*, 1971). Leaf tissue of bean and spinach exposed to 0.5 ppm O_3 (995 μg m^{-3}) for 1 h contained more sterol glycosides and acylated sterol glycosides and less free sterol than was found in leaf tissue from control plants (Tomlinson and Rich, 1973). As free sterols, being 'flatter', exert a greater influence on membrane permeability and integrity (Grunwald, 1971), the loss of free sterols and the synthesis of esterified derivatives have been considered the initial steps in membrane deterioration and the development of O_3 injury in leaves (Tomlinson and Rich, 1971, 1973). As injury progresses activities of acid phosphatase (Evans and Miller, 1972) and of cellulase (Dass and Weaver, 1972) increase. Visible symptoms of injury preceding necrosis can include the production of various pigments, such as anthocyanin (cyanidin glucoside) in curly dock (*Rumex crispus*) (Koukol and Dugger, 1967).

Sulphur dioxide

Bisulphite has been shown to inhibit α-glucan phosphorylase (α-1,4-glucan:ortho-phosphate glucosyltransferase) extracted from potato (*Solanum tuberosum*) by competing with phosphate at the binding site of phosphate (Kamogawa and Fukui, 1973). Experiments conducted by Kozioł (1980) on soyabean exposed to 63, 215 or 246 μg SO_2 m^{-3} (24, 81 or 93 ppb) for 34 days on a 14-h photoperiod suggested an inhibition of starch synthesis in mature and recently-matured (i.e. fully expanded) leaves; ratios of amylose:amylopectin could be significantly altered by exposure to SO_2 implying that the enzymes controlling their synthesis may exhibit different sensitivities to sulphite.

Nikolaevskii (1968) exposed six species to 125 ppm SO_2 (332 mg m^{-3}) for 17 h (length of photoperiod not given) and found that the activity of the pentose phosphate pathway was increased and of glycolysis decreased in ash leaved maple (*Acer negundo*), pentose phosphate pathway decreased and glycolysis increased in black poplar (*Populus canadensis*), balsam poplar (*P. balsamifera*) and tobacco (*Nicotiana affinis*), and the activities of both pathways decreased in silver birch (*Betula verrucosa*) and a salt bush (*Kochia trichophylla*). He has suggested that in SO_2-resistant species, the activity of the pentose phosphate pathway is either maintained or increased with a decrease in glycolysis and Krebs cycle activity, whereas in SO_2-sensitive species the reverse is found.

Pierre (1977) found that exposure of bean to SO_2 resulted in a general increase in the activities of enzymes associated with photosynthesis, glycolysis and the pentose phosphate pathway when extracted from pollutant-exposed leaves and then assayed *in vitro*, in contrast with a general decrease in their activities when assayed in the presence of HSO_3^- or SO_3^- *in vitro*. She has suggested, therefore, that such *in vitro* studies with the ionic species may not serve as adequate models for the action of SO_2 *in vivo*. The comment by Weinstein (1977), suggesting that the extraction and partial purification of enzymes from pollutant-treated material may remove the inhibitor so that activity *in vitro* may appear stimulated, may again be relevant.

Nitrogen oxides and ammonia

Little information is available on the effects of these gases on carbohydrate metabolism. In response to exposure to NO, tomato showed increased levels of nitrite reductase, glutamate dehydrogenase, glutamate oxaloacetate transaminase and glutamate pyruvate transaminase activities (Wellburn, Wilson and Aldridge, 1980). Increases in protein concentrations of the order of 10–20% were observed in pea and amino acids were found to increase in tomato exposed to NO_2 (Zeevart, 1976). If photosynthesis were also reduced, such changes in protein and amino acid concentrations might be reflected in corresponding decreases in the concentrations of reducing sugars in leaves.

Treatment with NH_3 increased the concentrations of amino acids, mostly at the expense of sucrose biosynthesis, in alfalfa (Platt, Plaut and Bassham, 1977) and in the alga *Chlorella pyrenoidosa* (Kanazawa, Kirk and Bassham, 1970; Kanazawa *et al.*, 1972). The reported increases in the activities of PEP carboxylase (Hammel, Cornwell and Bassham, 1979; Paul, Cornwell and Bassham, 1978; Platt, Plaut and Bassham, 1977) and of pyruvate kinase (Paul, Cornwell and Bassham, 1978; Platt, Plaut and Bassham, 1977) correlate with the increases in amino acids found in opium poppy (Paul, Cornwell and Bassham, 1978), alfalfa (Platt, Plaut and Bassham, 1977) and

Chlorella pyrenoidosa (Kanazawa, Kirk and Bassham, 1970) treated with NH_3. Enzymes of carbohydrate metabolism that have been reported to decrease in activity after treatment with NH_3 include sucrose synthetase and FBPase (Kanazawa, Kirk and Bassham, 1970), aldolase (Wakiuchi, Matsumoto and Takahashi, 1971) and starch synthetase (Matsumoto, Wakiuchi and Takahashi, 1971), whereas treatment with NH_3 was found to increase the activities of enolase (Kanazawa *et al.*, 1972), amylase (Rockwood, 1919), G6PD, phosphoglucoisomerase and PFK (Wakiuchi, Matsumoto and Takahashi, 1971). Increases in the activities of G6PD and PFK in plants exposed to NH_3 would imply increases in respiration, and Willis and Yemm (1955) indeed found that roots of barley (*Hordeum vulgare*) assimilating NH_3 showed an enhanced rate of uptake of O_2 and a depletion of reserve carbohydrates. In isolated cells of carrot (*Daucus carota*), NH_3 was found to stimulate glycolysis (Becarri *et al.*, 1969), although the pentose phosphate pathway had previously been shown by ap Rees and Beevers (1960a,b) to be the major pathway of induced respiration in carrot slices.

Carbohydrate pools

The next aspect of the effects of gaseous air pollutants on plant carbohydrate metabolism to be considered concerns the pollutant-induced changes in carbohydrate concentrations within the plant, summarized in *Table 18.2*. From these data, it is difficult to make any generalization for the effects of carbohydrate concentrations of one pollutant, let alone a generalization for all.

Studies by Kozioł (1980) on soyabean and by Cowling and Kozioł (1978) and Kozioł and Cowling (1980) on perennial ryegrass (*Lolium perenne*) exposed to SO_2 have shown that significant increases in the concentrations of free carbohydrates and decreases in the concentrations of storage carbohydrates can occur without any significant effects on growth or on dry matter production. Such alterations in free carbohydrate concentrations may, however, have important consequences regarding either the predisposition of plants to phytophagous insect attack or possible alterations in the sensitivity of plants to mixtures of pollutants, notably SO_2 in combination with photochemical oxidants.

Phaseolus vulgaris is the favoured host of the Mexican bean beetle (*Epilachna varivestris*), an insect also considered an economically important pest of soyabean although it is only marginally adapted to this alternative host (Hughes, Potter and Weinstein, 1982). Exposure of the preferrred host, *Phaseolus*, to $390\,\mu g\ SO_2\ m^{-3}$ (0.15 ppm) for seven days resulted in female beetles showing a preference for fumigated leaf tissue (Hughes, Potter and Weinstein, 1981); no differences in growth, development time or survival were found between larvae fed on control or SO_2-treated leaves, nor in the number of eggs laid by females over a ten-day period. When beetles were offered leaves of soyabean exposed to $524\,\mu g\ SO_2\ m^{-3}$ (0.2 ppm) for seven days, the larvae developed faster and grew larger and females showed a significant preference for SO_2-treated leaves to those from control plants (Hughes, Potter and Weinstein, 1982); further, females fed on SO_2-treated leaves were more fecund — not only were more eggs laid but their viability was also higher. The feeding response of the Mexican bean beetle is influenced by several factors including host volatiles and sugar concentrations (Hughes, Potter and Weinstein, 1981). Information in the literature on the effects of gaseous pollutants on a wide variety of plants confirms that exposure to pollutants can alter the concentrations of sugars, organic and amino acids and the emission of volatiles from plants, often without causing visible symptoms of injury or reductions in growth or yield. Where crop injury and yield reductions due to the direct effects of air pollutants are minimal, it becomes important to ascertain whether further potential reductions in yield might be sustained by making the crop more susceptible to insect attack.

There are numerous reports of the effects of leaf age influencing the amount of O_3 damage sustained, the consensus being that leaves expanded to 75–100% of their mature lamina area are most sensitive to damage (Glater, Solberg and Scott, 1962; Menser, Heggestad and Street, 1963; Ting and Dugger, 1968; Ting and Mukerji, 1971; Tingey, Fites and Wickliff, 1973a). Other studies have shown that light intensity and photoperiod can also affect a plant's response to O_3 or PAN (Dugger et al., 1963b; MacDowall, 1965; Menser et al., 1963). In combination, these studies seem to indicate an involvement of light-mediated metabolism in predisposing leaves to oxidant damage. Thus, rates of photosynthesis and the amount of photosynthate in sensitive leaves become likely candidates for further study, especially as it has been shown that leaf photosynthetic activity reaches a maximum when leaves have expanded to about 65–80% maximum lamina area, thereafter decreasing (Rawson and Woodward, 1976; Woodward and Rawson, 1976). Loading leaves of bean (Phaseolus vulgaris) with carbohydrate by submitting them to an extended (72 h) photoperiod enhanced damage caused by exposure to 0.9–1.0 ppm O_3 (1.74–1.93 mg m^{-3}), whereas plants depleted of carbohydrate by extended dark periods (48–72 h) were not damaged by exposure to 0.6 ppm O_3 (1.16 mg m^{-3}) for 30 min, even though it was shown that the stomata were open during exposure (Dugger et al., 1962). Additionally, Dugger and co-workers demonstrated that leaves of plants kept in the dark became susceptible to O_3 damage when fed sugars through petioles or even directly to the leaf by surface application. Dugger, Koukol and Palmer (1966) also demonstrated that the middle and lower leaves of the O_3-resistant variety of tobacco (Nicotiana tabacum), Bel-B had a lower sugar content than did the leaves of the sensitive variety, Bel-W3. In both varieties the top (developing) leaves had sugar concentrations about twice those of the lower leaves and were found to be resistant to damage. Lee (1965) extended the study of sugar concentrations and O_3 damage in tobacco and reported that leaves containing 5–14 mg sucrose or 5–35 mg reducing sugars g^{-1} dry weight were most sensitive to O_3 damage. However, no correlation between sugar concentrations and O_3 damage could be found in radish (Raphanus sativus) (Athanassious, 1980), bean (Phaseolus vulgaris) (Dunning, Heck and Tingey, 1974) or soyabean (Glycine max) (Dunning, Heck and Tingey, 1974; Tingey, Fites and Wickliff, 1973a). As sugar concentrations have been shown to influence the response of leaf flecking in the varieties of tobacco currently used as bio-indicators for O_3, attention should be paid to ambient SO_2 concentrations and the possible effects on leaf carbohydrate concentrations when interpreting the response of the bio-indicators.

Translocation

One area of research that has been neglected concerns the possible effect of gaseous pollutants on the translocation of carbohydrates from leaves, and yet as regards the growth and ultimate productivity of the plant this is of major importance, second perhaps only to photosynthesis. Noyes (1980) showed that translocation was inhibited by 39, 44 or 69% in bean exposed for 2 h to 0.1, 1.0 or 3.0 ppm SO_2 (0.26, 2.62 or 7.85 mg m^{-3}), respectively, while Teh and Swanson (1982) found exposure to 2.9 ppm

TABLE 18.2. Effects of air pollutants on carbohydrate pools

Plant species	Experimental conditions	Effect	Reference
HF:			
Pinus ponderosa	0.6 ppb, 8 h day^{-1}, 5 days wk^{-1}, 3 wks 1.8 ppb, 8 h day^{-1}, 5 days wk^{-1}, 3 wks 6.1 ppb, 8 h day^{-1}, 2 days wk^{-1}, 3 wks 12.2 ppb, 4 h day^{-1}, 2 days wk^{-1}, 3 wks	Decrease in starch, increase in non-starch polysaccharides Increase in starch towards end of experiment	Adams and Emerson (1961)
Phaseolus vulgaris, Zea mays, Lycopersicon esculentum	5.8–12.9 ppb, 4–12 days	No consistent effect on nucleotide pools	McCune, Weinstein and Mancini (1970)
Phaseolus vulgaris (effects on seed produced)	0.71 ppb, 69–92 days continuous 2.6 ppb, 69–92 days continuous 11.2 ppb, 69–92 days continuous 12.9 ppb, 69–92 days continuous	Slight increase in total and reducing sugars, decrease in starch Substantial decrease in starch	Pack (1971)
Phaseolus vulgaris Lycopersicon esculentum Phaseolus vulgaris Glycine max	17.3 ppb, 4 days 12.4 ppb, 6 days 12.4 ppb, 9 days 30 ppb, 3–5 days	No effect on phosphorylated sugars No effect on glucose, fructose or sucrose Decrease in glucose, fructose and sucrose Decrease in sucrose and non-reducing sugars, increase in glucose, fructose and total reducing sugars	Pack and Wilson (1967) Weinstein (1961) Yang and Miller (1963a)
O$_3$:			
Pinus strobus	0.05 ppm, 4, 9, 17, 22 wks	Increase in total soluble and reducing sugars in primary needles	Barnes (1972)
P. echinata P. elliottii P. serotina P. taeda	0.05 ppm, 11 or 20 wks	Increase in total soluble and reducing sugars in primary needles after 11 wks; increase in total soluble and reducing sugars in secondary needles after 20 wks	
All five Pinus species	0.05, 0.15 ppm, 5 wks	Increase in total soluble and reducing sugars in primary needles	
Ulmus americana	0.9 ppm, 2 h, assayed 24 h after exposure	Decrease in non-structural carbohydrates in all leaves, stems and roots	Constantinidou and Kozlowski (1979)
Citrus limon	0.15–0.25 ppm, 8 h day^{-1}, 5 days wk^{-1}, 9 wks	Increase in reducing sugars, decrease in starch	Dugger, Koukol and Palmer (1966) and Dugger and Palmer (1969)
Acer saccharum	0.2–0.3 ppm, 2 h or 0.1 ppm, 2 h day^{-1}, 14 days	No change in starch concentrations	Hibben (1969)

Species	Treatment	Effect	Reference
Pinus ponderosa	Ambient exposure in field	Decrease in soluble sugars in phloem of trees showing visible damage	Miller, Cobb and Zavarin (1968)
Pinus ponderosa	0.3 ppm, 9 h day^{-1}, 33 days	Increase in soluble sugars, decrease in polysaccharides	Miller *et al.* (1969)
Brassica oleracea	0.2 or 0.35 ppm, 6 h day^{-1}, 1.5 days wk^{-1}, 105 days	Increase in total carbohydrates	Pippen *et al.* (1975)
Zea mays	0.2 or 0.35 ppm, 2.5 h day^{-1}, 3 days wk^{-1}, 71 days	Decrease in total carbohydrates	
Lactuca sativa	0.2 or 0.35 ppm, 6 h day^{-1}, 3 days wk^{-1}, 58 days	No effect	
Fragaria chiloensis	0.2 or 0.35 ppm, 6.5 h day^{-1}, 1.5 days wk^{-1}, 198 days	No effect	
Lycopersicon esculentum	0.2 or 0.35 ppm, 2.5 h day^{-1}, 3 days wk^{-1}, 116 days	Decrease in total carbohydrates	
Glycine max	0.05 ppm, 2 h	Decrease in reducing sugars and starch	Tingey, Fites and Wickliff (1973b)
Pinus ponderosa	99 ppb, 6 h day^{-1}, 20 wks	Increase in soluble sugars and starch in tops; decrease in soluble sugars and starch in roots	Tingey, Wilhour and Standley (1976)
Phaseolus vulgaris	0.25 ppm, 2.5–3 h	Increase in sterol glycosides and acylated sterol glycosides	Tomlinson and Rich (1971)
Pinus strobus, P. taeda	0.1, 0.2, 0.3, 0.6, 1.0 ppm for 10 min, 7 or 21 days	Decrease in label from photoassimilated ^{14}C in soluble sugars, increase in label in sugar phosphates	Wilkinson and Barnes (1973)
PAN: *Avena sativa*	12–14 ppm, 4 h, excised coleoptiles exposed in medium, incubated with [^{14}C] glucose	Decrease in ^{14}C in sucrose, UDPG and cellulose, no effect on G1P or G6P	Gordon and Ordin (1972)
SO$_2$: *Picea abies, Larix leptolepis, L. decidua, Pinus contorta latifolia, P. sylvestris*	3 ppm, 13 h	Decrease in starch	Börtitz (1964)
	0.7–1.4 ppm, field conditions	Decrease in starch	Börtitz (1968)
Picea abies	Ambient conditions, Ore mountains (GDR)	Increase in total sugars in severely injured needles	Börtitz (1969)
Ulmus americana	2 ppm, 6 h, assayed 24 h after exposure	Decrease in non-structural carbohydrates in all leaves, stems and roots	Constantinidou and Kozlowski (1979)

266

TABLE 18.2 (cont.)

Plant species	Experimental conditions	Effect	Reference
Citrus limon	0.25 ppm, 8 h day^{-1}, 5 days wk^{-1}, 9 wks	Increase in reducing sugars, decrease in starch	Dugger, Koukol and Palmer (1966)
Taxus baccata	0.5, 50, 500 ppm, 20 or 72 h, field conditions in December and March	Increase in glucose, fructose and starch, decrease in saccharose and raffinose	Höllwarth (1977)
Pinus contorta	10, 25, 50, 100 ppm (in solution), 22 h	Change in glycolipid composition resulting in increased release of sugars from tissue	Khan and Malhotra (1977)
Pisum sativum	10000 ppm, 24, 48, 72, 96 h	Increase in glucose and fructose, decrease in sucrose	Koštir et al. (1970)
Glycine max	24, 82, 93 ppb, 34 days	Increase in free carbohydrates, decrease in starch (dependent upon leaf age)	Koziol (1980)
Lolium perenne	19, 148 ppb, 29 days, harvested, then additional 22 days	No effect on incorporation of ^{14}C into sugars at end of 29 days; increase in ^{14}C in fructose and sucrose at 148 ppb after additional 22 days	Koziol and Cowling (1978)
Lolium perenne	19, 148 ppb, 29 days, harvested, then additional 22 days	Increase in total free carbohydrates after 29 days, decrease in free carbohydrates after additional 22 days; decrease in fructosans after both periods	Koziol and Cowling (1980)
Phaseolus vulgaris	0.40, 0.77, 1.53, 3.06, 4.03 or 10.06 ppm for 24 h	Increase in reducing sugars and starch up to 3.06 ppm, decrease in sugars and starch at higher SO$_2$ concs.	Koziol and Jordan (1978)
Spinacia oleracea (chloroplasts)	1–5 mM SO$_3^{2-}$, 6 min	Decrease in PGA, sugar phosphates and saccharose, increase in glycolate	Libera, Ziegler and Ziegler (1974)
Pinus banksiana	0.34–0.51 ppm, 96 h	Increase in reducing sugars, decrease in non-reducing sugars	Malhotra and Sarkar (1979)
Papaver somniferum (cell suspension)	10, 20 mM HSO$_3^-$ or SO$_3^{2-}$	Increase sucrose and starch	Paul and Bassham (1978)
Hordeum vulgare	5 ppm, 15–60 min	Increase incorporation of ^{14}C in sugar phosphates, decrease in sucrose	Spedding and Thomas (1973)
NH$_3$:			
Cucumis sativus	20 or 200 mg l^{-1} NH$_3$, 5 days	Increase in UDPG, UDP, G1P, F6P, decrease in starch, G6P and free sugars	Matsumoto, Wakiuchi and Takahashi (1968, 1969)
Papaver somniferum	2 mM NH$_4$Cl, 2–60 min	Decrease in sucrose Bassham (1978)	Paul, Cornwell and Bassham (1978)

SO_2 (7.60 mg m^{-3}) for 2 h inhibited translocation in bean by 45%. In both studies it was concluded that phloem-loading was rate limiting.

Root growth

If exposure to gaseous pollutants were inhibiting translocation of assimilate from leaves, perhaps the most sensitive monitor for such an effect would be the restricted development of root systems. Mansfield (personal communication) has pointed out that very little attention has been paid to such effects of pollutants on root systems, especially in grasses. *Table 18.3* summarizes the results of some studies on the effects of O_3 or SO_2 on plant growth that included data for roots. In all but one case the weight of

TABLE 18.3. Effects of O_3 or SO_3 on root and shoot growth compared with control plants

Plant species	Experimental conditions	% change in growth (dry wt)		Reference
		Root	Shoot/leaf	
O_3:				
Daucus carota	0.19 ppm, 6 h intermittent, 108 days	−32%[a]	+13%	Bennett and Oshima (1976)
	0.25 ppm, 6 h intermittent, 108 days	−46%[a]	+2%	
Glycine max	0.10 ppm, 8 h day^{-1}, 5 days wk^{-1}, 3 wks	−21%[a]	−9%	Tingey, Fites and Wickliff (1973b)
SO_2:				
Ulmus americana	2 ppm, 6 h day^{-1}, 9 days, harvested 7 days later	−7%	(no data)	Constantinidou and Kozlowski (1979)
	2 ppm, 6 h day^{-1}, 9 days, harvested 35 days later	−22%[a]	(no data)	
Ailanthus altissima	0.1 ppm, 14 days	+16%	−33%[a]	Marshall and Furnier (1981)
	0.2 ppm, 14 days	−42%[a]	−68%[a]	
Pinus resinosa	0.2 ppm, 91 h, 32 °C	−22%[a]	+7%	Norby and Kozlowski (1981)
	0.2 ppm, 91 h, 22 °C	−20%[a]	−18%	

Per cent reduction (−%) or increases (+%) calculated from dry weight data presented by authors.
[a] Indicates authors having reported a significant effect on changes in dry weight in plants between experimental and control treatments.

the roots of plants exposed to these gases was lower than that of control plants. Such pollutant-induced reductions in root growth are likely to have important consequences in perennials such as grasses, in which regrowth after cutting or grazing is dependent upon the reserve assimilate stored in the roots.

Conclusions

Although there have been several studies concerned with the interactions of gaseous pollutants or their ionic species with enzymes of carbohydrate metabolism, it is difficult to predict what the alterations in carbohydrate concentrations in leaf tissue will be because of the complex interactions of environmental and edaphic factors modifying plant response. Some plants respond to exposure to air pollutants by retaining more free carbohydrate in leaves at the expense of storage carbohydrate synthesis; such an enhancement of free carbohydrate concentrations in leaves is likely to be important as regards attack by phytophagous insects or the predisposition to oxidant damage in those species known to be sensitive when leaf carbohydrate concentrations fall within a certain range. An area of research meriting further study is the effect of exposure to gaseous pollutants on the translocation of carbohydrates from leaves, since the development of roots, young leaves, flowers and fruit is dependent upon this supply of carbon.

Acknowledgement

The financial support of the Agricultural Research Council during the preparation of this chapter is gratefully acknowledged.

References

ADAMS, D. F. and EMERSON, M. T. Variations in the starch and total polysaccharide content of *Pinus ponderosa* needles with fluoride fumigation, *Plant Physiology* **36**, 261–264 (1961)

ap REES, T. and BEEVERS, H. Pathways of glucose dissimilation in carrot slices, *Plant Physiology* **35**, 830–838 (1960a)

ap REES, T. and BEEVERS, H. Pentose phosphate pathway as a major component of induced respiration of carrot and potato slices, *Plant Physiology* **35**, 839–847 (1960b)

ARRIGONI, O. and MARRÉ, E. Azione degli adenosinfosfati sull' ossidazione diretta del glucoso-6-fosfato in estratti vegetali, *Giornale di Biochemica* **4**, 1–9 (1955)

ATHANASSIOUS, R. Ozone effects on radish (*Raphanus sativus* L. cv. Cherry Belle): foliar sensitivity as related to metabolite levels and cell architecture, *Zeitschrift für Pflanzenphysiologie* **97**, 183–187 (1980)

AXELROD, B. and BANDURSKI, R. S. Phosphoglyceryl kinase in higher plants, *Journal of Biological Chemistry* **204**, 939–948 (1953)

BALLANTYNE, D. J. Fluoride inhibition of the Hill reaction in bean chloroplasts, *Atmospheric Environment* **6**, 267–273 (1972)

BARNES, R. L. Effects of chronic exposure to ozone on soluble sugar and ascorbic acid contents of pine seedlings, *Canadian Journal of Botany* **50**, 215–219 (1972)

BECCARI, E., D'AGNOLO, G., MORPUGO, G. and POCCHIARI, T. Glucose and pyruvate metabolism in *Daucus carota* cells: the effects of the ammonium ion, *Journal of Experimental Botany* **62**, 110–112 (1969)

BENNETT, J. P. and OSHIMA, R. J. Carrot injury and yield response to ozone, *Journal of the American Society for Horticultural Science* **101**, 638–639 (1976)

BONNER, J. and WILDMAN, S. G. Contributions to the study of auxin physiology, *Growth* **10** (6th Growth Symposium): 51–58 (as quoted in Ordin and Skoe, 1963, *see below*) (1946)

BÖRTITZ, S. Physiologische und biochemische Beiträge zur Rauchschadenforschung. I. Untersuchungen über die individuell unterschiedliche Wirkung von SO$_2$ auf Assimilation und einige Inhaltstoffe der Nadeln von Fichten (*Picea abies* [L.] Karst) durch Küvettenbegasung einzelner Zwiege im Freilandversuch, *Biologisches Zentralblatt* **83**, 501–513 (1964)

BÖRTITZ, S. Physiologische und biochemische Beiträge zur Rauchschadenforschung. VII. Einfluss letaler SO$_2$-Begasungen auf den Stärkehaushalt von Koniferennadeln, *Biologisches Zentralblatt* **87**, 63–70 (1968)

BÖRTITZ, S. Physiologische und biochemische Beiträge zur Rauchschadenforschung. XI. Analysen einiger Nadelinhaltsstoffe an Fichten unterschiedlicher individueller Rauchhärte aus einem Schadgebiet, *Archiv für Forstwesen* **18**, 123–131 (1969)

BURROUGHS, L. F. and SPARKS, A. H. The identification of sulphur dioxide-binding compounds in apple juices and ciders, *Journal of the Science of Food and Agriculture* **15**, 176–185 (1964)

BURROUGHS, L. F. and SPARKS, A. H. Sulphite-binding power of wines and ciders. III. Determination of carbonyl compounds in a wine and calculation of its sulphite-binding power, *Journal of the Science of Food and Agriculture* **24**, 207–217 (1973)

CAPRON, T. M. and MANSFIELD, T. A. Inhibition of net photosynthesis in tomato in air polluted with NO and NO$_2$, *Journal of Experimental Botany* **27**, 1181–1186 (1976)

CHANG, C. W. Effect of fluoride on nucleotides and ribonucleic acid in germinating corn seedling roots, *Plant Physiology* **43**, 669–674 (1968)

CHUNG, C. W. and NICKERSON, W. J. Polysaccharide synthesis in growing yeast, *Journal of Biological Chemistry* **208**, 395–407 (1954)

CONSTANTINIDOU, H. A. and KOZLOWSKI, T. T. Effects of sulfur dioxide and ozone on *Ulmus americana* seedlings. II. Carbohydrates, proteins and lipids, *Canadian Journa of Botany* **57**, 176–184 (1979)

COWLING, D. W. and KOZIOL, M. J. Growth of ryegrass (*Lolium perenne* L.) exposed to SO₂. I. Effects on photosynthesis and respiration, *Journal of Experimental Botany* **29**, 1029–1036 (1978)

DASS, H. C. and WEAVER, G. M. Enzymatic changes in intact leaves of *Phaseolus vulgaris* following ozone fumigation, *Atmospheric Environment* **6**, 759–763 (1972)

DE BENEDETTI, E., FORTI, G., GARLASCHI, F. M. and ROSA, L. On the mechanism of ammonia stimulation of photosynthesis in isolated chloroplasts, *Plant Science Letters* **7**, 85–90 (1976)

DEMOURA, J., LETOURNEAU, D. and WEISE, A. C. The response of potato (*Solanum tuberosum*) tuber phosphoglucomutase to fluoride, *Biochemie und Physiologie der Pflanzen* **164**, 228–233 (1973)

DUGGER, W. M., Jr., KOUKOL, J. and PALMER, R. L. Physiological and biochemical effects of atmospheric oxidants on plants, *Journal of the Air Pollution Control Association* **16**, 467–471 (1966)

DUGGER, W. M., Jr., KOUKOL, J., REED, W. D. and PALMER, R. L. Effect of peroxyacetyl nitrate on ¹⁴CO₂ fixation by spinach chloroplasts and pinto bean plants, *Plant Physiology* **38**, 468–472 (1963a)

DUGGER, W. M., Jr. and PALMER, R. L. Carbohydrate metabolism in leaves of rough lemon as influenced by ozone, *Proceedings of the First International Citrus Symposium* **2**, 711–715 (1969)

DUGGER, W. M. Jr. and TAYLOR, O. C. Interaction of light and smog components in plants, *Plant Physiology* **36** (Supplement) xliv (1961)

DUGGER, W. M., Jr., TAYLOR, O. C., CARDIFF, E. and THOMPSON, C. R. Relationship between carbohydrate content and susceptibility of pinto bean plants to ozone damage, *Proceedings of the American Society for Horticultural Science* **81**, 304–315 (1962)

DUGGER, W. M., Jr., TAYLOR, O. C., THOMPSON, C. R. and CARDIFF, E. The effect of light on predisposing plants to ozone and PAN damage, *Journal of the Air Pollution Control Association* **13**, 423–428 (1963b)

DUGGER, W. M. Jr. and TING, I. P. Air pollution oxidants — their effects on metabolic processes in plants, *Annual Review of Plant Physiology* **21**, 215–234 (1970)

DUNNING, J. A., HECK, W. W. and TINGEY, D. T. Foliar sensitivity of pinto bean and soybean to ozone as affected by temperature, potassium nutrition and ozone dose, *Water, Air and Soil Pollution* **3**, 301–313 (1974)

EVANS, L. S. and MILLER, P. R. Ozone damage to ponderosa pine: a histological and histochemical appraisal, *American Journal of Botany* **59**, 297–304 (1972)

FALK, M. and GIGUÈRE, P. A. On the nature of sulphurous acid, *Canadian Journal of Chemistry* **36**, 1121–1125 (1958)

GEHMAN, H. and OSMAN, E. M. The chemistry of the sugar-sulphite compound and its relationship to food problems, *Advances in Food Research* **5**, 53–96 (1954)

GLATER, R. B., SOLBERG, R. A. and SCOTT, F. M. A developmental study of the leaves of *Nicotiana glutinosa* as related to their smog-sensitivity, *American Journal of Botany* **49**, 954–970 (1962)

GORDON, W. C. and ORDIN, L. Phosphorylated and nucleotide sugar metabolism in relation to cell wall production in *Avena* coleoptiles treated with fluoride and peroxyacetyl nitrate, *Plant Physiology* **49**, 542–545 (1972)

GRUNWALD, C. Effects of free sterols, steryl esters and steryl glycoside on membrane permeability, *Plant Physiology* **48**, 653–655 (1971)

HALL, M. A., BROWN, R. L. and ORDIN, L. Inhibitory products of the action of peroxyacetyl nitrate upon indole-3-acetic acid, *Phytochemistry* **10**, 1233–1238 (1971)

HALL, M. A. and ORDIN, L. Subcellular location of phosphoglucomutase and UDPglucose pyrophosphorylase in *Avena* coleoptiles, *Physiologia Plantarum* **20**, 624–633 (1967)

HAMMEL, K. E., CORNWELL, K. L. and BASSHAM, J. A. Stimulation of dark CO₂ fixation by ammonia in isolated mesophyll cells of *Papaver somniferum* L., *Plant and Cell Physiology* **20**, 1523–1529 (1979)

HANSON, G. P. and STEWART, W. S. Photochemical oxidants: effect on starch hydrolysis in leaves, *Science* **168**, 1223–1224 (1970)

HIBBEN, C. R. Ozone toxicity to sugar maple, *Phytopathology* **59**, 1423–1428 (1969)

HILL, A. C. and BENNETT, J. H. Inhibition of apparent photosynthesis by nitrogen oxides, *Atmospheric Environment* **4**, 341–348 (1970)

HILL, A. C. and LITTLEFIELD, N. Ozone. Effect on apparent photosynthesis, rate of transpiration and stomatal closure in plants, *Environmental Science and Technology* **3**, 52–56 (1969)

HÖLLWARTH, M. Zum Verhalten einiger Kohlenhydrate aus Nadeln von *Taxus baccata* L. an städtischen Standorten unterschiedlicher Immissionsbelastung, *Angewandte Botanik* **51**, 277–285 (1977)

HORSMAN, D. C. and WELLBURN, A. R. Synergistic effect of SO_2 and NO_2 polluted air upon enzyme activity in pea seedlings, *Environmental Pollution* **8**, 123–133 (1975)

HUGHES, P. R., POTTER, J. E. and WEINSTEIN, L. H. Effects of air pollutants on plant:insect interactions: reactions of the Mexican bean beetle to SO_2-fumigated pinto beans, *Environmental Entomology* **10**, 741–744 (1981)

HUGHES, P. R., POTTER, J. E. and WEINSTEIN, L. H. Effects of air pollution on plant-insect interactions: increased susceptibility of greenhouse-grown soybeans to the Mexican bean beetle after plant exposure to SO_2, *Environmental Entomology* **11**, 173–176 (1982)

JIRÁČEK, V., MACHÁČKOVÁ, I. KOSTÍR, J. Nachweis der Bisulfit-Addukte (α-oxysulfonsäuren) von Carbonylverbindungen in den mit SO_2 behandeiten Erbsenkeimungen, *Experimentia* **28**, 1007–1008 (1972)

KAMOGAWA, A. and FUKUI, T. Inhibition of α-glucan phosphorylase by bisulfite competition at the phosphate binding site, *Biochimica et Biophysica Acta* **302**, 158–166 (1973)

KANAZAWA, T., KANAZAWA, K., KIRK, M. R. and BASSHAM, J. A. Regulatory effects of ammonia on carbon metabolism in *Chlorella pyrenoidosa* during photosynthesis and respiration, *Biochimica et Biophysica Acta* **256**, 656–669 (1972)

KANAZAWA, T., KIRK, M. R. and BASSHAM, J. A. Regulatory effects of ammonia on carbon metabolism in photosynthesizing *Chlorella pyrenoidosa*, *Biochimica et Biophysica Acta* **205**, 401–408 (1970)

KHAN, A. R. and MALHOTRA, S. S. Effects of aqueous sulphur dioxide on pine needle glycolipids, *Phytochemistry* **16**, 539–543 (1977)

KIELHÖFFER, E. Die Bindung der schwefligen Säure an Weinbestandteile, *Weinberg Keller* **5**, 461–476 (1958)

KIELLEY, W. W. and MEYERHOFF, O. Studies on adenosine-triphosphatase of muscle. II. A new magnesium-activated adenosinetriphosphatase, *Journal of Biological Chemistry* **176**, 591–601 (1948)

KING, M. E. *Biochemical Effects of Ozone*, PhD Thesis, Illinois Institute of Technology (quoted in Dugger and Ting, 1970, see above) (1961)

KOŠTÍR, J., MACHÁČKOVÁ, I., JIRÁČEK, V. and BUCHAR, E. Einfluss des Schwefeldioxids auf den Gehalt freier Saccharide und Aminosäuren in Erbsen-Keimpflanzen, *Experimentia* **26**, 604–605 (1970)

KOUKOL, J. and DUGGER, W. M. Jr. Anthocyanin formation as a response to ozone and smog treatment in *Rumex crispus* L., *Plant Physiology* **42**, 1023–1024 (1967)

KOUKOL, J., DUGGER, W. M. Jr. and BELSER, N. O. The inhibition of cyclic photophosphorylation by peroxyacetyl nitrate, *Plant Physiology* **38** (Supplement): xii (1963)

KOUKOL, J., DUGGER, W. M. Jr. and PALMER, R. L. Inhibitory effect of peroxyacetyl nitrate on cyclic photophosphorylation by chloroplasts from black valentine bean leaves, *Plant Physiology* **42**, 1419–1422 (1967)

KOZIOL, M. J. *Effects of Prolonged Exposure to SO_2 on the Growth and Carbohydrate Metabolism of Soyabean and Ryegrass*. DPhil Thesis, University of Oxford, Oxford, UK (1980)

KOZIOL, M. J. and COWLING, D. W. Growth of ryegrass (*Lolium perenne* L.) exposed to SO_2. II. Changes in the distribution of photoassimilated ^{14}C, *Journal of Experimental Botany* **29**, 1431–1439 (1978)

KOZIOL, M. J. and COWLING, D. W. Growth of ryegrass (*Lolium perenne* L.) exposed to SO_2. III. Effects on free and storage carbohydrate concentrations, *Journal of Experimental Botany* 1687–1699 (1980)

KOZIOL, M. J. and JORDAN, C. F. Changes in carbohydrate levels in red kidney bean (*Phaseolus vulgaris* L.) exposed to sulphur dioxide, *Journal of Experimental Botany* **29**, 1037–1043 (1978)

LEE, C-J., MILLER, G. W. and WELKIE, G. W. The effects of hydrogen fluoride and wounding on respiratory enzymes in soybean leaves, *International Journal of Air and Water Pollution* **10**, 169–181 (1966)

LEE, T. T. Sugar content and stomatal width as related to ozone injury in tobacco leaves, *Canadian Journal of Botany* **43**, 677–685 (1965)

LIBERA, W., ZIEGLER, I. and ZIEGLER, H. The action of sulfite on the HCO_3^--fixation and the fixation pattern of isolated chloroplasts and leaf tissue slices, *Zeitschrift für Pflanzenphysiologie* **74**, 420–433 (1975)

LÜTTGE, V., OSMOND, C. B., BALL, E., BRINKMAN, E. and KINZE, G. Bisulfite compounds as metabolic inhibitors and non-specific effects on membranes, *Plant and Cell Physiology* **13**, 505–514 (1972)

McCUNE, D. C. and WEINSTEIN, L. H. Metabolic effects of atmospheric fluorides on plants, *Environmental Pollution* **1**, 169–174 (1971)

McCUNE, D. C., WEINSTEIN, L. H., JACOBSON, J. S. and HITCHCOCK, A. E. Some effects of atmospheric fluoride on plant metabolism, *Journal of the Air Pollution Control Association* **14**, 465–468 (1964)

McCUNE, D. C., WEINSTEIN, L. H. and MANCINI, J. F. Effects of hydrogen fluoride on the acid-soluble nucleotide metabolism of plants, *Contributions from the Boyce Thompson Institute* **24**, 213–226 (1970)

MacDOWALL, F. D. H. Predisposition of tobacco to ozone damage, *Canadian Journal of Plant Science* **45**, 1–12 (1965)

MALHOTRA, S. S. and SARKAR, S. K. Effects of sulphur dioxide on sugar and free amino acid content of pine seedlings, *Physiologia Plantarum* **47**, 223–228 (1979)

MARSHALL, P. E. and FURNIER, G. R. Growth responses of *Ailanthus altissima* seedlings to SO₂, *Environmental Pollution, Series A* **25**, 149–153 (1981)

MASSART, L. and DUFAIT, R. Über die KCN⁻ un NaF-Hemmung der Garung mit besonderer Berücksichtigung der Metalle als aktivaturen der Fermente, *Zeitschrift für physiologische Chemie* **272**, 157–170 (1942)

MATSUMOTO, H., WAKIUCHI, N. and TAKAHASHI, E. Changes in sugar levels in cucumber leaves during ammonium toxicity, *Physiologia Plantarum* **21**, 1210–1216 (1968)

MATSUMOTO, H., WAKIUCHI, N. and TAKAHASHI, E. The suppression of starch synthesis and the accumulation of uridine diphosphoglucose in cucumber leaves due to ammonium toxicity, *Physiologia Plantarum* **22**, 537–545 (1969)

MATSUMOTO, H., WAKIUCHI, N. and TAKAHASHI, E. Changes of starch synthetase activity of cucumber leaves during ammonium toxicity, *Physiologia Plantarum* **24**, 102–105 (1971)

MELCHIOR, N. and MELCHIOR, J. B. Inhibition of yeast hexokinase by fluoride ion, *Science* **124**, 402–403 (1956)

MENSER, H. A., HEGGESTAD, H. E. and STREET, O. E. Response of plants to air pollutants. II. Effects of ozone concentrations and leaf maturity on injury to *Nicotiana tabacum*, *Phytopathology* **53**, 1304–1308 (1963)

MENSER, H. A., HEGGESTAD, H. E., STREET, O. E. and JEFFREY, R. N. Response of plants to air pollutants. I. Effects of ozone on tobacco plants preconditioned by light and temperature, *Plant Physiology* **38**, 605–609 (1963)

MILLER, G. W. Properties of enclase in extracts from pea seed, *Plant Physiology* **33**, 199–206 (1958)

MILLER, P. R., COBB, F. W. Jr. and ZAVARIN, E. Photochemical oxidant injury and bark beetle (*Coleoptera: Solytidae*) infestation of ponderosa pine. III. Effect of injury on oleoresin composition, phloem carbohydrates and phloem pH, *Hilgardia* **39**, 135–140 (1968)

MILLER, P. R., PARMETER, J. R., FLICK, B. H. and MARTINEZ, C. W. Ozone dosage response of Ponderosa pine seedlings, *Journal of the Air Pollution Control Association* **19**, 435–438 (1969)

MUDD, B. J. Enzyme inactivation by peroxyacetyl nitrate, *Archives of Biochemistry and Biophysics* **102**, 59–65 (1963)

MUDD, J. B. Response of enzyme systems to air pollutants, *Archives of Environmental Health* **10**, 201–206 (1965)

MUDD, J. B. and DUGGER, W. M. Jr. The oxidation of reduced pyridine nucleotides by peroxyaceyl nitrates, *Archives of Biochemistry and Biophysics* **102**, 52–58 (1963)

MUDD, J. B., McMANUS, T. T., ONGUN, A. and McCULLOGH, T. E. Inhibition of glycolipid biosynthesis in chloroplasts by ozone and sulfhydryl reagents, *Plant Physiology* **48**, 335–339 (1971)

MUKERJI, S. K. and YANG, S. F. Phosphoenolpyruvate carboxylase from spinach leaf tissue: inhibition by sulfite ion, *Plant Physiology* **53**, 829–834 (1974)

NICKERSON, W. J. and CHUNG, C. W. Reversal of fluoride inhibition of yeast growth with glucose-1-phosphate, *American Journal of Botany* **39**, 669–679 (1952)

NIELANDS, J. B. and STUMPF, P. K. *Outlines of Enzyme Chemistry*, p. 258, John Wiley and Sons, Inc., New York (1955)

NIKOLAEVSKII, V. S. (Role of some oxidative systems in the respiration and resistance to sulfur dioxide damage of plants.) *Fiziologiya rastenii* **15**, 110–115 (1955)

NORBY, R. J. and KOZLOWSKI, T. T. Response of SO₂-fumigated *Pinus resinosa* seedlings to post-fumigation temperature, *Canadian Journal of Botany* **59**, 470–475 (1981)

NOYES, R. D. The comparative effects of sulfur dioxide on photosynthesis and translocation in bean, *Physiological Plant Pathology* **16**, 73–79 (1980)

OGREN, W. L. Increasing carbon fixation by crop plants. In *Photosynthesis 77: Proceedings of the 4th International Congress on Photosynthesis* (Reading, UK, 4–9 Sept. 1977), (Eds. D. O. HALL, J. COOMBS and T. W. GOODWIN), pp. 721–733, The Biochemical Society, London (1978)

ORDIN, L. and ALTMAN, A. Inhibition of phosphoglucomutase activity in oat coleoptiles by air pollutants, *Physiologia Plantarum* **18**, 790–797 (1965)

ORDIN, L. and HALL, M. A. Studies on the cellulose synthesis by a cell-free oat coleoptile enzyme system: inactivation by airborne oxidants, *Plant Physiology* **42**, 205–212 (1967)

ORDIN, L., HALL, M. A. and KINDINGER, J. I. Oxidant-induced inhibition of enzymes involved in cell wall polysaccharide synthesis, *Archives of Environmental Health* **18**, 623–626 (1969)

ORDIN, L. and PROPST, B. Effect of fluoride on metabolism of oat coleoptile sections, *Plant Physiology* **37** (Supplement): lxviii (1962)

ORDIN, L. and SKOE, B. P. Inhibition of metabolism in *Avena* coleoptile tissue by fluoride, *Plant Physiology* **38**, 416–421 (1963)

OSMOND, C. B. and AVADHANI, P. N. Inhibition of the β-carboxylation pathway of CO₂ fixation by bisulfite compounds, *Plant Physiology* **45**, 228–230 (1970)

PACK, M. R. Effects of hydrogen fluoride on bean reproduction, *Journal of the Air Pollution Control Association* **21**, 133–137 (1971)

PACK, M. R. and WILSON, A. M. Influence of hydrogen fluoride fumigation on acid-soluble phosphorus compounds in bean seedlings, *Environmental Science and Technology* **1**, 1011–1013 (1967)

PAUL, J. S. and BASSHAM, J. A. Effects of sulfite on metabolism in isolated mesophyll cells from *Papaver somniferum*, *Plant Physiology* **62**, 210–214 (1978)

PAUL, J. S., CORNWELL, K. L. and BASSHAM, J. A. Effects of ammonia on carbon metabolism in photosynthesizing leaf-free mesophyll cells from *Papaver somniferum*, *Planta*, **142**, 49–54 (1978)

PIERRE, M. Action du SO_2 sur le métabolisme intermédaire. II. Effet de doses subnécrotiques de SO_2 sur les enzymes de feuilles de Haricot, *Physiologie végétale* **15**, 195–205 (1977)

PIPPEN, E. L., POTTER, A. L., RANDALL, V. G., NG, K. G., REUTER, F. W., MORGAN, A. I. and OSHIMA, R. J. Effect of ozone fumigation on crop composition, *Journal of Food Science* **40**, 672–676 (1975)

PLATT, S. G., PLAUT, Z. and BASSHAM, J. A. Ammonia regulation of carbon metabolism in photosynthesizing leaf discs, *Plant Physiology* **60**, 739–742 (1977)

RAPP, G. W. and SLIWINSKI, R. A. The effect of sodium monofluorophosphate on the enzymes phosphorylase and phosphatase, *Archives of Biochemistry and Biophysics* **60**, 379–383 (1956)

RAWSON, H. M. and WOODWARD, R. G. Photosynthesis and transpiration in dicotyledonous plants. I. Expanding leaves of tobacco and sunflower, *Australian Journal of Plant Physiology* **3**, 247–256 (1976)

RAY, W. J. Jr. and KOSHLAND, D. E. Jr. Identification of amino acids involved in phosphoglucomutase action, *Journal of Biological Chemistry* **237**, 2493–2505 (1962)

ROCKWOOD, E. W. The effect of neutral salts upon the activity of ptyalin, *Journal of the American Chemical Society* **41**, 228–230 (1919)

ROSA, L. Uncouplers affect light modulation of photosynthetic enzymes. In *Photosynthesis IV. Regulation of Carbon Metabolism* (Proceedings of the 5th International Photosynthesis Conference, Halkidiki, 1980), Balaban International Science Services, Philadelphia, pp. 405–414 (1981)

ROSS, C. W., WIEBE, H. H. and MILLER, G. W. Effect of fluoride on glucose catabolism in plant leaves, *Plant Physiology* **37**, 305–309 (1962)

SPEDDING, D. J. and THOMAS, W. J. Effect of sulphur dioxide on the metabolism of glycolic acid by barley (*Hordeum vulgare*) leaves, *Australian Journal of Biological Sciences* **26**, 281–286 (1973)

SPIKES, J. D., LUMRY, R. L. and RIESKE, J. S. Inhibition of the photochemical activity of chloroplasts. I. Salts. *Archives of Biochemistry and Biophysics* **55**, 25–37 (1955)

SRIVASTAVA, H. S., JOLLIFFE, P. A. and RUNECKLES, V. C. Inhibition of gas exchange in bean leaves by NO_2, *Canadian Journal of Botany* **53**, 466–474 (1975)

TANAKA, H., TAKANASHI, T. and YATAZAWA, M. Experimental studies on sulfur dioxide injuries in higher plants. I. Formation of glyoxylate-bisulfite in plant leaves exposed to sulfur dioxide, *Water, Air and Soil Pollution* **1**, 205–211 (1972)

TAYLOR, O. C., DUGGER, W. M. Jr., THOMAS, M. D. and THOMPSON, C. R. Effect of atmospheric oxidants on apparent photosynthesis in citrus trees, *Plant Physiology* **36** (Supplement): xxvi–xxvii (1961)

TEH, K. H. and SWANSON, C. A. Sulfur dioxide inhibition of translocation in bean plants, *Plant Physiology* **69**, 88–92 (1982)

TENG, J. and WHISTLER, R. L. Cellulose and chitin. In *Phytochemistry: The Processes and Products of Photosynthesis* **1**, (Ed. L. P. MILLER), pp. 249–269, Van Nostrand Reinhold Company, London (1973)

TEWFIK, S. and STUMPF, P. K. Carbohydrate metabolism in higher plants. IV. Observations on triose phosphate dehydrogenase, *Journal of Biological Chemistry* **192**, 519–525 (1951a)

TEWFIK, S. and STUMPF, P. K. Carbohydrate metabolism in higher plants. V. Enzymic oxidation of fructose diphosphate, *Journal of Biological Chemistry* **192**, 526–533 (1951b)

THOMAS, M. D., HENDRICKS, R. H. and HILL, G. R. Some chemical reactions of sulphur dioxide after absorption by alfalfa and sugar beets, *Plant Physiology* **19**, 212–226 (1944)

TING, I. P. and DUGGER, W. M. Jr. Factors affecting ozone sensitivity and susceptibility of cotton plants, *Journal of the Air Pollution Control Association* **18**, 810–813 (1968)

TING, I. P. and MUKERJI, S. K. Leaf ontogeny as a factor in susceptibility to ozone: amino acid and carbohydrate changes during expansion, *American Journal of Botany* **58**, 497–504 (1971)

TINGEY, D. T. Ozone induced alterations in the metabolite pools and enzyme activities of plants. In *Air Pollution Effects on Plant Growth, ACS Symposium Series*, **3**, (Ed. M. DUGGER), pp. 40–57, The American Chemical Society, Washington, DC (1974)

TINGEY, D. T., FITES, R. C. WICKLIFF, C. Activity changes in selected enzymes from soybean leaves following ozone exposure, *Physiologia Plantarum* **33**, 316–320 (1975)

TINGEY, D. T., FITES, R. C. and WICKLIFF, C. Foliar sensitivity of soybeans to ozone as related to several leaf parameters, *Environmental Pollution* **4**, 183–192 (1973a)

TINGEY, D. T., REINART, R. A., WICKLIFF, C. and HECK, W. W. Chronic ozone or sulfur dioxide exposures, or both, affect the early vegetative growth of soybean, *Canadian Journal of Plant Science* **53**, 875–879 (1973b)

TINGEY, D. T., WILHOUR, R. G. and STANDLEY, C. The effect of chronic ozone exposure on the metabolite content of Ponderosa pine seedlings, *Forest Science* **22**, 234–241 (1976)

TODD, G. W. Effect of low concentrations of ozone on the enzymes catalse, peroxidase, papain and urease, *Physiologia Plantarum* **11**, 457–463 (1958)

TOMLINSON, H. and RICH, S. Effect of ozone on sterols and sterol derivatives in bean leaves, *Phytopathology* **61**, 1404–1405 (1971)

TOMLINSON, H. and RICH, S. Anti-senescent compounds reduce injury and steroid changes in ozonated leaves and their chloroplasts, *Phytopathology* **63**, 903–906 (1973)

VAS, K. and INGRAM, M. Preservation of fruit juices with less SO_2, *Food Manufacture* **24**, 414–416 (1949)

WAKIUCHI, N., MATSUMOTO, H. and TAKAHASHI, E. Changes of some enzyme activities of cucumber during ammonium toxicity, *Physiologia Plantarum* **24**, 248–253 (1971)

WEIGL, J. and ZIEGLER, H. Die räumliche Verteilung von ^{35}S und die Art der markierten Verbindungen in Spinatblättern nach begasung mit $^{35}SO_2$, *Planta* **58**, 435–477 (1962)

WEINSTEIN, L. H. Effects of atmospheric fluoride on metabolic constituents of tomato and bean leaves, *Contributions from the Boyce Thompson Institute* **21**, 215–231 (1961)

WEINSTEIN, L. H. Fluoride and plant life, *Journal of Occupational Medicine* **19**, 49–78 (1977)

WEINSTEIN, L. H. and ALSCHER-HERMAN, R. Physiological responses of plants to fluorine. In *Effects of Gaseous Air Pollution in Agriculture and Horticulture*, pp. 139–167, (Eds. M. H. UNSWORTH and D. P. ORMROD), Butterworths, London (1982)

WELLBURN, A. R., WILSON, J. and ALDRIDGE, P. H. Biochemical responses of plants to nitric oxide polluted atmospheres, *Environmental Pollution, Series A* **22**, 219–228 (1980)

WILKINSON, T. G. and BARNES, R. L. Effects of ozone on $^{14}CO_2$ fixation patterns in pine, *Canadian Journal of Botany* **51**, 1573–1578 (1973)

WILLIS, A. J. and YEMM, E. W. The respiration of barley plants. VIII. Nitrogen assimilation and the respiration of the root system, *New Phytologist* **54**, 163–181 (1955)

WOODWARD, R. G. and RAWSON, H. M. Photosynthesis and transpiration in dicotyledonous plants. II. Expanding and senescing leaves of soybean, *Australian Journal of Plant Physiology* **3**, 257–267 (1976)

WOO, K. C. and CANVIN, D. T. Effect of ammonia on photosynthetic carbon fixation in isolated spinach cells, *Canadian Journal of Botany* **58**, 505–510 (1980a)

WOO, K. C. and CANVIN, D. T. Effect of ammonia, nitrite, glutamate, and inhibitors of N metabolism on photosynthetic carbon fixation in isolated spinach leaf cells, *Canadian Journal of Botany* **58**, 511–516 (1980b)

YANG, S. F. and MILLER, G. W. Biochemical studies on the effect of fluoride on higher plants. I. Metabolism of carbohydrates, organic acids and amino acids, *Biochemical Journal* **88**, 505–509 (1963a)

YANG, S. F. and MILLER, G. W. Biochemical studies on the effect of fluoride on higher plants. II. The effect of fluoride on sucrose-synthesizing enzymes from higher plants, *Biochemical Journal* **88**, 509–516 (1963b)

YANG, S. F. and MILLER, G. W. Biochemical studies on the effect of fluoride on higher plants. III. The effect of fluoride on dark carbon dioxide fixation, *Biochemical Journal* **88**, 517–522 (1963c)

ZEEVART, A. S. Some effects of fumigating plants for short periods with NO_2, *Environmental Pollution* **11**, 97–108 (1976)

ZELITCH, I. α-Hydroxysulfonates as inhibitors of the enzymatic oxidation of glycolic and lactic acids, *Journal of Biological Chemistry* **224**, 251–260 (1957)

ZIEGLER, I. The effect of SO_3^{2-} on the activity of ribulose-1,5-diphosphate carboxylase in isolated spinach chloroplasts, *Planta* **103**, 155–163 (1972)

ZIEGLER, I. The effect of air polluting gases on plant metabolism, *Environmental Quality and Safety* **2**, 182–208 (1973a)

ZIEGLER, I. Effect of sulphite on phosphoenolpyruvate carboxylase and malate formation in extracts of *Zea mays*, *Phytochemistry* **12**, 1027–1030 (1973b)

ZIEGLER, I. Malate dehydrogenase in *Zea mays*: properties and inhibition by sulphite, *Biochimica et Biophysica Acta* **364**, 28–37 (1974)

Chapter 19

Air pollutant effects on biochemicals derived from metabolism: organic, fatty and amino acids

R. L. Heath

DEPARTMENT OF BOTANY AND PLANT SCIENCES, UNIVERSITY OF CALIFORNIA AT RIVERSIDE, USA

Introduction

To summarize the data on the effects of air pollutants on the metabolism of organic, amino and fatty acids would be a truly exceptional task. There are few common patterns, or good hypotheses, on which to summarize the observed changes. Aside from interesting guesses on the involvement of SO_2 on the glutamate family of amino acids (Pahlich, 1975; Ziegler, 1975; Hällgren, 1978), most researchers end up simply listing the *in vivo* changes in metabolites after fumigation. In this discussion I will attempt to summarize briefly the data concerning amino/organic/fatty acids and discuss their metabolic origins. In doing so, I will limit my discussion to the two classes of air pollutants for which there is enough data to make coherent arguments — ozone and sulfur dioxide.

The metabolic interactions, which need to be understood at the onset of this discussion, are shown diagrammatically in *Figure 19.1* (after Waterman, 1968). All metabolic systems are linked by genetic and physiological controls and are affected by available nutrients, which interact with factors external to the plant. These processes acting together lead to a general response, which on the simplest level might be plant growth and the production of foodstuff. The controls on these processes are provided both by energy and carbon flow (synthesis and maintenance), and by informational flow (signals of response) involving ionic balance and hormones. These interactions dictate that a disturbance in one process necessarily affects the other processes; e.g. excess external abscissic acid will lead to stomatal closure and this will disrupt photosynthetic carbon metabolism. Unfortunately, most data on the effects of air pollutants are collected with a degree of disregard for these relationships. For the most part, pool sizes of metabolites are measured at a set time after the air pollutant stress. Under the best of conditions, only a minimal amount of information for very complex control systems is obtained. A pool size of a metabolite depends upon the balance between the flow of intermediates into and out of that pool. Further, many pools are closely regulated by the cell and a major shift in pool size might represent a grosser disruption of the controls rather than a simple change in rate of metabolism.

There must be a clear distinction made between alterations in the amount of a metabolite resulting from its decreased synthesis and/or the increased metabolism and chemical destruction of the metabolite itself. For example, several papers (Mudd *et al.*, 1969; Mudd, McManus and Ongun, 1971; Mudd, this volume) have shown that ozone

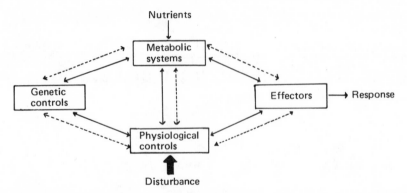

Figure 19.1. Schematic showing the interactions of metabolism and other tissue responses. Adapted from Waterman (1968). *See* text for a more complete discussion. ———— represents carbon and energy flows. ––– represents informational or control flows for regulation

chemically attacks amino acids, lipids and proteins. Aromatic and sulfur containing amino acids are the most easily oxidized by ozone (Mudd *et al.*, 1969). A chemically altered metabolite could be a potential inhibitor of the metabolic machinery or confound its normal operation. For instance, once an amino acid in a protein has been altered, the protein seems to be more susceptible to metabolic degradation (*see* Davies, 1979).

Finally, in studying any metabolic change induced by a stress, it must be emphasized that not only should the pool sizes of a given compound be monitored but the flow of metabolites into and out of that pool should also be examined. Very little of the latter has been done. Most investigators monitor only the relative amount of acids; for example, within plant tissue before and after a stress. Often pool size is measured only once following the stress and rarely is a complete time course followed (but *see* Pierre and Queiroz, 1981).

Amino acids

Metabolism is complicated. Compounds apparently far removed from each other chemically can alter each other's interaction with air pollutants. These possible interactions within the cell's milieu are demonstrated by the data of Pauls and Thompson (1980). Using microsomes from cotyledons of *Phaseolus vulgaris*, they found that exposure to ozone caused a loss in both lipid and protein (81 ± 2 and $17 \pm 3\%$, respectively), but by adding a cytosol extract to the preparation, this loss could be reduced to $62 \pm 3\%$ and $2 \pm 2\%$ respectively. Further, the production of malondialdehyde (MDA, a measure of lipid oxidation, Heath, 1975) was reduced by nearly 50% by adding the cytosol extract. This protection may have been due to sulfhydryls present in the cytosol extract, which competed for O_3, thus lessening the extent of the destruction of biochemicals and production of MDA (at least in part, *see* Mudd, McManus and Ongun, 1971a and Mudd *et al.*, 1971).

Closer relationships can be seen if we can break down intermediate metabolism into an interrelated set of subsystems as shown in *Figure 19.2* for amino acids (*see* Beevers, 1976). The 'backbone' for metabolic carbon flow includes glycolysis, the pentose shunt, and the tricarboxylic acid cycle (outlined area). From these simple sugars and organic acids the major families of amino acids (and as shown later, the major lipid families) can be derived. The major amino acid families are:

(1) *triose phosphate-derived family* of serine (SER), glycine (GLY) and cysteine (CYS);
(2) *pyruvate-derived family* of valine (VAL), alanine (ALA), and leucine (LEU);
(3) *phosphoenolpyruvate-derived family* of phenylalanine (PHE) and tyrosine (TYR) and the indole nucleus of tryptophan (TRY);
(4) *pentose shunt-derived family* of histidine (HIS) and the completion of TRY;
(5) *malate-derived (or late TCA cycle) family* of aspartate (ASP) and its amine (ASN), lysine (LYS), methionine (MET), threonine (THR) and isoleucine (ILE); and
(6) *oxoglutarate-derived (or early TCA cycle) family* of glutamate (GLU) and its amine (GLN), proline (PRO) and arginine (ARG)

As we will see later, the lipid families feed into and out of these pathways through acetyl CoA, which provides a link between all organic acids. A glutamate/glutamine shuttle also acts to trap amino groups and transfer them to other amino acceptors, such as ornithine (ORN), ASP, carbamyl phosphate, and nucleoside bases (Altman, Friedman and Levin, 1982; Beevers, 1976). Nothing will be said about the nucleoside bases. Very little information exists in the literature on ORN or carbamyl phosphate, though there is a growing awareness of the importance of 'storage amines' like putrescine and spermidine derived from those compounds (Priebe, Klein and Jäger, 1978).

With knowledge of these biochemical families, we can investigate their reactions with air pollutants and focus our attention on the more critical areas in the context of the larger picture. Yet, judgement must be used in interpreting the bewildering collections

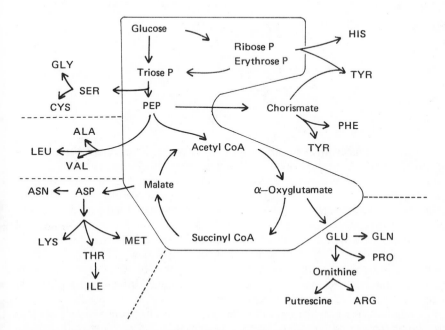

Figure 19.2. Pathways of amino acid metabolism. Outlined region represents the metabolic pathways of glycolysis, pentose-shunt, and the Tricarboxylic Acid Cycle. The six major families of amino acids are derived from these pathways and are described in the text (Beevers, 1976)

TABLE 19.1. Changes in organic acids under various conditions of air pollutant fumigation

Ref.[a]	Species	Fumigant	Dose (ppm)	Treatment	Organic acids	Ratio[b]
(1)	Pine	SO_2	0.35		VAL	0.52
			0.53			1.60
			0.35		TYR	0.81
			0.53			2.69
(2)	Ryegrass	SO_2	0.16	$-SO_4^{-2}$	VAL	4.68
				$+SO_4^{-2}$		1.09
				$-SO_4^{-2}$	TYR	1.10
				$+SO_4^{-2}$		1.52
(3)	Bean	SO_2	0.1	6 days	GLU	1.06
				12 days		1.20
				6 days	GLY	1.89
				12 days		1.21
(4)	Tomato	HF	12 ppb	No recovery	MAL	0.53
				12 h recovery		1.14
(5)	Spruce	Natural emission		Necrosis:		
				light-	GLY	0.06
				mid-		0.16
				heavy-		1.23
				light-	GABA[c]	2.04
				mid-		1.22
				heavy-		0.66

[a] Data were taken from the following references: (1) Malhotra and Sarkar, 1979; (2) Cowling and Bristow, 1979; (3) Pierre and Queiroz, 1981; (4) Weinstein, 1961; (5) Jäger and Grill, 1975.
[b] Data are expressed as the ratio of pool sizes (fumigated/control) for each organic acid listed.
[c] GABA, gamma-aminobutyric acid.

of data. For example, as shown in *Table 19.1*, changes in amino acids can depend upon:

(1) the level of the pollutant (ref. 1);
(2) the condition of the plant (ref. 2);
(3) the time of exposure (ref. 3);
(4) the extent of the recovery period (ref. 4, for HF);
(5) the acid measured (all refs.);
(6) the plant species involved (all refs.);
(7) the amount of injury (ref. 5); and
(8) the type of pollutant (all refs.).

The literature abounds with examples of apparently contradictory data (for excellent reviews, *see* Heck, 1977; Mudd, 1973 and Horsman and Wellburn, 1976).

In *Table 19.2*, I have summarized a large amount of data on amino acids pools by 'averaging' species, doses and times of observation after the stress. To handle this inconsistent mass of information, I have taken several liberties. I have used a ratio of pool size in exposed tissue to pool size in control tissue as a measure of the amount of change. This covers a wide variety of values, since the absolute pool sizes vary greatly with species, developmental age, and period of the day (Horsman and Wellburn, 1976). I then reduced this pool size ratio to a qualitative system of $+/0/-$ to symbolize increase, no change, or decrease, respectively. For extremely large changes (usually greater than 2 or less than 0.5), double symbols ($++$ or $--$) are used.

There is some consensus in the data that the triose phosphate-derived amino acids are increased by ozone treatment. The late TCA-derived amino acid family also shows some increase. Other groups have not been as well studied but show no obvious common pattern of change. LEU and HIS generally show increases. Many amino acids have not been carefully monitored and little attention has been paid to those at the beginning of each pathway.

With SO_2 fumigation, particularly at high doses, pyruvate-derived amino acids generally increase. With the possible exception of the triose-derived family, much variation exists. The changes in the pool sizes of the amino acids at the beginning of the TCA derived families — GLU and ASP — vary considerably. Yet, GLU and ASP, amide-containing amino acids, seem to vary in opposition to their precursors. The amine storage acid, ARG, seems to increase generally with SO_2 fumigation. As mentioned previously, the GLU pathway seems to be involved heavily with enzyme inactivation or inhibition (Pahlich, 1975; Ziegler, 1975).

Changes in amino acid pools appear to derive from prior metabolic events, especially with the TCA metabolites. As pointed out, more measurements are necessary in these studies but a few families seem to be affected more by fumigation than others. A further

TABLE 19.2. Amino acid pool size changes following fumigation with O_3 or SO_2

Amino acid families		Ratio of pool sizes (fumigated/control)									
		Reference: ($O_3 \rightarrow$)				Reference: ($SO_2 \rightarrow$)					
		1	2	3	4	5	6	7	8	9	10
A. Triosephosphate	CYS										+
	GLY	+	++	-(+)	+	+		--(+)	+		+(+)
	SER	+	+	++	+	+	+	0	+		-
B. Pyruvate	ALA	+	-	++	+	+	0	-	0	++	++
	LEU	+	++	++	+	++	++				--(+)
	VAL	+	++	0	+	0				+	-(+)
C. Pentose shunt	HIS		++	++		--			++	-	
	PHE	+	-	+		+					-(++)
	TRY					++					
	TYR		0	0		+					
D. Early TCA	ARG					+		++	-(+)		+
	GLU	-	+	-(+)	0	+	+	-	-	--	+(-)
	GLN	0			+	+		+(-)	+		
	ORN							+			
	PRO			0				+(0)			-
E. Late TCA	ASP	0	++	++	0	+			-	-	+
	ASN		++		+	--	++	+(-)		+	
	ILE		++	0		+				0	-(+)
	LYS		+	++		+					++
	MET	+		-		-					+
	THR		+	+		+	++		-(+)		+

The data were taken from the following references:

1. Tomlinson and Rich, 1967 O_3 fumigation   Bean
2. Ting and Mukerji, 1971   Cotton
3. Mumford et al., 1972   Corn pollen
4. Bennett, Heggestad and McNulty, 1977   Bean
5. Cowling and Bristow, 1979 SO_2 fumigation   Ryegrass
6. Arndt, 1970   Rye
7. Jäger and Grill, 1975   Spruce
8. Pierre and Queiroz, 1981   Bean
9. Godzik and Linskens, 1974   Bean
10. Malhotra and Sakhar, 1979

The data were evaluated as described in the text. The $+$, 0, $-$ represent increase, no change or decrease in the pool size by fumigation when compared to the control, respectively. No level of statistical significance is implied since the data vary greatly. ( ) represents data taken at highest dose where it is separate in the data from lower doses.

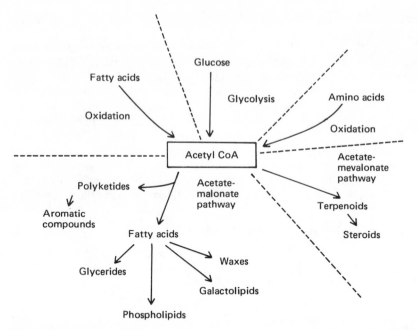

Figure 19.3. Pathways of lipid metabolism derived from acetyl Co–A. Adapted from Vickery and Vickery (1981). The three major oxidative pathways lead into Acetyl Co–A, which is the precursor for the two major pathways to polar lipids (acetate-malonate) or to terpenoids (acetate-mevalonate)

look at amine storage compounds might provide some valuable insight. Of particular note is the lack of changes in some of the TCA intermediates during SO_2 fumigation (Pierre and Queiroz, 1981). The most dramatic changes are for malate and malonate, occurring some 6 to 12 days after the beginning of the fumigation.

Lipids

Air pollutant interactions with lipids are less vague, though data show that the 'obvious' conclusion that ozone simply oxidizes double bonds in fatty acids may not be correct. *Figure 19.3* shows two general pathways of lipid metabolism, the *acetate-malonate* pathway leading to neutral and polar lipids or fatty acid synthesis, and the *acetate-mevalonate* pathway leading to steroids. There are other secondary pathways (Vickery and Vickery, 1981), but I would like to concentrate only on the first one. The major branch point of lipid metabolism and the precursors of the secondary products pathways is acetyl Co-A. Degradation of fatty acids through β-oxidation yields acetyl Co-A (Stumpf, 1980). Exogenous acetate can also be energized and easily converted into acetyl Co-A in most plant tissues; so, not surprisingly, many investigators have used exogenous acetate to label lipid intermediates. Even tracer amounts may either inflate the acetate pool or yield unknown specific activities of acetyl Co-A and its derivatives (for rarely is the amount of acetyl Co-A within the tissue known).

There is much uncertainty regarding specific pathways for all fatty acid synthesis, but a reasonable idea is offered in *Figure 19.4*. The fatty acids are either synthesized *de novo* by the fatty acid synthetase using malonyl Co-A as an intermediate or by chain-elongation from pre-existing shorter-chain fatty acids (by the 'elongase' reaction that uses shorter-chain fatty acid precursors, *see* Stumpf, 1980). Obviously, using ^{14}C-acetate as a tracer, the specific activity of the fatty acids produced by the two systems will be quite different, since fewer ^{14}C units are added to the fatty acids via the elongase reaction under short-term labeling. Linolenic acid (C18:3) has always represented a difficult problem in plant biochemistry, since it is labeled so slowly by exogenous acetate. This is no small problem since a large proportion of fatty acids in plants (60–80%) are polyunsaturated fatty acids (PUSFA).

For polar lipid biosynthesis, the fatty acids are attached to a glycerol backbone using glycerol-3-phosphate to form a diacylglyceride phosphate (or PA, phosphatidic acid, as in *Figure 19.5*; *see* Vickery and Vickery, 1981). PA always occurs in low concentrations in plant tissue; high concentrations usually indicate the activation of a lipase that degrades other phospholipids to PA. *Figure 19.5* also shows the biosynthetic pathway to other important polar lipids. There are two major branches; the diacylglyceride branch gives rise to the galactolipid series, monogalactosidyl diglyceride (MGDG) and digalactosidyl diglyceride (DGDG) as well as to sulfolipids (from SO_4^{2-}, not shown), triglycerides (for storage lipids) and phosphatidyl choline (PC, a major component of membranes); and the cytidine diphosphate (CDP)-diglyceride branch, using nucleotides to activate the polar moieties, produces phosphatidyl serine (PS), phosphatidyl ethanolamine (PE), and phosphatidyl glycerol (PG). To complicate investigations, PE can be methylated to form PC and, thus, interconnect both branches.

It has long been believed that ozone injury to plants is due to the oxidation of the unsaturated fatty acids of the membrane (Tomlinson and Rich, 1969; Menzel, 1976), since ozone in chemical systems rapidly reacts with double bonds leading to oxidation products (Criegee, 1975; Heath, 1978; Heath and Tappel, 1976); a pathway sometimes mistakenly referred to as 'lipid peroxidation'. This misnomer is due in part to the use of an oxidation and peroxidation product of polyunsaturated fatty acids — malondialdehyde (MDA) — as a measure of the degree of ozone attack. Ozone oxidation of double bonds is not a peroxidation — the reaction and products are quite different (*see* later paragraphs). Yet, papers still continue to suggest that peroxidation causes injury (Pauls and Thompson, 1980).

Mudd *et al.* (1971) have shown that the amount of MDA produced per amount of O_3 absorbed chemically is about 2% for chloroplasts made up of 89% PUSFA (63%, C18:3), although the ratio of MDA/O_3 can be a little higher with pure PC. There are factors which can modify this reaction especially in an aqueous chemical system — for example, 20 μmol GSH will reduce MDA production from 5 μmol of C18:2 by only 5%

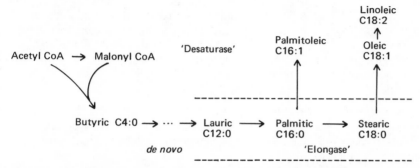

Figure 19.4. Biosynthetic pathways of fatty acids. Adapted from Stumpf (1980). The 'normal' or *de novo* pathway is that found through much experimentation in which acetate is fed to plant tissues

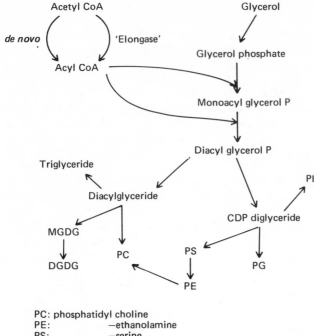

PC: phosphatidyl choline
PE: —ethanolamine
PS: —serine
PG: —glycerol
PI: —inositol

MGDG: monogalactosyl diglyceride
DGDG: digalactosyl diglyceride

Figure 19.5. Pathways of polar lipid biosynthesis. Adapted from Vickery and Vickery (1981). Abbreviations are listed on the figure and in the text

even though 5 μmol ozone is added and 8 μmol GSH is used up (Mudd *et al.*, 1971). In this case, the production of water-soluble peroxides (measured by I_2 release from I^-) was reduced to zero. A greater reduction in the amount of MDA produced by glutathionine can be noted if pure PC is used instead of PUSFA in membranes (Mudd, McManus and Ongun, 1971). Heath (1978) more closely examined the amount of MDA produced per amount of O_3 consumed and found that the ratio of MDA/O_3 was nearer $6.5 \pm 1.2\%$ for pure PUSFA in Pi buffer, but that the amount of MDA measured by the thiobarbituric acid (TBA) test (involving heating) could be three to seven times that of MDA measured by its UV absorption. In an algal system, Heath, Chimiklis and Frederick (1974) found that the amount of MDA/O_3 used was close to 5%. Further, MDA production and ozone decomposition did not really begin until the cell's viability (measured by plating) declined. It is clear that:

(1) MDA measures only part of the reaction (2–6%);
(2) the amount of MDA produced is very dependent upon the lipid species;
(3) water solutions can produce mostly hydrogen peroxide (which can be destroyed by sulfhydryls and catalase); and
(4) MDA concentrations measured by the TBA test are probably overestimated due to heating during the test.

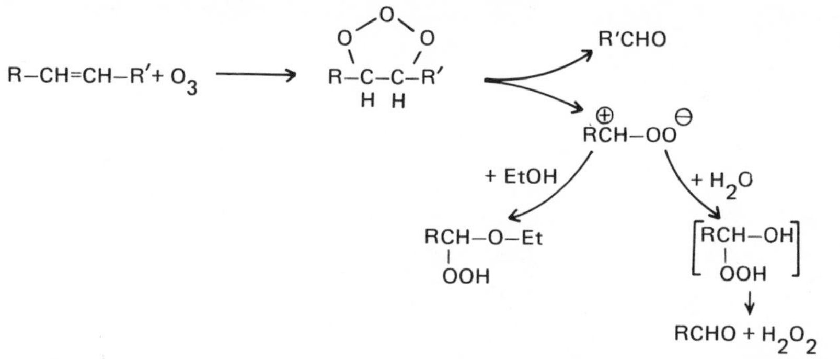

$$R-CH=CH-R' + O_3 \longrightarrow R-\underset{H}{C}-\underset{H}{C}-R' \text{ (ozonide)}$$

$$\longrightarrow R'CHO$$

$$\longrightarrow \overset{\oplus}{R}CH-OO^{\ominus}$$

$$+ EtOH \longrightarrow R\underset{OOH}{CH}-O-Et$$

$$+ H_2O \longrightarrow \left[R\underset{OOH}{CH}-OH \right] \longrightarrow RCHO + H_2O_2$$

Figure 19.6. The Creigee mechanism of the ozonolysis of the double bonds in poly-unsaturated fatty acids. Adapted from Heath and Tappel (1976). The abbreviation for ethanol is EtOH (80–100%) and the bracket stands for a hypothetical unstable intermediate

The role of undecomposed hydrogen peroxide in the production of HO˙ by the Weiss Reaction (Heath, 1979) has not been explored.

These apparent contradictions become less severe if one closely examines the chemistry of the reactions of O_3 with double bonds of organic molecules *in water* via the Creigee Mechanism (*Figure 19.6*). Heath and Tappel (1976) clearly demonstrated that, in water, all the 'peroxide' released by ozone attack of C18:2 was hydrogen peroxide by using a new, selective test. Yet, if a polar organic solvent (95% ethanol was used), the peroxide formed then was a lipid hydroperoxide (*Figure 19.6*).

The work by Tomlinson and Rich (1969) has been used to suggest that lipid oxidation is the fundamental mechanism of ozone damage. That work had three major findings:

(1) lipid emulsions applied to leaves prevented much of the ozone injury (50–90% inhibition);
(2) tobacco varieties which varied in their susceptibility to visible injury had varied ratios of unsaturated to saturated fatty acids; and
(3) severe visible injury reduced the amounts of extractable fatty acids.

There are other and perhaps better interpretations of their data. The emulsions could have easily reacted with the ozone in the air before its entry into the leaf. Although the stomata were open, no measurements of ozone interaction at the surfaces were made. If Table 1 of Tomlinson and Rich (1969), involving varieties, is reformulated as in my *Table 19.3*, no correlation between degree of visible injury and ratio of unsaturated to

TABLE 19.3. Comparison of the fatty acid content of various cultivars of tobacco with susceptibility to ozone injury

Cultivar	% Visible injury	Relative unsaturation of fatty acids
SS Bell-W-3	80+	1.21
S Bell-W-3	70–80	1.65
Conn 49	60–80	1.00
R Bell-W-3	10–30	2.00
68116	<10	1.45

Data were taken from Tomlinson and Rich (1969), from their Table 1 and their text. Relative unsaturation is the ratio of polyunsaturated fatty acids to saturated fatty acids, normalized to Conn 49.

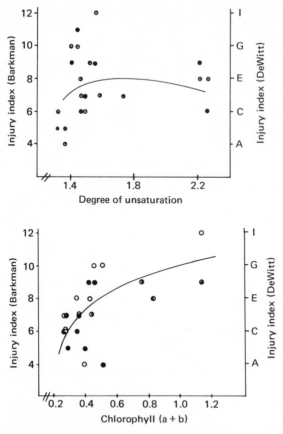

Figure 19.7. Comparison between injury level and unsaturated fatty acids in varied lichen species. Data were taken from Beltman *et al.* (1980). Each data point indicates one species of epiphytic lichen. The correlation between the degree of unsaturation of the fatty acids and two injury indices Barkman (○) and Dewitt (●) is shown in the top portion, while the bottom portion shows the correlation of injury with chlorophyll content of the lichen

saturated fatty acids is seen. Further, the data of Beltman *et al.* (1980) on the effects of air pollution on a wide variety of epiphytic lichens show (*Figure 19.7*) that while there is a correlation between chlorophyll content and injury score (by two methods), there is no correlation between unsaturated fatty acids and visible injury. In other studies, Cameron *et al.* (1970) and Cameron (1975) showed that two varieties of corn (*Zea mays*) differed in their susceptibility to ozone injury in the field. Yet, Heath (unpublished data) could find only a small difference in fatty acid content of the polar lipids, as shown in *Table 19.4*. The difference in the ozone sensitivity between the two varieties of corn disappears when the corn is grown in greenhouses; yet, the small lipid difference remains. Harris and Heath (1981) have traced this differential sensitivity to stomatal control of pollutant entry and to differential water status of the cultivars rather than any obvious biochemical difference.

Tomlinson and Rich (1969) used an extremely high dose of ozone (1 ppm) which possibly resulted in gross tissue damage. They suggested in a later paper that lower doses did not cause such extreme differences. Later work by Swanson, Thomson and

Mudd (1973) and Fong and Heath (1981) showed little change in lipids under more normal ozone stress (0.3–0.5 ppm for 1–2 h); certainly not of the magnitude suggested by the more quoted work of Tomlinson and Rich (1969). With extreme injury, the relatively mild solvent (20 ml of 1:1, $CHCl_3$: MeOH solvent g^{-1} tissue) may not have extracted all the lipid, leading to an apparent observed loss of lipid, especially if there were extensive oxidation of the protein (*see* Mudd, this volume).

There may be some smaller changes, however. As Swanson, Thomson and Mudd (1973) pointed out, although there were no significant changes in fatty acid content, there were more free fatty acids apparent on TLC separation following fumigation of pinto beans (*Phaseolus vulgaris*) with O_3. Fong and Heath (1981) saw no significant change in fatty acid levels or in total lipid phosphate (to within 5–8%), though there was an apparent increase in PE and decrease of PG, which increased with the dose (0.3–0.5 ppm). A larger change was observed with the MGDG to DGDG ratio, occurring several hours after the end of the fumigation. This agrees well with Mudd *et al.* (1971), who observed that ozone inhibited the formation of DGDG more than that of MGDG. Both observations fit the general pattern well, if steps between DGDG and MGDG and leading from CDP-diglyceride to PG were inhibited. The inhibition of PG would then appear as a stimulation of PE (*Figure 19.5*).

Trevathen, Moore and Orcutt (1979) reported similar findings, with the exception that the total extractable lipid apparently increased with ozone exposure (0.3 ppm) by about 15–25%; there was, however, an 18–35% general decrease in total fatty acids in the triglyceride (storage) fraction. Unfortunately, the decline in fatty acid and in linolenic acid (C18 : 3) concentrations did not follow the pattern of injury of the various leaves at different developmental stages. The leaves with uniform fleck had the least decline in fatty acid concentrations for two varieties of tobacco (*see Table 19.5*).

Using *Chlorella* as a model system, Frederick and Heath (1975) found that while MDA increased as the cell's viability declined, the loss of fatty acids could be shown only statistically on a percentage abundance basis; triunsaturates declined while saturates increased. As more recently shown (Heath and Frederick, 1979), this occurs long after many changes in membrane permeability occur and about the time of cellular death.

TABLE 19.4. Lipid content of two cultivars of corn. Fatty acid comparison of field grown corn (*Zea mays*)

		% of major fatty acids			
	C16:0	C16:1	C18:0	C18:2	C18:3
1. Bonanza	13.2 ± 0.8	1.0 ± 0.4	2.0 ± 0.1	4.5 ± 0.2	77.2 ± 1.6
Monarch advance	16.1 ± 1.6	4.6 ± 0.7	2.3 ± 0.3	4.2 ± 0.5	72.8 ± 3.3
2. Bonanza	9.7 ± 1.8	1.6 ± 0.4	1.3 ± 0.2	6.8 ± 1.0	80.1 ± 3.5
Monarch advance	15.4 ± 1.0	2.4 ± 0.3	1.7 ± 0.1	9.5 ± 1.0	70.5 ± 3.7
	Ratio (Monarch advance/bonanza)				
1.	1.22	4.6	1.15	0.93	0.94
2.	1.59	1.5	1.31	1.40	0.88
	Standard deviation of ratio				
1.	± 0.14	± 1.8	± 0.15	± 0.11	± 0.04
2.	± 0.28	± 0.4	± 0.18	± 0.25	± 0.06

Monarch advance is the ozone-sensitive cultivar, while Bonanza is ozone-resistant. The $\pm$ value represents 1 standard deviation of 4 repeated trials of the GLC determination of fatty acids.

Corn was grown in the field at Riverside under the conditions described by Cameron *et al.* (1970) and Harris and Heath (1981). Lipids were separated and analyzed according to Fong and Heath (1981) and Frederick and Heath (1975). Preparation no. 1 is data from a 60 day old tip section while no. 2 is data from a 68 day old mid section (unpublished data, R. L. Heath).

TABLE 19.5. Variation in lipid content of tobacco with amounts of necrosis

Material		Amounts from fumigated plants	
		Amounts from controls	
		NC88 (%)	Coker 86 (%)
Lipid total	T	97	171
	M	121	153
	B	121	153
Total FA of	T	73	52
triglycerides	M	30	26
	B	26	33
C18:3	T	104	58
	M	33	28
	B	37	49
C16:0	T	64	67
	M	38	27
	B	29	40

Data were taken from Trevathan, Moore and Orcutt, 1979. Plants were exposed to 0.25 ppm O_3.

The symbols represent: youngest leaves (T) with ozone injury only at leaf apex/margins; recent mature (M) uniform ozone-induced fleck over leaf; and oldest leaves (B), ozone-induced fleck only at base of leaf.

Acetate labeling patterns of lipids were studied using this algal system (*Table 19.6*). Cells were exposed to a short pulse of radioactive acetate following the ozone exposure. The control cells took up acetate very rapidly reaching steady-state within 30 min, while ozonated cells took acetate up more slowly but reached about the same steady-state level. It is not known what products were present in the aqueous phase after $CHCl_3$:MeOH separation, but it was likely to be amino and organic acids. Again, the ozone-stressed cells had a lower level of radioactivity compared to control cells. The level within the control cells rapidly rose to a maximum and then fell, while the level in the ozone-treated cells gradually decreased to a lower level. This indicated a slower movement of radioactivity from the aqueous phase to the organic phase in ozone-

TABLE 19.6. Movement of ^{14}C-acetate into Chlorella after ozone exposure

Time (min)	O_3-exposed/control (%)		
	Total acetate	*Aqueous acetate*	*Organic acetate*
15	—	—	—
30	43/39	32	29
60	82/83	59	39
90	89/144	158	49

^{14}C-acetate group into various lipids classes

O_3-exposed/control (%)

PC	PE	PG	PI	MGDG	DGDG
62 ± 37	260 ± 45	91 ± 30	172 ± 51	49 ± 15	42 ± 12

Chlorella sorokiniana grown in autotrophic medium were harvested and resuspended in phosphate buffer. Ozone (72 ppm) at 60 ml/min was introduced into a cell suspension (3×10^8 cells ml^{-1}; 20 ml) for 45 min before labeling (*see* Heath and Frederick, 1979). Cells were labeled with 2 mM Na-Acetate (6×10^{11} DPM/mol, s.a.) for the indicated time. Control cells were gassed with just the oxygen carrier gas. At various times, cells (5 ml) were removed and extracted according to Frederick and Heath (1975). The aqueous phase and lipid (or $CHCl_3$) phases were counted. An aliquot of cells (50 μl) was filtered on a millipore, washed with 5 ml cold solution and counted for the total acetate. The aqueous phase was acidified, evaporated to near dryness, resuspended in EtOH:H$_2$O (1:1), and counted. The lipid phase was spotted on TLC (silica gel; activated), which was developed in two dimensions, as described in Fong and Heath (1981). Each spot was then counted and the phosphate or galactose content determined. PC = phosphatidyl choline; PI = phosphatidyl inositol; PE = phosphatidyl ethanolamine; PG = phosphatidyl glycerol; MGDG = monogalactosyl diglyceride; DGDG = digalyactosyl diglyceride.

treated cells. The radioactivity in the lipid phase rose slowly in both control and ozone-treated cells, not reaching a steady-state level even after 90 min. After TLC separation, an inhibition of specific activity (s.a.) in PC and a lesser inhibition in s.a. in PG, was observed while PI and PE showed a stimulation in s.a. The s.a. of both MGDG and DGDG was lowered by ozone stress. This is similar to what was suggested in a previous paragraph. Acetate uptake is probably not influenced by ozone-induced changes in membrane permeability since acetate appears to move passively as its uncharged form.

During the extraction of lipids from cells labeled for various times with ^{14}C-acetate, the radioactivity in the non-lipid or aqueous phase suggested that this pool of compounds is depleted gradually and argues for a rapid equilibrium between acetate and this aqueous pool followed by a slower flow of acetate into the lipid pool. Thus, a general model can be proposed as follows.

(1) Acetate penetrates the cell very rapidly with no ozone effect.
(2) Acetate is converted into acetyl CoA; from acetyl CoA, metabolites flow into the TCA cycle. Ozone could slow this step due to a limitation of energy (Pell and Brennan, 1973)
(3) Lipid synthesis, requiring NADPH and ATP, proceeds at a relatively low rate. Ozone, by lowering the total energy of the cell, slows this step greatly.
(4) Only a few key steps of lipid metabolism are affected by ozone (see previous paragraphs); the fluxes are greatly lowered but the pool size is maintained, in part, by control mechanisms.

With SO_2 exposure, the picture seems to be similar with respect to fatty acids. There is little in the literature to indicate gross oxidation of polar lipids although chlorophyll seems to be sensitive to SO_2 and bleaches relatively easily (Malhotra, 1977). Khan and Malhotra (1977) have looked in more detail at biosynthetic pathways of fatty acids and have found a 20–50% decline in all glycolipids after treatment of sections of pine needles (*Pinus contorta*) with aqueous SO_2. Loss of glycolipids was matched by a loss of C18:3 and a gain in C16:O on a % relative abundance measure. A later biosynthetic study by Malhotra and Khan (1978) showed a general loss of all polar lipids with both aqueous SO_2 (treatment of segments of needles) and gaseous SO_2 (treatment of whole seedlings). In both cases, the pools of PG and PE showed a slightly larger drop than those of other polar lipids, but MGDG/DGDG ratio rose slightly as noted with ozone (their Tables 2 and 3 and my summary — *Table 19.7*). However, they suggest that concentrations of free fatty acids decrease with SO_2 treatment. Their data show a more total 'impaired cellular metabolism' by SO_2. Sulphur dioxide injury can also be reversed by a recovery period indicating metabolic recovery (*Table 19.7b* and *c*).

TABLE 19.7. Effect of gaseous SO_2 on lipid biosynthesis in Jack pine seedlings

	Conc. (ppm)	Duration (h)	Recovery (h)	Average of polar lipids (% of control)
A.	0.18	24	0	$74 \pm 8\%$
	0.35	24	0	$32 \pm 9\%$
B.	0.20	24	0	$76 \pm 10\%$
	0.20	24	48	$90 \pm 14\%$
C.	0.37	1	0	$41 \pm 4\%$
	0.37	1	24	$69 \pm 19\%$

Data were taken from Malhotra and Khan (1978).
Small sections of pine seedlings were incubated for 1–3 h with 1-[^{14}C]-acetate after a period of exposure and a period of recovery (as listed) and the polar lipid radioactivity was determined.

Similarly, Grunwald (1981) found that the amounts of free fatty acid were lowered 75–85% by exposure of soybeans (*Glycine max*) to 0.8 ppm of SO_2 for 20 days. The polar lipids were reduced by only 25–35% and the neutral lipids were increased by 50–90%. This would indicate a blockage of carbon movement into the polar lipid fraction. However, the decline in free fatty acid concentrations is disturbing. Generally the presence of free fatty acids is indicative of lipase activation during lipid isolation. Grunwald (1981) took samples in the field and transported them on ice. He did not use boiling isopropanol to inactivate lipases but ground his tissue in chloroform/methanol. Beans, unfortunately, are known to have active lipases (Quinn and Williams, 1978). It may be that lipases are inactivated by SO_2 exposure and so he observed a decline in the release of free fatty acids by SO_2. Along this line, the presence of more free fatty acids in the ozone studies of Swanson, Thomson and Mudd (1973) could have been due to activation of lipase by ozone or a general release of cellular material by loss of cell permeability.

Based on the available data, one can determine that fatty acids, especially in polar lipids, are little changed after a low dose of air pollutant. At high levels, the cell structure is disrupted and fatty acids become more vulnerable to oxidative reactions. Storage lipids may prove to be quite a different story. The few studies conducted so far (*see Table 19.5*) suggest that these lipids may suffer more alterations even at low doses, probably due to metabolic changes. Little work has been done on lipid turnover, since it is a very difficult area to investigate. No doubt, more surprises await us there.

Conclusions

Too often a major motivation for a particular study is that it is technically feasible; a clear statement of a specific scientific question is lacking. It is a simple matter to look at all free amino acids in a cell sap preparation, and technically, the work is relatively easy. But the results, consisting of a simple listing of the observations, contribute little to understanding metabolism and stress effects. Metabolism is much too closely regulated and interactive for this type of 'shot-gun' approach. Too often the chemistry between an air pollutant and a specific biochemical is not fully understood. This leads to misleading or confusing ideas, as with the term lipid peroxidation.

What is needed in these studies are precise and relevant hypotheses based on a good conceptual grasp of the pertinent physiological interrelationships that exist at several cell and tissue levels. This, plus good communication among chemists, physiologists, and ecologists will provide the basis for solutions to air pollution problems and insight into general stress physiology.

References

ALTMAN, A., FRIEDMAN, R. and LEVIN, N. Arginine and ornithine decarboxylases, the polyamine biosynthetic enzymes of mung bean seedlings, *Plant Physiology* **69**, 876–879 (1982)

ARNDT, U. Concentration changes of free amino acids in plants under the effect of HF and SO_2, *Staub-Reinhalt. Luft.* **30**, 28–32 (1970)

BEEVERS, L. *Nitrogen Metabolisn in Plants*, Edward Arnold, W. Clowers & Sons, Ltd, London, Chap. 5 (1976)

BELTMAN, I. H., DEKOK, L. J., KUIPER, P. J. C. and VAN HASSELT, P. R. Fatty acid composition and chlorophyll content of epiphytic lichens and a possible relation to their sensitivity of air pollution, *Oikos* **35**, 321–326 (1980)

BENNETT, J. H., HEGGESTAD, H. E. and McNULTY, I. B. Ozone and leaf physiology. In *Proceedings of the Fourth Annual Meeting of the Plant Growth Regulator Working Group*, pp. 323–330, Hot Springs, Ark. (1977)

CAMERON, J. W. Inheritance in sweet corn for resistance to acute ozone injury, *Journal of The American Society for Horticultural Science* **100**, 577–579 (1975)

CAMERON, J. W., JOHNSON, H. Jr., TAYLOR, O. C. and OTTO, H. W. Differential susceptibility of sweet corn hybrids to field injury by air pollution, *Hortscience* **5**, 217–219 (1970)

COWLING, D. W. and BRISTOW, A. W. Effects of SO_2 on sulphur and nitrogen fractions and on free amino acids in perennial ryegrass, *Journal of the Science of Food and Agriculture* **30**, 354–360 (1979)

CRIEGEE, R. Mechanism of ozonolysis, *Angewandte Chemie (International Edition)* **14**, 745–760 (1975)

DAVIES, D. D. Factors affecting protein turnover in plants. In *Nitrogen Assimilation in Plants*, (Eds. E. J. HEWITT and C. V. CUTTING), pp. 369–396, Academic Press, London (1979)

FONG, F. and HEATH, R. L. Lipid content in the primary leaf of bean (*Phaseolus vulgaris*) after ozone fumigation, *Zeitschrift für Pflanzenphysiologie* **104**, 109–115 (1981)

FREDERICK, P. E. and HEATH, R. L. Ozone-induced fatty acid and viability changes in *Chlorella*, *Plant Physiology* **55**, 15–19 (1975)

GODZIK, S. and LINSKENS, H. F. Concentration changes of free amino acids in primary bean leaves after continuous and interrupted SO_2 fumigation and recovery, *Environmental Pollution* **7**, 25–38 (1974)

GRUNWALD, C. Foliar fatty acids and sterols of soybean field fumigated with SO_2, *Plant Physiology* **68**, 868–871 (1981)

HÄLLGREN, J-E. Physiological and biochemical effects of sulfur dioxide on plants. In *Sulfur in the Environment, Part II, Ecological Impacts*, (Ed. J. O. NRIAGU), pp. 164–209, John Wiley & Sons, New York (1978)

HARRIS, M. J. and HEATH, R. L. Ozone sensitivity in sweet corn (*Zea mays* L.) plants: a possible relationship to water balance, *Plant Physiology* **68**, 885–890 (1981)

HEATH, R. L. Ozone. In *Responses of Plants to Air Pollution*, (Eds. J. B. MUDD and T. T. KOZLOWSKI), pp. 23–55, Academic Press, New York (1975)

HEATH, R. L. The reaction stoichiometry between ozone and unsaturated fatty acids in an aqueous environment, *Chemistry and Physics of Lipids* **22**, 25–37 (1978)

HEATH, R. L. Breakdown of ozone and formation of hydrogen peroxide in aqueous solutions of amine buffers exposed to ozone, *Toxicology Letters* **4**, 449–453 (1979)

HEATH, R. L., CHIMIKLIS, P. and FREDERICK, P. E. Role of potassium and lipids in ozone injury to plant membranes. In *Air Pollution Related to Plant Growth*, (Ed. W. M. DUGGER, Jr.), pp. 58–75, American Chemical Society Symposium Series, New York (1974)

HEATH, R. L. and FREDERICK, P. E. Ozone alteration of membrane permeability in *Chlorella* I: permeability of potassium ion as measured by ^{86}Rb tracer, *Plant Physiology* **64**, 455–459 (1979)

HEATH, R. L. and TAPPEL, A. L. A new sensitive assay for the measurement of hydroperoxides, *Analytical Biochemistry* **76**, 184–191 (1976)

HECK, W. W. Plants and microorganisms. In *Ozone and Other Photochemical Oxidants*, (Ed. S. K. FRIEDLANDER), Chap. 11, National Academy of Science, Washington, DC (1977)

HORSMAN, D. C. and WELLBURN, A. R. Guide to the metabolic and biochemical effects of air pollutants on higher plants. In *Effects of Air Pollutants on Plants*, (Ed. T. A. MANSFIELD), pp. 185–199, Cambridge University Press, Cambridge (1976)

JÄGER, H. J. and GRILL, D. Einfluss von SO_2 und HF auf freier Aminosäuren der Fichte (*Picea abies* L. Karsten), *European Journal of Forest Pathology* **5**, 279–286 (1975)

KHAN, A. A. and MALHOTRA, S. S. Effects of aqueous sulfur dioxide on pine needle glycolipids, *Phytochemistry* **16**, 539–543 (1977)

MALHOTRA, S. S. Effects of aqueous sulfur dioxide on chlorophyll destruction in *Pinus contorta*, *New Phytologist* **78**, 101–109 (1977)

MALHOTRA, S. S. and KHAN, A. A. Effects of sulfur dioxide fumigation on lipid biosynthesis in pine needles, *Phytochemistry* **17**, 241–244 (1978)

MALHOTRA, S. S. and SARKAR, S. K. Effects of sulfur dioxide on sugar and free amino acid content of pine seedlings, *Physiologia Plantarum* **47**, 223–228 (1979)

MENZEL, D. B. Role of free radicals in the toxicity of air pollutants (NO_2 and O_3). In *Free Radicals in Biology*, Vol. II, (Ed. W. A. PRYOR), pp. 181–202, Academic Press, New York (1976)

MUDD, J. B. Biochemical effects of some air pollutants on plants. In *Air Pollution Damage to Vegetation*, (Ed. J. A. NAEGELE), *Advances in Chemistry Series*, **122**, pp. 31–47, American Chemical Society, Washington, DC (1973)

MUDD, J. B., LEAVITT, R., ONGUN, A. and McMANUS, T. T. Reaction of ozone with amino acids and proteins, *Atmospheric Environment* **3**, 669–682 (1969)

MUDD, J. B., McMANUS, T. T. and ONGUN, A. Inhibition of lipid metabolism in chloroplasts by ozone, *Proceedings of the Second International Clean Air Congress*, pp. 256–260, Academic Press, New York (1971)

MUDD, J. B., McMANUS, T. T., ONGUN, A. and McCULLOGH, T. E. Inhibition of glycolipid biosynthesis in chloroplasts by ozone and sulfhydryl reagents, *Plant Physiology* **48**, 335–339 (1971)

MUMFORD, R. A., LIPKE, H., LAUFER, D. A. and FEDER, W. A. Ozone-induced changes in corn pollen, *Environmental Science and Technology* **6**, 427–430 (1972)

PAHLICH, E. Effect of sulfur dioxide-pollution on cellular regulation. A general concept of the mode of action of gaseous air contamination, *Atmospheric Environment* **9**, 261–263 (1975)

PAULS, K. P. and THOMPSON, J. E. *In vitro* simulation of senescence-related membrane damage by ozone-induced lipid peroxidation, *Nature* **283**, 504–506 (1980)

PELL, E. J. and BRENNAN, E. Changes in respiration, photosynthesis, ATP, and total adenylate content of ozonated pinto bean foliage as they relate to symptom expression, *Plant Physiology* **51**, 378–381 (1973)

PIERRE, M. and QUEIROZ, O. Enzyme and metabolic changes in bean leaves during continuous pollution by subnecrotic levels of SO_2, *Environmental Pollution, Series A* **25**, 41–51 (1981)

PRIEBE, A., KLEIN, H. and JÄGER, H. J. Role of polyamines in SO_2-polluted pea plants, *Journal of Experimental Botany* **29**, 1045–1050 (1978)

QUINN, P. J. and WILLIAMS, W. P. Plant lipids and their role in membrane function, *Progress in Biophysics and Molecular Biology* **34**, 109–179 (1978)

STUMPF, P. K. Biosynthesis of saturated and unsaturated fatty acids. In *The Biochemistry of Plants: Lipids, Structure and Function* Vol. 4 (Ed. P. K. STUMPF), pp. 177–204, Academic Press, New York (1980)

SWANSON, E. S., THOMSON, W. W. and MUDD, J. B. The effect of ozone on leaf cell membranes, *Canadian Journal of Botany* **51**, 1213–1219 (1973)

TING, I. P. and MUKERJI, S. K. Leaf ontogeny as a factor in susceptibility to ozone: amine acid and carbohydrate changes during expansion, *American Journal of Botany* **58**, 497–504 (1971)

TOMLINSON, H. and RICH, S. Metabolic changes in free amino acids of bean leaves exposed to ozone, *Phytopathology* **57**, 972–974 (1967)

TOMLINSON, H. and RICH, S. Relating lipid content and fatty acid synthesis to ozone injury in tobacco leaves, *Phytopathology* **59**, 1284–1286 (1969)

TREVATHAN, L. E., MOORE, L. D. and ORCUTT, D. M. Symptom expression and free sterol and fatty acid composition of flue-cured tobacco plants exposed to ozone, *Phytopathology* **69**, 582–585 (1979)

VICKERY, M. L. and VICKERY, B. *Secondary Plant Metabolism*, Chap. 3, University Park Press, Baltimore, MD (1981)

WATERMAN, T. H. System theory and biology. In *Systems Theory and Biology*, (Ed. M. D. MESAROVIC), pp. 11–39, Springer-Verlag, New York (1968)

WEINSTEIN, L. H. Effects of atmospheric fluoride on metabolic constituents of tomato and bean leaves, *Contributions from the Boyce Thompson Institute* **21**, 215–237 (1961)

ZIEGLER, I. The effects of SO_2 pollution on plant metabolism, *Residue Reviews* **56**, 79–105 (1975)

Chapter 20

Biosynthesis and emission of hydrogen sulfide by higher plants

P. Filner*

ARCO PLANT CELL RESEARCH INSTITUTE, DUBLIN, CALIFORNIA, USA

and

H. Rennenberg†, J. Sekiya‡, R. A. Bressan§, L. G. Wilson, L. Le Cureux and T. Shimei

MSU-DOE PLANT RESEARCH LABORATORY, MICHIGAN STATE UNIVERSITY, USA

Introduction

Sulfur dioxide is rapidly oxidized after entry into plants, with a half-life measured in minutes or hours at most (Miller and Xerikos, 1979; Garsed and Read, 1977a, 1977b, 1977c). Sulfate is the principal oxidation product (Ziegler, 1975; De Cormis, 1969), and it accumulates to high concentrations during chronic exposure of plants to SO_2 (Thomas et al., 1943; Thomas, Hendricks and Hill, 1950; Thomas, 1951). It is not surprising, therefore, that at sublethal concentrations, gaseous SO_2 can satisfy the nutritional sulfur requirements of plants (Thomas et al., 1943). The inhibition of plant growth by chronic exposure to SO_2 may be largely, perhaps entirely, due to effects of the resultant high internal sulfate concentrations (Thomas, Hendricks and Hill, 1950). Injury from acute exposure to a high concentration of SO_2, on the other hand, appears to involve a more specific response to SO_2. The development of waterlogged, then necrotic regions on leaves, is not elicited by sulfate or any other common sulfur compound besides SO_2 or bisulfite. Furthermore, the symptoms of acute injury are quite similar in many plant species (Barrett and Benedict, 1970).

In addition to the attraction of being a more specific response to SO_2, the acute injury response lends itself to experimentation more readily than the chronic injury response in several other important ways. The response can be elicited within hours, so it is unnecessary to grow plants for most or all of their life-span in atmospheres with precisely controlled levels of SO_2. Research on acute injury is better suited to a modest budget and modest facilities because it requires less space, less controlled environment equipment and less time than research on chronic injury.

Regardless of the precise mechanism of acute injury, SO_2 must participate in an interaction within the plant which leads to injury, either directly or perhaps through a

* Former address: MSU-DOE Plant Research Laboratory, Michigan State University, East Lansing, Michigan 48824, USA
† Present address: Botanisches Institut der Universität Köln, Gyrhofstrasse 15, D-5000 Köln 41, DFR
‡ Present address: Department of Agricultural Chemistry, Yamaguchi University, Yamaguchi 753, Japan
§ Present address: Department of Horticulture, Purdue University, West Lafayette, Indiana 47907

certain sequence of intermediate reactions. If one or more of these reactions involves a chemical reaction of the sulfur in SO_2, it can be found by following the fate of ^{35}S in $^{35}SO_2$. The problem is how to distinguish the chain of reactions leading to injury in a maze of other reactions. The key is to have the right control — a plant or plant tissue which is less susceptible to injury by SO_2, but with a metabolism which is otherwise as similar as possible, to that of the plant or tissue exhibiting a high level of injury.

We chose to search for mechanisms of acute injury and of resistance to injury by SO_2 in cucurbits, as wide differences in susceptibility to injury by SO_2 were reported to exist in this family in an old survey of susceptibility under field conditions (*see* Thomas and Hendricks, 1956).

Results

Genetically determined resistance

We found that genetic differences in susceptibility to acute injury existed between genera in the Cucurbitaceae (*Figure 20.1*; Bressan, Wilson and Filner, 1978; Bressan *et al.*, 1979a; Bressan *et al.*, 1981). Cultivars of *Cucumis sativus* (cucumber) were more

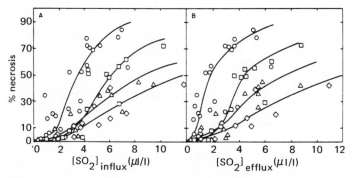

Figure 20.1. A: Dependence of injury of the four cucurbit cultivars on SO_2 exposure. Injury is given as per cent of the leaf area which was necrotic 24 h after the end of the fumigation. (The SO_2 exposure is given as the concentration of SO_2 entering the chamber, multiplied by the duration of the exposure in min and divided by 1000 min.) Chipper (formerly SC 25) cucumber (○); National Pickling (□); Prolific Straightneck (△); Small Sugar (◇). B: Same as A except that the time normalized SO_2 concentrations shown are the average concentrations of the gas leaving the chamber. (Bressan, Wilson and Filner, 1978)

sensitive than cultivars of *Cucurbita pepo* (squash and pumpkin). We found differences in susceptibility between cultivars of the same species as well. This genetic difference in resistance to SO_2 injury in cucumber behaved like a simple Mendelian trait, with resistance being dominant (Bressan *et al.*, 1981).

GENETIC DIFFERENCES IN SO_2 UPTAKE ARE NOT ATTRIBUTABLE TO STOMATAL BEHAVIOR

This genetically determined resistance to SO_2 injury was attributable to differences in SO_2 absorption from air. The four cultivars exhibited clearcut resistance differences

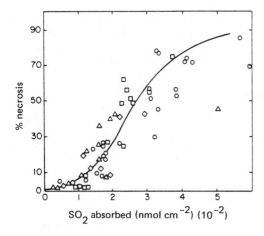

Figure 20.2 Dependence of injury sustained by the four cultivars on the amount of SO_2 absorbed. Injury is given as the per cent of leaf area which was necrotic 24 h after the end of the fumigation. The amount of SO_2 absorbed was calculated from the difference between the rates of SO_2 influx and efflux. The data are from the same experiments as the data in *Figure 20.1*. Chipper (formerly SC 25) (○); National Pickling (□); Prolific Straightneck (△); Small Sugar (◇). (Bressan, Wilson and Filner, 1978)

when exposed to the same external concentrations of SO_2 for the same length of time (Bressan, Wilson and Filner, 1978). However, injury per unit of SO_2 absorbed was very similar in the four cucurbit cultivars (*Figure 20.2*). It follows that the more resistant a cultivar is to injury by SO_2 at a particular concentration the lower the rate of SO_2 absorption at that concentration. Differences in resistance due to differences in uptake of a gas are usually presumed to reflect differences in stomatal numbers or apertures. The small differences in number of stomata per unit leaf area were insufficient to account for the differences in the uptake of SO_2 among the four cultivars. Similarly, rates of CO_2 fixation per unit leaf area among the cultivars differed little, regardless of whether or not SO_2 was present; this makes it unlikely that stomatal status alone could account for uptake differences (P. Filner, unpublished observations). Furthermore, the same four cultivars exhibited a very similar pattern of differences in resistance when leaf discs were floated on bisulfite solutions (Bressan *et al.*, 1979a). Uptake by the discs does not proceed via the stomata under these conditions. This was evident from the fact that injury decreased with distance from the cut edges, indicating that uptake probably proceeded via those edges. We therefore believe that the genetic differences in uptake of gaseous SO_2 reflect differences in permeability of a barrier to SO_2 entry other than the stomatal aperture. A likely candidate for this barrier is the plasma membrane.

GENETIC TOLERANCE OF SO_2 IS ASSOCIATED WITH TOLERANCE OF OTHER STRESSES

In the course of this work, we happened to observe that the cucurbit cultivars we used differed in cold tolerance in the same order as they differed in SO_2 tolerance (Bressan *et al.*, 1979b). This observation brought to mind the idea of Levitt and Kozlowski (1972) that there is a common denominator of stress tolerance — probably some property of a membrane. One can think of the plasma membrane as a sort of Achilles heel of the plant cell. When the membrane fails, nothing inside can save the cell. It is therefore quite plausible that, during evolution, there would have been no opportunity for natural selection to have favored a variant with a cell component possessing higher stress tolerance than the plasma membrane. It follows that sequences of sublethal failures of cell function caused by stresses would all tend to converge on a common lethal failure: that of the plasma membrane. To evaluate this hypothesis, we looked at the response of leaves and leaf discs to a series of environmental insults.

ETHANE EMISSION IS A CONVENIENT ASSAY OF INJURY

We chose to use ethane emission to assay injury rather than estimating necrotic leaf area by eye. Wounded living tissue produces ethylene by enzymatic processes (Boller and Kende, 1980), which actually declines at high levels of injury (Elstner and Konse, 1976). Ethane emission, on the other hand, is thought to result from reactions of radicals with linolenic acid in membranes (Konze and Elstner, 1978), something which probably never happens in a healthy intact cell. Green chloroplasts and light appear to be required for ethane production, probably to produce the radicals (*Figure 20.3*; Wilson *et al.*, 1979). Thus, ethylene can be considered the living plant cell's cry of 'ouch', while ethane is a manifestation of the dead or dying cells giving up the ghost. Since both ethylene and ethane can be measured by the same gas chromatographic procedure, why favor one over the other? First, backgrounds of ethane are usually much lower than those of ethylene. Second, ethylene rises, then falls with increasing injury, so at least two injury levels must be used to decide on which side of the injury curve a given ethylene emission level is, and hence which injury level the ethylene emission indicates. Actually, the most attractive assay is the ethane:ethylene ratio because it increases

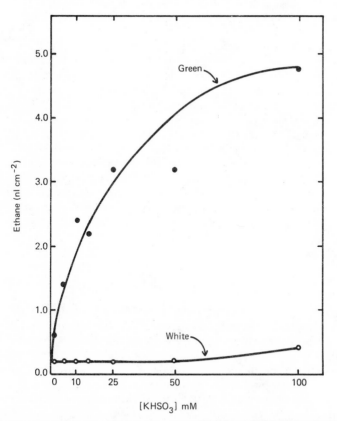

Figure 20.3. Emission of ethane by green and white leaf discs of variegated *Coleus*. Leaf discs in groups of ten were incubated in light for 2 h with the indicated bisulfite concentrations. They were then washed and replaced in the flasks with 3 ml of water with chloramphenicol to suppress bacterial growth. The flasks were sealed with rubber serum bottle caps. After 24 h, ethane in the headspace of each flask was determined by gas chromatography

monotonically with injury and is dimensionless (Bressan *et al.*, 1979a). Therefore, it does not require measurement of the leaf area, which must be done in order to compare emission rates of either ethylene or ethane by different samples. Another major advantage of the ethane/ethylene assay is that very little tissue is required. As little as one to five leaf discs, less than 10 mm in diameter are sufficient for the assay. Thus, a leaf or even a particular zone of a leaf may be 'biopsied', leaving the rest of the organ or organism intact for further study.

IS THE PLASMA MEMBRANE THE ACHILLES HEEL OF THE STRESSED PLANT CELL?

Using the ethane emission assay, we found that leaf discs from cucurbits with differing susceptibility to injury by bisulfite and SO_2 respond to cold, heat, salt, and alcohols in the same order (*Table 20.1*). These treatments cannot be described as survivable

TABLE 20.1. Comparisons of ethane emissions by cucurbit cultivars in response to various stresses

Treatment [a]	Post treatment (h)	Ethane (nl cm^{-2} chipper)	Ethane (% of emission by Chipper)			
			Chipper (cucumber)	National Pickling (cucumber)	Prolific straight-neck (squash)	Small sugar (pumpkin)
Bisulfite (15 mM, 2 h)	24	3.4±0.8	100±23.5	88.2±16.7	23.5±4.1	17.6±8.3
Alcohol (10%, 2.5 h)	23	7.2±0.1	100±1.4	73.6±0.01	37.5±0.01	30.6±0.02
Heating (65°C, 3 h)	24	2.4±0.05	100±2.1	75.0±8.3	25.0±8.0	8.3±8.3
Chilling (2-4°C, 66 h)	24	8.6±2.6	100±30.2	69.8±23.2	6.9±2.3	6.4±2.9
KCl (3 mM, 22 h)	0	5.7±1.0	100±16.9	87.7±11.6	74.1±7.0	70.6±5.3

[a] Leaves 110–150 cm^2 were selected, usually the fourth leaf from the apex of a plant 30–35 days old. In the case of heating, internodes and leaves down to leaf 5 from the apex collapsed when the whole plant was heated, so discs were punched from leaf 7. In the case of alcohol treatment, plants 50–55 days old were used and discs were punched from leaf 14 from the base. Ethane emitted after KCl treatment was calculated from triplicate samples. Variance is reported as standard error.

'stresses' because the leaf discs were not checked for viability after stressing. Rather, the treatments of the discs should be looked upon as fairly extreme environmental insults. The discs lose their ability to fix CO_2 more or less concomitantly with increase in ethane emission. We believe that the observed constancy of order of responses by the four cucurbit cultivars to various environmental insults, including bisulfite, supports the concept of the plasma membrane as the Achilles heel of the plant cell, and conversely, that which must be toughened in order for a plant to gain tolerance of acute stresses.

Developmentally determined resistance

We noticed in fumigation of cucurbits that young leaves were injured far less than mature leaves on the same plant (*Figure 20.4*; Bressan, Wilson and Filner, 1978). When SO_2 uptake by young and mature leaves was measured we found that the resistant

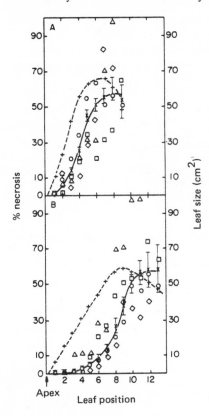

Figure 20.4. Dependence of injury on leaf position relative to the apex: node no. 1 is nearest the apex. Data from the experiments presented in *Figure 20.3* were divided into two sets regardless of differences in SO_2 exposure or absorption: A, data from slowly growing plants of mean age 40 days with five to nine leaves; B, data from rapidly growing plants of mean age 43 days with 10–13 leaves. Mean per cent necrosis of Chipper (formerly SC 25) (○); National Pickling (□); Prolific Straightneck (△); Small Sugar (◇). Mean area of leaves of all four cultivars (+ --- +). (Bressan, Wilson and Filner, 1978)

young leaves took up far more SO_2 (Bressan, Wilson and Filner, 1978). Thus it became clear that young leaves had to have a biochemically-based resistance mechanism which functioned after SO_2 entered the plant. This gave us the opportunity we were seeking: a chance to look for a difference in SO_2 metabolism in resistant versus sensitive, but otherwise closely similar, plant material.

Fumigations were performed with $^{35}SO_2$ in a closed system, and the distribution of ^{35}S in metabolites was determined. The kinetics of labeling of the different fractions were followed by snipping pieces from leaves inside the closed system at various times and removing them via an acidified 'water gate'. This enabled samples to be removed without disturbing the $^{35}SO_2$ atmosphere within the closed chamber (*Figure 20.5*).

YOUNG, SO_2 RESISTANT LEAVES EMIT MUCH MORE H_2S THAN MATURE SENSITIVE LEAVES

Only small differences in amounts of the ^{35}S metabolites of $^{35}SO_2$ were found in fractions obtained from homogenates of young and mature leaves, regardless whether from genetically resistant or sensitive cucurbits. About 60% of the absorbed $^{35}SO_2$ was oxidized to $^{35}SO_4^{2-}$ (*Table 20.2*; Sekiya, Wilson and Filner, 1980; Sekiya, Wilson and Filner, 1982). Virtually as a last resort, we looked at volatile sulfur emitted by young, resistant leaves, and mature, sensitive leaves on the same plant. To our surprise and delight, we found a very large difference: the young leaves emitted a lot of H_2S, while the mature leaves emitted very little (*Table 20.3*; Sekiya, Wilson and Filner, 1982). In some experiments the rates of H_2S emission per leaf area differed by more than 100-

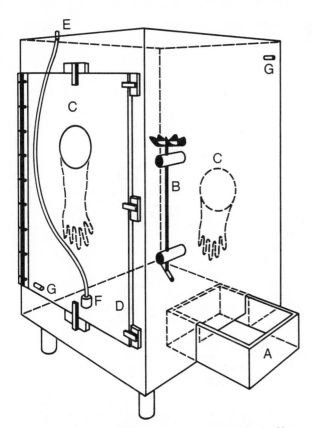

Figure 20.5. Chamber for closed system fumigations with $^{35}SO_2$. The chamber was constructed of Plexiglas. Its major features were: A, a water gate, which when filled with dilute lactic acid provided a gas-tight closure through which leaf samples in vials could be passed; B, a fan to circulate air in the chamber, which was driven by a magnetic stirrer below the chamber; C, gloves which permitted working inside the chamber without opening it; D, a door for initially placing in the chamber plants, vials with caps, forceps, scissors, and alkaline $^{35}SO_3^{2-}$ solution; E, a port for introducing acid to convert $^{35}SO_3^{2-}$ solution in beaker (F) to $^{35}SO_2$ gas; and G, inlet and outlet ports for flushing the chamber atmosphere into a trap at the end of the experiment. A word of caution — when the gloves are used, overpressure in the chamber tends to push out the solution in the water gate, so the water gate should not be filled to excess, or the outside part should be covered with a glass or Plexiglass plate

fold. The H_2S emitted by young leaves was equal to more than 10% of the SO_2 absorbed. When $^{35}SO_2$ was used, the specific radioactivity of the entire $H_2^{35}S$ was about 73% that of the absorbed $^{35}SO_2$.

We believe that the positive correlation of H_2S emission with SO_2 resistance is a meaningful clue to the developmentally regulated, biochemically based mechanism of resistance encountered in cucurbits. Resistance differences between young and mature leaves have been observed in other species (Guderian, 1977) so SO_2 uptake and H_2S emission should be determined in other species, to see how general the phenomenon is.

SO$_2$ PROBABLY CAUSES INJURY INDIRECTLY, THROUGH A NON-INJURIOUS PRIMARY REACTION

Is SO_2 itself causing injury directly, or must it be transformed to something else? We believe for the following reasons that it probably has to be transformed. A higher level

TABLE 20.2. ^{35}S distribution of fractions of cucumber leaves exposed to $^{35}SO_2$

Time (min)	^{35}S distribution (%)				
	SO_4^{2-}	SO_3^{2-}	S-organic and sulfide	Lipid	Residue
Young leaves					
30	22.9	52.0	13.4	3.1	1.5
180	66.7	13.3	14.8	3.6	1.6
Mature leaves					
30	40.0	37.0	16.8	3.8	2.3
180	60.5	19.6	14.7	3.1	2.3

A cucumber plant, about five weeks old with about 500 cm^2 of leaf area, was exposed in a closed 40 l Plexiglas chamber to a pulse of $^{35}SO_2$ (1.67 μCi μmol^{-1}) which reached a maximum concentration of 15 μl l^{-1} and was almost completely absorbed by the plant within 30 min. The chamber was illuminated with fluorescent lamps which produced 0.8 mW cm^{-2}. The chamber temperature was about 28 °C.

The leaves were extracted with 80% ethanol containing N-ethylmaleimide to react with thiols, sulfide and sulfite, and the ethanol-soluble fraction was extracted with chloroform to obtain the lipids, and leave the polar organic sulfur compounds in ethanol. The residue was extracted with 1% trichloroacetic acid to obtain sulfate, The N-ethylmaleimide derivatives were separated by thin layer chromatography and electrophoresis.

TABLE 20.3. Differences in the emission of H_2S between resistant and sensitive cucurbit leaves in response to SO_2. Emission of volatile ^{35}S from cucumber leaves fumigated with $^{35}SO_2$

	Wt (g)	Total ^{35}S uptake (cpm g fresh wt^{-1})	Volatile ^{35}S (cpm g fresh wt^{-1})	Volatile ^{35}S (%)
Young leaves	1.85	6668×10^3	858×10^3	12.9
Mature leaves	5.10	1060×10^3	2×10^3	0.2

Immediately after fumigating a cucumber plant with $^{35}SO_2$ for 30 min in light, leaves were detached and placed in individual chambers through which an airstream was drawn; the airstream was then passed into N-ethylmaleimide and Zn acetate traps in series. After 3 h, total radioactivity remaining in leaf, and trapped volatile radioactivity were determined.

of SO_2 is found inside young leaves than inside mature leaves throughout an acute exposure to SO_2, but the mature leaves are injured far more severely (*Table 20.2*; Bressan, Wilson and Filner, 1978; Wilson *et al.*, 1979). This could mean that SO_2 is not itself the toxic substance; however, it could also mean that young leaves do not have injurable sites accessible to the absorbed SO_2. Virtually all of the absorbed SO_2 is metabolized within about three hours (*Table 20.2*; Wilson *et al.*, 1979) in light or dark (Wilson *et al.*, unpublished observations). In experiments with bisulfite and leaf discs, we found that bisulfite can be absorbed in darkness without injury, but a later exposure to light results in injury, as manifested by necrosis and ethane production, in proportion to duration of the light exposure (*Table 20.4*). Injury therefore appears not to require the persistence of bisulfite. These results may have another meaning, however: there may be a primary injury event caused directly by SO_2/HSO_3^-. A (reduced?) metabolite of SO_2/HSO_3^- could form and cause injury which is latent, i.e. storable in darkness. The injury would remain latent until the light comes on and secondary consequences begin to occur. The difference between a reaction which is itself injurious, but in a way not manifested in darkness, as opposed to a reaction which is not injurious, but makes the tissue photosensitive, may be little more than semantic.

SO_2 is oxidized to SO_4^{2-} to about the same extent, ca. 60%, in resistant and sensitive cucurbit leaves. Assuming that oxidation in resistant and sensitive leaves proceeds via

TABLE 20.4. Effect of bisulfite concentration and length of incubation in the light on ethane and ethylene emission

I Dark with $KHSO_3$ (h)	II Light (h)	III Dark (h)	(mM)	Light in Period II × $KHSO_3$ in Period I (h × mM)	Ethane after Period I+II+III (nl cm^{-2})	Ethylene (nl cm^{-2})
2	30	0	0	0	0.8 ± 0.1	0.8 ± 0.1
2	30	0	5	150	2.4 ± 0.3	2.0 ± 0.1
2	20	10	7.5	150	6.0 ± 0.2	1.4 ± 0.2
2	11.5	18.5	15	173	5.9 ± 0.3	1.0 ± 0.1
2	5	25	30	150	7.7 ± 0.6	0.5 ± 0.0
2	3	27	50	150	7.2 ± 0.2	0.4 ± 0.0
2	2	28	75	150	6.3 ± 0.4	0.5 ± 0.0
2	1.5	28.5	100	150	7.6 ± 0.3	0.4 ± 0.0

Leaf discs of cucumber, Chipper (formerly SC25) were treated in the dark for 2 h in various concentrations of $KHSO_3$. They were then washed and incubated in the light to give 150 mM × h. At the end of the light period the discs were placed in darkness and then after 30 h, all were sampled for ethane and ethylene (nl cm^{-2} ± standard error).

the same pathway, it appears unlikely that such oxidation is the crucial detoxification process. Thus, the only striking difference between the fates of SO_2 in the leaf types is conversion of about 10% of the absorbed SO_2 to the reduced product, H_2S, in young leaves, but not in mature leaves (*see Tables 20.3 and 20.5*). This pattern suggests that a product of reduction of SO_2 causes injury, and that either further reduction of this product, or a diversionary reduction to H_2S, of a precursor to the injurious product, protects the leaf from injury (*Figure 20.6*). Still another possible explanation is that SO_2/HSO_3^- causes injury in a certain compartment, in which reduction rather than oxidation occurs. This compartment could be the chloroplast.

TABLE 20.5. Lack of dilution by $^{32}SO_4$ of ^{35}S in $H_2^{35}S$ made from $^{35}SO_2$

	Young leaf	Mature leaf
Na_2SO_4—$^{35}SO_2$		
Fresh weight (two leaves) (g)	2.07	6.07
Total uptake of ^{35}S (A) (cpm g fresh wt^{-1})	3.563×10^6	2.467×10^6
H_2S trapped (B) (cpm g fresh wt^{-1})	0.252×10^6	0.073×10^6
(nmol g fresh wt^{-1})	224	52.9
(B/A) × 100 (%)	7.1	2.9
Specific radioactivity of H_2S (mCi/mmol)	0.505	0.629
Specific radioactivity of SO_2 (mCi/mmol)	0.696	0.696
$Na_2^{35}SO_4$—SO_2		
Fresh weight (g)	1.40	2.70
Total uptake of ^{35}S (A) (cpm g fresh wt^{-1})	18.17×10^6	8.56×10^6
H_2S trapped (B) (cpm g fresh wt^{-1})	0.032×10^6	0.003×10^6
(nmol g fresh wt^{-1})	157	31.3
(B/A) × 100 (%)	0.18	0.04
Specific radioactivity of H_2S (mCi/mmol)	0.092	0.047
Specific radioactivity of SO_4^{-2} (mCi/mmol) (estimated)	0.8–1.5	

The conditions were the same as in *Table 20.3*.

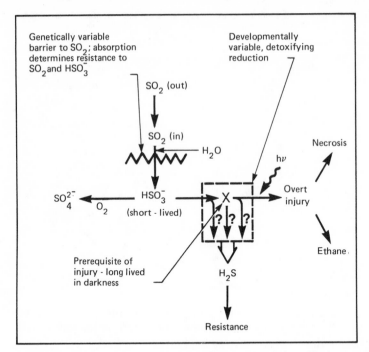

Figure 20.6. Resistance to SO_2 injury by synthesis and emission of hydrogen sulfide

There may be more than one path of oxidation of SO_2 to SO_4^{2-} in leaves. Ream and Wilson (1982) have obtained some preliminary evidence for the existence in leaves of two such systems. Extracts of cucurbit leaves contain a light-dependent, partly particulate system and a light-independent soluble system. The light-independent, soluble enzyme is not inhibited by DCMU or exogenous superoxide dismutase. The light-dependent system is inhibited by DCMU or superoxide dismutase, which suggests that it involves superoxide formed by transfer of electrons from chloroplast Photosystem II to molecular oxygen. However, it is doubtful that this reaction would occur to a significant extent in a healthy intact chloroplast, or if it did, that the superoxide would be left untouched by endogenous superoxide dismutase. Therefore, the light-dependent oxidation should be viewed as most likely a test-tube artifact. Nevertheless, it may be significant that the light-independent oxidation reaction is the predominant one in extracts of young leaves, i.e. the leaves which are most resistant to acute injury by SO_2.

Paths of H_2S synthesis

SULFATE IS NOT AN INTERMEDIATE IN H_2S SYNTHESIS FROM SO_2

Three pathways for the synthesis of H_2S in response to SO_2/HSO_3^- have to be considered (*Figure 20.7*). Since much of the SO_2/HSO_3^- absorbed by leaf tissue of cucurbits is oxidized to sulfate, light-dependent reduction of sulfate to sulfide may be part of the path of H_2S synthesis from SO_2/HSO_3^-; subsequent to reduction, sulfide may be split off carrier-bound sulfide, and released as H_2S. Alternatively, carrier-bound sulfide may be incorporated into cysteine, from which H_2S may be released by the

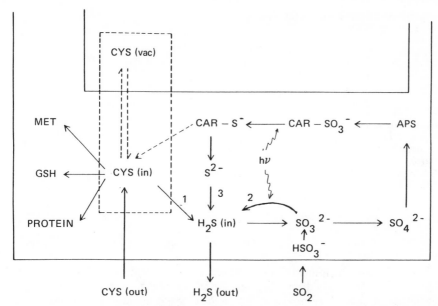

Figure 20.7. Paths of hydrogen sulfide production in leaves. (1) Light-independent H₂S synthesis from
L-cysteine. (2) Light-dependent H₂S synthesis from SO₂. (3) Light-dependent H₂S synthesis from sulfate

action of cysteine desulfhydrase. A third path of H_2S synthesis may proceed via direct
reduction of SO_2/HSO_3^- (*Figure 20.7*). This path of H_2S synthesis is not only supported
by the developmentally determined resistance to SO_2 in cucurbits discussed above, but
also by labeling experiments with $^{35}SO_2$ and $^{35}SO_4^{2-}$ (*Table 20.5*). When cucurbit
plants were exposed to unlabeled Na_2SO_4 and fumigated with $^{35}SO_2$, about 10% of the
^{35}S absorbed was incorporated into H_2S. The H_2S released showed a specific
radioactivity comparable to the $^{35}SO_2$ used for the fumigation. However, when
cucurbit plants were exposed to $Na_2{}^{35}SO_4$ and fumigated with unlabeled SO_2, only
0.2% of the ^{35}S in the leaves was incorporated into H_2S, although 85% of the label was
present as $^{35}SO_4^{2-}$. The specific radioactivity of the H_2S released under these
conditions was one order of magnitude smaller than the specific radioactivity of the
$^{35}SO_4^{2-}$ inside the cells (*Table 20.5*). Thus, direct reduction of SO_2/HSO_3^- to H_2S rather
than reduction subsequent to an oxidation to sulfate seems to be the biosynthetic path
leading to H_2S emission in response to SO_2/HSO_3^-.

THREE PRECURSORS, THREE PATHWAYS

We found that H_2S emission from cucurbit leaves can occur not only in response to
SO_2/HSO_3^- (De Cormis, 1968, 1969; Wilson, Bressan and Filner, 1978; Sekiya, Wilson
and Filner, 1982), but also in response to SO_4^{2-} (Spaleny, 1977; Wilson, Bressan and
Filner, 1978; Sekiya *et al.*, 1982a; Winner *et al.*, 1981) and L-cysteine (Rennenberg *et al.*,
1982; Sekiya *et al.*, 1982b). A number of other sulfur compounds by themselves did not
cause H_2S emission (*Table 20.6*). This does not necessarily mean that volatile sulfur is
not produced in response to these sulfur compounds. Upon exposure to L-methionine,
for example, methylmercaptan is emitted by cucurbit leaf discs after a lag-period of
6–14 h (Schmidt *et al.*, in preparation). The sulfur of dithioerythritol does not serve as a
precursor of H_2S, but dithioerythritol causes H_2S emission in response to sulfate in

TABLE 20.6. Sulfur compounds which do not cause cucurbit leaves to emit H_2S

D-Cysteine
L-Cystine
O-Methyl-L-cysteine
Cysteamine
3-Mercaptopropionic acid
L-Homocysteine
S-Methyl-L-cysteine
L-Cysteic acid
DL-Cysteic acid
Taurine
L-Serine
L-Methionine

darkness. Without dithioerythritol H_2S emission in response to SO_4^{2-} in darkness is virtually undetectable (Sekiya *et al.*, 1982a).

The H_2S emission phenomenon in response to each precursor/effector (SO_4^{2-}, SO_2/HSO_3^-, L-cysteine) could be distinguished by inhibitor studies. The light-dependent H_2S emission in response to SO_2/HSO_3^- could be inhibited by cyanazine (*Figure 20.8*), a triazine herbicide inhibiting the electron flow between Photosystems II and I (Brewer, Arntzen and Slite, 1979), but was not inhibited by amino-oxyacetic acid. Amino-oxyacetic acid is an inhibitor of pyridoxal phosphate dependent enzyme reactions. Both cysteine synthase and cysteine desulfhydrase are likely to require pyridoxal phosphate (Kredich, Keeman and Foote, 1972; Giovanelli, Mudd and

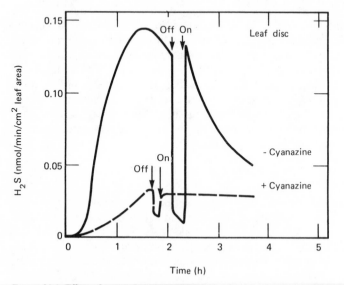

Figure 20.8. Effect of cyanazine on H_2S emission in response to bisulfite. Leaf discs from each half of one leaf of cucumber (Chipper) were placed in a pair of 250 ml flasks with 10 mM potassium bisulfite, one with cyanazine and one without. The flasks were illuminated with 8 mW cm-2 of light and air was drawn through the flasks to a flame photometric sulfur analyzer. The light was turned off then back on at the indicated times

TABLE 20.7 Effect of DCMU, atrazine, cyanazine on H₂S emission in cucumber leaf

Substrate	Inhibitor	Volatile S (nmol 4 h⁻¹)		Radioactive activity (%)
25 mM SO₄⁻²	—		54.6	100
	0.1 mM cyanazine		3.93	7.2
25 mM SO₄⁻²	—		35.0	100
	0.1 mM atrazine		6.86	19.6
25 mM SO₄⁻²	—		21.3	100
	0.1 mM DCMU		3.01	14.1
10 mM cysteine	—		52.7	100
	0.1 mM cyanazine		40.3	76.5
10 mM KHSO₃	—	detached leaf	5.13	100
	0.1 mM cyanazine	whole leaf	0.57	11.1

Leaf discs were prepared and used with procedures similar to those described in the legend for *Figure 20.10*.

Datko, 1980). Light-dependent H₂S emission in response to SO₄²⁻ was also inhibited by cyanazine (*Table 20.7*), but stimulated rather than inhibited by amino-oxyacetic acid (Sekiya *et al.*, 1982a). Sulfate uptake was sometimes inhibited by amino-oxyacetic acid, sometimes not, depending upon the developmental state of the leaf. When sulfate uptake was inhibited, H₂S production was, of course, secondarily inhibited. Light-independent H₂S emission from L-cysteine is not affected by cyanazine, but is strongly inhibited by amino-oxyacetic acid (*Figure 20.9*; Sekiya *et al.*, 1982b).

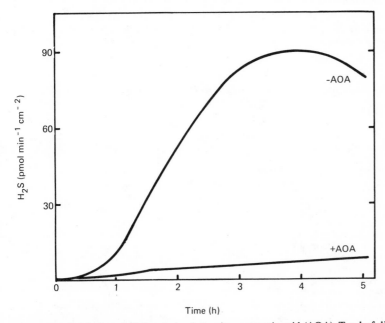

Figure 20.9. Inhibition of H₂S emission by amino-oxyacetic acid (AOA). Ten leaf discs from a half of a young cucumber leaf were floated on 10 mM L-cysteine solution and ten leaf discs from the other half of the same leaf were floated on a 10 mM L-cysteine solution containing 1 mM amino-oxyacetic acid. Light intensity was 8 mW cm⁻². H₂S emission was monitored by a sulfur analyzer

L-CYSTEINE IS PROBABLY NOT AN INTERMEDIATE IN H₂S PRODUCTION FROM SULFATE

Inhibition of H_2S emission in response to sulfate by cyanazine and other herbicides which inhibit photosynthetic electron transport (Sekiya *et al.*, 1982a) suggests that the same photosystem which functions in photosynthetic CO_2 reduction is responsible for light-dependent sulfate reduction and the generation of H_2S. Again, direct release of H_2S from carrier-bound sulfide or incorporation of carrier-bound sulfide into cysteine and subsequent desulfhydration are possible paths of H_2S synthesis from sulfate (*Figure 20.7*). Stimulation of H_2S emission by amino-oxyacetic acid, however, suggests that cysteine desulfhydrase is not participating in synthesis of H_2S in response to sulfate. Therefore, either direct release of H_2S from carrier-bound sulfide or a path of cysteine degradation that does not include cysteine desulfhydrase activity may be responsible for H_2S emission in response to sulfate.

Evidence is accumulating that the H_2S emitted in response to sulfate is synthesized without entering the cysteine pool(s) of the cucurbit cells. When glutathione synthesis, a major path of sulfur assimilation in cucurbits (Sekiya *et al.*, 1982a), is inhibited, sulfate reduction remains unaffected; whereas the incorporation of ^{35}S into cysteine is reduced under these conditions, emission of H_2S is stimulated by up to 80% (Rennenberg and Filner, 1982; *Table 20.8*). Therefore it appears that the excess of sulfate reduced in the presence of inhibitors of glutathione synthesis is released as H_2S into the atmosphere.

AVAILABILITY OF O-ACETYLSERINE PROBABLY DETERMINES THE RATE OF H₂S PRODUCTION

The rate of H_2S emission from sulfate can be inhibited by *o*-acetylserine, or by the precursors of this compound: acetyl coenzyme A and coenzyme A (*Table 20.9*). The lowered rate of H_2S emission is accompanied by an enhanced incorporation of ^{35}S from $^{35}SO_4^{2-}$ into cysteine. From these observations it appears that cysteine synthesis in curcurbit leaves is limited by the availability of *o*-acetylserine. We believe that the extent to which the amount of sulfate reduced exceeds the amount of *o*-acetylserine

TABLE 20.8. Influence of inhibitors of glutathione synthesis on H₂S emission from mature pumpkin leaves

SO_4^{2-} conc. (mM)	Inhibitor	Conc. (mM)	Relative H_2S emission (%±SE)
25	—	—	100±9.5
25	MSX	0.1	181±8.9
0	MSX	0.1	nd[a]
25	BSX	0.05	126±9.3
0	BSX	0.05	nd
25	cystamine	5	153±9.9
0	cystamine	5	nd
25	γ-methyl-glutamate	5	122±8.6
0	γ-methyl-glutamate	5	nd
25	D-glutamate	5	128±8.7
0	D-glutamate	5	7±4.3
25	GSH	1	183±9.6
0	GSH	1	nd

From Rennenberg and Filner, 1982.
The data were obtained in three replicates of 12 separate experiments, each replicate employing discs from a different leaf and a different plant. Eight leaf discs (21 cm² leaf area) of one-half of a mature pumpkin leaf were floated for 3 h in the light (4 mW cm⁻²) in 10 ml of a treatment solution containing one of the inhibitors indicated and either 25 mM or no sulfate. Eight leaf discs from the other half of the same leaf were floated under the same conditions in 10 ml of a 25 mM sulfate solution as a control. Controls emit 16–40 pmol H_2S × cm⁻² leaf area × min⁻¹.
[a] nd: H₂S emission not detected.

TABLE 20.9. **Effect of *o*-acetylserine and its precursors on H₂S emission from pumpkin leaf discs**

Compound added	Concentration (mM)	H₂S emission (%)
—	—	100
o-Acetylserine	5	27
L-serine	5	59
Acetyl coenzyme A	2	20
Na acetate	2	95
Pyruvate	2	90
3-Fluoro-pyruvate	0.05	137
Coenzyme A	2	22

Substrate: 25 mM sulfate, light 8 mW cm⁻², emission for 3 h. (From Rennenberg, in preparation.)
For each compound tested, two groups of leaf discs (18 cm² leaf area), each from one-half of an expanding leaf, were placed in 250-ml flasks. One group of leaf discs was floated on 10-ml 25 mM K_2SO_4 containing one of the compounds indicated; the other group of leaf discs was floated on 10 ml of a 25 mM K_2SO_4 solution without additions. Volatile emission in the light (8 mW cm⁻²) was continuously monitored by a flamephotometric sulfur analyzers and the emission without addition was taken as 100%.

available determines how much H_2S is emitted. This idea is supported by the observation that 3-fluoropyruvate, an inhibitor of the synthesis of the *o*-acetylserine precursor, acetyl coenzyme A, by pyruvate dehydrogenase (Bisswanger, 1981), stimulates H_2S emission while inhibiting the synthesis of cysteine (*Table 20.9*).

Hydrogen sulfide emission in response to sulfate in continuous light goes through a series of transients. First it rises to a peak within about 1 h, then declines gradually. If we shifted from light to dark, then back to light, an overshoot peak followed by oscillating transients occurred (*Figure 20.10*; Sekiya *et al.*, 1982a). We found that this peak can be eliminated by adding *o*-acetylserine to leaf discs. Acetate does not work, nor does serine. This result suggests that H_2S is emitted when (bound?) sulfite is produced in excess of the rate at which *o*-acetylserine is formed. We suggest that immediately following a shift of leaf tissue from dark to light, acceleration of *o*-acetylserine production lags behind the acceleration of sulfide production, but eventually photosynthetic energy and photosynthetically generated carbon compounds increase the rate of *o*-acetylserine synthesis, so less sulfide excess exists to be released as H_2S.

LEAVES HAVE A SEEMINGLY FUTILE SULFUR CYCLE

The observation that amino-oxyacetic acid inhibits light-independent emission of H_2S in response to L-cysteine in the dark indicates that the path of synthesis of H_2S in this instance may be the degradation of L-cysteine by cysteine desulfhydrase. However, what was initially light-independent H_2S emission in response to L-cysteine, in the light became a partially light-dependent, and cyanazine sensitive process after a lag-period of up to 90 min (Rennenberg *et al.*, 1982; *Figure 20.11*). The length of this lag-period decreased with increasing leaf age. These results suggested the existence of a seemingly futile sulfur cycle (*Figure 20.7*), which was confirmed experimentally by pulse chase experiments with ³⁵S-cysteine, in which the emission of radiolabeled and unlabeled H_2S as well as the labeling pattern of sulfur compounds inside the cells were analyzed (Rennenberg *et al.*, 1982). These experiments showed that a wave of radiolabel passed through sulfide and H_2S, then sulfite and sulfate, after which a second wave of radiolabel in sulfide and H_2S was detected. The second wave of label in sulfide and H_2S

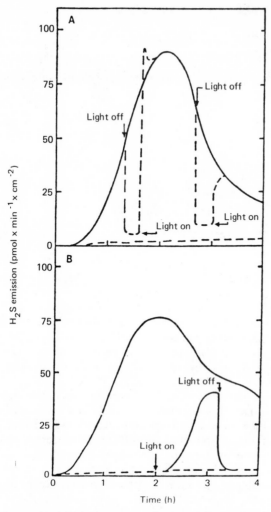

Figure 20.10. Light dependence of H$_2$S emission by discs from young cucumber leaves in response to 25 mM K$_2$SO$_4$. Three groups of eight discs (18 cm^2 leaf area) were obtained from a single leaf and each group of discs was floated on 10 ml of K$_2$SO$_4$ (pH 6) in a 125-ml Erlenmeyer flask coupled to sulfur analyzer through an automatic channel selector. One group of discs was kept in darkness; one group was kept in light (8 mW cm^{-2}) continuously; the third group was illuminated, but the illumination was interrupted for brief periods of darkness

was light-dependent and inhibited by cyanazine or by dilution of the sulfate pool of the cells with exogenous sulfate, while the first wave was not.

One would expect the sulfide in the second wave of emitted H$_2$^{35}S to be further metabolized, with reincorporation of some of the H$_2$^{35}S into L-cysteine. This was not observed, however (Rennenberg *et al.*, 1982). One possible explanation is that the large expansion of the unlabeled L-cysteine pool, as a result of adding 10 mM unlabeled L-cysteine as a chase, blocked further synthesis of labeled L-cysteine. The fact that no further increase in labeling of protein with ^{35}S occurred after the first wave of emission of H$_2$^{35}S is consistent with this interpretation (Rennenberg *et al.*, 1982).

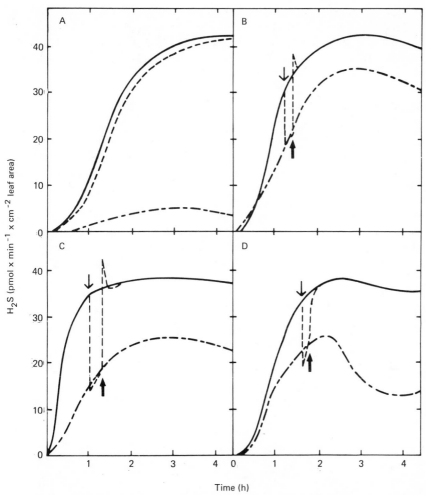

Figure 20.11. Effect of amino-oxyacetic acid and cyanazine on H_2S emission by cucumber leaf discs in response to L-cysteine. The data were obtained in three replicates of eight separate experiments, each replicate employing leaf discs from a different leaf and a different plant. Eight leaf discs (21 cm^2 leaf area) from one-half of a cucumber leaf were floated for 3 h in the dark (a) or in the light (b–d; 4 mW cm^{-2}) in 10 ml 10 mM L-cysteine and used as control; eight leaf discs from the other half of the leaf were exposed to 10 ml 10 mM L-cysteine under the conditions listed below. H_2S emission was measured continuously by a flame photometric sulfur analyzer. Light on. (A) Expanding leaf, continuous dark. —— control; – – – + 100 µM cyanazine; — — — — + 100 µM AOA. (B) Expanding leaf. —— light, control; — — — — — dark. (C) Mature leaf. —— light, control; — — — — — dark. (D) Expanding leaf, continuous light. —— control; — — — — — + 100 µM cyanazine

From these data it can be concluded that the H_2S emitted by cucurbit cells in response to L-cysteine in the light is derived from two different processes (*Figure 20.7*). One part of the H_2S emitted is produced by a light-dependent process that is inhibited by cyanazine; the balance of the H_2S emitted is derived from a light-independent process that is inhibited by amino-oxyacetic acid. The light-independent process appears to be the desulfhydration of L-cysteine, the light-dependent process appears to be the light-dependent reduction step in the sulfate assimilation pathway. We believe that in the light L-cysteine is degraded to sulfide, but only part of the sulfide produced this way is released into the atmsophere as H_2S. The balance of the H_2S produced would be oxidized to sulfate via sulfite. Subsequently, the sulfate synthesized via this pathway would be reduced so that the sulfur from L-cysteine via sulfide, sulfite and sulfate would enter the H_2S pool of the cells again.

FIELD PLANTS SPONTANEOUSLY EMIT VOLATILE SULFUR IN A LIGHT-DEPENDENT MANNER

An important question which should be addressed is whether H_2S emission by plants is solely a laboratory artefact or does it occur in nature? We have invested a great deal of effort in seeking the answer to this question. In the course of our investigations single leaves of several plant species grown under field conditions were placed in leaf chambers and air, cleansed of background sulfur by passage through charcoal and Purafil filters, was pulled through the leaf chambers. The amount of volatile sulfur emitted by the leaves under these conditions was too small to be measured in a continuous flow system. Using a stainless steel trap cooled with liquid oxygen, we were able to detect volatile sulfur emissions from intact leaves on untreated plants growing under field conditions (*Table 20.10*). However, in addition to volatile S, CO_2 was also

TABLE 20.10. Emission of volatile sulfur from field crops in 1981

Crop	Volatile sulfur emission (pmol h^{-1})		
	In light	*In dark*	*Chamber control*
Cantaloupe	42 (2)	0 (1)	0 (1)
Okra	49 (2)	0 (2)	2 (2)
Sugar beet	21 (8)	0 (2)	32 (4)
Apple	38 (8)	2 (6)	1 (5)
Grape	39 (11)	2 (2)	76 (7)
Black bean	21 (5)	6 (4)	59 (6)
White bean	38 (4)	2 (3)	226 (4)
Swiss chard	23 (3)		149 (1)
Collards	343 (7)	35 (3)	91 (8)

Plants grown in a field on the Michigan State University Campus were used. A leaf about 50 cm^2 was placed in a leaf chamber while still attached to the plant. The leaf chamber was thermostatted by circulating water in a water jacket. Air was drawn through a Purafil filter, then through the chamber and into a stainless steel trap cooled with liquid nitrogen. A second chamber contained a leaf in darkness; a third chamber had no leaf. After 1–4 h of trapping the trapped volatiles were analyzed by a CO_2 analyzer and a flamephotometric sulfur analyzer. The data are corrected for CO_2 interference (numbers in parentheses) with the sulfur analyzer. Usually the leaves were undamaged by the procedure.

trapped at the temperature of liquid oxygen, and in amounts sufficient to interfere with the flamephotometric determination of volatile S. Therefore, the content of the traps was first passed through a non-destructive CO_2 analyzer and then through the flamephotometric sulfur detector; the apparent amount of S measured was corrected for the interference by the CO_2 in the sample by means of a standard curve. The corrected values for the emission of volatile sulfur from leaves of nine species are given in *Table 20.10*. In the light, substantial amounts of volatile S were trapped with all the species tested, whereas only minute amounts were trapped in the dark. But even in the light the amount of volatile sulfur trapped was insufficient to allow gas chromatographic analysis. Thus, we do not yet know whether H_2S is the principal sulfur compound emitted under field conditions.

In the data gathered during the summer of 1981, only in four of the nine species did the amount of volatile S trapped provide a clear proof that volatile S was emitted from the leaves in the light. In the experiments with the other species the average amount of S trapped was smaller than the average amount of S trapped from air which passed through control chambers without leaves. High amounts and high variation in the sulfur content of the controls, however, seem to be due to insufficient purification of the air before it enters the leaf chambers. The Purafil traps used for this purpose, although trapping H_2S, CH_3SH and $(CH_3)_2S$, do not remove COS, CS_2 or $(CH_3)_2S_2$ from the air (Adams et al., 1980). These compounds, however, may be emitted from the soil (Farwell et al., 1979). As the amount of volatile S trapped in experiments with leaves kept in the dark was consistently lower than the amount of volatile S trapped in controls without leaves, volatile S that is not removed by the Purafil traps seems to be absorbed by the leaves. This absorption may cause an under-estimation of the sulfur emitted. Although the present data clearly show that volatile sulfur is emitted by leaves in the light under field conditions, our test system needs further improvement.

LIVING PLANTS APPEAR TO BE A MAJOR SOURCE OF NATURAL ATMOSPHERIC SULFUR

It is admittedly rather risky to extrapolate from data obtained on a handful of species. Nevertheless, we wish to do so because it is instructive. The average rates of emission under field conditions observed, if extrapolated to leaves of all plants in tropic and temperate zones, could account for 7.4 million tons of sulfur per year, or about 7% of the approximately 100 million metric tons of atmospheric sulfur thought to enter the atmosphere by natural processes each year (cf. Meyer, 1977). It would not surprise us if the contribution of plants to natural atmospheric sulfur proved to be a far higher percentage of the total, when sulfur emissions from the major species which cover the earth's productive land masses have been measured in the various environments in which they grow. We know that in the laboratory, leaves are capable of stimulated emission rates 1000-fold greater than the spontaneous rates we have detected in the field.

Conclusion

The most intriguing question about H_2S synthesis and emission by plants is 'why?'. What is the function of this pathway? One remote possibility is that the futile sulfur cycle might be a way for plants to get rid of excess photosynthetic reductant. However, the rate of sulfate reduction or cysteine desulfhydration is about 1/1000th that of CO_2 fixation. It is therefore difficult to envisage this as a significant way to burn off electrons. A more plausible function of H_2S synthesis and emission is to maintain at safe levels pools of pathway-specific reductants and carriers which function in sulfate reduction. Suppose, when there is no possibility to incorporate sulfide into organic sulfur compounds, photosynthetic energy could elevate the concentration of sulfate pathway reductants to levels which produce harmful side reactions. Then it would be to the plant's advantage to use the futile sulfur cycle to burn off the excess reductants. In much the same way, the photorespiratory pathway can be considered to be a futile cycle which prevents photosynthetic energy from doing damage when CO_2 is scarce. Still another possible function of the intracellular sulfur cycle and the emission of H_2S out of this cycle that has to be considered is to buffer the cell's thiol concentration. Changes in light intensity, to which plants are continually exposed under natural conditions, will cause tremendous changes in sulfate reduction. Comparable fluctuations of the thiol/disulfide status of a plant cell do not appear to be tolerable, as important functions such as protein synthesis (Fahey *et al.*, 1980), the catalytic activities of major proteins (*cf.* Giovanelli, Mudd and Datko, 1980, Rennenberg, 1982; Jocelyn, 1972b), and detoxification phenomena (*cf.* Halliwell, 1981; Lamoureux and Russness, 1981; Rennenberg, 1982), are highly susceptible to even minor changes in the thiol concentration. Therefore, a means of precise regulation of the cell's content of low molecular weight thiols would be expected to be an essential tool for a plant to survive in a continually changing environment.

Although the steady state concentration of cysteine in plant cells has not been extensively studied, the available data indicate that the cysteine concentration is maintained at a very low level (Smith, 1975; Giovanelli, Mudd and Datko, 1978, 1980; Brunhold and Schmidt, 1978). Almost all the cell's low molecular weight thiol seems to be present as glutathione (*cf.* Rennenberg, 1982), which apparently functions as a storage and transport form of cysteine (*cf.* Rennenberg, 1982). The glutathione pool of green cells seems to be carefully regulated, as excess cysteine incorporated into this peptide is translocated to other parts of the plant (Rennenberg, Schmitz and Bergmann, 1979; Bonas *et al.*, 1982). It is obvious that long-distance transport of glutathione can be used for the regulation of the cells' thiol/disulfide status in the short-time range only. Translocation of glutathione to other parts of the plant at a rate which substantially exceeds the rate of glutathione degradation, would only be translocation of the plant's problem of ridding itself of excess reduced sulfur.

High amounts of excess reduced sulfur, however, could be handled by the emission of H_2S from the intracellular sulfur cycle. Irrespective of whether plants are exposed to excess SO_2/HSO_3^-, SO_4^{2-}, or cysteine, the intracellular sulfur cycle would allow a plant to use as much of these compounds as necessary for its own growth and development. It would simultaneously give a plant the possibility to maintain the cysteine and thiol concentration under a critical concentration by releasing excess sulfur into the atmosphere. The emission of H_2S may be compared with a pressure valve. Through it, excess sulfur is released out of the intracellular sulfur cycle.

Exposure of a plants to high amounts of sulfur may not only occur under manmade circumstances, such as SO_2 fumigation from industrial sources or the use of sulfur containing fertilizers, but also may result from geological activity, e.g. volcanos, hot springs, etc. As these geological activities seem to have covered most parts of the earth in former times, the ability to get rid of excess sulfur may have been far more important for survival of plants in ancient times than today. Therefore, today's manmade exposure of plants to dangerously high concentrations of sulfur may not be a new environmental situation for plants. The emission of H_2S by plants may be the old answer to an old problem.

Acknowledgement

This research was supported by the United States Department of Energy under Contract DE-AC02-ER0-1338. H.R. was a recipient of a Deutsche Forschungsgemeinschaft fellowship.

References

ADAMS, A., TARWELL, S. O., ROBINSON, E. and PACK, M. A. *Biogenic sulfur emission in the SURE region*, Electric Power Research Institute Report EA 1516, Washington State University, Air Pollution Research Station, Pullman, Washington 99164 (1980)

BARRET, T. W. and BENEDICT, H. M. Sulfur dioxide. In *Recognition of Air Pollution Injury to Vegetation: A Pictorial Atlas*, pp. C1–C10, National Air Pollution Control Association, Pittsburgh (1970)

BISSWANGER, H. Substrate specificity of the pyruvate dehydrogenase complex from *Escherichia coli*, *Journal of Biological Chemistry* **256**, 815–822 (1981)

BOLLER, T. and KENDE, H. Regulation of wound ethylene synthesis in plants, *Nature* **286**, 259–260 (1980)

BONAS, H., SCHMITZ, K., RENNENBERG, H. and BERGMANN, L. Phloem transport of sulfur in *Ricinus*, *Plants* **155**, 82–88 (1982)

BRESSAN, R. A., WILSON, L. G. and FILNER, P. Mechanisms of resistance to sulfur dioxide in the Curcurbitaceae, *Plant Physiology* **61**, 761–767 (1978)

BRESSAN, R. A., LECUREUX, L., WILSON, L. D. and FILNER, P. Emission of ethylene and ethane by leaf tissue exposed to injurious concentrations of sulfur dioxide or bisulfite ion, *Plant Physiology* **63**, 924–930 (1979a)

BRESSAN, R. A., LECUREUX, L., WILSON, L. G. and FILNER, P. Release of ethane by leaf tissue in response to various environmental stresses, *Plant Physiology* **63**, S-327 (1979b)

BRESSAN, R. A., LECUREUX, L., WILSON, L. G., FILNER, P. and BAKER, L. R. Inheritance of resistance to sulfur dioxide in cucumber, *Hort Science* **16**, 332–333 (1981)

BREWER, P. E., ARNTZEN, C. J. and SLITE, F. W. Effects of atrazine, cyanazine and procyazine in the photochemical reaction of isolated chloroplasts, *Weed Science* **27**, 300–308 (1979)

BRUNHOLD, C. and SCHMIDT, A. Regulation of sulfate assimilation in plants. 7. Cysteine inactivation of adenosine 5'-phosphosulfate sulfotransferase in *Lemna minor* L., *Plant Physiology* **61**, 342–347 (1978)

DECORMIS, L. Dégagement d'hydrogene sulfuré par des plantes soumises à une atmosphère contenant de l'anhydride sulfureux, *Comptes Rendus de l'Académie des Sciences* **266D**, 682–685 (1968)

DECORMIS, L. Quelques aspects de l'absorption du soufre par les plants soumises à une atmosphère contenant du SO$_2$. In *Proceedings of the 1st European Congress on the Influence of Air Pollution on Plants and Animals*, pp. 75–78, Wageningen, The Netherlands, Centre for Agricultural Publishing and Documentation, Wageningen (1969)

ELSTNER, E. F. and KONZE, J. R. Effect of point freezing on ethylene and ethane production by sugar beet leaf disks, *Nature* **263**, 351–352 (1976)

FAHEY, R. C., DISTEFANO, D. L., MEIER, G. P. and BRYAN, R. N. Role of hydration state and thiol-disulfide status in the control of thermal stability and protein synthesis in wheat embryo, *Plant Physiology* **65**, 1062–1066 (1980)

FARWELL, S. O., SHARRARD, A. E., PACU, M. R. and ADAMS, D. F. Sulfur compounds volatilized from soil at different moisture contents, *Soil Biology and Biochemistry* **11**, 411–415 (1979)

GARSED, S. G. and READ, D. J. SO$_2$ metabolism in soybeans (*Glycine max* var. Biloxi). I. Effects of light and dark on uptake and translocation of ^{35}SO$_2$, *New Phytologist* **78**, 111–119 (1977a)

GARSED, S. G. and READ, D. J. SO$_2$ metabolism in soybean (*Glycine max* var. Biloxi). II. Biochemical distribution of ^{35}SO$_2$ products, *New Phytologist* **79**, 583–592 (1977b)

GARSED, S. G. and READ, D. J. Uptake and metabolism of ^{35}SO$_2$ in plants differing in sensitivity to SO$_2$, *Environmental Pollution* **13**, 173–186 (1977c)

GIOVANELLI, J., MUDD, S. H. and Datko, A. H. Homocysteine biosynthesis in green plants, *Journal of Biological Chemistry* **253**, 5665–5677 (1978)

GIOVANELLI, J., MUDD, S. H. and DATKO, H. A. Sulfur amino acids in plants. In *The Biochemistry of Plants*, Vol. 5, (Ed. J. B. MIFLIN), pp. 453–505, Academic Press, New York (1980)

GUDERIAN, R. *Air pollution. Phytotoxicity of Acidic Gases and its Significance in Air Pollution Control*, Springer-Verlag, Heidelberg (1977)

HALLIWELL, B. Ascorbic acid and the illuminated chloroplast, *Advances in Chemistry* **200**, 263–274 (1981)

JOCELYN, P. C. *Biochemistry of the SH group*, pp. 100–101, Academic Press, London (1972a)

JOCELYN, P. C. *Biochemistry of the SH group*, pp. 190–221, Academic Press, London (1972b)

KONZE, J. R. and ELSTNER, E. I. Ethane and ethylene formation by mitochondria as indication of aerobic lipid degradation in response to wounding of plant tissue, *Biochimica et Biophysica Acta* **528**, 213–221 (1978)

KREDICH, N. M., KEEMAN, B. S. and FOOTE, L. J. The purification and subunit structure of cysteine desulfhydrase from *Salmonella typhimurium*, *Journal of Biological Chemistry* **247**, 7157–7162 (1972)

LAMOUREUX, G. L. and RUSNESS, D. G. Catabolism of glutathione conjugates of pesticides in higher plants. In *Sulfur in Pesticide Action and Metabolism*, (Eds. J. D. ROSEN, P. S. MAGEE and J. E. CASIDA), ACS Symposium Series, **158**, 133–164 (1981)

LEVITT, J. and KOZLOWSKI, T. T. *Responses of Plants to Environmental Stresses*, Academic Press, New York (1972)

MEYER, B. *Sulfur, Energy, and Environment*, pp. 142–168, Elsevier Publishing Company, Amsterdam (1977)

MILLER, J. E. and XERIKOS, P. B. Residence times of sulfite in SO_2 sensitive and tolerant soybeans, *Environmental Pollution* **18**, 259–264 (1979)

REAM, J. E. and WILSON, L. G. Possible significance to SO_2 injury of dark and light-dependent bisulfite oxidizing activities in *Cucumis sativa* leaves, *Plant Physiology* **69**, S-16 (1982)

RENNENBERG, H., SCHMITZ, K. and BERGMANN, L. Long-distance transport of sulfur in *Nicotiana tabacum*, *Planta* **147**, 57–62 (1979)

RENNENBERG, H. Glutathione metabolism and possible biological roles in higher plants, *Phytochemistry* **21**, 2771–2781 (1982)

RENNENBERG, H. and FILNER, P. Stimulation of H_2S emission from pumpkin leaves by inhibition of glutathione synthesis, *Plant Physiology* **69**, 766–770 (1982)

RENNENBERG, H., SEKIYA, J., WILSON, L. G. and FILNER, P. Evidence for an intracellular sulfur cycle in cucumber leaves, *Planta* **154**, 516–524 (1982)

SEKIYA, J., WILSON, L. G. and FILNER, P. Positive correlation between H_2S emission and SO_2 resistance in cucumber, *Plant Physiology* **62**, S-407 (1980)

SEKIYA, J., WILSON, L. G. and FILNER, P. Resistance to injury by sulfur dioxide: correlation with its reduction to, and emission of hydrogen sulfide in Cucurbitaceae, *Plant Physiology* **70**, 437–441 (1982)

SEKIYA, J., SCHMIDT, A., RENNENBERG, H., WILSON, L. G. and FILNER, P. Hydrogen sulfide emission by cucumber leaves in response to sulfate in light and dark, *Phytochemistry* **21**, 2173–2178 (1982a)

SEKIYA, J., SCHMIDT, A., WILSON, L. G., and FILNER, P. Emission of hydrogen sulfide by leaf tissue in response to L-cysteine, *Plant Physiology* **70**, 430–436 (1982b)

SMITH, I. K. Sulfate transport in cultured tobacco cells, *Plant Physiology* **55**, 303–307 (1975)

SPALENY, J. Sulphate transformation to hydrogen sulphide in spruce seedlings, *Plant & Soil* **48**, 557–563 (1977)

THOMAS, M. D. Gas damage to plants, *Annual Review of Plant Physiology* **2**, 293–322 (1951)

THOMAS, M. D., HENDRICKS, R. H., COLLIER, G. R. and HILL, R. G. Utilization of SO_4^{2-} and SO_2 for the nutrition of alfalfa, *Plant Physiology* **18**, 345–371 (1943)

THOMAS, M. D., HENDRICKS, R. H. and HILL, G. R. Sulfur metabolism of plants. Effect of SO_2 on vegetation, *Industrial and Engineering Chemistry* **42**, 2231–2235 (1950)

THOMAS, M. D. and HENDRICKS, R. H. Effects of air pollution on plants. In *Air Pollution Handbook*, (Eds. P. L. MAGILL, F. R. HOLDEN and C. ACKLEY), pp. 9-1 to 9-44, McGraw-Hill, New York (1956)

WILSON, L. G., BRESSAN, R. A. and FILNER, P. Light-dependent emission of hydrogen sulfide from plants, *Plant Physiology* **61**, 184–189 (1978)

WILSON, L. G., BRESSAN, R., LECUREUX, L., REAM, J. and FILNER, P. Destruction of bisulfite by young and mature leaves, *Plant Physiology* **63**, 5–329 (1979)

WINNER, W. F., SMITH, C. L., KOCH, G. W., MOONEY, H. A., BEWLEY, J. E. and KROUSE, H. R. H_2S emission rates from plants and patterns of stable sulfur, *Nature* **289**, 672–674 (1981)

ZEIGLER, I. Effects of SO_2 pollution on plant metabolism, *Residue Reviews* **56**, 79–105 (1975)

Chapter 21

Modification of plant cell buffering capacities by gaseous air pollutants

E. Nieboer, J. D. MacFarlane

DEPARTMENTS OF BIOCHEMISTRY AND BIOLOGY, McMASTER UNIVERSITY, HAMILTON, CANADA

and

D. H. S. Richardson

SCHOOL OF BOTANY, TRINITY COLLEGE, DUBLIN, IRELAND

Introduction

Evaluations of buffer capacities of plant samples and tree bark homogenates have been made in environmental assessment and exposure studies, although the number of published reports are few (Thomas, Hendricks and Hill, 1944; Skye, 1968; Johnsen and Søchting, 1973; Hällgren, 1978). Because of the current interest in cellular pH gradients and proton-motive forces in cell biochemistry and bioenergetics (*see* Nicholls, 1982), new methods for the determination of intracellular pH and buffer capacities have become available. It appears opportune therefore to review these advances and to assess their merit in studies of the uptake and toxicity of gaseous air pollutants.

There appears to be some confusion about the exact meaning of buffer capacity. For this reason, the basic principles defining this concept are outlined, as well as the integrally linked phenomenon of pH and its measurement. The more reliable and promising methods of intracellular pH and buffer capacity determinations are described in enough detail to serve as a primer for interested researchers. A compilation is provided of the commonly observed intracellular pH values, and their dependence on experimental conditions is reviewed. The few estimates of intracellular buffer capacities reported for plants are also critically examined. After reviewing the known effects on these parameters of exposures to acidic and basic gaseous substances, the potentially damaging repercussions of alterations in intracellular pH and buffer capabilities are considered. It is hoped that the background and scope provided in this article will serve to challenge and stimulate researchers to design experiments to test the fundamental importance of such induced intracellular pH-related changes.

Basic concepts

Buffer capacity

Qualitatively, a buffer is defined as a solution which resists changes in pH despite the addition of substantial quantities of acid or base. More formally, the buffer index or

313

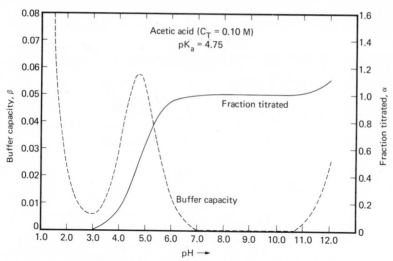

Figure 21.1. Demonstration of the expression in equation (1) relating the buffer capacity β to the slope of a potentiometric titration curve. To facilitate the comparison, the axes of the titration curve have been reversed since, by convention, pH is plotted *versus* volume of base added ($\equiv \alpha$, the fraction titrated). C_T denotes the sum of the concentrations of the protonated and unprotonated forms

capacity is referred to as the 'inverse slope'. This term originates from its relation to the slope of an acid/base titration curve as illustrated in *Figure 21.1*. Thus

$$\beta = \frac{dC_b}{dpH} \quad \text{or} \quad \beta = -\frac{dC_a}{dpH} \tag{1}$$

with β the buffer index, dC_b and dC_a respectively the increments of strong base or acid added, and dpH the corresponding incremental change in pH. For a monoprotic weak acid, the expression summarized in (2) may be derived from first principles (Van Slyke, 1922; Butler, 1964; Bates, 1973).

$$\beta = 2.303 \left[\frac{K_w}{[H^+]} + [H^+] + \frac{CK_a[H^+]}{(K_a + [H^+])^2} \right] \tag{2}$$

K_w is the ion-product of water, K_a is the weak acid dissociation constant, and C is the total buffer concentration (sum of protonated and unprotonated forms). In this derivation, all activity coefficients were assigned values of unity, and pH is defined as $-\log[H^+]$ (*see* comments on pH below). The first two terms within the square brackets on the right-hand side of equation (2), describe the buffering action of water, which increases with increasing concentration of hydrogen ions or hydroxyl ions ($[OH^-] = K_w/[H^+]$). Effectively, these contributions correspond to the innate buffering capacity of the hydrogen ion (*see* curve for HCl in *Figure 21.2*) or hydroxyl ion (NaOH curve in *Figure 21.2*). It is evident from the third term in equation (2), and the curves for acetic acid, hydrogen phosphate and hydrogen borate, in *Figure 21.2*, that the buffering capacity depends on the total buffer concentration, and that the magnitude of K_a largely determines the pH range in which useful buffering is possible.

For a mixture of weak acid buffers, the index at a specific pH corresponds to the sum of the separate effects of each constituent buffer (Van Slyke, 1922). This is expressed quantitatively in equations (3) and (4).

$$\beta = 2.303\left[\frac{K_w}{[H^+]} + [H^+] + \sum_{i=1}^{n}\frac{C_i K_{ai}[H^+]}{(K_{ai}+[H^+])^2}\right] \tag{3}$$

$$\beta = 2.303\left[\frac{K_w}{[H^+]} + [H^+] + \frac{C_T \bar{K}_a[H^+]}{(\bar{K}_a+[H^+])^2}\right] \tag{4}$$

In equation (4), C_T is summed over all buffer components (i.e. concentration of weak acid plus concentration of its conjugate base for each buffer), while $\bar{K}_a$ denotes an average dissociation constant describing the weak-acid properties of the functional groups involved in the buffering action over a narrow pH range, such as that expected for cell cytosol at intracellular pH values.

pH concept

Formally, pH is defined in terms of the activity of the hydrogen ion, a_{H^+} (Bates, 1973). Recalling that

$$a_{H^+} = [H^+]\gamma_{H^+} \tag{3}$$

with γ_{H^+} the hydrogen ion activity coefficient, we may write

$$pH = -\log a_{H^+} = -\log [H^+]\gamma_{H^+} \tag{6}$$

In an electrochemical cell, such as that consisting of the calomel and glass electrodes, a_{H^+} may be related to the measured e.m.f. E of the cell, the standard cell potential $E^{0\prime}$, and the potential E_j across the liquid junction between the calomel electrolyte and the test solution.

$$pH = -\log a_{H^+} = \frac{(E - E^{0\prime} - E_j)F}{2.303RT} \tag{7}$$

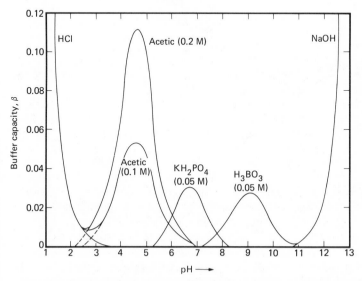

Figure 21.2. Plots of the buffer capacity based on equation (2) for various solutions of strong acid, strong base and buffers prepared from the indicated weak acid and the strong base NaOH. The concentrations specified refer to the total concentration of the buffer components (the acid and its conjugate base). (After Bates, 1973)

Of the quantities in equation (7), E, R (the gas constant), T (the absolute temperature) and F (the Faraday constant) are physically defined or measurable. This is in contrast to $E^{0\prime}$ and E_j, which are not obtainable experimentally. To overcome this impasse, pH has been defined operationally.

$$pH = pH(S) + \frac{(E_x - E_s)\,F}{2.303\,RT} \qquad (8)$$

In this expression, pH(S) is the pH value of a standard buffer, E_x is the cell e.m.f. for the unknown solution and E_s that for the standard buffer solution. Even with this operational approach to pH measurements, no changes in the reference potential ($E^{0\prime}$) and liquid junction potential (E_j) may occur between the pH-meter standardization step and the actual measurement for the unknown sample. E_j is known to change if the composition, concentration of buffer components, and ionic strength in the unknown sample are substantially different from the standard buffer solution(s). $E^{0\prime}$ also has components that may alter during use. In addition, contributions to the cell e.m.f. from cations other than the proton also occur. Consequently, considerable uncertainty may be associated with pH measurements, even if good laboratory practices are employed (Skoog and West, 1980; Willard et al., 1981). We will refer to these inherent complicating factors when intracellular pH measurements and values are discussed.

Intracellular pH

Methods of measurement

Direct evaluations of intracellular buffer capacities are few, and these will be discussed in the next section. However, a rearrangement of the expression summarized in equation (1) indicates that intracellular pH changes (ΔpH) in response to the uptake of an acid or base (ΔC_a or ΔC_b) reflect the intracellular buffer index, and may thus be employed in its assessment.

$$\Delta pH = \beta^{-1}\,\Delta C_a \qquad (9)$$

Of course, a knowledge of intracellular pH values is required. A review of the techniques available for pH measurements inside cells is thus pertinent. Waddell and Bates (1969) have provided a comprehensive account of this topic, although recently nuclear magnetic resonance spectrometry has allowed a new and very promising approach (Roberts et al., 1981a; Jardetsky and Roberts, 1981).

MICROELECTRODES

Microelectrodes sensitive to proton activity are available for direct insertion into plant cells. In spite of the uncertainties associated with pH measurements, the response of glass and other microelectrodes to pH is not in doubt if proper standardization and measurement techniques are employed. Waddell and Bates (1969) and Spanswick and Miller (1977) have pointed out some potential difficulties in their use. Insertion of the microelectrode is crucial since the pH-sensitive tip must be totally submerged in the intracellular fluid to circumvent erroneous readings. Plant cell walls are often prepunctured to avoid electrode damage. Of course, cell disturbance and damage must be minimized. Correct placement of the reference microelectrode is also important. E.m.f. corrections are required when it is placed in a different cellular compartment from the pH-sensitive electrode, such as in the vacuole. Not surprisingly (*cf.* E_j in equation (7)), substantial liquid junction potentials can develop when a reference electrode with a very small opening is employed. In order to minimize E_j, it is also important that the calibrating buffers contain concentrations of the major ions approximating those of the cell. And finally, protein adsorption can poison the indicator electrode; its occurrence may be verified by checking the electrode response in a standard solution before and after the intracellular readings.

DISTRIBUTION OF WEAK ACIDS AND BASES

The most widely employed method for intracellular pH evaluations is based on the distribution of weak acids or bases between the cellular compartments and the incubation medium. It is based on the premise that at equilibrium the concentration of the undissociated form is the same on both sides of the intervening membrane. A knowledge of the pK_a of the indicator compound, and measurement of the pH and volume of the incubation medium or extracellular volume, the intracellular volume, and the amounts of the indicator compound in the medium or extracellular space and inside the cell, permit an assessment of the intracellular hydrogen ion activity (Waddell and Butler, 1959; Waddell and Bates, 1969; Heldt *et al.*, 1973). The principle of this approach is best illustrated with the aid of the schematic in *Figure 21.3* depicting the chloroplast structure. The volume contained by the inner membrane is accessible to ^3H_2O but inaccessible to $[^{14}C]$-sucrose. In contrast, the intermembrane space is permeable to both radiolabelled molecules. This application of both permeable and non-permeable molecules allows an evaluation of the volumes of both the internal and external water spaces (Addanki, Cahill and Sotos, 1968; Heldt and Sauer, 1971; Heldt *et al.*, 1973). To complete the measurements, the intra/extra-compartment distribution of the weak acid 5,5-dimethyl$[2-^{14}C]$oxazolidine-2,4-dione (DMO) must be determined. Heldt *et al.* (1973) have extended this approach to evaluate both the pH of the stroma and the thylakoid spaces of intact chloroplasts. Planimetric estimation from electron micrographs of the thylakoid space and the stroma space was required, as well as the additional use of $[^{14}C]$methylamine as a weak base pH probe. Mathematical manipulation of the experimental data yielded pH estimates for both internal chloroplast spaces. In principle, this approach should permit the simultaneous evaluation of an average cytoplasmic pH value (including that of the chloroplast) and

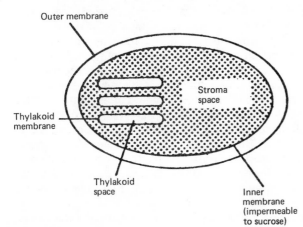

Figure 21.3. Schematic diagram of the chloroplast structure. (Courtesy of Heldt *et al.*, 1973)

the vacuolar pH in plant cells. Spanswick and Miller (1977) have determined the pH of both compartments for *Nitella translucens*, but evaluated the amount of radiolabel in the vacuole and cytoplasm after physical separation.

DMO appears to possess the attributes suitable for a weak acid pH probe. It has a pK_a value of 6.2 at 22 °C and appears relatively insensitive to ionic strength changes (Addanki, Cahill and Sotos, 1968; Bakker, Rottenberg and Caplan, 1976). The DMO molecule is not metabolized, and apparently does not adsorb onto tissues and proteins. As required, its anion is non-permeant. In contrast, there is evidence that protonated methylamine is taken up by plant cells, although considerably higher concentrations appear to be required than for NH_4^+ (Smith and Walker, 1978; Smith, Raven and Jaysuriya, 1978).

Although the assumptions involved in the DMO pH evaluation have been thoroughly tested, some uncertainty in its use remains. It is also important to emphasize that conventional pH measurements involving the pH meter are still inherent in this technique, both of the external medium and in the assignment of the pK_a value for the DMO indicator molecule.

NUCLEAR MAGNETIC RESONANCE SPECTROMETRY (NMR)

NMR is now firmly established as one of the most versatile instrumental techniques in chemistry, and more recently in biology and medicine (Dwek *et al.*, 1977; Shulman, 1979; Jardetzky and Roberts, 1981). It is routinely employed in qualitative and quantitative analysis, in the evaluation of protein structure in solution (including conformational changes due to substrate and inhibitor binding), in membrane dynamics studies, and in the dynamics of ATP production and usage in whole organs. Intracellular pH determinations have also been reported for erythrocytes (Moon and Richards, 1973), and yeast cells (Navon *et al.*, 1979). Roberts *et al.* (1980, 1981a,b) have demonstrated that NMR is also suitable as a non-invasive method for the determination of cytoplasmic and vacuolar pH in higher plant cells.

Orientation of nuclear magnetic dipoles in a magnetic field results in energy characterized by the nuclear spin quantum number, I. For nuclei such as ^1H and ^{31}P, $I = \frac{1}{2}$ and two energy states result. The exact energy separation depends on the type of nucleus and the magnitude of the applied field (generated by electromagnets and superconducting magnets), but is in the MHz radiofrequency range of the electromagnetic spectrum. Atoms in different chemical environments in a molecule absorb energy at slightly different frequencies, and thus a spectrum serves as a fingerprint for the molecule. Proton (^1H, 100% natural abundance) NMR, as well as carbon (^{13}C, 1.1% abundance) and phosphorus (^{31}P, 100% abundance) NMR have become routine with modern instrumentation. The ^{31}P spectrum of maize root tips is reproduced in *Figure 21.4*. The position of the peaks on the abscissa (the chemical shift) is measured relative to a suitable reference compound also present in the sample tube (*see* legend to *Figure 21.4*), and is expressed in parts per million. This is a dimensionless quantity and corresponds to: (chemical shift in Hz relative to reference compound) $\times 10^6$/(spectrometer frequency in Hz). ^{31}P chemical shifts depend on the chemical environment of the phosphorus and the pH. Consequently, inorganic phosphate shows up in a different position from phosphate bound in glucose-6-phosphate or ATP. Furthermore, inorganic phosphate located in intracellular compartments of different pH also show dissimilar chemical shifts. These types of observations, combined with metabolic experiments which alter the relative concentrations of phosphate containing molecules, allow identification of the various

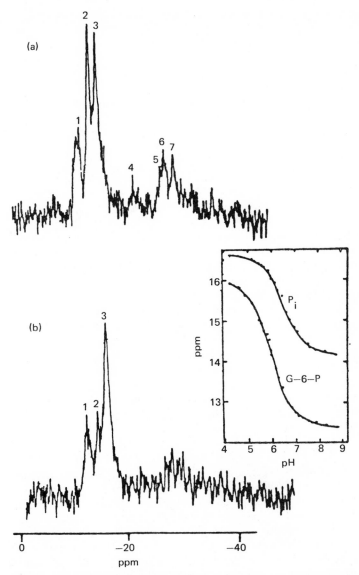

Figure 21.4. (a) 145.7 MHz ^{31}P-NMR spectrum of maize root tips ($\sim$900 tips or $\sim$1.2 g) 1.5 mm long. Maize (*Zea mays* L.) WF9 $\times$ Bear 38 were soaked for 6 h in tap water, and grown for two days on moist tissue in the dark, at 25 °C. Plants were collected in the light, root tips 2–3 mm long being excised with a razor blade, and then put on ice. Before being placed in the NMR tube, they were washed three times with ice-cold distilled water. (Washing with 1 mM EDTA did not affect the quality of the spectra recorded). Spectra were obtained at 5°C with $\sim$400 scans at 0.409 s per scan without proton irradiation and without lock using a modified Bruker HXS-360 spectrometer. Shifts are referenced to 0.05 M methylene diphosphonic acid (MDP) in pH 8.9 Tris buffer, located in a coaxial capillary tube. (b) Spectrum of root tips 2.5 mm long, grown and prepared as (a). Peak assignments: peak 1, glucose-6-phosphate; 2, cytoplasmic inorganic phosphate; 3, vacuolar inorganic phosphate; 4 and 5, ATP; 6 and 7, uridine diphosphoglucose. Rationale for peak identification is provided in the original publication. The inset illustrates the pH dependence of the chemical shift for standard solutions of inorganic phosphate and glucose-6-phosphate. It is this dependence that permits pH measurements. (Courtesy of Roberts *et al.*, 1980)

TABLE 21.1. Observed pH values

Compartment	Common range[a]	References[b]
Cytoplasmic	6.5–7.5	Davis, 1974; Komor and Tanner, 1974; Spanswick and Miller, 1977; Roberts et al., 1981a.
Vacuolar	4.0–5.5	Roa and Pickard, 1976; Roberts et al., 1981a.
Chloroplast		
Thylakoid space	5.0–6.5	Heldt et al., 1973; Falkner et al., 1976.
Stromal space	7.0–7.5	

[a] Exact value depends on plant species, method of measurement, external pH and whether observations are in the light or dark.
[b] An excellent secondary reference is Smith and Raven (1979)

peaks, also called resonances. As the chemical shift of ^{31}P resonances yield titration curves when plotted against pH (inset in *Figure 21.4*), pH determination from chemical shifts is possible. Roberts, Wade-Jardetzky and Jardetzky (1981b) conclude that a knowledge of intracellular ionic strength and free Mg^{2+} concentration in the sample are required if the determination of intracellular pH by ^{31}P-NMR is to be considered accurate within ± 0.05 to 0.1 pH unit. However, this limitation is no more serious than ionic strength and competing ion effects in microelectrode measurements, and of course can be minimized by suitable selection of the calibrating matrix. There is no doubt that ^{31}P-NMR, and for that matter also ^{1}H- and ^{13}C-NMR (Radda and Seeley, 1979), will be extremely useful for testing whether intracellular pH changes occur during physiological events and translocation phenomena. The advantages of the NMR approach include: it is non-invasive and non-destructive, it is relatively rapid, it is possible to measure not only mean pH values but also pH values for intracellular compartments, and most importantly, the dynamics of cell metabolism may be monitored (Dwek et al., 1977; Radda and Seeley, 1979).

Observed values

Typical intracellular pH values are summarized in *Table 21.1* and *Figure 21.5*. The cytoplasmic pH appears to be near 7, and is similar to that reported for the stromal space of chloroplasts. Both the thylakoid space of chloroplasts and cell vacuoles are considerably more acidic. As indicated in *Table 21.1*, the exact intracellular pH value depends on the plant species, method of measurement, external pH and whether the observations are made in the light or dark. Cytoplasmic pH is relatively insensitive to moderate external pH perturbations, as illustrated in *Figure 21.5*. The response appears to be linear for the plant species for which data is shown. This apparent resistance is interpreted as reflecting the existence of internal pH regulation. Corroborating evidence is provided by the work of Roberts et al. (1981a). They have demonstrated for maize root tips that neither the cytoplasmic pH nor vacuolar pH is altered during active H^+ extrusion. Light has been demonstrated to effect reciprocally the pH of the thylakoid space which is decreased, while that of the stroma space is increased.

The data for *Nitella translucens* plotted in *Figure 21.5* demonstrate good agreement between pH measurements by glass microelectrodes and the DMO weak-acid method. The significantly lower value, obtained with a plastic-insulated antimony microelectrode, is attributed to imperfections in the insulation in the region of the microelectrode outside the cell (Spanswick and Miller, 1977); antimony electrodes are also known to be oxygen sensitive (Roa and Pickard, 1976). ^{31}P-NMR pH determinations also appear to agree with DMO evaluations (Smith and Raven, 1979).

Buffer capacities

Titration of tissue homogenates

A standard approach to the evaluation of buffer capacities of plant tissue is to titrate homogenates with strong acid or base. The reciprocal of the slopes of curves of pH *versus* volume of titrant evaluated at specific pH values corresponds to the buffer capacity β (*cf.* equation (1)). Two examples are give in *Figures 21.6* and *21.7* for ground bark samples. The more closely the curves shift along the abscissa toward that of distilled water, the smaller the value of $|\Delta C_b|/\Delta pH$ between pH 3 and 8, and thus also of the buffer capacity. For the curves in *Figure 21.6*, the pH changes by about one-third to 1 pH unit ($\beta = 50$ to 150 μmol H$^+$ g^{-1} air dry wt pH unit^{-1}) with the addition of 1 ml of base to the sample while those in *Figure 21.7* are altered by 1–2 pH units ($\beta = 25$ to 50 μmol H$^+$ g^{-1} air dry wt pH unit^{-1}). Titration of ground lichen tissue in a similar manner showed species variation in buffer capacities roughly ranging from 25 to 150 μmol H$^+$ g^{-1} air dry wt pH unit^{-1} (Skye, 1968).

Roberts *et al.* (1981a) have proposed a modification of the homogenate titration technique which also permitted the evaluation of the cytoplasmic buffer capacity. Their results are summarized in *Table 21.2*. The homogenate buffer capacity found for maize root tips of about 20 μmol H$^+$ g^{-1} fresh wt pH unit^{-1} corresponds to that of the more weakly buffered lichen samples. Note that, as expected, β varies with pH. Roberts *et al.* (1981a) designate the reported cytoplasmic buffer capacity to be an upper limit, since

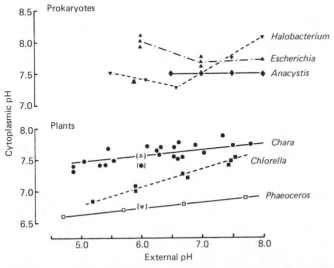

Figure 21.5. Cytoplasmic pH as a function of external pH. Symbols in parentheses are for *Nitella*; other organisms are identified on the figure. Values for *Nitella* are a comparison of measurements with antimony microelectrode ($\triangledown$), glass microelectrode ($\triangle$) and DMO (*). Values for *Phaeoceros* were measured with antimony microelectrodes; other values were obtained with DMO or methylamine (*Escherichia* at pH 8.0). (Courtesy of Smith and Raven, 1979.)

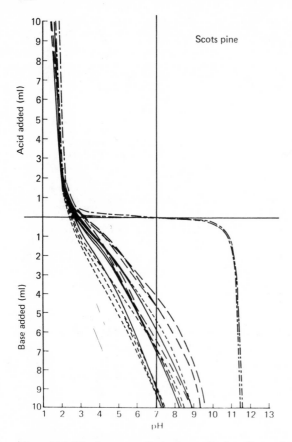

Figure 21.6. Titration curves for bark samples from Scots pine; ----, samples collected in city centres; —, samples from the transition zone; – – –, samples from rural control areas located at the city periphery (*see* text). The strong acid/base titration of water is depicted by –·–·–. Two gram samples of ground air-dry samples were titrated with 0.1 M NaOH or 0.1 M HCl in the presence of water twice the sample volume. (Courtesy of Skye, 1968)

TABLE 21.2. Buffer capacity of maize root tip tissue[a,b]

Sample	Buffer capacity designation	Apparent buffer capacities (μmol H$^+$ g^{-1} fresh wt pH unit^{-1})		
		pH 5–7	pH 6–7	pH 7–8
A. Homogenate	Whole root tips	21	20	18
B. HClO$_4$ extract (cold)	Root-cap slime (mucigel) plus small metabolites	18	14	11
C. 80% Hot ethanol extract	Small metabolites	7	6.5	6.5
D. A—(B—C)	Cytoplasmic	10	12.5	13.5

[a] From Roberts *et al.*, 1981a.
[b] Evaluated by potentiometric titration; for details *see* original paper.

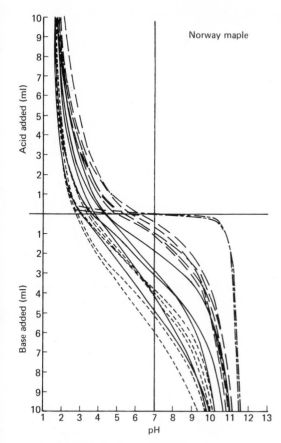

Figure 21.7. Titration curves for Norway maple. The symbolism employed and the experimental details are as in *Figure 21.6.* (Courtesy of Skye, 1968)

some part of the buffering substances must have been derived from the vacuoles. They also emphasize that the procedure outlined does not correct for the buffer capacity of cell walls, which occur in extract A of *Table 21.2*, but this is probably not a significant error for pH values exceeding 5.5.

A proper evaluation of this approach, or some modified form, is warranted as it may provide a convenient procedure for the evaluation of plant cell intracellular pH buffer capacities.

As mentioned in the section on pH measurements, separation of vacuolar and cytoplasmic fluids has been reported (Spanswick and Miller, 1977). Consequently, β values could be evaluated by titrating potentiometrically such isolated components. Another approach would be to measure ΔpH (equation (9)) by, for example, NMR in translocation studies when the influx of a labelled acid or base (ΔC_a) can also be assessed.

Calculations based on buffer component concentration

For pH values between 3 and 11, the $[OH^-]$ and $[H^+]$ terms in equations (2) to (4) contribute in a minor manner to the buffer capacity. Consequently, β may be simplified to

$$\beta = \frac{2.303 \, C_T \bar{K}_a [H^+]}{(K_a + [H^+])^2} \tag{10}$$

For a $p\bar{K}_a$ value of 6.50, the following relationships are valid:

$$\beta(pH \; 6) = \beta(pH \; 7) = 0.42 \, C_T \tag{11}$$

$$\beta(pH \; 6.5) = 0.58 \, C_T \tag{12}$$

Raven and Smith (1976) and Smith and Raven (1979) have calculated that the cytoplasmic buffer capacity in the pH range 6.0–8.0 for algae, like that for animal cells, is at most 20 mmol H^+ l^{-1} (pH unit)$^{-1}$. From equation (11), we see that this corresponds to about a concentration of 50 mM for the cytoplasmic buffering components.

There are a number of natural buffers which are effective in the cytoplasmic pH range of 6.5–7.5. It is readily demonstrated using equation (2) that the relative buffer capacities of 100%, 33% and 4% occur at pH values corresponding to the pK_a value, $pK_a \pm 1$ and $pK_a \pm 2$, respectively. Thus suitable buffers should have $pK_a = 7 \pm 1$ and therefore would include bicarbonate ($pK_a = 6.4$) and phosphate compounds ($pK_a = 6.8 - 7.0$) (Smith and Raven, 1979). Interestingly, Raven (1974) has determined that for the alga *Hydrodictyon africanum* the concentration of inorganic phosphate in both vacuole and cytoplasm was 1–2 mM, and that the cytoplasm contained 35 mM phosphate as organic and polyphosphates. Thus, based on the estimate that the total cytoplasmic buffer concentration is about 50 mM, 15 mM of buffering equivalents may perhaps be assigned to free amino acids and proteins. This appears a little high, based on the analytical data for *Chlorella* reported by Kanazawa (1964). Intracellular membranes, including that of internal structures such as thylakoids, appear to possess considerable buffering ability (Heldt *et al.*, 1973). In the cytosol, free amino acids such as histidine ($pK_a = 6$, imidazole-N), cysteine ($pK_a = 8.3$, —SH) and cystine ($pK_a = 8.0$, —NH$_2$) may be expected to contribute to the buffering capacity (Smith and Raven, 1979). In proteins, the imidazole moiety of histidine has been reported to have pK_a values of 5–7 (Markley, 1975), while the α-amino group of the N-terminal residues has pK_a values near 7 (e.g. Brown and Bradbury, 1975). As with histidine, micro-environments in proteins (Shrager *et al.*, 1972; Rabenstein, 1973) are also responsible for the wide range of values ($pK_a = 2$–6) observed for the side chain carboxylates of aspartic and glutamic acid residues (Dobson and Williams, 1977). Potentially, all these protonation centres should add to the total intracellular buffer index.

The organic acids that accumulate in vacuoles account for their lower pH values (Matile, 1978). Analogously, Heldt *et al.* (1973) have explained pH differences between the stroma and the thylakoid space in terms of a thylakoid buffer with pK_a of 5.5 and a concentration four times greater than the stroma buffer (assigned pK_a value of 6.8). It is difficult to compare the buffer capacities expressed in μmol H^+ g^{-1} (pH unit)$^{-1}$ of the previous section with those discussed here in mM H^+ l^{-1} (pH unit)$^{-1}$. In principle, it should be possible to establish experimentally suitable conversion factors.

The studies by Roberts *et al.* (1980; 1981a) have shown for maize root-tip cells incubated with external aqueous ammonia concentrations of 6–35 mM, that both the vacuolar and cytoplasmic buffer capacities could be exceeded, even during short-lived incubations (5 min). Permeant weak acids might also be expected to accomplish this providing the protonated species bears no charge and is present in adequate supply (*see* earlier discussions of the weak acid DMO). It may be significant that reductions in photosynthetic fixation rates have been observed in lichens on incubation at pH 4 in acetate buffer (Boileau, Nieboer and Richardson, unpublished results for *Cladonia rangiferina*), and at pH values of 2–3 in phthalate buffer (Türk and Wirth, 1975; for *Hypogymnia physodes* and *Xanthoria parietina*). At the experimental pH values, significant amounts of the uncharged acid occurred in both buffers.

Modification of cell pH by gaseous pollutants

Speciation and uptake

As pointed out by Mansfield and Freer-Smith (1981), a gaseous pollutant entering the cells of a leaf (or thallus) must diffuse through the extracellular water contained in the cell walls. This dissolution process must therefore involve acidification of this extracellular fluid by acidic substances. Some of the more common proton generating reactions are summarized in *Table 21.3*. From the weak acid dissociation equilibrium

TABLE 21.3. Acid/base properties of gaseous air pollutants

Dissolution or dissociation equilibrium	pK_a
$HCl \rightleftharpoons Cl^- + H^+$	—
$SO_3 + H_2O \rightleftharpoons HSO_4^- + H^+$	—
$2NO_2 + H_2O \rightleftharpoons HNO_2 + H^+ + NO_3^-$	—
$SO_2 \cdot H_2O \rightleftharpoons HSO_3^- + H^+$	1.91
$HSO_4^- \rightleftharpoons SO_4^{2-} + H^+$	1.99
$HNO_2 \rightleftharpoons NO_2^- + H^+$	3.15
$HF \rightleftharpoons F^- + H^+$	3.17
$H_2S \rightleftharpoons HS^- + H^+$	7.02
$HSO_3^- \rightleftharpoons SO_3^{2-} + H^+$	7.18
$NH_3 + H_2O \rightleftharpoons NH_4^+ + OH^-$	9.24 (pK_b)

pK_a values are for zero ionic strength and 25 °C (from Smith and Martell, 1976).

expression, it is readily deduced that 50%, 9% and 1% of the undissociated acid exists at pH values equal to the pK_a, $pK_a + 1$ and $pK_a + 2$, respectively. At pH 4, the concentration of $SO_2 \cdot H_2O$ and HSO_4^- is about 1%; while this occurs at pH 5 for HNO_2 and HF. At pH 5, H_2S and HSO_3^- only dissociate to the degree of 1%. Consequently, under field conditions, significant permeant neutral weak acid concentrations may be expected. Such uncharged acids enter the cell and penetrate organelles easily (Spedding *et al.*, 1980a,b; Heath, 1980; Richardson and Nieboer, 1983). Heath (1980) explains that intracellular accumulation occurs because of high cytoplasmic pH and that this induces water flow into the cell. Experiments with DMO have shown this preferential accumulation. For *Nitella translucens*, at an external pH of 6 and an internal pH of 7.5, the internal accumulation factor for DMO was 9 (Spanswick and Miller, 1977). As already discussed, the uptake of significant amounts of weak acid or base are expected and known to lower or increase intracellular pH values (*see Table 21.4*).

Strong acids will dissociate completely in the extracellular moisture. The decreased pH may be expected to be reflected in lower intracellular pH values as demonstrated by the data in *Figure 21.5*. Although membranes tend to be impermeable to anions, there are pumps for SO_4^{2-}, Cl^- and HPO_4^{2-} which are also known to translocate other species derived from pollutants. Examples are arsenate by the phosphate translocator (Sanders and Windom, 1980; Richardson *et al.*, 1984), and sulphite by the sulphate

TABLE 21.4. Known effects of exposure on buffer capacity and pH

Plant	Pollutant	Component examined	β (pH < 7)	pH	References
Spruce needles	$SO_2(g)$	Homogenates	Reduced	—	Hällgren, 1978
Alfalfa	$SO_2(g)$	Homogenates	Reduced	—	Thomas, Hendricks and Hill, 1944
Lichens[a]	$SO_2(g)$	Thallus moisture	—	− Parmelia + Xanthoria	Türk, Wirth and Lange, 1974
Maize root tips	$NH_3(aq)$	Cytoplasm and vacuole	—	Increased	Roberts et al., 1981a

[a] SO_2 exposure decreased the thallus pH of all species tested (10), except specimens of Xanthoria parietina.

transport system (Hampp and Ziegler, 1977). Such anion uptake processes most likely involve the concomitant uptake of protons (Beever and Burns, 1980), and presumably would thus be pH dependent.

Effect of exposure on pH and β values

It may be concluded from the discussion in the previous section, that both weak acid and strong acid pollutants tend to reduce intracellular pH values. Additional evidence for acidification is summarized in *Table 21.4* in response to $SO_2(g)$ exposure, and the increase in intracellular alkalinity due to the permeant uncharged ammonia molecules is also noted. As indicated in the footnote to *Table 21.4*, the pH increase observed for *Xanthoria parietina* was anomalous. However, this species often grows on basic substrata or nutrient-enriched bark (Hawksworth and Rose, 1976). It is conceivable that the absorbed SO_2 interacted with basic particulates trapped in the thallus.

Implications of changes in pH and β values

Evidence is growing in support of the critical role of pH gradients across membranes and proton-related membrane energization in driving ATP synthesis, in the translocation of cations, anions and neutral metabolites, and even in regulating enzyme activity (Mitchell, 1977; Williams, 1978a,b; Williams, 1979; Smith and Raven, 1979; Beever and Burns, 1980; Wellburn, this volume). Evidence for the existence of pH gradients in plants has already been presented. It would seem essential therefore for a plant to regulate these pH gradients, and to be able to generate proton electrochemical potential (free energy) differences for membrane energization. The mentioned relative insensitivity of intracellular pH to changes in external pH, and the invariance of cytoplasmic and vacuolar pH during H^+ extrusion by maize root tips provide support for the existence of such internal control. The exact mechanism for this pH-stat is not resolved (Smith and Raven, 1979; Roberts et al., 1981a). Significant alterations in cytoplasmic pH and β values are likely to be inhibitory and detrimental to plants.

Since protonmotive forces assume such importance in plant physiology and metabolism, the reduced β values reported in *Table 21.4* would most likely be damaging. It is thus of some significance that Skye (1968) was able to conclude for epiphytic lichens growing in the vicinity of Stockholm, Sweden: 'Of the investigated species, it is thus with a few exceptions the species having the lowest buffering capacity for acid substances that disappear first when we proceed from the normal area in towards the lichen-free areas' (*see Figure 21.6* and *21.7*). He collected samples in three zones: a control area with normal, or almost normal, lichen vegetation at the periphery of the area studied; a 'lichen desert' in the vicinity of the urban centre; and an intervening transitional zone. The buffering capacities were determined for lichen samples collected in the control zone. It is also of interest that the β values of tree bark samples increased on moving toward the urban centre (*see Figures 21.6* and *21.7*). Johnsen and Søchting (1973) have reported a similar finding. This observation was interpreted as indicating an accumulation of weak acids in the bark. Lower bark pH values were indeed found by Skye (1968) for trees in the transitional zone compared to the control area. Similarly, Johnsen and Søchting (1973) found a linear inverse relationship between bark pH and average winter SO_2 concentrations.

Concluding remarks

Although fragmentary, the available data indicates that gaseous air pollutants with acidic or basic solution properties can alter both the intracellular pH and buffering capacity. Evidence has been reviewed which suggests that such changes are likely to be detrimental to a plant. Data for lichens indicate that plant distribution may depend on the inherent buffering capacity of a species.

   It is clear that the technology for intracellular pH and buffer capacity measurements is now reliable and generally available. The pH of cellular compartments can be monitored concurrently with an examination of cell dynamics during the uptake of toxic substances. The sophistication of the instrumentation and techniques calls for interdisciplinary cooperation, but this adds to the potential of making genuine gains in our understanding of intracellular uptake and toxicity mechanisms. Direct measurements of intracellular pH values are badly needed in exposure studies to confirm current hypotheses that intracellular pH changes contribute to plant toxicity. Such studies will have the added benefit of enhancing our knowledge of fundamental processes such as internal pH regulation, and would permit the testing of specific hypotheses concerning metabolic control by cytoplasmic pH and cellular and intracellular proton gradients and fluxes.

Addendum

During the symposium at Oxford we had most fruitful discussions with Loughman and Williams. The Oxford group has recently reviewed the historical aspects of the application of NMR spectrometry to pH measurements, have examined potato and maize tissues in detail including the optimization of experimental and instrumental conditions, and have recorded spectra of whole deseeded plants (Kime *et al.*, 1982a). Intracellular changes resulting from the uptake of inorganic phosphate, D-mannose, Mn^{2+} and 2,4-dinitrophenol by potato and maize root tissues are described in a second paper (Kime, Loughman and Ratcliffe, 1982b). It is concluded that NMR provides a powerful, direct method for following intracellular changes in plant tissues. Additional indications of the full potential of NMR in pollution research are provided by model studies with lipid membranes of ion transport, of intravesicular pH control and intravesicular precipitation of inorganic hydroxides and oxides (Mann and Williams, 1982; Mann *et al.*, 1982). Finally, a book on the measurement and regulation of intracellular pH has just been published (Nuccitelli and Deamer, 1982); it treats in considerable detailed some of the topics examined in this review.

References

ADDANKI, S., CAHILL, F. D. and SOTOS, J. F. Determination of intramitochondrial pH and intramitochondrial–extramitochondrial pH gradient of isolated heart mitochondria by the use of 5,5-dimethyl-2,4-oxazalidinedione, *Journal of Biological Chemistry* **243**, 2337–2348 (1968)

BAKKER, E. P., ROTTENBERG, H. and CAPLAN, S. R. An estimation of the light-induced electrochemical potential difference of protons across the membrane of *Halobacterium halobium*, *Biochimica et Biophysica Acta* **440**, 557–572 (1976)

BATES, R. G. *Determination of pH. Theory and Practice*, 2nd edn, pp. 31–111, Wiley-Interscience, New York (1973)

BEEVER, R. E. and BURNS, D. J. W. Phosphorus uptake, storage and utilization by fungi. In *Advances in Botanical Research*, (Ed. H. W. WOOLHOUSE), pp. 127–219, Academic Press, London (1980)

BROWN, L. R. and BRADBURY, J. H. Proton-magnetic-resonance studies of the lysine residues of ribonuclease A, *European Journal of Biochemistry* **54**, 219–227 (1975)

BUTLER, J. N. *Ionic Equilibrium. A Mathematical Approach*, pp. 149–152, Addison-Wesley, Reading, Massachusetts (1964)

DAVIS, R. F. Photoinduced changes in electrical potentials and H^+ activities of the chloroplast cytoplasm and vacuole of *Phaeoceros laevis*. In *Membrane Transport in Plants*, (Ed. U. ZIMMERMANN and J. DAINTY), pp. 197–201, Springer-Verlag, New York (1974)

DOBSON, C. M. and WILLIAMS, R. J. P. Nuclear magnetic resonance studies of the interaction of lanthanide cations with lysozyme. In *Metal-Ligand Interactions in Organic Chemistry and Biochemistry*, part 1, (Eds. B. PULLMAN and N. GOLDBLUM), pp. 255–282 (1977)

DWEK, R. A., CAMPBELL, I. D., RICHARDS, R. E. and WILLIAMS, R. J. P. *NMR in Biology*, Academic Press, London (1977)

FALKNER, G., HORNER, F., WERDAN, K. and HELDT, H. W. pH changes in the cytoplasm of the blue-green alga *Anacystis nidulans* caused by light-dependent proton flux into the thylakoid space, *Plant Physiology* **58**, 717–718 (1976)

HALLGREN, J. E. Physiological and biochemical effects of sulphur dioxide on plants. In *Sulfur in the Environment*, (Eds. J. O. NRIAGU), *Part 2. Ecological Impacts*, pp. 164–209, J. Wiley and Sons, New York (1978)

HAMPP, R. and ZIEGLER, I. Sulfate and sulfite translocation via the phosphate translocator of the inner envelope membrane of chloroplasts, *Planta* **137**, 309–312 (1977)

HAWKSWORTH, D. L. and ROSE, F. *Lichens as Pollution Monitors*, Edward Arnold, London (1976)

HEATH, R. L. Initial events in injury to plants by air pollutants, *Annual Review of Plant Physiology* **31**, 395–431 (1980)

HELDT, H. W. and SAUER, F. The inner membrane of the chloroplast envelope as the site of specific metabolite transport, *Biochimica et Biophysica Acta* **234**, 83–91 (1971)

HELDT, H. W., WERDAN, K., MILOVANCEV, M. and GELLER, G. Alkalization of the chloroplast stroma caused by light-dependent proton flux into the thylakoid space, *Biochimica et Biophysica Acta*, **314**, 224–241 (1973)

JARDETZKY, O. and ROBERTS, G. C. K. *NMR in Molecular Biology*, Academic Press, New York (1981)

JOHNSEN, I. and SØCHTING, U. Influence of air pollution on the epiphytic lichen vegetation and bark properties of deciduous trees in the Copenhagen area, *Oikos* **24**, 344–351 (1973)

KANAZAWA, T. Changes of amino acid composition of *Chlorella* cells during their life cycle, *Plant and Cell Physiology* **5**, 333–353 (1964)

KIME, M. J., RATCLIFFE, R. G., WILLIAMS, R. J. P. and LOUGHMAN, B. C. The application of ^{31}P nuclear magnetic resonance to higher plant tissue. 1. Detection of spectra, *Journal of Experimental Botany* **33**, 656–669 (1982a)

KIME, M. J., RATCLIFFE, R. G. and LOUGHMAN, B. C. The application of ^{31}P nuclear magnetic resonance to higher plant tissue. 2. Detection of intracellular changes, *Journal of Experimental Botany* **33**, 670–681 (1982b)

KOMOR, E. and TANNER, W. The hexose-proton cotransport system of *Chlorella*, *Journal of General Physiology* **43**, 568–581 (1974)

MANN, S. and WILLIAMS, R. J. P. Precipitation within unilamellar vesicles. I. Studies of silver (I) oxide formation, *Journal of the Chemical Society, Dalton Transactions*, 311–316 (1982)

MANN, S., KIME, M. J., RATCLIFFE, R. G. and WILLIAMS, R. J. P. Precipitation within unilamellar vesicles. II. Membrane control of ion transport, *Journal of the Chemical Society, Dalton Transactions*, 771–774 (1982)

MANSFIELD, T. A. and FREER-SMITH, P. H. Effects of urban air pollution on plant growth, *Biological Reviews* **56**, 343–368 (1981)

MARKLEY, J. L. Observation of histidine residues in proteins by means of nuclear magnetic resonance spectroscopy, *Accounts of Chemical Research* **8**, 70–80 (1975)

MATILE, P. Biochemistry and function of vacuoles, *Annual Review of Plant Physiology* **29**, 193–213 (1978)

MITCHELL, P. A commentary on alternative hypotheses of protonic coupling in the membrane systems catalysing oxidative and photosynthetic phosphorylation, *Federation of European Biochemical Societies Letters* **78**, 1–20 (1977)

MOON, R. B. and RICHARDS, J. H. Determination of intracellular pH by ^{31}P magnetic resonance, *Journal of Biological Chemistry* **248**, 7276–7278 (1973)

NAVON, G., SHULMAN, R. G., YAMANE, T., ECCLESHALL, T. R., LAM, K-B., BARONOFSKY, J. J. and MARMUR, J. Phosphorous-31 nuclear magnetic resonance studies of wild-type and glycolytic pathway mutants of *Saccharomyces cerevisiae*, *Biochemistry* **18**, 4487–4499 (1979)

NICHOLLS, D. G. *Bioenergetics, An Introduction to the Chemiosmotic Theory*, Academic Press, London (1982)

NUCCITELLI, R. and DEAMER, D. W. *Intracellular pH: Its Measurement, Regulation and Utilization in Cellular Functions*, Kroc Foundation Series, Vol. 15, Alan R. Liss, Inc., New York (1982)

RABENSTEIN, D. L. Nuclear magnetic resonance studies of the acid-base chemistry of amino acids and peptides. 1. Microscopic ionization constants of glutathione and methyl-mercury-complexed glutathione, *Journal of the American Chemical Society* **95**, 2797–2803 (1973)

RADDA, G. K. and SEELEY, P. J. Recent studies on cellular metabolism by nuclear magnetic resonance, *Annual Review of Physiology* **41**, 749–769 (1979)

RAVEN, J. A. Phosphate transport in *Hydrodictyon africanum*, *New Phytologist* **73**, 421–432 (1974)

RAVEN, J. A. and SMITH, F. A. Cytoplasmic pH regulation and electrogenic H^+ extrusion, *Current Advances in Plant Science* **8**, 649–661 (1976)

RICHARDSON, D. H. S. and NIEBOER, E. Ecophysiological responses of lichens to sulphur dioxide, *Journal of the Hattori Botanical Laboratory* **54**, 331–351 (1983)

RICHARDSON, D. H. S., NIEBOER, E., LAVOIE, P. and PADOVAN, D. Anion accumulation by lichens. 1. The characteristics and kinetics of arsenate uptake by *Umbilicaria muhlenbergii*, *The New Phytologist* (in press) (1984)

ROA, R. L. and PICKARD, W. F. The vacuolar pH of *Chara braunii*, *Journal of Experimental Botany* **27**, 853–858 (1976)

ROBERTS, J. K. M., RAY, P. M., WADE-JARDETZKY, N. and JARDETZKY, O. Estimation of cytoplasmic and vacuolar pH in higher plant cells by ^{31}P NMR, *Nature* **283**, 870–872 (1980)

ROBERTS, J. K. M., RAY, P. M., WADE-JARDETZKY, N. and JARDETZKY, O. Extent of intracellular pH changes during H^+ extrusion by maize root-tip cells, *Planta* **152**, 74–78 (1981a)

ROBERTS, J. K. M., WADE-JARDETZKY, N., JARDETZKY, O. Intracellular pH measurements by ^{31}P nuclear magnetic resonance. Influence of factors other than pH on ^{31}P chemical shifts, *Biochemistry* **20**, 5389–5394 (1981b)

SANDERS, J. G. and WINDOM, H. L. The uptake and reduction of arsenic species by marine algae, *Estuarine and Coastal Marine Science* **10**, 555–567 (1980)

SHRAGER, R. I., COHEN, J. S., HELLER, S. R., SACHS, D. H. and SCHECHTER, A. N. Mathematical models for interacting groups in nuclear magnetic resonance titration curves, *Biochemistry* **11**, 541–547 (1972)

SHULMAN, R. G. (Ed.). *Biological Applications of Magnetic Resonance*, Academic Press, New York (1979)

SKOOG, D. A. and WEST, D. M. *Principles of Instrumental Analaysis*, 2nd edn, pp. 539–556, Saunders, Philadelphia (1980)

SKYE, E. Lichens and air pollution, A study of cryptogamic epiphytes and environment in the Stockholm region, *Acta Phytogeographica Suecica* **52**, 8–123 (1968)

SMITH, R. M. and MARTELL, A. E. *Critical Stability Constants*, Volume 4: *Inorganic Complexes*, Plenum Press, New York (1976)

SMITH, F. A. and RAVEN, J. A. Intracellular pH and its regulation, *Annual Review of Plant Physiology* **30**, 289–311 (1979)

SMITH, F. A., RAVEN, J. A. and JAYSURIYA, H. D. Uptake of methylammonium ions by *Hydrodictyon africanum*, *Journal of Experimental Botany* **29**, 121–133 (1978)

SMITH, F. A. and WALKER, N. A. Entry of methylammonium and ammonium ions in *Chara* internodal cells, *Journal of Experimental Botany* **29**, 107–120 (1978)

SPANSWICK, R. M. and MILLER, A. G. Measurement of the cytoplasmic pH in Nitella translucens. Comparison of values obtained in microelectrode and weak acid methods, *Plant Physiology* **59**, 664–666 (1977)

SPEDDING, D. J., ZEIGLER, I., HAMPP, R. and ZEIGLER, H. Effect of pH on the uptake of ^{35}S-sulfur from sulfate, sulfite and sulfide by isolated spinach chloroplasts, *Zeitschrift für Pflanzenphysiologie* **96**, 351–364 (1980a)

SPEDDING, D. J., ZIEGLER, I., HAMPP, R. and ZEIGLER, H. Effect of pH on the uptake of ^{35}S-sulfur from sulfate, sulfite and sulfide by *Chlorella vulgaris, Zeitschrift für Pflanzenphysiologie* **97**, 205–214 (1980b)

THOMAS, M. D., HENDRICKS, R. H. and HILL, G. R. Some chemical reactions of sulphur dioxide after absorption by alfalfa and sugar beets, *Plant Physiology* **19**, 212–226 (1944)

TÜRK, R. and WIRTH, V. The pH dependence of SO_2 damage to lichens, *Oecologia (Berlin)* **19**, 285–291 (1975)

TÜRK, R., WIRTH, V. and LANGE, O. L. CO_2-Gaswechsel-Untersuchungen zur SO_2-Resistenz von Flechten, *Oecologia (Berlin)* **15**, 33–64 (1974)

VAN SLYKE, D. D. On the measurement of buffer values and on the relationship of buffer value to the dissociation constant of the buffer and the concentration and reaction of the buffer solution, *Journal of Biological Chemistry* **52**, 525–570 (1922)

WADDELL, W. J. and BATES, R. G. Intracellular pH, *Physiological Reviews* **49**, 285–329 (1969)

WADDELL, W. J. and BUTLER, T. C. Calculation of intracellular pH from the distribution of 5,5-Dimethyl-2,4-oxazolidinedione (DMO). Application to skeletal muscle of the dog, *Journal of Clinical Investigation* **38**, 720–729 (1959)

WILLARD, H. H., MERRITT, L. L., DEAN, J. A. and SETTLE, F. A. *Instrumental Methods of Analysis*, 6th edn, pp. 652–663, Van Nostrand, New York (1981)

WILLIAMS, R. J. P. The history and the hypotheses concerning ATP-formation by energized protons. *Federation of European Biochemical Societies Letters* **85**, 9–19 (1978a)

WILLIAMS, R. J. P. Energy states of proteins, enzymes and membranes, *Proceedings of the Royal Society of London, Series B* **200**, 353–389 (1978b)

WILLIAMS, R. J. P. Some unrealistic assumptions in the theory of chemi-osmosis and their consequences, *Federation of European Biochemical Societies Letters* **102**, 126–132 (1979)

Part VI

Relationship of biochemical effects to environmental problems

Chapter 22

Biochemical diagnostic tests for the effect of air pollution on plants

N. M. Darrall

CENTRAL ELECTRICITY RESEARCH LABORATORIES, LEATHERHEAD, UK

H. J. Jäger

INSTITUT FÜR PFLANZENÖKOLOGIE DER JUSTUS LIEBIG-UNIVERSITÄT, GIESSEN, FRG

Introduction

The protection of vegetation against effects of air pollution is dependent on the development of reliable methods of assessing injury, whether it be the appearance of visible symptoms, effects on growth and ontogony or changes in physiological and biochemical processes. During that period of air pollution research when it was assumed that yield losses did not occur in the absence of visible injury, much effort was concentrated on the description and quantification of visible symptoms as reductions in growth were reported to be proportional to the area of leaf tissue destroyed (Hill and Thomas, 1933). With the improvement of techniques, effects on yield have also been found in the absence of visible injury, but such effects can be quantified in experiments of several weeks duration.

There have been considerable interest amongst some research groups in the development of biochemical diagnostic tests to permit rapid screening of crops and natural vegetation for air pollution damage without time-consuming measurements of growth parameters (Keller, 1982; Jäger, 1982). Some assays have been designed to give an early indication of effects (usually visible injury), for example, by measurements of chlorophyll content; other assays have been used either to provide a rapid quantitative estimate of effects by measuring the production of ethylene, or to map the extent of areas damaged by pollutants by measuring changes in peroxidase activity in leaf tissues. Arndt (1970) has suggested the following guidelines for the selection of biochemical indicators:

(1) rapid and unambiguous response to low pollutant concentrations,
(2) very specific reaction with the individual pollutant components,
(3) possibility of detecting metabolic changes by simple methods,
(4) high reproducibility of results.

The aim of this chapter is to discuss the parameters which have been used as diagnostic tests for air pollution injury and to assess their usefulness in the field.

Photosynthesis

Several authors have suggested that measurements of changes in the rate of photosynthesis would provide a reliable and sensitive method for determining pollutant effects in plants. Thomas (1956), Keller (1958) and Taylor, Cardiff and Mersereau (1965) reported that this parameter was suitable for quantifying the effects of SO_2, and Keller (1973) also used the method for assessing the effects of HF. There are now many reports of inhibition of photosynthesis by SO_2, O_3 and NO_x at concentrations causing growth reductions in the absence of visible injury. The response time is rapid and inhibition can be detected within 30 min of the commencement of an injurious fumigation, for example above 200–300 ppb SO_2, in barley (*Hordeum vulgare*) and alfalfa (*Medicago sativa*) (Bennet and Hill, 1973) and 35 ppb in broad bean (*Vicia faba* cv Dylan) (Black and Unsworth, 1979). Reductions in net photosynthesis were found to be proportional to the pollutant concentration; residual effects persisted for only 2–3 h after the cessation of fumigation where the level of inhibition did not exceed 20–30%. Because of the rapid response time of this parameter, diagnostic assays could only be performed during fumigation and would not give an integrated value for the effect of the episode. Rates of photosynthesis will, however, change in response to a wide variety of other parameters, such as light intensity, temperature, stomatal aperture, and the physiological age of the tissue, which will complicate the interpretation of the data in field pollution studies.

Chlorophyll fluorescence

In 1972, Arndt suggested the use of chlorophyll fluorescence characteristics as a diagnostic test for air pollutants. The emission of fast fluorescent transients on illumination of dark adapted tissue, the Kautsky effect (Kautsky and Hirsch, 1931), was found to be modified by heavy metals and sulphur dioxide and indicates disturbances within the chloroplasts to photosystem II (Arndt, 1974). Synchronized cultures of *Chlorella pyrenoidosa* were used, and the direction of the response to SO_2 (as sulphite in solution) was found to be concentration dependent. At concentrations of 0.1 mM sulphite or above a decrease in fluorescent transients was found, and at lower concentrations a slight increase occurred. Changes in emissions of slow fluorescent transients were also reported, and these were thought to be related to temporal alterations in the thylakoid membrane ultrastructure and pigments associated with ionic transport and photophosphorylation (Papageorgio, 1975). In general, an increase in fluorescence indicates a poorer use of available light energy and a reduction in fluorescence and increased utilization. Changes in both fast and slow transients in response to gaseous pollutants were first reported by Chimiklis and Heath (1975). An initial increase and subsequent decay of transients was found on fumigation of *Chlorella* cultures with 26 μM O_3. Schreiber *et al.* (1978) modified the technique for use with intact leaves, and measured both transients and stationary fluorescence yield. The reduction in transients was correlated with visual assessments of injury which developed subsequently. Stationary fluorescence yield was also recorded during the fumigation. Whilst not providing the same amount of information as an examination of fluorescent transients from dark-adapted plants, the method is much simpler, and continuous recordings can be made during a fumigation. Both alterations in the rate of photosynthetic processes and the destruction of chlorophyll will account for changes in stationary fluorescence yield. The advantages of measurements of these various fluorescence characteristics are that a large number of determinations can be made with high precision in intact plants. Despite these features, the method is not readily applicable as a diagnostic tool in the field. Firstly, the changes in fluorescent characteristics are not pollutant specific, and the direction of response can vary between different concentrations of the same pollutant (Arndt, 1974). Secondly, other factors can alter fluorescence yield, for example light intensity, light quality, cell water content and the developmental stage of the tissue (Goedheer, 1972).

The potential usefulness of measurements of fluorescence transients for screening for SO_2 tolerance has been investigated by G. B. Wilson (personal communication). Ryegrass populations known to be of differing tolerances were screened and significant differences found. However, no relationship was apparent between tolerance and fluorescence characteristics.

Leaf pigments

Numerous studies have used the measurement of leaf pigment content as an assay for air pollutant effects (*see Table 22.1*). Reports of decreased chlorophyll concentrations in plants fumigated with SO_2, O_3 and HF and increased concentrations in plants fumigated with NO_2 are listed by Horsman and Wellburn (1976). In a number of surveys around point sources, the degree of visible injury was related to the loss of chlorophyll attributed to SO_2 fumigations (Dörries, 1932; Katz and Shore, 1955), and Jamrich (1968) related loss of chlorophyll to the fluorine content of leaves around a point source. Knudson, Tibbits and Edwards (1977) used measurements of chlorophyll content to quantify visible injury from laboratory fumigation experiments with O_3, as it was found to be less subjective than a visual assessment of injury. Whilst the assay may

TABLE 22.1. Biochemical diagnostic tests using leaf pigments

Pigment	Plant species	Pollutant(s)	Reference
Chlorophyll a:b ratio	Common silver fir	0.57 ppm SO_2	Müller (1957)
Chlorophyll a + b	Conifers	Preponderantly SO_2 Samples from field sites in Detroit River Industrial area and Sudbury Ni–Cu smelting district	Katz and Shore (1955)
Chlorophyll a and b	Hybrid black poplar Common elder Lilac	Preponderantly F from aluminium smelter Samples taken from field site	Gronebaum-Turk and Mathé (1975)
Chlorophyll	Pinto bean	O_3	Knudsen, Tibbits and Edwards (1977)
Carotenes	Barley Rye Oat	7 ppb HF or 190 ppb SO_2 or 307 ppb HCl	Arndt (1971)
Phenolics	Norway Spruce (young trees in pots)	Predominantly F at field sites various distances from an industrial source	Yee-Meiler (1974)
Phenolics	Spruce sp.	SO_2	Grill, Esterbauer and Beck (1975)
Phenolics	Bean	160 ppb $O_3 \times 3$ h	Howell and Kremer (1973)

provide an accurate measure of injury in fumigations performed in controlled environments, chlorophyll formation is very sensitive to almost any factor which disturbs metabolic processes, for example light intensity, temperature, mineral nutrient status, water stress and waterlogging as well as many pathogens. The presence of NO_2 with SO_2 in a fumigation episode may reduce the sensitivity of the assya, as NO_2 can cause greening of leaf tissue (Taylor and Eaton, 1966; Horsman and Wellburn, 1976). The amount of chlorophyll degradation as a result of SO_2 fumigations also varies with the time of year. Lauenroth and Dodd (1981) have reported maximum effect on total chlorophyll levels in western wheatgrass (*Agropyron smithii*) during August. A more specific assay is therefore required. Müller (1957) found a preferential degradation of chlorophyll a over chlorophyll b in seedlings of common silver fir (*Abies alba*) exposed to SO_2, and he suggested that the change in ratio of these two forms could be used specifically to determine the effects of acid gases. A similar preferential degradation of chlorphyll a in plants exposed to SO_2 has also been reported in alfalfa (*Medicago sativa*) (Katz and Shore, 1955), in Norway spruce (*Picea abies*) (Börtitz, 1964), in sessile oak (*Quercus petraea*) (Ricks and Williams, 1975) and in western wheatgrass (*Agropyron smithii*) (Lauenroth and Dodd, 1981). There are, however, other reports of similar rates of degradation of chlorophyll a and b in plants exposed to SO_2 (Dässler, 1972; Ricks and Williams, 1975; Bell and Mudd, 1976; Rabe and Kreeb, 1980) and one report of an increase in chlorophyll a concentration in the alga *Euglena gracilis* on fumigation with SO_2 (5 ppm) (de Koning and Jegier, 1970). These results indicate that the preferential degradation of chlorophyll a to b in plants exposed to SO_2 is either restricted to certain species or particular experimental conditions and is not suitable as a diagnostic test. There are also a few reports of preferential breakdown of chlorophyll a in the presence of other pollutants, for example ozone (Nouchi and Odaira, 1973; Knudson, Tibbits and Edwards, 1977) and fluorine (Gronebaum-Turck and Mathé, 1975).

Changes in the content of a number of pigments other than chlorophyll have also been used as indicators of air pollutant effects by assaying extractable concentrations. Carotenes are the second most abundant pigments in leaf tissues after chlorophylls. Arndt (1971) suggested the use of β-carotene concentrations as a reliable early indication of injury from SO_2, HF and HCl fumigations. Leaf tissue from oat (*Avena sativa*), barley (*Hordeum sativum*) and rye (*Secale cereale*) fumigated with 7 ppb HF, 190 ppb SO_2 or 307 ppb HCl for one month showed reduced concentrations of β-carotene whilst there were no consistent trends in total chlorophyll content or the ratio of chlorophyll a to b. Changes in the concentration of phenolic compounds have also been used as indicators of air pollutant effects. Yee-Meiler (1974) found an increasing concentration of phenolics with decreasing distance from a source of fluoride emissions, simultaneous measurements of fluoride concentrations in the leaves were not, however, made. Grill, Esterbauer and Beck (1975) reported increased levels of phenolics in spruce needles exposed to SO_2 and Howell and Kremer (1973) found a similar response in bean (*Phaseolus vulgaris*) leaves from plants fumigated with 160 ppb O_3 for 3 h. The development of phenolic pigments is therefore not a specific diagnostic test and it is not known whether the concentration of extractable phenolics can be correlated with the degree of injury from exposure to pollutants.

Leaf reflectance characteristics

A more recently developed method for detecting injury from air pollutants is the measurement of changes in reflectance and transmission characteristics of leaves. These can largely be attributed to changes in the concentrations of leaf pigments, 65–75% of which are chlorophylls. The method was however found to be more sensitive than a comparison of extractable chlorophyll content and was found to be applicable to assessments of chronic or mildly acute injury from ozone fumigations where the diffuse nature of the injury precludes a visual assessment (Runeckles and Resh, 1975). Concurrent rust infections were found to interfere minimally with the assay, but as leaf water content was found to alter the reflectance characteristics appreciably, the authors recommended that care should be taken to ensure maximum turgor in the leaf material used. Nowak and Lautman (1974) also advocated the use of spectral reflectivity measurements for diagnosing sub-visual changes following fumigations with SO_2 or O_3 from laboratory experiments using bean and tobacco leaves. Changes were recorded between the wavelengths 475 and 750 nm although certain cyclic patterns complicated the interpretation of results in plants fumigated with SO_2. It was suggested that *in situ* continuous measurements of narrower wavebands might enable specific identification of pollutants in the field but this remains to be demonstrated. Omasa *et al.* (1980) have, however, distinguished between visible injury from SO_2 and O_3 fumigations in the laboratory. Reflectance characteristics were recorded photographically under constant lighting conditions using an intereference filter in the 661–681 nm waveband. From an analysis of the spectral images, injury from the two pollutants could be identified and quantified. Using the difference in reflectance characteristics of leaves above a white and a black background, Schubert, MacLeod and Howell (1972) were able to distinguish between two forms of visible injury in ozone-stressed soybeans, bleaching and browning. It must be noted however that the above results were obtained in the laboratory under constant environmental conditions in which all other stress factors were eliminated. The possibility of using reflectance measurements as a diagnostic test in screening for tolerance to pollutants has been investigated in ryegrass (Wilson, personal communication). Significant differences were found between phenotypes when fumigated with SO_2 but no relationship between chronic tolerance and leaf reflectance was found.

Extension of the technique to field studies, including remote sensing by aerial photography, has indicated the need for a carefully integrated programme of analysis (van Genderen, 1974). Colour infrared film was found to be the most suitable to detect vegetation that was affected (van Genderen, 1974) but other stress factors were also listed which would similarly cause a loss of reflectance in the near infrared region, namely pathogens, insects, frost injury, mineral deficiency, drought and flooding. Areas of vegetation subject to pollutant stress could only be delineated after recourse to an inventory of emission sources. Sapp (1979) similarly encountered problems in distinguishing between SO_2 damage, drought stress and the effects of herbicides and weed infested areas in field work, whilst in laboratory experiments (Sapp, 1980) it was possible by the use of visible reflectance characteristics to distinguish between three classes of chlorosis and four classes of necrosis in wheat (*Triticum aestivum*) fumigated with SO_2. Results from a second species, soyabean (*Glycine max*) were, however, less conclusive. There are a number of studies in which reflectance measurements from remote sensing have been used to delineate areas of pollutant stress in forests. This method has the advantage that it is more efficient and less costly than ground survey methods over large land areas and also provides a permanent record of trends. To date specific areas affected by SO_2 emissions (Wert, 1969; Murtha and Trerise, 1977; Williams, 1978; Murtha, 1979) and O_3 emissions (Larsh, Miller and Wert, 1970; Weber and Polcyn, 1972) have been delineated in this way. Other surveys have covered areas of mixed vegetation near industrial areas (van Genderen, 1974), grasslands exposed to

SO$_2$ (Taylor, Leininger and Osberg, 1979) and in areas exposed to ozone episodes (Olson, 1977). Barber, Horler and Ferns (1981) have used spectral reflectance measurements and obtained correlations between certain wavelengths and shoot growth in laboratory experiments. In the field, however, it was only possible to demonstrate qualitatively the differences in reflectance between fumigated plots and the surrounding area by aerial photography. Other factors which were found to alter the reflectance were differences in plant development and injury from storm damage. Attempts have also been made to diagnose areas of pollutant stress using satellite data, but it was found that there was insufficient resolution for quantitative work, and only large areas of stressed vegetation could be delineated (Fritz and Pennypacker, 1975).

There is considerable potential for the use of reflectance measurements to delineate areas of vegetation subject to pollutant stress and also to quantify effects in laboratory experiments. *Table 22.2* lists examples in which this technique has been used as a diagnostic test. At present, however, changes in reflectance characteristics reported are not sufficiently specific to be used as a diagnostic test to distinguish air pollutant stress.

TABLE 22.2. Biochemical diagnostic techniques using leaf reflectance characteristics

Plant species	Pollutant	Measurement	Reference
Pine forests	Oxidants at field site in California	Remote sensing	Wert (1969)
Soyabean Maize	O$_3$	Laboratory study	Schubert, MacLeod and Howell (1972)
Mixed vegetation	Mixed field sites in industrial NE England	Remote sensing	van Genderen (1974)
Bean Tobacco	SO$_2$ 5 ppm × 16 h or O$_3$ 880 ppb × 25 h or O$_3$ 75 ppb × 292 h	Laboratory study	Nowak and Lautman (1974)
Bean	50 ppb O$_3$ × 6 h daily	Laboratory study	Runeckles and Resh (1975)
Mixed hardwoods and conifers	SO$_2$ acute dose	Remote sensing	Murtha and Trerise (1977)
Mixed vegetation	SO$_2$ acute dose	Remote sensing	Williams (1978)
Grasslands	SO$_2$ from field fumigation site (ZAPS)	Remote sensing	Taylor, Leininger and Osberg (1979)
Sunflower	1.5 ppm SO$_2$ 3 h or 800 ppb O$_3$ 2 h	Laboratory fumigation	Omasa *et al.* (1980)

Enzyme assays

There are numerous reports in the literature of changes in enzyme activity associated with fumigation episodes (for reviews *see* Horsman and Wellburn, 1976; Hällgren, 1978; Mudd, 1973) and a number of these responses have been used as diagnostic tests for air pollutants (*Table 22.3*). The first enzyme which was used as an indicator of pollutant stress was polyphenol oxidase. Godzik (1967) sampled trees of several species from sites in Upper Silesia, Poland in the vicinity of a coking plant where ambient air pollutants included SO$_2$, NO$_2$ and hydrocarbons.

Increased levels of enzyme activity of up to eightfold were found in the leaves of horse chestnut (*Aesculus hippocastanum*) that developed visible injury symptoms. Marked changes in other deciduous tree species including oak (*Quercus robus*) and common elder (*Sambucus nigra*) were not found. Peroxidase activity has been claimed to be a

TABLE 22.3. Biochemical diagnostic techniques using enzyme assays

Enzyme	Plant species	Pollutant(s)	Reference
Polyphenoloxidase	Horse chestnut	Mixed including SO_2, NO_2 and hydrocarbons	Godzik (1967)
Peroxidase	Mixed hardwoods and conifers	Preponderantly F at field sites various distances from aluminium plant	Keller and Schwager (1971)
Peroxidase	Mixed hardwoods and conifers	Preponderantly F at field sites various distances from aluminium plant	Keller (1974)
Peroxidase	Norway spruce (young trees)	200 ppb SO_2 for five months over winter continuously	Keller, Schwager and Yee-Meiler (1976)
Peroxidase	Soyabean cuu York and Wye	350 ppb $O_2 \times 2$ h	Curtis, Howell and Kremer (1976)
Peroxidase	Cowberry Crowberry	Mixed from chemical plant includes SO_2 and F	Mikkonen and Huttunen (1981)
Peroxidase	Norway spurce Scots pine	Mixed fumigation from local chemical plant includes SO_2 and F	Huttenen (1981)
Glutamate dehydrogenase	Pea	900 μg m^{-3} SO_2 for <14 days in laboratory conditions	Pahlich, Jäger and Steubing (1972)
Glutamate dehydrogenase	Pea	100, 150 or 200 ppb SO_2 18 days in laboratory conditions	Jäger and Klein (1977)
Glycosidases	Norway spruce	50, 100 or 200 ppb SO_2 for several weeks in outdoor chambers	Bucher-Wallin (1979)
Several enzymes simultaneously	Alfalfa Broad bean Tulip Beet Tobacco Barley Tomato	Mixed, in slight to medium pollution levels near Stuttgart, Germany	Rabe and Kreeb (1979)

sensitive assay to indicate pollutant doses which will subsequently give rise to visible injury (Curtis and Howell, 1971). Elevated enzyme activity has been reported after exposure to SO_2, fluorides, ozone and traffic exhaust. Curtis, Howell and Kremer (1976) and Dass and Weaver (1972) both using concentrations of 350 ppb O_3 reported elevated peroxidase activities after short-term fumigations of soyabean (*Glycine max*) and bean (*Phaseolus vulgaris*) respectively. In fumigations of pea (*Pisum sativum*) with 0.2–2.0 ppm SO_2 and 0.1 or 1.0 ppm NO_2 for three weeks, SO_2 alone stimulated the enzyme activity whilst NO_2 alone had no effect. In combination with SO_2 however, NO_2 lowered the threshold concentration of SO_2 at which the stimulation of activity was first seen and increased the level of stimulation above that from SO_2 alone (Horsman and Wellburn, 1975). Fumigation of soyabean (*Glycine max*) with 0.1 ppb fluoride for between 24 and 144 h also stimulated activity (Lee, Miller and Welkie, 1966). Keller and Schwager (1971) reported that leaf fluoride content and peroxidase activity both increase with decreasing distance from an aluminium smelter; the relationship between these two parameters was plotted for various months of the year.

It can be seen from the above data that peroxidase activity could not be used as a specific diagnostic test for a particular pollutant as more than one gas stimulated activity of the enzyme. Also it is not known whether the assay could be used quantitatively as a biochemical indicator of effects from fumigation episodes since in none of the reports had changes in enzyme activity been correlated with the level of injury sustained, whether visible injury or effects on growth. The stimulation in enzyme activity has, however, been used as a qualitative diagnostic test in field and laboratory studies (Keller and Schwager, 1971; Keller, 1974; Curtis, Howell and Kremer, 1976; Keller, Schwager and Yee-Meiler, 1976). Keller (1974) used the assay as a diagnostic test to survey and map areas of trees exposed to fluoride emissions. A fairly close correspondence was found between areas in which elevated enzyme activity was detected and the areas in which visible symptoms of injury subsequently developed.

Certain glycosidases were also found to be reliable indicators of the later development of visible injury after fumigations of SO_2. The β-glucosidase and β-1,3-glucanase activities of beech (*Fagus sylvatica*) and the β-galactoside and β-1,3-glucanase of Norway spruce (*Picea abies*) were significantly raised in plants in which necrosis occurred several days later (Bucher-Wallin, Bernhard and Bucher, 1979). No correlations were however made between the two parameters to determine if the assay could be used as a quantitative indicator of pollutant effects.

Changes in glutamate dehydrogenase activity have been claimed to be the most sensitive enzymatic indicator of air pollution effects (Pahlich, Jäger and Steubing, 1972, Horsman and Wellburn, 1976; Jäger and Klein, 1977). Jäger and Klein suggested that it be used to diagnose severe SO_2 injury before the appearance of visible symptoms in conjunction with measurements of inorganic sulphur and the buffering capacity of the leaf tissue. Their data however, were, for only one species, pea (*Pisum sativum*) and Rabe and Kreeb (1979) working with several cultivated species found that the direction of the changes in the activity of glutamate dehydrogenase was dependent on the species fumigated. When exposed to winter urban levels of pollutants in Stuttgart (1975–1976) that included a winter mean concentration of 50 ppb SO_2, significant increases of activity were found in tulip (*Tulipa*) (red emperor), broad bean (*Vicia faba*) and tobacco (*Nicotiana tabacum*), no change in barley (*Hordeum distichon*), beet (*Beta vulgaris*) or tomato (*Lycopersicon esculentum*) and either no change (fumigation in late summer) or a significant inhibition of activity (fumigation in spring) in alfalfa (*Medicago sativa*). Ozone has also been reported to increase peroxidase activity. Tingey, Fites and Wickliff (1976a) reported effects on soybean cultivars exposed to 490 ppb O_3 for 2 h. Wellburn *et al.* (1981), who looked at the effects of SO_2 and NO_2 alone and in combination on several enzymes in perennial ryegrass and pea, expressed changes in glutamate dehydrogenase activity as a ratio to glutamine synthetase activity rather than expressing the enzyme on the basis of tissue dry weight. Glutamine synthetase activity is unaffected by fumigation with SO_2 or NO_2 and so it functions as an internal standard for the assay. This method of expressing the results by reference to glutamine synthetase considerably reduced the variability of the data. Rabe and Kreeb (1979) assayed the activity of several enzymes including glutamate dehydrogenase and concluded that several assays must be used simultaneously to diagnose air pollution stress in the absence of visible injury. The assays used by the authors were chlorophyll a and soluble protein levels, the activity of the enzymes glucose-6-phosphate dehydrogenase, isocitrate dehydrogenase, glutamate dehydrogenase, asparagine aminotransferase and alanine aminotransferase. The authors, however, suggested that such technically-demanding analyses should be dispensed with where measurements of visible injury could be made.

In conclusion it can be said that enzyme assays may be useful diagnostic tests where they can be carried out more quickly and easily than measurements of growth reductions or visible injury. They have proved useful indicators of areas subject to pollutant stress where the source of emissions is already known and an early indication of visible injury is required. As pointed out by Wellburn *et al.* (1976), however, this is only a valid method where the source of pollution is of recent origin, as natural selection of resistant genotypes is likely to take place in areas with a long history of fumigation episodes. The technique has not been extended to diagnose areas in which growth reductions are likely to take place or make quantitative estimates of effect.

Endogenous metabolites

Levels of endogenous metabolites have also been suggested as useful diagnostic parameters for air pollution effects, notably amino acids (Jäger, Pahlich and Steubing, 1972; Godzik and Linskens, 1974; Jäger, 1975; *see also Table 22.4*). There are many

TABLE 22.4. Biochemical diagnostic tests using levels of endogenous metabolites

Metabolite	Plant species	Pollutant(s)	Reference
Amino acids	Pea	340 ppb SO_2 for 7, 14 or 21 days	Jäger, Pahlich and Steubing (1972)
Amino acids	Bean	700 ppb SO_2 for up to 96 h	Godzik and Linskens (1974)
Amino acids	Pea Maize Lettuce	300 ppb SO_2 for 5, 8 or 16 days	Jäger (1975)
Ethylene	A range of herbaceous and woody plants	250–750 ppb O_3	Tingey, Standley and Field (1976)
Ethylene and ethane	Cucumber Pumpkin	5.5 ppm SO_2 for 16 h	Bressan *et al.* (1979)
Ethylene	Common beech Larch Norway spruce Scots pine	<225 ppb SO_2 for several weeks in outdoor chambers	Bucher (1979)
Ethylene	Mixed herbaceous and woody species	Concentration up to 1 ppm O_3 for 2 h	Tingey, Wilhour and Taylor (1978)

reports of changes in the levels of free amino acids in response to fumigations with SO_2, O_3 and $SO_2 + HF$ (Arndt, 1970; Jäger and Grill, 1975; Horsman and Wellburn, 1976; Malhotra and Sarkar, 1979; Pierre and Queiroz, 1981). The authors report an increase in the levels of most amino acids and amines (Priebe, Klein and Jäger, 1978) after fumigation treatments. Glutamate and aspartate, important in the synthesis of proteins and other amino acids are, by contrast, lower in fumigated material (Godzik and Linskens, 1974; Pierre and Queiroz, 1981). Proline levels, which rise substantially in response to water stress (Steward, 1972; Hsiao, 1973), only increased by a factor of two. These results indicate that pollutant stresses alter the rate of protein metabolism and there are several reports of decreased concentrations of protein in fumigated plants (Craker, 1972; Jäger, Pahlich and Steubing, 1972; Constantinidou and Kozlowski, 1979). Craker (1972) suggested that the immediate initiation of catabolic enzymes accounts for the fall in protein concentrations and the work of Tomlinson and Rich (1969) on ^{14}C incorporation by plants fumigated with ozone would substantiate this suggestion. Tingey, Wilhour and Taylor (1978) proposed that changes in the rate of protein synthesis may provide a good indicator of pollutant response; it would, however, be a nonspecific indicator of stress since many factors affect the rate of this process in plants.

Ethylene and ethane production

The production of stress ethylene and ethane have also been suggested as useful biochemical indicators of air pollution (Tingey, Standley and Field, 1976; Bressan et al., 1979; Bucher, 1979; see also Table 22.4). Ethylene appeared to be formed primarily from stressed but functional leaves and ethane from visibly damaged leaves (Bressan et al., 1979). Ethylene production as a response to an air pollutant was first reported by Craker (1971) who exposed tomatoes to 90 ppb O_3 for 2 h and measured the subsequent accumulation of ethylene from the plants, which were placed in a sealed chamber for up to 4 h after the fumigation treatment. Tingey, Standley and Field (1976) similarly recorded ethylene accumulation over a period of 22 h after exposure to various concentrations of ozone between 250 and 750 ppb. Twenty-nine species, including annuals, herbaceous perennials and woody perennials, were used and good correlations were found between ethylene levels and visible injury. Increased ethylene production of up to 20-fold was found at the highest concentration of air pollutants. Ethylene production has more recently been correlated with reductions in plant growth in response to ozone fumigations. Tingey (1980) found a highly significant negative correlation between the two parameters in *Phaseolus* sp. Ethylene production has also been reported in response to SO_2 fumigations (Bressan et al., 1979; Peiser and Yang, 1979; Bucher, 1979, 1981). Bressan et al. (1979) fumigated cucumber (*Cucumis sativus*) and pumpkin (*Cucurbita pepo*) with SO_2 concentrations of up to 40 ppm, and measured both ethylene and ethane evolution from excised leaf discs of fumigated plants in the 24-h period following exposure. Strong linear correlations were found between the concentration of SO_2 to which the plants were exposed and the amount of ethylene or ethane produced. Similar correlations were found between ethylene production and uptake of SO_2 or visible injury. By contrast Bucher (1981) working with clonal material from trees found no correlation between the degree of visible injury and ethylene emissions. Clonal material (two to three years old) from common beech (*Fagus sylvatica*), larch (*Larix decidua*), Norway spruce (*Picea abies*) and Scots pine (*Pinus sylvestris*) was fumigated with constant concentrations of SO_2 ranging between 25 and 225 ppb. An increase of up to tenfold in ethylene emissions was detected in beech after eight days, in larch after five to seven days and in Scots pine after five days. A second peak occurred after about 15 days and the cyclic pattern was still evident after six weeks' exposure in Scots pine. The oscillations in ethylene evolution make the assay rather unsuitable for quantitiative indications of the effects of SO_2. There are a number of possible reasons for the absence of any correlation between injury and ethylene emissions. Firstly, ethylene emissions were assayed on plant material fumigated for up to 25 days rather than for a few hours as in the experiments previously described. It is likely that the metabolic changes during a long-term fumigation are rather different to those during the period of hours after a short-term fumigation (less than one day). Secondly, the variability in response of the material may be so great as to mask the effect of the experimental treatment. It is possible that the production of ethylene could be used as a quantitative indicator of visible injury and growth reductions, resulting from fumigations with air pollutants, but only when the episode is comparatively short (less than one day). Bressan *et al.* (1979) used a correlation between visible injury and ethane evolution to screen for resistance in cucurbits. A difference in ethane evolution of up to 24-fold was found between leaf discs cut from sensitive and resistance cultivars fumigated with SO_2 for several hours under controlled environment conditions. It must, however, be noted that other stresses, namely water stress, insect damage, temperature extremes, drought, γ-irradiation, disease and mechanical wounding are also known to stimulate ethylene production (Abeles, 1973), and would interfere with the use of such assays in the field.

Härtel turbidity test

Two assays of a more general biochemical nature which have been used as diagnostic tests for air pollution effects are the Härtel turbidity test, and tissue buffering capacity, as summarized in *Table 22.5*. Härtel has used the turbidity test to map areas affected by

TABLE 22.5. Biochemical diagnostic tests of a general nature

Test	Plant species	Pollutant	Reference
Härtel turbidity test	Norway spruce	SO_2	Härtel (1953)
Härtel turbidity test	Norway spruce	SO_2	Härtel and Papesch (1955)
Härtel turbidity test	Norway spruce	Predominantly SO_2 at field sites in industrial areas	Härtel (1972)
Buffering capacity	Norway spruce 2–8 m tall	Predominantly SO_2 at field sites in industrial areas	Grill (1971)
Buffering capacity	Pea	100, 150 and 250 ppb SO_2 for 18 days	Jäger and Klein (1977)

pollutants and to classify injury zones (Härtel, 1953, Härtel and Papesch, 1955; Härtel, 1972). The test is based on the observation that aqueous extracts of conifer needles, especially Norway spruce, are more turbid if taken from plants exposed to SO_2 (Härtel, 1953). The phenomenon was reported at concentrations which do not result in visible injury, and a correlation of $r = 0.86$ was reported between SO_2 concentration and the turbidity of the extracts in a survey of plants sampled at various distances from an emission source (Härtel, 1953; Härtel and Papesch, 1955). A correlation was also found between maximum turbidity values and the decline in annual growth. This assay has also been used by Keller, Schwager and Yee-Meiler (1976) to map areas of vegetation affected by emissions from an aluminium smelter in the Rhone Valley in Switzerland, and by Steinhübel (1957) and Materna and Ryskova (1961) in Czechoslovakia. Keller, Schwager and Yee-Meiler (1976) found the method less tedious than the assay for peroxidase activity but also less sensitive. Its use is also restricted to coniferous trees. The increase in turbidity is thought to result from the higher solubility of certain substances, e.g. polysaccharides, proteins, polyphenols, pectins and tannins with the fall in tissue pH and also from the increasing concentrations of phenolic compounds in the needles (Materna and Ryskova, 1961). The effect of other gaseous pollutants on this assay are not known.

Tissue buffering capacity

Changes in the buffering capacity of tissues exposed to acid gases such as SO_2 have been suggested as useful diagnostic tests when used in conjunction with the estimation of the sulphur content of the tissues and possibly another biochemical test (Jäger and Klein, 1977). The principle behind the assay is that in the detoxification of acid pollutants such as SO_2, the buffering capacity of the cellular fluids is depleted. Lower buffering capacities of leaf extracts plants exposed to SO_2 have been reported by Grill (1971), Grill and Härtel (1972) and Jäger and Klein (1977) in Norway spruce and pea respectively. Keller, Schwager and Yee-Meiler (1976) however failed to detect differences in buffering capacity of another species of Norway spruce (*Picea excelsa*) fumigated with SO_2 in outdoor chambers during a winter period. From the positive results obtained using this assay, it can be seen that it is more widely applicable than the turbidity test previously described which could only be used on conifers. Like the turbidity test the effects of other pollutants on the assay are not known and these may interfere with quantitative assessments. In mapping areas of vegetation damaged by pollutants it is, therefore, imperative to know all the contributing factors such as SO_2, NO_2, O_3, HCl, organic compounds and heavy metal contaminants.

Summary

There are a large number of biochemical changes which occur as a result of fumigations with air pollutants that have been suggested as suitable biochemical indicators of effects on vegetation. A list of these assays and the authors who suggested their use can be found in *Table 22.1*. Despite the variety of tests used there are several features common to all which limit their usefulness as diagnostic tests.

Firstly, none of the assays are specific to a particular pollutant, which means that whilst an effect can be demonstrated, the source and nature of the pollutant must be ascertained from monitoring data or emission inventories. It is also known that none of the assays are specific indicators of pollutants; in many cases known environmental factors can elicit the same response. This means that it is important when using such assays that control samples are included, and that these should be subjected to the same environmental stresses in the absence of the pollutants. This is a criterion that is very difficult to achieve. An alternative approach is the use of an internal standard, the levels of which are affected by environmental stress but not by air pollutants. This has been done by Wellburn *et al.* (1981) who were investigating the effect of SO_2 and NO_2 fumigations on glutamate dehydrogenase activity. Glutamin synthetase is unaffected by these air pollutants and their results were therefore expressed on the basis of this second enzyme rather than on a dry weight basis.

Secondly, few of the diagnostic tests have been developed sufficiently to be applied quantitatively, exceptions being the use of leaf reflectance measurements and stress ethylene evolution. This means that, apart from these, it is not known whether the threshold concentration of pollutant(s) which causes injury is the same as the threshold for effects on the diagnostic test. This may lead to inaccuracies in delimiting affected areas. Another factor that may lead to inaccuracies in the use of assays that have been described is that there is little information available on the interactions between pollutants, and as they rarely occur singly in the environment this is an important consideration for field studies.

Thirdly, biochemical studies on the effect of air pollutants on plants have shown that the direction and degree of response is dependent on the species examined (*see* Rabe and Kreeb, 1979). In making quantitative assessments of injury using biochemical indicators, therefore, each species has to be assessed individually and the work involved in this may outweight the advantage of using a biochemical diagnostic test rather than measurements of effects on yield or visible damage.

Fourthly, many of the biochemical tests have been developed in the laboratory on plants fumigated for short periods of a few hours. Whilst these assays may be useful in this situation, e.g. screening of cultivars for resistance in breeding programmes (Bressan *et al.*, 1979), extrapolation to the field may not be possible. Fumigation episodes in the field are often long term and it is likely that the cumulative biochemical changes will be rather different. An example of this is the difference in ethylene evolution in response to short-term and long-term fumigations to SO_2 (Bressan *et al.*, 1979; Bucher, 1981, respectively).

The potential usefulness of biochemical diagnostic techniques is therefore restricted by many factors. In the absence of other variables, like environmental stresses, such assays could be useful as qualitative or quantitative diagnostic tests as part of a screening programme in plant breeding.

In the field, however, the situation is very complex and detailed studies of all stress factors involved should be made before the results of a biochemical diagnostic test can be attributed to the effects of a particular air pollutant.

Acknowledgements

The work was carried out at the Central Electricity Research Laboratories and is published by permission of the Central Electricity Generating Board.

References

ABELES, F. B. *Ethylene in Plant Biology*, Academic Press, New York (1973)

ARNDT, U. Konzentrationsänderungen bei freien Aminosäuren in Pflanzen unter dem Einfluss von Fluorwasserstoff und Schwefeldioxid, *Staub-Reinhalf Luft* **36**, 256–259 (1970)

ARNDT, U. Konzentrationsänderungen bei Blattfarbstoffen unter dem Einfluss von Luftveruntreinigungen, Ein Diskussionsbeitrag zur Pigmentanalyse, *Environmental Pollution* **2**, 37–48 (1971)

ARNDT, U. The Kautsky-effect as a sensitive proof for air pollution effects on plants, *Chemosphere* **1**, 187–190 (1972)

ARNDT, U. The Kautsky-effect: A method for the investigation of air pollutants in chloroplasts, *Environmental Pollution* **6**, 181–194 (1974)

BARBER, J., HORLER, D. N. H. and FERNS, D. C. Remote sensing of effects of air pollutants on plants, *Final Report to the Commission of European Community Contract No. ENV-398-UK(N)* (1981)

BELL, N. J. B. and MUDD, C. H. Sulphur dioxide resistance in plants: a case study of *Lolium perenne*. In *effects of Air Pollutants on Plants*, (Ed. T. A. MANSFIELD), pp. 87–103, Cambridge University Press, Cambridge (1976)

BENNETT, J. H. and HILL, A. C. Inhibition of apparent photosynthesis by pollutants, *Journal of Environmental Pollution* **2**, 526–530 (1973)

BLACK, V. J. and UNSWORTH, M. H. Effects of low concentrations of sulphur dioxide on net photosynthesis and dark respiration of *Vicia faba*, *Journal of Experimental Botany* **30**, 473–483 (1979)

BÖRTITZ, S. Physiologische und biochemische Beiträge zur Rauchschaden-Forschung. 1. Mitteilung, *Biologische Zentralblatt* **83**, 501–513 (1964)

BRESSAN, R. A., LECUREUX, L., WILSON, L. G. and FILNER, P. Emission of ethylene and ethane by leaf tissue exposed to injurious concentrations of sulphur dioxide or bisulphite ion, *Plant Physiology* **63**, 924–930 (1979)

BUCHER, J. B. SO_2-induziertes Stress-Athylen in den Assimilations organen von Waldbäumen. Mitteilungen aus dem Institut für Forst- und Holzwirtschaft, Ljubljana. IUFRO-Bericht der X. Fachtagung in Ljubljana, Sept. 1978, Fachgrupe S 2.09, pp. 93–101 (1979)

BUCHER, J. B. SO$_2$-induced ethylene evolution of forest tree foliage, *European Journal of Forest Pathology* **11**, 369–373 (1981)

BUCHER-WALLIN, I. K., BERNHARD, L. and BUCHER, J. B. Einfluss niedriger SO$_2$-Konzentrationen auf die Aktivität einiger Glykosidasen der Assimilationsorgane verkonter Waldbäume, *European Journal of Forest Pathology* **9**, 6–15 (1979)

CHIMIKLIS, P. and HEATH, R. Fluorescence transients of O$_3$ gassed *Chlorella*, *Plant Physiology* **56** (Supplement): 5, abstract no. 23 (1975)

CONSTANTINIDOU, H. A. and KOZLOWSKI, T. T. Effects of sulphur dioxide and ozone on *Ulmus americana* seedlings. II. Carbohydrates, proteins, and lipids, *Canadian Journal of Botany* **57**, 176–184 (1979)

CRAKER, L. E. Ethylene production from ozone injured plants, *Environmental Pollution* **1**, 299–304 (1971)

CRAKER, L. E. Influence of ozone on RNA and protein content of *Lemna minor* L., *Environmental Pollution* **3**, 319–323 (1972)

CURTIS, C. R. and HOWELL, R. K. Increases in peroxidase isoenzyme activity in bean leaves exposed to low doses of ozone, *Phytopathology* **61**, 1306–1307 (1971)

CURTIS, C. R., HOWELL, R. K. and KREMER, D. F. Soybean peroxidases from ozone injury, *Environmental Pollution* **11**, 189–194 (1976)

DASS, H. C. and WEAVER, G. M. Enzymatic changes in intact leaves of *Phaseolus vulgaris* following ozone fumigation, *Atmospheric Environment* **6**, 759–763 (1972)

DÄSSLER, H. G. Zur Wirkungsweise der Schadstoffe und der Einfluss von SO$_2$ auf Blattfarbstoffe, *Mitteilungen aus der Forstlichen Bundesversuchsanstalt, Wien* **97**, 353–366 (1972)

DÖRRIES, W. Uber die Brauchbarkeit der spektroskopischen Phäophytinprobe in der Rauchschaden-Diagnostik, *Zeitschrift für Pflanzenkrankheiten und Pflanzenschutz* **42**, 257–273 (1932)

FRITZ, E. L. and PENNYPACKER, S. P. Attempts to use satellite data to detect vegetative damage and alteration caused by air and soil pollutants, *Phytopathology* **65**, 1056–1060 (1975)

VAN GENDEREN, J. L. Remote sensing of environmental pollution on Teeside, *Environmental Pollution* **6**, 221–234 (1974)

GODZIK, S. Polyphenol oxidase activity in vegetation injured by industrial air pollution, *Biuletyn Zakladu Badan Nauk* **10**, 103–113 (1967)

GODZIK, S. and LINSKENS, H. F. Concentration changes of free amino acids in primary bean leaves after continuous and interrupted SO$_2$ fumigation and recovery, *Environmental Pollution* **7**, 25–37 (1974)

GOEDHEER, J. C. Fluorescence in relation to photosynthesis, *Annual Review of Plant Physiology* **23**, 87–112 (1972)

GRILL, D. Pufferkapazität gesunder und rauchgeschädigter Fichtennadeln, *Zeitschrift für Pflanzenkrankheiten und Pflanzenschutz* **78**, 612–622 (1971)

GRILL, D., ESTERBAUER, H. and BECK, G. Untersuchungen an phenolischen Substanzen und Glucose in SO$_2$-geschädigten Fichtennadeln, *Phytopathologische Zeitschrift* **82**, 182–184 (1975)

GRILL, D. and HÄRTEL, O. Zellphysiologische und biochemische Untersuchungen an SO$_2$-begasten Fichtennadeln. Resistenz und Pufferkapazität, *Mitteilungen aus der Forstlichen Bundesversuchsanstalt, Wien* **97**, 367–384 (1972)

GRONEBAUM-TURK, K. and MATHÉ, P. Der Einfluss von Fluorverbindungen auf die Chloroplastenpigmente von Pappel-, Holunder- und Fliederblättern im Freiland bei verschiedener Belastung, *European Journal of Forest Pathology* **5**, 183–184 (1975)

HÄLLGREN, J. E. Physioligical and biochemical effects of sulphur dioxide on plants. In *Sulphur in the Environment. Part II. Ecological Impacts*, (Ed. J. O. NRIAGU), pp. 163–209, Wiley and Sons Inc., New York (1978)

HÄRTEL, O. Eine neue Methode zur Erkennung von Raucheinwirkungen an Fichten, *Zentralblatt für die gesante Forst- und Holzwirtschaft* **72**, 12–31 (1953)

HÄRTEL, O. and PAPESCH, E. Über die Wachsausscheidung von Koniferen, *Berichte der Deutschen botanischen Gesellschaft* **68**, 133–142 (1955)

HÄRTEL, O. Langjährige Messreihen mit dem Trübungstest an abgasgeschädigren Fichten, *Oecologia* **9**, 103–111 (1972)

HILL, G. R. and THOMAS, M. D. Influence of leaf destruction by sulphur dioxide and by clipping on yield of alfalfa, *Plant Physiology* **8**, 223–245 (1933)

HORSMAN, D. C. and WELLBURN, A. R. Synergistic effect of SO$_2$ and NO$_2$ polluted air upon enzyme activity in pea seedlings, *Environmental Pollution* **8**, 123–133 (1975)

HORSMAN, D. C. AND WELLBURN, A. R. Guide to the metabolic and biochemical effects of air pollutants on higher plants. In *Effects of Air Pollutants on Plants*, (Ed. T. A. MANSFIELD), pp. 185–199, Cambridge University Press, Cambridge (1976)

HOWELL, R. K. and KREMER, D. F. The chemistry of physiology of pigmentation in leaves injured by air pollution, *Journal of Environmental Quality* **2**, 434–438 (1973)

HSIAO, T. C. Plant responses to water stress, *Annual Review of Plant Physiology* **24**, 519–570 (1973)

HUTTUNEN, S. Seasonal variation of air pollution stress in conifers, *Mitteilungen aus der Forstlichen Bundesversuchsanstalt, Wien* **137**, 103–112 (1981)

JÄGER, H. J. Wirkung von SO_2-Begasung auf die Aktivität von Enzymen des Aminosäurestoffwechsels und den Gehalt frier Aminosäuren in unterschiedlich resistenten Pflanzen, *Zeitschrift für Pflanzenkrankheiten und Pflanzenschutz* **82**, 139–148 (1975)

JÄGER, H. J. Biochemical indication of an effect of air pollution on plants. In *Monitoring of Air Pollutants by Plants: Methods and Problems* (Proceedings of an International Workshop, Osnabrück (FRG), September 1981), (Eds. L. STEUBING and H. J. JAGER), pp. 99–108, Dr W. Junk, The Hague (1982)

JÄGER, H. J. and GRILL, D. Einfluss von SO_2 und HF auf freie Aminosäuren der Fichte (*Picea abies* [L.] Karstem), *European Journal of Forest Pathology* **5**, 279–286 (1975)

JÄGER, H. J. and KLEIN, H. Biochemical and physiological detection of sulphur dioxide injury to pea plants (*Pisum sativum*), *Journal of the Air Pollution Control Association* **27**, 464–466 (1977)

JÄGER, H. J., PAHLICH, E. and STEUBING, L. Die Wirkung von Schwefeldioxid auf den Aminosäure und Proteingehalt von Erbsenkeimligen, *Angewandte Botanik* **46**, 199–211 (1972)

JAMRICH, V. Je stabilita chlorofylu faktorom odolnost; prot; dymu? *Zb Ved Pr. Lesn. Fak Vys. Sk Lesn. Drevarsk Zvolene* **10**, 7 (1968)

KATZ, M. and SHORE, V. C. Air pollution damage to vegetation, *Journal of the Air Pollution Control Association* **5**, 144–150 (1955)

KAUTSKY, H. and HIRSCH, A. Neue Versuche zur Kohlensäure assimilation, *Naturwissenschaften* **19**, 964 (1931)

KELLER, H. Beiträge zur Erfassung der durch schweflige Säure hervorgerufenen Rauchschäden an Nadelhölzern, *Forstwiss. Forsch. (Beihefte zum Forstw. Cbl.)* **10**, 5–63 (1958)

KELLER, Th. Über die schädigende Wirkung des Fluors, *Schweizerische Zeitschrift für Forstwesen* **124**, 700–706 (1973)

KELLER, Th. The use of peroxidase activity for monitoring and mapping air pollution areas, *European Journal of Forest Pathology* **4**, 11–19 (1974)

KELLER, Th. and SCHWAGER, H. Der Nachweis unsichtbarer ('physiologischer') Fluorimmissionsschädigungen an Waldbäumen durch eine einfache kolorimetrische Bestimmung der Peroxidase-Aktivität, *European Journal of Forest Pathology* **1**, 6–18 (1971)

KELLER, Th., SCHWAGER, H. and YEE-MEILER, D. Der Nachweis winterlicher SO_2-Immissionen an jungen Fichten, *European Journal of Forest Pathology* **6**, 244–249 (1976)

KELLER, Th. Physiological bioindications on an effect of air pollution on plants. In *Monitoring of Air Pollutants by Plants: Methods and Problems* (Proceedings of an International Workshop Osnabrück (FRG) September 1981), (Eds. L. STEUBING and H. J. JÄGER), pp. 85–96, Dr W. Junk, The Hague (1982)

KNUDSON, L. L., TIBBITS, T. W. and EDWARDS, G. E. Measurements of ozone injury by determination of leaf chlorophyll concentration, *Plant Physiology* **60**, 606–608 (1977)

DE KONING, H. W. and JEGIER, Z. Effects of sulphur dioxide and ozone on *Euglena gracilis, Atmospheric Environment* **4**, 357–361 (1970)

LARSH, R. N., MILLER, R. P. and WERT, S. L. Aerial photography to detect and evaluate air pollution damage to Ponderosa pine, *Journal of the Air Pollution Control Association* **20**, 289–292 (1970)

LAUENROTH, W. K. and DODD, J. L. Chlorophyll reduction in western wheatgrass (*Agropyron smithii* Rydb.) exposed to sulphur dioxide, *Water, Air and Soil Pollution* **15**, 309–315 (1981)

LEE, C.-J., MILLER, G. W. and WELKIE, G. W. The effects of hydrogen fluoride and wounding on respiratory enzymes in soybean leaves, *International Journal of Air and Water Pollution* **10**, 169–181 (1966)

MALHOTRA, S. S. and SARKAR, S. K. Effects of sulphur dioxide on sugar and free amino acid content of pine seedlings, *Physiologia Plantarum* **47**, 223–228 (1979)

MATERNA, J. and RYSKOVA, L. Prispevek k poznani biochemickych zakladu Hartelova zakaloveho testu, *Lesnictvi* **7**, 389–400 (1961)

MIKKONEN, H. and HUTTUNEN, S. Dwarf shrubs as bioindicators, *Silva Fennica* **15**, 475–480 (1981)

MUDD, J. B. Biochemical effects of some air pollutants on plants. In *Air Pollution Damage to Vegetation. Advances in Chemistry Series.* No. 122, (Ed. J. A. NEAGLE), pp. 31–47, American Chemical Society, Washington, DC (1973)

MURTHA, P. A. Modelling tree damage-type patterns for photointerpretation of SO_2 injury, *International Archives of Photogrammetry* **13**, 14 (1979)

MURTHA, P. A. and TRERISE, R. Four years after: photointerpretation of the residual effects of SO_2 damage to conifers and hardwoods. *Proceedings of the 6th Biennial Workshop, Aerial Colour Photography in Plant Sciences*, pp. 25–30, American Society for Photogrammetry, Colorado State University (1977)

MÜLLER, J. Spezifischer Nachweis von SO_2-Rauchschäden an Pflanzen mit Hilfe von Blattpigmentanalysen, *Naturwissenschaften* **44**, 453 (1957)

NOUCHI, I. and ODAIRA, T. Influence of ozone on plant pigments, *Journal of The Japanese Society Air Pollution* **8**, 120–125 (1973)

NOWAK, M. B. and LAUTMAN, D. A. *Study of air pollutant signatures for remote sensing*. Final report NASA Research Grant, NGR 22-01-072, Northeastern University, Boston, Massachusetts, 46 pp. (1974)

OLSEN, C. E. Jr. Pre-visual detection of stress in pine forests. *Proceedings of The 11th International Symposium on Remote Sensing of Environment* **2**, 933–944 (1977)

OMASA, K., ABO, T., HASHIMOTO, Y. and AIGA, I. Evaluation of air pollution injury to plants by image processing. *Research Report from The National Institute for Environmental Studies No. 11*, Ibaralki, Japan, pp. 249–254 (1980)

PAHLICH, E., JÄGER, H. J. and STEUBING, L. Beeinflussung der Aktivitäten von Glutamatdehydrogenase and Glutaminsynthetase aus Erbsenkeimlingen durch SO₂, *Angewandte Botanik* **46**, 183–197 (1972)

PAPAGEORGIO, G. Chlorophyll fluorescence: an intrinsic probe of photosynthesis. In *Bioenergetics of Photosynthesis*, (Ed. GOVINDJEE), pp. 319–371, Academic Press, New York (1975)

PEISER, G. D. and YANG, S. F. Ethylene and ethane production from sulphur dioxide-injured plants, *Plant Physiology* **63**, 142–145 (1979)

PIERRE, M. and QUEIROZ, O. Enzymic and metabolic changes in bean leaves during continuous pollution by subnecrotic levels of SO₂, *Environmental Pollution, Series A* **25**, 41–53 (1981)

PRIEBE, A., KLEIN, H. and JÄGER, H. J. Role of polyamines in SO₂-polluted pean plants, *Journal of Experimental Botany* **29**, 1045–1050 (1978)

RABE, E. and KREEB, K. H. Enzyme activities and chlorophyll and protein content in plants as indicators of air pollution, *Environmental Pollution* **19**, 119–137 (1979)

RABE, R. and KREEB, K. H. Bioindication of air pollution by chlorophyll destruction in plant leaves, *Oikos* **34**, 163–167 (1980)

RICKS, G. R. and WILLIAMS, R. J. H. Effects of atmospheric pollution on deciduous woodland. Part 3. Effects on photosynthetic pigments of leaves of *Quercus petraea* (Mattuschka Leibl.), *Environmental Pollution* **8**, 97–106 (1975)

RUNECKLES, V. C. and RESH, H. M. The assessment of chronic ozone injury to leaves by reflectance spectrophotometry, *Atmospheric Environment* **9**, 447–452 (1975)

SAPP, C. D. Remote sensing of sulphur dioxide effects on vegetation. Photometric analysis of aerial photographs. USA Environmental Protection Agency Report EPA-600/7-79-138; TVA/ONR-79/01, 31 pp. (1979)

SAPP, C. D. Remote sensing of sulphur dioxide effects on vegetation. Spectroradiometry, USA Environmental Protection Agency Report EPA-600/7-80-159; TVA/ONR-80/11, 33 pp. (1980)

SCHREIBER, U., VIDAVER, W., RUNECKLES, V. C. and ROSEN, P. Chlorophyll fluorescence assay for ozone injury in intact plants, *Plant Physiology* **61**, 80–84 (1978)

SCHUBERT, J. S., MACLEOD, N. H. and HOWELL, R. K. Changes in optical properties of stressed and diseased leaves, *Agronomy Abstracts* **64**, 164 (1972)

STEINHÜBEL, G. Pouzitie Härtlovho testu Pri diagnoze skod symovymi plynmi, *Biologia (Bratislava)* **12**, 611–617 (1957)

STEWARD, C. R. Proline content and metabolism during rehydration of wilted excised leaves in the dark, *Plant Physiology* **50**, 678–681 (1972)

TAYLOR, J., LEININGER, W. C. and OSBERG, T. R. The use of remote sensing in evaluating SO₂ damage to grasslands. In *Bioenvironmental Impact of a Coal-fired Power Plant: Fourth Interim Report, Colstrip, Montana, USA*. US Environmental Protection Agency, EPA 600/3-79-004 Section **24**, 899–923 (1979)

TAYLOR, O. C., CARDIFF, E. A. and MERSEREAU, J. D. Apparent photosynthesis as a measure of air pollution damage, *Journal of the Air Pollution Control Association* **15**, 171–173 (1965)

TAYLOR, O. C. and EATON, F. M. Suppression of plant growth by nitrogen dioxide, *Plant Physiology* **41**, 132–135 (1966)

THOMAS, M. D. The invisible theory of plant damage, *Journal of the Air Pollution Control Association* **5**, 205–208 (1956)

TINGEY, D. T. Stress ethylene production — a measure of plant response to stress, *Hortscience* **15**, 630–633 (1980)

TINGEY, D. T., FITES, R. C. and WICKLIFF, C. Differential foliar sensitivity of soybean cultivars to ozone associated with differential enzyme activities, *Physiologia Plantarum* **37**, 69–72 (1976)

TINGEY, D. T., STANDLEY, C. and FIELD, R. W. Stress ethylene evolution: a measure of ozone effects on plants, *Atmospheric Environment* **10**, 969–974 (1976)

TINGEY, D. T., WILHOUR, R. G. and TAYLOR, G. C. The measurement of plant responses. In *Handbook of Methodology for the Assessment of Air Pollution Effects on Vegetation*, (Eds. W. W. HECK, S. V. KRUPA and S. N. LINZON), pp. 7-1 to 7-35, Air Pollution Control Association, Upper Midwest Section (1978)

TINGEY, D. T., PETTIT, N. and BARD, L. Effect of chlorine on stress ethylene production, *Environmental and Experimental Botany* **18**, 61–66 (1978)

TOMLINSON, H. and RICH, S. Relating lipid content and fatty acid synthesis to ozone injury of tobacco leaves, *Phytopathology* **59**, 1284–1296 (1969)

WEBER, F. P. and POLCYN, F. C. Remote sensing to detect stress in forests, *Photogrammetric Engineering* **38**, 163–175 (1972)

WELLBURN, A. R., CAPRON, T. M., CHAN, H.-S. and HORSMAN, R. C. Biochemical effects of atmospheric pollutants on plants. In *Effects of Air Pollutants on Plants*, (Ed. T. A. MANSFIELD), pp. 105–114, Cambridge University Press, Cambridge (1976)

WELLBURN, A. R., HIGGINSON, C., ROBINSON, D. and WALMSLEY, C. Biochemical explanations of more than additive inhibitory effects of low atmospheric levels of sulphur dioxide plus nitrogen dioxide upon plants, *New Phytologist* **88**, 223–237 (1981)

WERT, S. L. A system for using remote sensing techniques to detect and evaluate air pollution effects in forest stands, *Proceedings of the 6th International Symposium on Remote Sensing of the Environment*, University of Michigan, Ann Arbor, Michigan, Oct. 13–16, 1969, pp. 1169–1178 (1969)

WILLIAMS, D. R. Applications of remote sensing to vegetation injury caused by air pollution, *Proceedings of a Symposium on Vegetation Damage Assessment*, Seattle, Feb. 1978, American Society of Photogrammetry, 529–543 (1978)

YEE-MEILER, D. Über den Einfluss fluorhaltiger Fabrikabgase auf den Phenolgehalt von Fichtennadeln, *European Journal of Forest Pathology* **3**, 214–221 (1974)

Chapter 23

Changes in the nutritional quality of crops

L. Skärby

SWEDISH WATER AND AIR POLLUTION RESEARCH INSTITUTE, GÖTEBORG, SWEDEN

Introduction

Recently, it was declared that in addition to assessment of crop yield losses, there is a demand for more knowledge about changes in crop quality due to air pollutants (Heck *et al.*, 1982). However, one of the first times crop quality was an issue in discussions on the effects of air pollutants on plants was in the 1960s, when grape farmers in California and later in the eastern USA reported a change in the quality of grapes exposed to ozone. Grape juice from vineyards located in areas affected by photochemical oxidants had a more sour taste than juice from grapes grown elsewhere. Since changes in quality are often more subtle and more difficult to detect or prove, it is understandable that, to date, the effects of air pollutants on crops have been associated with visual symptoms and/or yield losses. These effects are not only more easily detected but also result in direct economic losses.

The quality of a crop comprises three broad aspects:

(1) hygenic and toxicological conditions;
(2) taste and aesthetic values;
(3) relative composition of nutrients, especially proteins, fats, carbohydrates, metals and vitamins.

Table 23.1 provides examples of different changes in the quality of crops exposed to gaseous air pollutants. In this chapter the primary concern will be the changes in the nutritional quality of crops. Since toxic substances, reported to be induced in plants by air pollutants, may drastically change the nutritional value of the crops, reports dealing with such toxicological aspects will also be included.

Literature review

The literature on the subject of air pollutants causing different changes in quality is quite scarce, and reports that deal specifically with changes in nutritional quality are still rarer. In *Table 23.2* a short review is presented on recent research reports which deal with the changes in nutritional quality and nutritional composition in crops due to exposure to gaseous air pollutants.

TABLE 23.1. Examples of changes in the quality of crops exposed to gaseous air pollutants

	Example
(1) Hygenic and toxicological changes	Elevated levels of fluoride in plants
	Elevated levels of pesticides
	Elevated levels of toxic phenolic compounds caused by ozone, changing the metabolism in plants
(2) Taste and aesthetic changes	Taste changes in grapes exposed to ozone
	Necrotic spots on spinach exposed to ozone
	Abscission of flowers in ornamentals exposed to ethylene
(3) Nutritional changes	Changes in nutrient composition due to ozone and SO_2

TABLE 23.2. Literature on changes in nutritional quality and nutritional composition of crops caused by gaseous air pollutants

Plant species	Air pollutant	Quality aspect	Reference
Cabbage Carrot Corn Lettuce Strawberry Tomato	O_3	Nutritional composition	Pippen et al. (1975)
Soybean	SO_2	Protein content	Sàrdi (1981)
Soybean	SO_2	Nutritional composition	Sprugel et al. (1980)
Soybean	O_3	Oil and protein content	Howell and Rose (1980)
Alfalfa	O_3	Nutritional composition	Thompson and Kats (1976)
Alfalfa	O_3	Rate of cell wall digestion (in vitro) Nutritional composition	Howell and Smith (1977)
Alfalfa	O_3	Nutritional quality (Coumestrol, 4',7-DHF)	Hurwitz, Pell and Sherwood (1979)
Alfalfa	O_3	Nutritional quality (4',7-DHF)	Skärby and Pell (1979)
Alfalfa	O_3	Nutritional quality (daidzein, genistein, formononetin)	Vendryes Jones and Pell (1981)
Soybean	O_3	Nutritional quality (coumestrol, daidzein, hydroxyphaseollin, sojagol)	Keen and Taylor (1975)
Potato	O_3	Nutritional quality Nutritional composition	Pell, Weissberger and Speroni (1980)
Potato	O_3	Nutritional quality Nutritional composition	Pell and Pearson (1981)

One of the first scientific reports on effects of air pollutants on crop quality with a focus on nutritional composition was published by Pippen et al. (1975). The study included a survey of the composition of cabbage (*Brassica oleracea capita*), carrot (*Daucus carota*), corn (*Zea mays*), lettuce (*Lactuca sativa*), strawberry (*Fragaria* sp.) and tomato (*Lycopersicon esculentum*), plants grown in carbon-filtered air, 200 ppb O_3, and 350 ppb O_3, respectively. Items determined quantitatively included five vitamins, nitrogen, fibre, solids, ash, carbohydrates and up to nine metals. A number of consistent

TABLE 23.3. Effects of ozone on crop quality (nutritional composition)

Crop	Solids	Nitrogen	Fibre	Ash	Carbo-hydrate	β-Caro-tene	Vit. C	Thiamin	Ribo-flavine	Niacin
Cabbage	↑	↑ NS[a]	↑	↑	↑	Same	↑	↑	↑ NS	↑ NS
Carrot	↓	Same[c]	ND[b]	ND	ND	Same	↓ NS	Same	↓ NS	↑
Corn	↓	↑	ND	ND	↓	↓		↑	ND	ND
Lettuce	Same	↑	Same	Same	Same	Same	Same	↑	Same	Same
Strawberry	Same	↑ ↓	Same	Same	Same	ND	↑ NS	Same	Same	↑
Tomato	↓	Same	↓	↑	↓	Same	↓	↓	↓ NS	↓ NS

Some results from the work of Pippen et al. (1975)
↑ = Significant increase.
↓ = Significant decrease.
[a] NS = Not significant.
[b] ND = Not determined.
[c] Same as control.

trends were evident from the data. For example, O_3 reduced the content of solids in three out of the six crops; some plants seemed to react to O_3 by producing increased amounts of the reducing agent ascorbic acid. *Table 23.3* summarizes the most interesting results. In conclusion, the results provide important indications that ozone does not have a major and general deleterious impact on crop composition but may nevertheless have a serious impact on certain susceptible species or varieties.

The main determining factor for quality in leguminous plants is the protein content of the seeds. Three different studies on the effects of air pollutants on soybeans (*Glycine max*) have been reported recently. Sárdi (1981) noted significant decreases (96%, 87% of control) in the protein content of seeds of soybean and peas (*Pisum sativum*) grown in exposure chambers and exposed to 0.22 ppm SO_2. It was further reported that low levels (0.06 ppm) of SO_2 increased the protein contents (104%, 106% of control) in seeds of both species. The author claims that yields of plants grown in a polluted environment show considerable losses, both in quality and quantity in most cases; therefore, a study of change in the rate of production and final concentration of essential amino acids would be important in assessing the food value of the crops.

Sprugel et al. (1980) reported on the effects of SO_2 on seed yield and quality in field-grown soybeans. Content of nutrients in soybean seeds exposed to SO_2 was compared to control seeds. When exposed to 0.36 and 0.79 ppm SO_2 the concentration of protein in the seed was significantly lower, as seen in *Table 23.4*. The content of oil, nitrogen, phosphorus and potassium did not change significantly. The authors conclude that seed quality, although modified by the SO_2 fumigation treatment, was much less affected than seed yield.

In 1980, Howell and Rose reported effects of ambient ozone concentrations on soybean seed quality. The oil and protein content in the seed was studied and it was found that protein concentrations were significantly higher and the percentage of oil significantly lower in seed produced in unfiltered air compared to filtered air. These data were said to be typical of oil seed crops in that when the percentage of protein increases in the seed, the percentage of oil decreases.

In a study from California, Thompson and Kats (1976) reported various effects of photochemical air pollution on susceptible ('Hayden') and tolerant ('Eldorado') varieties of alfalfa (*Medicago sativa*). The green forage yield of plants grown in the non-filtered air chambers was 42% and 33% less for the susceptible and tolerant variety, respectively, when compared to the forage yield of plants in the filtered air chambers. Chemical analyses of dried forage samples showed that the chemical composition was

TABLE 23.4. Determination of nutrient content of soybean seeds exposed to SO₂

Nutrient	SO₂ 0.36 ppm	SO₂ 0.79 ppm
	(% of control)	
Protein	98	97
Sulphur	—[a]	127
Calcium	—	107
Magnesium	95	92
Copper	—	130
Zinc	107	111
Manganese	—	88
Boron	92	94

(After Sprugel et al., 1980).
[a] = No significant difference.

altered in both varieties grown in the non-filtered air. The result was a plant forage which was lower in fibre, β-carotene, vitamin C and, in some cases, nitrogen, while niacin concentrations were significantly increased. Due to the lower fibre and carbohydrate levels in 'Hayden' grown in non-filtered air, the concentrations of eight out of ten elements analysed were higher. Calcium was the only metal that consistently occurred at higher concentrations in plants of both varieties when grown in filtered air.

In 1977, Howell and Smith also reported quality changes in greenhouse-grown alfalfa exposed to ambient ozone levels and filtered air. A decrease in the *in vitro* rate of cell wall digestion after ozone exposure was observed. It was suggested that intake of forage diets by ruminants might be adversely affected in that ozone-injured alfalfa is more slowly digested and consequently might enlarge ruminal fill. Conversely, the alfalfa exposed to ozone was nutritionally equal or superior to the alfalfa exposed to filtered air. For example, alfalfa in the non-filtered environment had higher nitrogen levels than plants from a filtered air environment. Finally, the authors emphasize the importance of studying the effects of air pollution on crop quality.

Quality studies on alfalfa and potato

In 1977, Dr Eva Pell initiated a study in which the objective was to determine the impact of ozone on qualitative parameters of alfalfa foliage and potato tubers. (*Solanum tuberosum*). It was hypothesized that there was a correlation between the expression of ozone-induced foliar injury and the qualitative status of the plant organ. Flavonoids, when present in excessively high concentrations, can alter the quality of crops. Flavonoids are phenolic compounds which represent a widespread group of water-soluble compounds, mainly coloured red, purple or yellow. These compounds are natural constituents of all vascular plants and they normally exist as phenolic glycosides. In healthy plants, these are present only in trace amounts, while in plants stressed by biotic or abiotic factors, flavonoids can exist in high concentrations. The first qualitative parameter to be considered was coumestrol in alfalfa (Hurwitz, Pell and Sherwood, 1979). Coumestrol belongs to the isoflavonoid group and can exist as an aglycone in trace amounts in healthy tissue of alfalfa (Bickoff *et al.*, 1967). However, if alfalfa becomes infected with pathogenic fungi, coumestrol production is induced

(Sherwood *et al.*, 1970). In contrast to other flavonoids, which on the whole are harmless substances, isoflavonoids have oestrogenic, insecticidal and fungicidal properties. Coumestrol has been found in foliage of soybean, another legume, in response to ozone-induced injury (Keen and Taylor, 1975). Coumestrol as an oestrogenic flavonoid with oestrogenic properties is reported to have caused fertility dysfunctions in sheep (Braden, Hart and Lamberton, 1967). Thus, coumestrol, if found in ozone-exposed alfalfa, would change the nutritional quality of the crop.

In the alfalfa cultivar ('Buffalo' exposed to 0, 0.20 and 0.30 ppm ozone in exposure chambers for 2.5 h, coumestrol was never detected. However, seven other fluorescent compounds accumulated in the ozonized leaves; one of these was identified as 4',7-dihydroxyflavone (4',7-DHF). The concentration of 4',7-DHF increased as the severity of foliar injury increased. Furthermore, foliage exhibited more ozone-induced injury and accumulated higher levels of 4',7-DHF in trials conducted during December, January and February than in trials conducted in August (*Table 23.5*). This variation

TABLE 23.5. Concentration of 4',7-dihydroxyflavone (ppm dry weight) in alfalfa foliage from 'Buffalo' cultivar 48 h after exposure to 580 μg m^{-3} (0.30 ppm)

Time of experiment	4',7-DHF (ppm)
January	110
February	199
August	7

(After Hurwitz, Pell and Scherwood, 1979).

might be attributable to differences in growth, temperature, light quality and quantity that are known to be factors influencing the ozone sensitivity of plants. Very little is known about the function of 4',7-DHF in plants. However, it has been shown to accumulate as an aglycone in response to pathogenic fungal infection in alfalfa (Olah and Sherwood, 1971). The production of 4',7-DHF was also induced in detached alfalfa shoots treated with abiotic stresses such as copper chloride or cadmium chloride (Sherwood *et al.*, 1970). The next step in this work was to establish whether this was a response unique to 'Buffalo' or characteristic of other alfalfa cultivars (Skärby and Pell, 1979). Four additional cultivars ('Ladak', 'Sonora', 'Moapa' and 'Vernal') were exposed to 0 and 0.40 ppm O_3 for 3 h after which the presences and quantity of coumestrol and 4',7-DHF were determined; coumestrol, however, was never detected. Injured foliage of all four cultivars contained elevated levels of 4',7-DHF, but no significant cultivar $\times O_3$ interaction was detected (*Table 23.6*).

A third study included a determination of whether biosynthesis of the oestrogenic isoflavonaids daidzein, genistein and formononetin was induced in alfalfa by O_3 treatment (Vendryes Jones and Pell, 1981). The cultivar 'Moapa' was exposed to 0.30 ppm O_3 for 2 h. Neither of these three compounds was detected in O_3-injured foliage. It is remarkable, then, that much greater concentrations of 4',7-DHF occur both in O_3-injured foliage of alfalfa and in foliage infected by plant pathogens. However, not much is known about the function of 4',7-DHF or its possible modification of the quality of alfalfa.

The possibility that O_3 might induce flavonoid synthesis has been studied in a few other plant species. Keen and Taylor (1975) found that coumestrol, daidzein and sojagol accumulated in soybean foliage in response to O_3-induced injury (*Table 23.7*).

TABLE 23.6. Concentration of 4',7-DHF (ppm dry weight) in alfalfa foliage from four cultivars, 48 h after exposure of 773 μg O$_3$ m^{-3} (0.40 ppm). Each value is the mean of four samples of tissue analysed

Control	Trial			Mean
(L, S, V, M)	I	II	III	
	Not detected	Not detected	Not detected	
Ladak (L)	60 ($\pm$16)	50 ($\pm$19)	42 ($\pm$17)	51
Sonora (S)	41 ($\pm$16)	64 ($\pm$28)	35 ($\pm$11)	47
Vernal (V)	43 ($\pm$18)	47 ($\pm$3)	46 ($\pm$19)	45
Moapa (M)	44 ($\pm$20)	53 ($\pm$15)	47 ($\pm$20)	48

(After Skärby and Pell, 1979.) Standard deviations in parentheses.

TABLE 23.7. Levels of coumestrol, daidzein, sojagol and hydroxyphaseollin (ppm fresh weight) in soybean leaves fumigated with 0.7 ppm ozone for 3 h or inoculated with an incompatible race of *Pseudomonas glycinea*

Treatment	Coumestrol	Daidzein	Hydroxyphaseollin	Sojagol
Control	5	50	<5	<1
Ozone-exposed	500	800	<5	35
Inoculated with P. glycinea	350	2000	810	40

(After Keen and Taylor, 1975.)

TABLE 23.8. Hydroxyphaseollin and coumestrol (ppm fresh weight) levels in soybean leaves exposed 2 h to various air pollutants or inoculated with an incompatible race of *Pseudomonas glycinea*

Treatment	Coumestrol	Hydroxyphaseollin
Control	5	<5
Ozone 0.6 ppm	690	<5
PAN 0.4 ppm	11	<5
NO$_2$ 18 ppm	50	<5
SO$_2$ 2.4 ppm	25	<5
P. glycinea	400	1200

(After Keen and Taylor, 1975.)

The production of coumestrol was also induced by NO$_2$, SO$_2$ and PAN but to a much lesser degree (*Table 23.8*). An accumulation of anthocyanins has been reported in morning glory (*Ipomoea* sp.), curly dock (*Rumex crispus* L) and poinsettia (*Euphorbia pulcherrima* Wild.) after exposure to ozone (Nouchi and Odaira, 1973; Koukol and Dugger, 1967; Craker and Feder, 1972). A number of studies have also been carried out on the influence of air pollution on the accumulation of other phenolic compounds in plants (Howell, 1974; Howell and Kremer, 1973). Howell, Devine and Hansen (1971) reported an increased production of caffeic acid in beans. Menser and Chaplin (1969) studied the effects on phenol and alkaloid content in tobacco leaves (*Nicotiana tabacum*

L.) after exposure to O_3 and found higher concentrations of polyphenols in leaves injured by O_3 than in healthy tobacco leaves.

The mechanisms of the induction of the biosynthesis of flavonoid compounds, as a response to different environmental stresses, have been discussed but they are not yet clearly understood. Olah and Sherwood (1971) have suggested that fungal-induced accumulation of flavonoid aglycones may involve conversion of pre-existing flavonoid glycosides to aglycones. Possibly glycosidases, necessary for the conversion, are produced by the host in response to the pathogen or are actually provided by the plant pathogen itself. The host may also synthesize aglycones immediately. The final step, the glycosylation, might be impaired by environmental stresses such as O_3. In parallel with this, Keen and Taylor (1975) suggested that O_3-induced foliar injury in soybeans operates through invocation of the hypersensitive disease-resistance response. This means that the toxicity associated with stippling and necrosis in O_3-damaged plants may be due to the post-treatment production of flavonoids and other phenolic compounds by the plant. Others have reported that hormones such as ethylene and cytokinins control the synthesis of flavonoids (Dedio and Clark, 1971; Craker, 1971). Craker, Standley and Starbuck (1971) and Tingey, Standley and Field (1976) reported an increased ethylene production from O_3-injured plants. One hypothetical explanation concerning the mechanisms of induction of flavonoid synthesis in plants exposed to O_3 might therefore be an increased ethylene production.

In 1977, Dr Eva Pell and co-workers started experiments to test the impact of O_3 on the quality and quantity of tubers from four different cultivars of potato known to be either susceptible or tolerant to O_3. They postulated that if potato plants were exposed to O_3, the tubers would be reduced in weight and number and that qualitative parameters including dry matter, starch and sugar content would be altered (*Table*

TABLE 23.9. Quality parameters studied in potato tubers exposed to ozone

Starch	
Nitrogen	Nutritional quality
Glycoalkaloids	
Total solids	Mealiness
Reducing sugars	Chip colour

(After Pell *et al.*, 1980; Pell and Pearson, 1981)

23.9). Total solids have been evaluated because of their direct correlation with mealiness; sugar status was determined because of its role in chip-colour. Finally, the concentration of glycoalkaloids was determined since high levels of alkaloids, especially the toxic substances solanine and chaconine in potatoes, will drastically change the quality of the potato. The number and weight of tubers were reduced in all cultivars when exposed to 0.20 ppm O_3 for 3 h once every other week for three months. Starch content was not affected; results on nitrogen content have not yet been published. Total solids were also significantly reduced by O_3 in all cultivars. The concentration of glycoalkaloids in tubers decreased significantly in three of the four cultivars. An increase was observed in tubers from 'Cherokee', a susceptible cultivar, which was a new finding. Finally, concentrations of reducing sugars in tubers increased in three of the four cultivars which showed that O_3 has the potential to increase sugar levels (Pell, Weissberger and Speroni, 1980, Pell and Pearson, 1981). The conclusion from these experiments is that foliar response as a criterion for identifying susceptible and tolerant cultivars is not always reliable. Ozone has the potential to alter tuber quality as regards content of total solids, sugars and total glycoalkaloids without necessarily inducing foliar injury.

Discussion

Although the biochemical effects of air pollutants do not always result in changes in the nutritional quality of the crop, it is evident from the literature reviewed here that such a potential to alter crop quality does, in fact, exist. Until now, relatively little interest has been focused on research into the quality of the harvested crop. There may be two reasons for there being only a few laboratories working on the effects of air pollutants on crop quality parameters. The first is that, from an economic point of view, changes in the nutritional quality of crops is not yet regarded to be as 'important' as changes in the taste, aesthetic quality, absence of toxic substances and the useful life of the harvested crop. The second reason is that assessing changes in crop quality is complicated by the fact that such changes are not necessarily connected with foliar injury or yield losses; further, the biochemical and physiological mechanisms behind specific changes in quality are not yet fully understood. An increased awareness of the nutritional value of different crops warrants further research on changes in nutritional quality as an important parameter in studying the effects of air pollutants on crops. It is evident that such research activities constitute an important initiative in a field where consistent information on the effects of crop quality is otherwise lacking.

To solve one part of the environmental problem of the effects of air pollutants on plants, one strategy might be to test breeding lines for resistance, but such a strategy aimed at avoiding adverse effects of pollutants on crop quality is far from being realized without considerable additional research. A discussion of such control strategies lies beyond the scope of this chapter.

The research on nutritional composition of plants is quite a complicated issue since this is dependent on many environmental factors other than air pollutants. In the western countries, however, a joint FAO/WHO Codex Alimentarius Commission exists to standardize acceptable levels of different nutrients and other substances in food products. Detailed requirements on nutritive quality will be available in the near future. This work is important as basic information for scientists working with air pollutants that give rise to alterations in nutritive composition of crops.

The present chapter has attempted to sum up existing knowledge about effects on nutritional quality of inorganic air pollutants. When it comes to effects of organic air pollutants such as ethylene, polyaromatic hydrocarbons, etc., such knowledge is, to judge from the literature, clearly lacking. The crops that have been studied so far are mainly leguminous plants and potatoes. We have very little consistent information on the effects of air pollutants on nutritional composition and nutritional quality of important products such as cereals, corn, vegetables, fruits and berries.

References

BICKOFF, E. M., LOPER, G. M., HANSON, C. H., GRAHAM, J. H., WITT, S. C. and SPENCER, R. R. Effect of common leafspot on coumestans and flavones in alfalfa, *Crop Science* 7, 259–261 (1967)
BRADEN, A. W. H., HART, N. K. and LAMBERTON, J. A. The oestrogenic activity and metabolism of certain isoflavones in sheep, *Australian Journal of Agricultural Research* 18, 335–348 (1967)
CRAKER, L. E. Ethylene production from ozone injured plants, *Environmental Pollution* 1, 229–304 (1971)

CRAKER, L. E. and FEDER, W. A. Development of the inflorescence in petunia, geranium and poinsettia under ozone stress, *Horticultural Science* **7**, 59–60 (1972)

CRAKER, L. E., STANDLEY, L. A. and STARBUCK, M. J. Ethylene control of anthocyanin synthesis in Sorghum, *Plant Physiology* **48**, 349–352 (1971)

DEDIO, W. and CLARK, K. W. Influence of cytokinins on isoflavone and anthocyanin synthesis in red clover seedlings, *Pesticide Science* **2**, 65–68 (1971)

HECK, W. W., TAYLOR, O. C., ADAMS, R., BINGHAM, G., MILLER, J., PRESTON, E. and WEINSTEIN, L. Assessment of crop loss from ozone, *Journal of Air Pollution Control Association* **32**, 353–361 (1982)

HOWELL, R. K. Phenols, ozone and their involvement in pigmentation and physiology of plant injury, *ACS Symposium Series No. 3, Air Pollution Effects on Plant Growth*, (Ed. M. Dugger), pp. 94–105, American Chemical Society, Washington, DC (1974)

HOWELL, R. K., DEVINE, T. E. and HANSON, C. H. Resistance of selected alfalfa strains to ozone, *Crop Science* **11**, 114–115 (1971)

HOWELL, R. K. and KREMER, D. F. The chemistry and physiology of pigmentation in leaves injured by air pollution, *Journal of Environmental Quality* **2**, 434–438 (1973)

HOWELL, R. K. and ROSE, L. P. Residual air pollution effects on soybean seed quality, *Plant Disease* **64**, 385–386 (1980)

HOWELL, R. K. and SMITH, L. W. Effects of ozone on nutritive quality of alfalfa, *Journal of Dairy Science* **60**, 924–928 (1977)

HURWITZ, B., PELL, E. and SCHERWOOD, R. T. Status of coumestrol and 4′,7-di-hydroxyflavone in alfalfa foliage exposed to ozone, *Phytopathology* **69**, 810–813 (1979)

KEEN, N. T. and TAYLOR, O. C. Ozone injury in soybean. Isoflavonoid accumulation is related to necrosis, *Plant Physiology* **55**, 731–733 (1975)

KOUKOL, J. and DUGGER, W. M. Jr. Anthocyanin formation as a response to ozone and smog treatment in *Rumex crispus* L., *Plant Physiology* **42**, 1023–1024 (1967)

MENSER, H. A. and CHAPLIN, J. F. Air pollution: Effects on the phenol and alkaloid content of cured tobacco leaves, *Tobacco* **169**, 73–74 (1969)

NOUCHI, I. and ODAIRA, T. Influence of ozone on plant pigments, *Chemical Abstracts* **79**, 122381 (1973)

OLAH, A. F. and SHERWOOD, R. T. Flavones, isoflavones and coumestans in alfalfa infected by *Ascochyta imperfecta*, *Phytopathology* **61**, 65–69 (1971)

PELL, E. J., WEISSBERGER, W. C. and SPERONI, J. Impact of ozone on quantity and quality of greenhouse-grown potato plants, *Environmental Science & Technology* **14**, 568–571 (1980)

PELL, E. J. and PEARSON, N. S. The impact of ozone on tuber characteristics of greenhouse-grown 'Norchip' and 'Cherokee' potato cultivars. (Abstract), *74th APCA Annual Meeting and Exhibition*, June 21–26, 1981, Philadelphia, PA, USA (1981)

PIPPEN, E. L., POTTER, A. L., RANDALL, V. G., NG, K. G., REUTER III, F. W., MORGAN, A. G. Jr. and OSHIMA, R. J. Effect of ozone fumigation on crop composition, *Journal of Food Science* **40**, 672–676 (1975)

SÁRDI, K. Changes in the soluble protein content of soybean *Glycine max* L. and pea *Pisum sativum* L. under continuous SO_2 and soot pollution, *Environmental Pollution* **25**, 181–186 (1981)

SHERWOOD, R. T., OLAH, A. F., OLESEN, W. H. and JONES, E. E. Effect of disease and injury on accumulation of a flavonoid estrogen, coumestrol, in alfalfa, *Phytopathology* **60**, 684–688 (1970)

SKÄRBY, L. and PELL, E. J. Concentrations of coumestrol and 4′,7-dihydroxyflavone in four alfalfa cultivars after exposure to ozone, *Journal of Environmental Quality* **8**, 285–286 (1979)

SPRUGEL, D. G., MILLER, J. E., MULLER, R. N., SMITH, H. J. and XERIKOS, P. B. Sulfur dioxide effects on yield and seed quality in field-grown soybeans, *Phytopathology* **70**, 1129–1133 (1980)

THOMPSON, C. R. and KATS, G. Effect of photochemical air pollution on two varieties of alfalfa, *Environmental Science & Technology* **10**, 1237–1241 (1976)

TINGEY, D. T., STANDLEY, C. and FIELD, R. W. Stress ethylene evolution: a measure of ozone effects on plants, *Atmospheric Environment* **10**, 969–974 (1976)

VENDRYES JONES, J. and PELL, E. J. The influence of ozone on the presence of isoflavones in alfalfa foliage, *Journal of Air Pollution Control Association* **31**, 885–886 (1981)

Chapter 24

Relationship of biochemical effects of air pollutants on plants to environmental problems: insect and microbial interactions

P. R. Hughes and J. A. Laurence

BOYCE THOMPSON INSTITUTE AT CORNELL UNIVERSITY, ITHACA, NY, USA

Introduction

Studies of effects of air pollution on ecosystems generally have not taken into account the possible interactions with other factors such as plant parasites (i.e. insects and pathogenic micro-organisms). Direct losses in plant productivity due to these pests have been studied intensively over many years and are estimated to exceed $30 billion per year in the USA alone (James, 1980). It has been recognized recently that if only a fraction of these losses were induced by interactions with air pollutants, the impact of the interaction in terms of economic loss and ecological perturbation could far exceed that due to air pollution alone. This realization has led to an increased interest in investigation of the effects of air pollutant-induced physiological and biochemical changes in plants on plant–pest interactions. The importance of this subject has been recognized by the National Crop Loss Assessment Network program in the USA (Heck *et al.*, 1982).

To examine the possible consequences of such changes in plants on the parasites, one should begin with an understanding of the basic plant-parasite relationship. In order to coexist, plants and their parasites must be in a state of equilibrium in which neither totally dominates the other. The pest's ability to utilize the plant must be counterbalanced by the plant's ability to defend against excessive herbivory or parasitism. The evolutionary result of this mutual antagonism is thought to be the various defenses of the plant, including antimicrobial phytoalexins and post-inhibitors (reviewed by Harborne and Ingham, 1978), antiherbivore allelochemics (Rosenthal and Janzen, 1979), glandular hairs (Tingey and Gibson, 1978; Johnson, Sorensen and Horber, 1980) and toxic non-protein amino acids (reviewed by Rosenthal and Bell, 1979). The parasites in return have evolved a variety of mechanisms for avoiding or overcoming the defenses of a plant species or group of plants (Rosenthal, Dahlman and Janzen, 1976; Levin, 1976; Harborne and Ingham, 1978; Brattsen, 1979; Bernays and Woodhead, 1982a,b). In general, neither the plant defenses nor the parasite countermeasures are completely effective, maintaining the balance in which both are able to exist but neither realizes its full potential.

The result of this coadaptation is a dependency of plant parasites on their hosts for both nutrition and for many chemical and physical cues that regulate their feeding, reproduction, and in the case of pathogenic microbes, infection behavior. Plants are often only marginally adequate nutritionally for insects (Southwood, 1973), and insects are known to be strongly affected by slight changes in the quality of their hosts (Hukusima, 1958; House, 1961; Southwood, 1973). The dependency of insects on chemical cues of plant origin has been the subject of many publications (Beroza, 1970; Wood, Silverstein and Nakajima, 1970; Jermy, 1976; Shorey and McKelvey, 1977; Hedin, Jenkins and Maxwell, 1977), and Dethier (1970) presents an overview. These chemical messengers are generally controlling steps in the sequence of behavior leading to feeding or oviposition and include attractants, repellents, stimulants, deterrents and arrestants. Pathogenic micro-organisms also depend on plant compounds for chemical and physical cues (*see* discussion by Preece, 1976; Dodman, 1979; Sequeira, 1980).

These chemical and physical properties of the plant are determined by the specific plant biochemistry and are subject to all the variations imposed by the plant's interaction with its environment (Sokolov, 1959; Steward *et al.*, 1959; Flück, 1963). Thus, soil characteristics, including chemical and physical factors, affect the production of plant compounds, as do climatic conditions, such as light intensity and quality, day length, temperature, relative humidity and rainwater or dew.

Air pollutants constitute another class of environmental factors that induce biochemical changes in plants. They can affect stomatal function, photosynthesis, respiration, translocation and many specific biochemical processes (*see* reviews by Ziegler, 1973; Wellburn *et al.*, 1976; Hällgren, 1978).

Such effects on the plant could be expected to upset the equilibrium between the plants and parasites. Indeed, one can find a vast literature relating plant exposure to air pollutants with subsequent changes in plant parasitism. As early as 1923, a relationship between pollution in the forest and outbreaks of insect pests was reported (Evenden, 1923). Since that time numerous reports concerning both insects and diseases have appeared, particularly from North America and Europe (Alstad, Edmunds and Weinstein, 1982; Laurence, 1981; Hay, 1977). If, in the context of this chapter, the term 'environmental problem' is defined as a change in the density or species composition of plant parasites, then these reports constitute considerable circumstantial evidence that biochemical changes induced in plants by air pollutants have a significant impact on environmental problems.

In this chapter, we will first examine more closely the evidence that a relationship exists between plant exposure to air pollutants and changes in plant-parasite populations. Then we will examine the changes known to occur in plants exposed to pollutants and discuss their relevance to the success of plant parasites.

Associations between air pollution and changes in plant parasitism

Insects

There are several hundred reports concerning effects of air pollutants on the plant–insect relationship. These have been partially reviewed by Alstad, Edmunds and Weinstein (1982) and Heagle (1973). The bibliography by Hay (1977) contains a number of additional references on the subject.

All but five of these papers have been descriptions of correlations between a measure of insect presence (e.g. damage, numbers, etc.) and a measure of pollutant presence (e.g. foliar concentration, damage, distance to the source). Almost all of these studies have been conducted in forest ecosystems, but a few have examined effects in cultivated plants (Przybylski, 1967, 1968, 1974, 1979) or along roadways (Braun, Flückiger and

Handwritten margin note (top): include in crop extensions that SO₂ could be studied in relation to insect preference — relationships, i.e. viruses, or fungus etc.

Oertli, 1981; Port and Thompson, 1980; Flückiger, Oertli and Baltensweiler, 1978). In general, no correlation or positive correlations have been found for most insect species at low or intermediate pollutant concentrations while negative correlations have been found at very high levels. Frequently, increased populations of so-called 'secondary' pests, those that are restricted primarily to weak hosts, were noted (Pfeffer, 1963; Donaubauer, 1966; Stark *et al.*, 1968; Bosener, 1969; Sierpinski, 1972). Changes in populations of these insects might be more accurately interpreted as reflections of changes in the proportion of weak trees in the forest rather than of a change in the plant–insect relationships.

The effects of gaseous air pollutants on the growth, development, fecundity, longevity and feeding behavior of Mexican bean beetles (*Epilachna varivestis* Mulsant) also have been examined, using either bean (*Phaseolus vulgaris* L.) or soybean (*Glycine max* [L.] Merrill) as the hosts. Beetles fed on bean foliage fumigated with HF developed more slowly and weighed less than those fed on control foliage (Weinstein *et al.*, 1973). The treated plants in these experiments contained about 1000 ppm F in the unifoliate leaves, which is a high level for an herbaceous plant. When bean foliage was fumigated with SO_2, no major effects were observed on beetle growth, development, fecundity or longevity, but adults exhibited a marked feeding preference for the fumigated leaves (Hughes, Potter and Weinstein, 1981). In similar experiments with soybean, the beetles grew larger, developed faster and were more fecund when fed on SO_2-fumigated foliage than on control leaves; again, adult females showed a distinct feeding preference for fumigated leaves (Hughes, Potter and Weinstein, 1982). These results on soybean have been confirmed recently in a field study in which the plants were exposed to a seasonal average of 43 μg SO_2 m^{-3} (i.e. an average of 367 μg m^{-3} for 191 h in 68 days) (Hughes *et al.*, in press). Benepal, Rangappa and Dunning (1979) reported in an abstract that feeding damage by this insect on ozone-fumigated beans (1568 μg m^{-3} for 2 h) was approximately double that on control plants. However, the full details of this work are not yet available.

Handwritten margin note (right): beetles prefer SO₂ fumigated leaves.

McNary *et al.* (1981) examined large areas of native mixed grass prairie that had been exposed to monthly median SO_2 levels of 73, 134 or 228 μg m^{-3} and found a tendency for grasshopper density to decrease with increasing SO_2 concentration. However, the experimental design did not permit a conclusion to be drawn as to whether this decrease was due to a change in the emigration–immigration ratio between plots or due to a reduction in reproduction.

The evidence available demonstrates that biochemical changes in plants induced by gaseous air pollutants can affect insect success. The numerous correlative-type studies have provided needed information but fail to establish a cause–effect relationship or to elucidate which of the many components present in a polluted area or 'fume zone' are responsible for the changes. However, the controlled experiments with HF, SO_2 and O_3 establish the cause–effect relationship in at least a few cases. Furthermore, these effects can be elicited under field conditions with pollutant levels well within those occurring naturally (Hughes *et al.*, in press).

Pathogens

Air pollution-induced modifications of the development of plant disease have been recognized for some time, but carefully controlled studies designed to quantify these effects are relatively new and very few studies have been performed under field conditions. The subject has been reviewed and the documented relationships between pollutants, pathogens and plants discussed (Heagle, 1973; Laurence, 1981). Although specifics of these interactions are available in those references, an overview of known effects is appropriate and will provide background for a discussion of the possible metabolic causes of observed relationships.

Fungal diseases have been reported to be stimulated or inhibited by exposure of host plants to air pollutants. Diseases caused by obligate parasites are almost always inhibited, but some caused by non-obligate pathogens, such as *Armillaria mellea*, *Fomes annosus* and *Botrytis* spp., increase in severity on pollutant-stressed plants. The diseases that are stimulated are those that are considered to be secondary problems but occur on weakened trees or plants. The increased success of *Botrytis* spp. is probably related to the production of infection courts (i.e. foliar lesions) by the pollutant.

The development of virus diseases is usually greater in plants exposed to air pollutants. Very few of these diseases have been investigated and there is little direct evidence as to the cause of the stimulation. Perhaps virus replication is increased as the plant repairs damage resulting from the pollutant. In addition to pollutant-induced increases in disease development, virus infection has often been shown to protect plants from ozone (Bisessar and Temple, 1977; Davis and Smith, 1974; Vargo, Pell and Smith, 1978); however, no definitive work has been done to explain this phenomenon.

Bacterial diseases are almost always inhibited in plants that have been exposed to air pollutants. Work with these pathogens has shown that not only are lesions smaller on plants exposed to pollutants, but latent periods of the pathogen may be increased, possibly resulting in changes in the epidemiology of the disease (Laurence and Reynolds, in press; Laurence and Aluisio, 1981). However, field experiments have not been performed to assess these effects on disease development in a population of plants.

Physiological bases for alterations of plant-parasite relations by air pollutants

Although a cause–effect relationship between air pollutant-induced changes in plants and altered parasite success has been demonstrated in many studies, especially with pathogens, the biochemical bases for these effects have been totally neglected. However, many biochemical changes in plants exposed to specific pollutants have been described, including alterations in free amino acids, simple sugars, lipids, steroids, flavonoids and other secondary plant compounds such as terpenes. Much is known about the importance of many of these compounds or groups of compounds to the nutrition and behavior of pathology of the parasites. Therefore, although the implications are somewhat speculative at this time, a consideration of specific pollutant-induced changes and the relevance of these to parasite success would seem useful, especially in stimulating ideas for future research.

Insects

Since the success of phytophagous insects is dependent on plant metabolites to guide feeding and oviposition behavior and to meet nutritional needs for growth and reproduction, air pollutant-induced changes can be grouped for convenience into two categories:

(1) those that affect host vulnerability to discovery by insects (i.e. host apparency, *see* Feeny, 1976), and
(2) those that affect the nutritional quality of the host.

EFFECTS ON PLANT APPARENCY

Changes in plant physical characteristics (e.g. color) or in secondary metabolites used by insects in host location and recognition would alter host appparency. Many color changes in plants have been reported in conjunction with injurious levels of pollutants (Malhotra and Blauel, 1980). Since orientation by many insects such as aphids and day-flying lepidoptera is known to be strongly affected by color, pollutant-related color changes are likely to be influencing host selection and subsequent colonization by some pest species. However, to our knowledge this possibility has not been investigated to date.

A few reports describe changes in secondary plant metabolites that are known to affect insect orientation. Emission of volatile terpenes from balsam fir (*Abies balsamea*) trees increased following exposure to SO_2 (5240 μg m^{-3}, 5 h day^{-1} for three days) (Renwick and Potter, 1981). Similarly with spruce, the volatile oil content was lower in foliage from smoke-damaged (Cvrkal, 1959) or SO_2-fumigated (Dässler, 1964) trees, as one would expect if exposure to the pollutants were causing increased loss to the atmosphere. Volatile terpenes are known to serve as chemical messengers for many pests of conifers, including many species of bark beetles and foliage-feeding lepidoptera (*see* review by Wood, 1982; Städler, 1974). In addition to affecting insect orientation, volatile terpenes emanating from plants are known to stimulate sexual maturation in locusts (Carlisle, Ellis and Betts, 1965; Ellis, Carlisle and Osborne, 1965).

Injury by SO_2 causes the production of ethane by many plants (Bressan *et al.*, 1978; Peiser and Yang, 1979). Recently, Kimmerer and Kozlowski (1982) showed that injury by SO_2 (786 μg m^{-3} continuously for three to five days) increased ethane production by red pine (*Pinus resinosa*) and paper birch (*Betula papyrifera*). Ethane acts as a strong repellent to *Monochamus alternatus*, a wood-boring beetle (Sumimoto, Shiraga and Kondo, 1975).

These few examples demonstrate that air pollutants cause biochemical changes that can affect chemical cues regulating insect behavior. It is important to note, however, that the SO_2 concentrations in all of these studies were quite high and in excess of the current US secondary air quality standard.

EFFECTS ON HOST QUALITY

Pollutants can affect the nutritional quality of plants as hosts for insects in several ways, including altering plant nutrition, causing changes in primary or secondary plant metabolites, or changing plant water balance.

Plant nutrition
Plant nutrition can be strongly affected by the lowered soil pH caused by prolonged wet or dry deposition of pollutants such as oxides of sulfur or nitrogen or HCl vapor (Ziegler, 1973; Nyborg, 1978 and references therein; National Academy of Sciences, 1977; Evans, 1982). Toxic levels of elements such as iron, manganese and aluminum can occur as a result of their increased solubility at low pH. Other elements, such as phosphorus, calcium, magnesium and molybdenum can become limited in supply at the low pH. Plant nutrients can also be rendered unavailable by complexing with pollutants such as fluoride (Weinstein and Alscher-Herman, 1982). Acidic precipitation caused by oxides of sulfur and nitrogen can increase leaching of nutrients from foliage (Wood and Bormann, 1974; Fairfax and Lepp, 1975).

Pollutants might also affect the availability of plant nutrients by reducing litter breakdown, fragmentation, decomposition, and chemical deterioration, thus interfering with nutrient cycling. Several studies have shown that some soil arthropods, which are responsible for much of the litter breakdown and fragmentation, are generally reduced in contaminated areas (Strojan, 1978 and references therein; Bromenshenk, 1979). Likewise, microbes which are largely responsible for litter decomposition and chemical deterioration can be affected by pollutants, especially heavy metals and acidic deposition (Smith, 1975; National Academy of Sciences, 1977; Hågvar and Amundsen, 1981). In addition to microbes involved in litter decomposition, those involved in nitrogen fixation can be very sensitive to pollutants (Nyborg, 1978 and references therein). The effect of pollutants on soil arthropods and microbes varies, depending on the conditions and the species, but the final step of showing that litter decomposition and subsequent nutrient cycling are being depressed has yet to be taken.

Pollutants can affect the plants in a positive manner by serving as nutrients. This has been best documented for oxides of sulfur (Nyborg, 1978; Cowling and Lockyer, 1976; Faller, 1971, 1972) and N-containing gases (Faller, 1972; Troiano and Leone, 1977). Nyborg (1978) presents calculations to demonstrate that the benefits of SO$_2$ emissions for overcoming sulfur deficiencies in agricultural crops are minor. The impact on other ecosystems, especially lower producing systems, may be much greater, but not all plants in deficient soils seem to respond (Elkiey and Ormrod, 1981; Bell and Clough, 1973).

Alteration of insect success by changes in plant quality associated with varied plant nutrition is well established. Although the mechanisms are unknown, changes in nitrogen or any of the macronutrients (K, Ca, Mg, P or S) or in any of several micronutrients examined affected the growth and/or reproduction of a wide range of phytophagous insects (Barker and Tauber, 1951; Allen and Selman, 1955; Todd, Parker and Gaines, 1972; Semtner, Rasnake and Terrill, 1980, and citations in these references).

Primary metabolites

Gaseous air pollutants induce many biochemical changes in plants that influence primary metabolite concentrations. The carbohydrate content of leaves, especially the concentrations of reducing sugars, appears to increase at low level exposures to O$_3$ or SO$_2$ and then decrease with prolonged exposures (Koštíř et al., 1970; Koziol and Jordan, 1978; Pell, Weissberger and Speroni, 1980; Prasad and Rao, 1981; Blum, Smith and Fites, 1982). Acute or chronic exposure of a variety of plants to fluoride caused increases in total free amino acids, organic acids, and reducing sugars, while the sucrose content decreased (Weinstein, 1961; Yang and Miller, 1963a,b; Arndt, 1970). The free amino acid content, especially γ-aminobutyric acid and alanine, of beans (*Phaseolus*), beets (*Beta*), corn (*Zea*), barley (*Hordeum*) and rye (*Secale*) increased when the plants were exposed to acute or chronic concentrations of O$_3$ (Tomlinson and Rich, 1967). Free amino acids and soluble protein also increased in tissues of bean (*Phaseolus*), soybean (*Glycine*) and pea (*Pisum*) exposed to low concentrations of SO$_2$, but exposure to higher doses resulted in a decreased concentration of proteins (Godzik and Linskens, 1974; Grill, Esterbauer and Beck, 1975; Priebe, Klein and Jäger, 1978; Sárdi, 1981). In field studies, Flückiger, Flückiger-Keller and Oertli (1978), Flückiger, Oertli and Baltensweiler (1978) showed that levels of free amino acids and reducing sugars in foliage of birch (*Betula pendula*) and hawthorn (*Crataegus monogyna*) increased with decreasing distance from a motorway. The extent to which these differences depend on pollutants from automobiles or microclimatic changes as-sociated with the motorway has yet to be determined.

Lipid biosynthesis in needles of lodgepole pine (*Pinus contorta*) and jack pine (*P. banksiana*) was markedly reduced by exposure of plants to SO_2 (472–917 μg m^{-3} for 24 h) (Malhotra and Khan, 1978). Sulfur dioxide-induced alterations in the surface of conifer needles have been observed, with the implication that SO_2 stimulates the production of waxes (*see* Hällgren, 1978 and references therein). However, such a stimulation remains to be demonstrated unequivocally.

The relevance of such changes in primary plant metabolites to phytophagous insects is manifold. Any of these compounds can serve in the regulation of insect feeding behavior, depending on the species (Augustine *et al.*, 1964; Mittler, 1970; Bernays and Chapman, 1978), and quantitative or qualitative changes can strongly affect insect growth and/or reproduction (Johansson, 1964; McNeil and Southwood, 1978). Surface wax components are involved in host acceptance (Bernays and Chapman, 1978) and recent evidence suggests a relationship between host plant lipids and the susceptibility of the Mexican bean beetle to desiccation (Wilson, 1981).

Secondary metabolites
Also subject to air pollutant-induced changes are many so-called 'secondary plant metabolites' that affect the acceptability or quality of plants as hosts for insects. SO_2, NO_2 or O_3 can cause the accumulation of isoflavonoids in leaves, although this might vary with plant species (Jones and Pell, 1981; Hurwitz, Pell and Sherwood, 1979). SO_2 fumigation increased the phenolic content of spruce (*Picea*) needles (Grill, Esterbauer and Beck, 1975) and O_3 appears to permit enzymatic oxidation of phenols by impairing cell membranes (Howell, 1974). Thalenhorst (1974) reported a positive correlation between spruce exposure to atmospheric pollutants (predominantly fluoride compounds) and the density and survival rate of *Sacchiphantes abietis* (L.) pseudo-fundatrices. Furthermore, he found that increased aphid survival was correlated with a diminished capacity of the exposed trees to produce a specific defensive reaction in response to host penetration by the aphids; the biochemical basis of this defense reaction has not been described but it very likely involves secondary compounds such as phenolics. Although plant phenols have generally been regarded as deterrent or deleterious to insects (Isman and Duffey, 1982; Levin, 1976) more recent reports have shown that they may also serve a nutrient role (Bernays and Woodhead, 1982a,b; Schopf, Mignat and Hedden, 1982). The phenolic, coumarin, has been suggested to be involved in the host specificity of several insects whose feeding is deterred by the chemical (Mansour, Dimetry and Rofaeel, 1982 and references therein). Many examples of such compounds affecting insect behavior or development can be found in the literature (*see* reviews by Beck and Reese, 1976; Hedin, Jenkins and Maxwell, 1977; Duffey, 1980).

Plant water balance
Another important way in which air pollutants, notably SO_2, can affect the plant–insect relationship is by altering the plant water balance. The effects of SO_2 on plant water loss are summarized by Hällgren (1978); even very low concentrations can greatly increase loss, especially under drought conditions. There are numerous reports hypothesizing that insect outbreaks can be related to drought stress of the host plants or that demonstate increased success of insects on water-stressed plants (Kennedy and Booth, 1959; Wearing and van Emden, 1967; White, 1974, 1976; Lewis, 1979). In general, the insects studied to date tended to do better on stressed plants, presumably

due to increased nutritional value of the plant resulting from higher concentrations of such materials as free amino acids, inorganic acids, and simple sugars. However, Scriber (1978) and Scriber and Slansky (1981) have summarized evidence that if the leaf water content becomes too low, insect success can be negatively affected.

Hormesis and hormonal changes
The stimulation of plant growth by low levels of pollutants has been reported for HF, SO_2 and O_3 (*see* brief review of Bennett, Resh and Runeckles, 1974). This apparent enhancement of physiological processes by low doses of a toxicant, termed hormesis (Stebbing, 1979) or hormoligosis (Luckey, 1968), is well known in animals. Considering the sensitivity of insects to differences in host quality and the improved quality of faster growing plants (Mattson, 1980), insect success may be greater on plants stimulated in this manner.

Little is known about the effects of pollutants on the biosynthesis, translocation or activity of plant growth hormones. However, there is evidence that SO_2 affects some of the hormones (Hällgren, 1978). Sulfite has been shown to enhance oxidation of indole-3-acetic acid *in vitro* (Meudt, 1971; Yang and Saleh, 1973), and O_3 as well as SO_2 caused increased production of ethylene in a great number of plants (Tingey, Standley and Field, 1976; Bressan *et al.*, 1978; Peiser and Yang, 1979; Kimmerer and Kozlowski, 1982). In addition, a number of *in vivo* effects of SO_2 on plant germination, flowering, leaf abscission, and senescence might be explained on the basis of pollutant interaction with plant hormones (Hällgren, 1978).

The effect of plant hormones on insect success has received scattered attention with contradictory results (*see* references in Neumann-Visscher, 1982). Recent studies by Neumann-Visscher and co-workers have been directed towards resolving these differences and clarifying the relationship between plant hormones and insect growth, fecundity and longevity. Information to date indicates that, at least for some grasshoppers and one aphid species, growth inhibiting hormones such as abscisic acid and ethylene can cause a depression of insect growth and reproduction while growth stimulating hormones such as indoleacetic acid, gibberellic acids, and cytokinins are associated with increased longevity and fecundity (Neumann-Visscher, 1982). Whether these effects are due to direct action of the hormones on insects or to hormone-related changes in the host tissue is not known. However, the precise mechanism would not be expected to change the impact of pollutant-induced hormonal shifts on the phytophagous insects.

The decreased grasshopper populations observed by McNary *et al.* (1981) in SO_2-fumigated plots of prairie grass might represent an example of a pollutant affecting insects by causing a shift in plant hormones. Accelerated plant senescence due to exposure to SO_2 has been reported many times, and the low gibberellin content of senescent vegetation is known to depress locust maturation and reproduction (Ellis, Carlisle and Osborne, 1965).

Pathogens

McCune *et al.* (1973) proposed three possible ways in which pollutants could alter pathogen success:

(1) a direct effect of the pollutant on the pathogen;
(2) an indirect effect on the pathogen through modification of the physiology or biochemistry of the host; or

(3) an alteration in the suitability of the host's organs as habitats for the pathogen (such as more or fewer infection sites).

It is generally believed that micro-organisms are tolerant to the direct action of air pollutants, particularly at ambient levels (Couey, 1965; Couey and Uota, 1961; Harding, 1968); however, it is possible that pathogens are sensitive to pollutants at certain times in their development (Krause and Weidensaul, 1978; Shriner, 1980). In many experiments (Laurence and Wood, 1978; Laurence et al., 1979; Laurence and Aluisio, 1981), significant inhibition of disease development has been observed when exposure to a pollutant occurred prior to inoculation, and therefore a direct effect on the pathogen was impossible.

Alterations in the suitability of a host, such as changes in the number of infection sites available, have been reported (Laurence, 1981). Whereas these effects on infection may be important, it is likely that the most important pollution-induced alterations in disease development occur after the pathogen is established and are related to changes in the physiology or biochemistry of the host.

Since most reports in the literature deal with reductions in disease incidence and severity of disease in plants exposed to pollutants, most of the discussion will pertain to possible mechanisms behind those observations. It is probable that, with the exception of virus-induced disease, increases in disease occur as a result of a pathogen's ability to invade weakened plants or take advantage of necrotic tissue as an infection court.

The plant surface is the site of the first interaction of host, pathogen and pollutant. In the development of many diseases, the plant surface is extremely important in determining the success of a pathogen as it represents the first physical barrier to infection. It is also important in pathogen recognition phenomena and as a substrate for the growth of some pathogens. Shriner (1980) has discussed the effects of pollutants on plant surfaces and the importance of that interaction in relation to the pathogen's success. There is little information, however, on the physical or chemical mechanisms by which surfaces of vegetation are altered in the presence of air pollutants.

Air pollutants are also known to modify the water relations and stomatal function of plants (Black and Unsworth, 1980; Majernik and Mansfield, 1971). Although the water status of plants is known to affect disease development, no work has been done to relate pollutant-induced changes in water relations to the success of pathogens. Once the pathogen is established, pollutant-modified stomatal behavior would probably be of minor importance in the development of disease, but modifications of turgor pressure, solute potential, or gaseous diffusion would likely affect pathogen success.

Plants that are exposed to air pollutants produce many compounds, either as secondary products or associated with detoxification. Kuć (1972) comments that 'It would be incredible if these compounds did not influence the growth and development of infections agents in tissues where they are produced or accumulate.' However, the production or accumulation of a particular metabolite and the concommitant observation of a change in pathogen success does not prove a cause–effect relationship. At this time, we can only speculate on what may be the cause of observed interactions.

Plants often respond to pathogen infection or wounding by producing phenolic compounds or phytoalexins, including terpenoids, naphthaldehydes, polyacetylenes and isoflavonoids (Kosuge, 1969; Kuć, 1972; Deverall, 1977). In some cases, these compounds are at least partly responsible for resistance to the pathogen, and the infection fails. In other cases, they may only slow down the disease cycle. If the pathogen, for some reason, does not stimulate the production of these chemicals and other conditions are favorable, the disease proceeds normally.

The production of isoperoxidases (Curtis and Howell, 1971; Curtis, Howell and Kremer, 1976), caffeic acid (Howell, 1970) and isoflavonoids (Keen and Taylor, 1975) have been reported following exposure of plants to O_3. The effects of other major pollutants on the formation of such compounds are poorly documented although it is known that peroxidases are generally increased following exposure to air pollutants (Keller, 1980; Keller and Schwager, 1971). These studies have not examined the antimicrobial potential of the altered metabolite pool.

Perhaps the best way to discuss potential effects of air pollutants on host metabolism is to consider an example. *Xanthomonas phaseoli* causes leaf spots on both bean (*Phaseolus vulgaris*) and soybean (*Glycine max*), (*X. phaseoli* var. *sojensis*). In both cases, exposure of the plants to SO_2 results in smaller lesions and delayed development when compared to control plants (Laurence and Aluisio, 1981; Laurence and Reynolds, (1982). Plants used in these studies were susceptible to the pathogens and so one would expect significant quantities of antimicrobial compounds to be formed.

Urs and Dunleavy (1974a,b) reported that horseradish peroxidase was toxic to *X. phaseoli* var. *sojensis* and peroxidase may be responsible for host resistance to the pathogen. It is also probable that SO_2 can stimulate production of phenolics and peroxidases in soybean, as has been demonstrated for O_3; thus, one might conclude that these plant-produced compounds could have been associated with the reduction in disease development that was observed.

Keen and Taylor (1975) demonstrated that several isoflavonoids accumulated in leaves of soybean following exposure of the plants to ozone, PAN, NO_2 or SO_2. Although the pollutant treatments did not stimulate the production of the soybean phytoalexin, there was a measurable increase in the concentration of coumestrol in leaves. Lyon and Wood (1975) found that coumestrol was produced by a resistant cultivar of bean (*Phaseolus vulgaris*) in response to inoculation with a race of *Pseudomonas phaseolicola*. The quantities of coumestrol they report are similar to those found by Keen and Taylor. Furthermore, 10 to 30 ppm coumestrol prevented growth of the bacterium for 36 to 48 h *in vitro*. The results of these studies would suggest that lesion formation would be delayed in plants exposed to SO_2, and the delay might result in smaller lesions. In fact, Laurence and Reynolds (1982) report smaller lesions and latent periods (the time from inoculation to visible symptom) of 24 to 48 h longer in plants exposed to SO_2.

Although these are only two examples of how pollutants might modify plant-pathogen interaction, they are probably illustrative of what occurs in many situations. It is obvious that the subject is poorly understood and that a great deal of research is necessary to establish the mechanisms involved in the modification of host-parasite relationships by air pollutants.

Conclusion

The relationship between biochemical effects of air pollutants on plants and environmental problems caused by insects and microbes can be examined as a sequence of steps. A biochemical change in the plant can produce a physiological change that, in turn, can affect the success of an insect or micro-organism. This altered success could cause a fluctuation in the population level that might subsequently produce an economic or ecological impact.

Clearly, the available data establish that air pollutant-induced changes in plants can affect the success of plant parasites; this has been documented in a few laboratory studies with insects and in many studies with plant pathogens. The biochemical bases for the observed effects on pest success are not known, and apparently no studies have been undertaken for their elucidation. Similarly, the physiological bases have been neglected, but a synthesis of results from many independent studies, although somewhat speculative, has provided useful and stimulating insight into the probable relationships involved. These relationships appear to be complex, since several changes, each of which could affect the plant–parasite relation, can be induced simultaneously by a single pollutant; furthermore, plants in the field are usually exposed to a combination of pollutants rather than to single compounds. Therefore, it is unlikely that only one mechanism is involved in any particular case. This complexity makes establishing either biochemical or physiological bases for pollutant-induced alterations of parasite success very difficult. Even so, much can be learned by a properly integrated combination of laboratory and field studies, and such investigations would contribute to a better understanding of the basic plant–parasite relationship as well as of the pollutant–plant–pest interaction.

The impact of the altered parasite success on populations of pest or 'non-pest' insects and microbes is not known. Certainly, many correlative-type field studies provide a large body of circumstantial evidence that such changes can result in devastating pest outbreaks under some conditions. However, data are lacking on the frequency with which such problems occur near sources of pollutants, i.e. cases in which parasite problems are not changed near a source are not reported. In many cases the pollutant undoubtedly represents just one of many environmental stresses, such as a temporary drought or an exceptionally hot or cold period, causing transient fluctions in the ecological balance and population levels that are partially buffered by density-dependent factors. At the other extreme, pollutant stress obviously can be more traumatic to the environment, causing direct death of both flora and fauna. What concentrations and exposures of specific pollutants or combinations of pollutants can the homeostatic mechanisms of the ecosystem tolerate and under what conditions is the pollutant-caused perturbation excessive in terms of the ability of the ecosystem to recover? These questions need to be addressed by a combination of controlled laboratory and field studies coupled with observations of natural systems, and their answer requires an integration of many scientific disciplines. Until such progress is made, however, it seems reasonable to assume that pollutants are currently causing significant changes in some plant–pest relationships with both economic and ecological consequences.

Acknowledgements

The authors would like to thank Dr L. H. Weinstein for his many valuable suggestions during the preparation of this chapter. We are also grateful to Ms Greta Colavito for her efforts in obtaining the references used and ensuring their proper citation, and to Ms Ellen Stoeber for typing the manuscript.

References

ALLEN, M. D. and SELMAN, I. W. Egg-production in the mustard beetle, *Phaedon cochleariae* (F.) in relation to diets of mineral-deficient leaves, *Bulletin of Entomological Research* **46**, 393–397 (1955)

ALSTAD, D. N., EDMUNDS, G. F. Jr. and WEINSTEIN, L. H. Effects of air pollutants on insect populations, *Annual Review of Entomology* **27**, 369–384 (1982)

ARNDT, U. Concentration changes of free amino acids in plants under the effect of hydrogen fluoride and sulfur dioxide, *Staub-Reinhaltung der Luft* **30**, 28–32 (1970)

AUGUSTINE, M. G., FISK, F. W., DAVIDSON, R. H., LA PIDUS, J. B. and CLEARY, R. W. Host–plant selection by the Mexican bean beetle, *Epilachna varivestis*, *Annals of the Entomological Society of America* **57**, 127–134 (1964)

BARKER, J. S. and TAUBER, O. E. Fecundity of and plant injury by the pea aphid as influenced by nutritional changes in the garden pea, *Journal of Economic Entomology* **44**, 1010–1012 (1951)

BECK, S. D. and REESE, J. C. Insect–plant interactions: Nutrition and metabolism. In *Biochemical Interaction Between Plants and Insects*, (Eds. J. W. WALLACE and R. L. MANSELL), pp. 41–92, Plenum Publishing Co., New York (1976)

BELL, J. N. B. and CLOUGH, W. S. Depression of yield in ryegrass exposed to sulphur dioxide, *Nature* **241**, 47–49 (1973)

BENEPAL, P. S., RANGAPPA, M. and DUNNING, J. A. Interaction between ozone and Mexican bean beetle feeding damage on beans [Abstract], *Annual Report of the Bean Improvement Cooperative* **22**, 50 (1979)

BENNETT, J. P., RESH, H. M. and RUNECKLES, V. C. Apparent stimulations of plant growth by air pollutants, *Canadian Journal of Botany* **52**, 35–41 (1974)

BERNAYS, E. A. and CHAPMAN, R. F. Plant chemistry and acridoid feeding behavior. In *Biochemical Aspects of Plant and Animal Coevolution*, (Ed. J. B. HARBORNE), pp. 99–141, Academic Press, Inc., London (1978)

BERNAYS, E. A. and WOODHEAD, S. Plant phenols utilized as nutrients by a phytophagous insect, *Science* **216**, 201–203 (1982a)

BERNAYS, E. A. and WOODHEAD, S. Incorporation of dietary phenols into the cuticle in the tree locust *Anacridium melanorhodon*, *Journal of Insect Physiology* **28**, 601–606 (1982b)

BEROZA, M. (Ed.) *Chemicals Controlling Insect Behaviour*. Academic Press, Inc., New York, 170 pp. (1970)

BISESSAR, S. and TEMPLE, P. J. Reduced ozone injury on virus-infected tobacco in the field, *Plant Disease Reporter* **61**, 961–963 (1977)

BLACK, V. J. and UNSWORTH, M. H. Stomatal responses to sulphur dioxide and vapor pressure deficit, *Journal of Experimental Botany* **31**, 667–677 (1980)

BLUM, U., SMITH, G. R. and FITES, R. C. Effects of multiple O_3 exposures on carbohydrate and mineral contents of ladino clover, *Environmental and Experimental Botany* **22**, 143–154 (1982)

BOSENER, R. Occurrence of bark-breeding forest pests in fume-damaged pine and spruce-stands [in German], *Archiv für Forstwesen* **18**, 1021–1026 (1969)

BRATTSEN, L. B. Biochemical defense mechanisms in herbivores against plant allelochemicals. In *Herbivores, Their Interaction with Secondary Plant Metabolites*, (Eds. G. A. ROSENTHAL and D. H. JANZEN), pp. 199–270, Academic Press, Inc., New York (1979)

BRAUN, S., FLÜCKIGER, W. and OERTLI, J. J. Einfluss einer Autobahn auf den Befall von Weissdorn (*Crataegus monogyna*) mit *Aphis pomi*, *Mitteilungen der Deutschen Gesellschaft für Allgemeine und Angewandte Entomologie* **3**, 138–139 (1981)

BRESSAN, R. A., WILSON, L. G., LE CUREUX, L. and FILNER, P. Use of ethylene and ethane emissions to assay injury by SO_2, *Plant Physiology* **61** (supplement): 93, Abstract no. 509 (1978)

BROMENSHENK, J. J. Responses of ground-dwelling insects to sulfur dioxide. In *Bioenvironmental Impact of a Coal-Fired Power Plant*, Fourth Interim Report, Colstrip, Montana, (Eds. E. M. PRESTON and T. L. GULLETT), pp. 673–703, EPA-600/3-79-044, US Environmental Protection Agency, Washington, DC (1979)

CARLISLE, D. B., ELLIS, P. E. and BETTS, E. The influence of aromatic shrubs on sexual maturation in the desert locust *Schistocerca gregaria*, *Journal of Insect Physiology* **11**, 1541–1558 (1965)

COUEY, H. M. Inhibition of germination of *Alternaria* spores by sulfur dioxide under various moisture conditions, *Phytopathology* **55**, 525–527 (1965)

COUEY, H. M. and UOTA, M. Effect of concentration, exposure time, temperature and relative humidity on the toxicity of sulfur dioxide to the spores of *Botrytis cinerea*, *Phytopathology* **51**, 739–814 (1961)

COWLING, D. W. and LOCKYER, D. R. Growth of perennial ryegrass (*Lolium perenne* L.) exposed to a low concentration of sulphur dioxide, *Journal of Experimental Botany* **27**, 411–417 (1976)

CURTIS, C. R. and HOWELL, R. K. Increases in peroxidase isoenzyme activity in bean leaves exposed to low doses of ozone, *Phytopathology* **61**, 1306–1307 (1971)

CURTIS, C. R., HOWELL, R. K. and KREMER, D. F. Soybean peroxidases from ozone injury, *Environmental Pollution* **11**, 189–194 (1976)

CVRKAL, H. Biochemische Diagnose an Fichten in Rauchgebieten. *Berichte aus der tscheshoslowakischen, Akademie der Landwirtschaftswissenschaften, Sektion Forstwesen* **5**, 1033–1048 (1959)

DÄSSLER, H. G. The effect of SO_2 on the terpene content of spruce needles [in German], *Flora* **154**, 376–382 (1964)

DAVIS, D. D. and SMITH, S. H. Reduction of ozone-sensitivity of pinto bean by bean common mosaic virus, *Phytopathology* **64**, 383–385 (1974)

DETHIER, V. G. Some general considerations of insects' responses to the chemicals in food plants. In *Control of Insect Behavior by Natural Products*, (Eds. D. L. WOOD, R. M. SILVERSTEIN and M. NAKAJIMA), pp. 21–28, Academic Press, Inc., New York (1970)

DEVERALL, B. J. *Defence Mechanisms of Plants*, Cambridge University Press, Cambridge, 110 pp. (1977)

DODMAN, R. L. How the defenses are breached. In *Plant Disease: An Advanced Treatise*, Vol. IV, (Eds. J. G. HORSFALL and E. B. COWLING), pp. 135–153, Academic Press, Inc., New York (1979)

DONAUBAUER, E. Secondary damages of forests caused by industrial exhaust fumes [in German, English summary], *Mitteilungen der Forstlichen Bundesversuchsanstalt* **73**, 101–110 (1966)

DUFFEY, S. S. Sequestration of plant natural products by insects, *Annual Review of Entomology* **25**, 447–477 (1980)

ELKIEY, T. and ORMROD, D. P. Sulphur and nitrogen nutrition and misting effects on the response of bluegrass to ozone, sulphur dioxide, nitrogen dioxide or their mixture, *Water, Air and Soil Pollution* **16**, 177–186 (1981)

ELLIS, P. E., CARLISLE, D. B. and OSBORNE, D. J. Desert locusts: Sexual maturation delayed by feeding on senescent vegetation, *Science* **149**, 546–547 (1965)

EVANS, L. S. Biological effects of acidity in precipitation on vegetation : A review, *Environmental and Experimental Botany* **22**, 155–169 (1982)

EVENDEN, J. C. *Smelter Killed Timber and Insects*. United States Department of Agriculture, Bureau of Entomology, Forest Insect Investigations, Typewritten report filed at R-1, Missoula, Montana (1923)

FAIRFAX, J. A. W. and LEPP, N. W. Effect of simulated 'acid rain' on cation loss from leaves, *Nature (London)* **255**, 324–325 (1975)

FALLER, N. Effects of atmospheric SO₂ on plants, *Sulphur Institute Journal* **6**, 5–7 (1971)

FALLER, N. Sulphur dioxide, hydrogen sulphate, nitrous gases, and ammonia as sole source of S and N for higher plants [in German, English summary], *Zeitschrift für Pflanzenernährung und Bodenkunde* **131**(2), 120–130 (1972)

FEENY, P. Plant apparency and chemical defense. In *Biochemical Interaction Between Plants and Insects*, (Eds. J. W. WALLACE and R. L. MANSELL), pp. 1–40, Plenum Publishing Co., New York (1976)

FLÜCK, H. Intrinsic and extrinsic factors affecting the production of secondary plant products. In *Chemical Plant Taxonomy*, (Ed. T. SWAIN), pp. 167–186, Academic Press, Inc., London (1963)

FLÜCKIGER, W., FLÜCKIGER-KELLER, H. and OERTLI, J. J. Biochemical changes in young birches in the vicinity of a highway [in German, English summary], *European Journal of Forest Pathology* **8**, 154–163 (1978)

FLÜCKIGER, W., OERTLI, J. J. and BALTENSWEILER, W. Observations of an aphid infestation on hawthorn in the vicinity of a motorway, *Naturwissenschaften* **65**, 654–655 (1978)

GODZIK, S. and LINSKENS, H. F. Concentration changes of free amino acids in primary bean leaves after continuous and interrupted SO₂ fumigation and recovery, *Environmental Pollution* **7**, 25–38 (1974)

GRILL, D., ESTERBAUER, H. and BECK, G. Phenols and glucose in SO₂-damaged needles of spruce [in German, English summary], *Phytopathologische Zeitschrift* **82**, 182–184 (1975)

HÅGVAR, S. and AMUNDSEN, T. Effects of liming and artificial acid rain on the mite (*Acari*) fauna in coniferous forest, *Oikos* **37**, 7–20 (1981)

HÄLLGREN, J.-E. Physiological and biochemical effects of sulfur dioxide on plants. In *Sulfur in the Environment. Part II. Ecological Impacts*, (Ed. J. O. NRIAGU), pp. 163–209, John Wiley & Sons, New York (1978)

HARDING, P. R. Jr. Effect of ozone on *Penicillium* mold decay and sporulation, *Plant Disease Reporter* **52**, 245–247 (1968)

HARBORNE, J. B. and INGHAM, J. L. Biochemical aspects of the coevolution of higher plants with their fungal parasites. In *Biochemical Aspects of Plant and Animal Coevolution*, (Ed. J. B. HARBORNE), pp. 343–405, Academic Press, Inc., London (1978)

HAY, C. J. *Bibliography on Arthropoda and Air Pollution*, USDA Forest Service General Technical Report NE-24 (1977)

HEAGLE, A. S. Interactions between air pollutants and plant parasites, *Annual Review of Phytopathology* **11**, 365–388 (1973)

HECK, W. W., TAYLOR, O. C., ADAMS, R., BINGHAM, G., MILLER, J., PRESTON, E. and WEINSTEIN, L. Assessment of crop loss from ozone, *Journal of the Air Pollution Control Association* **32**, 353–361 (1982)

HEDIN, P. A., JENKINS, J. N. and MAXWELL, F. G. Behavioral and developmental factors affecting host plant resistance to insects. In *Host Plant Resistance to Pests*, (Ed. P. A. HEDIN) pp. 231–275, American Chemical Society Symposium Series 62 (1977)

HOUSE, H. L. Insect nutrition, *Annual Review of Entomology* **6**, 13–26 (1961)

HOWELL, R. K. Influence of air pollution on quantities of caffeic acid isolated from leaves of *Phaseolus vulgaris*, *Phytopathology* **60**, 1626–1629 (1970)

HOWELL, R. K. Phenols, ozone, and their involvement in pigmentation and physiology of plant injury. In *Air Pollution Effects on Plant Growth*, (Ed. M. DUGGER), pp. 94–105, American Chemical Society, Washington DC (1974)

HUGHES, P. R., DICKIE, A. I. and PENTON, M. A. Increased success of the Mexican bean beetle on field-growth soybeans exposed to SO_2, *Journal of Environmental Quality* (in press) (1983)

HUGHES, P. R., POTTER, J. E. and WEINSTEIN, L. H. Effects of air pollutants on plant-insect interactions: Reactions of the Mexican bean beetle to SO_2-fumigated pinto beans, *Environmental Entomology* **10**, 741–744 (1981)

HUGHES, P. R., POTTER, J. E. and WEINSTEIN, L. H. Effects of air pollution on plant-insect interactions: Increased susceptibility of greenhouse-grown soybeans to the Mexican bean beetle after plant exposure to SO_2, *Environmental Entomology* **11**, 173–176 (1982)

HUKUSIMA, S. The effect of varying nitrogen levels in nutrition upon the arthropod fauna in young apple trees, with special reference to the increase of the mite and aphid populations, *Bulletin of the Faculty of Agriculture, Hirosaki University* **4**, 72–79 (1958)

HURWITZ, B., PELL, E. J. and SHERWOOD, R. T. Status of coumestrol and 4',7-di-hydroxyflavone in alfalfa foliage exposed to ozone, *Phytopathology* **69**, 810–813 (1979)

ISMAN, M. B. and DUFFEY, S. S. Phenolic compounds in foliage of commercial tomato cultivars as growth inhibitors to the fruitworm, *Heliothis zea*, *Journal of the American Society for Horticultural Science* **107**, 167–170 (1982)

JAMES, C. Economic, social and political implications of crop losses: A holistic framework for loss assessment in agricultural systems. In *Crop Loss Assessment*, (Eds. P. S. TENG and S. V. KRUPA), Miscellaneous Publication 7, Agricultural Experiment Station, University of Minnesota, 327 p. (1980)

JERMY, T. (ed.). *The Host-Plant in Relation to Insect Behaviour and Reproduction*, Akadémiai Kiadó, Budapest, Hungary, 322 pp. (1976)

JOHANSSON, A. S. Feeding and nutrition in reproductive processes in insects, *Symposia of the Royal Entomological Society of London* **2**, 43–55 (1964)

JOHNSON, K. J. R., SORENSON, E. L. and HORBER, E. K. Resistance of glandular-haired *Medicago* species to oviposition by alfalfa weevils (*Hypera postica*), *Environmental Entomology* **9**, 241–244 (1980)

JONES, J. V. and PELL, E. J. The influence of ozone on the presence of isoflavones in alfalfa foliage, *Journal of the Air Pollution Control Association* **31**, 885–886 (1981)

KEEN, N. T. and TAYLOR, O. C. Ozone injury in soybeans: Isoflavonoid accumulation is related to necrosis, *Plant Physiology* **55**, 731–733 (1975)

KELLER, Th. Bestimmungsmethoden für die Einwirkung von Luftverunreinigungen, *Schweizerische Zeitschrfft für Forstwesen* **131**, 239–253 (1980)

KELLER, Th. and SCHWAGER, H. Der Nachweis unsichtbarer ('physiologischer') Fluor-immissionsschadigungen an Waldbaremen durch eine einfache kolorimetrische Bestimmung der Peroxidase-Aktivitat, *European Journal of Forest Pathology* **1**, 6–18 (1971)

KENNEDY, J. S. and BOOTH, C. O. Responses of *Aphis fabae* Scop. to water shortage in host plants in the field, *Entomologia Experimentalis et Applicata* **2**, 1–11 (1959)

KIMMERER, T. W. and KOZLOWSKI, T. T. Ethylene, ethane, acetaldehyde, and ethanol production by plants under stress, *Plant Physiology* **69**, 840–847 (1982)

KOŠTÍR, J., MACHÁČKOVÁ, I., JIRÁČEK, V. and BUCHAR, E. Effect of sulfur dioxide on the content of free sugars and amino acids in pea seedlings [in German, English summary], *Experientia* **26**, 604–605 (1970)

KOSUGE, T. The role of phenolics in host response to infection, *Annual Review of Phytopathology* **7**, 195–222 (1969)

KOZIOL, M. J. and JORDAN, C. F. Changes in carbohydrate levels in red kidney bean (*Phaseolus vulgaris* L.) exposed to sulphur dioxide, *Journal of Experimental Botany* **29**, 1037–1043 (1978)

KRAUSE, C. R. and WEIDENSAUL, T. C. Ultrastructural effects on the sporulation, germination, and pathogenicity of *Botrytis cinerea*, *Phytopathology* **68**, 195–198 (1978)

KUĆ, J. Phytoalexins, *Annual Review of Phytopathology* **10**, 207–232 (1972)

LAURENCE, J. A. Effects of air pollutants on plant-pathogen interactions, *Zeitschrift für Pflanzenkrankheiten und Pflanzenschutz* **88**, 156–173 (1981)

LAURENCE, J. A. and ALUISIO, A. L. Effects of sulfur dioxide on expansion of lesions caused by *Corynebacterium nebraskense* in maize and by *Xanthomonas phaseoli* var. *sojensis* in soybean, *Phytopathology* **71**, 445–448 (1981)

LAURENCE, J. A. and REYNOLDS, K. L. Effects of concentration of SO_2 and other characteristics of exposure on the development of *Xanthomonas phaseoli* lesions in red kidney bean, *Phytopathology* **72**, 1243–1246 (1982)

LAURENCE, J. A., WEINSTEIN, L. H., McCUNE, D. C. and ALUISIO, A. L. Effects of sulfur dioxide on southern corn leaf blight of maize and stem rust of wheat, *Plant Disease Reporter* **63**, 975–978 (1979)

LAURENCE, J. A. and WOOD, F. A. Effects of ozone on infection of soybean by *Pseudomonas glycinia*, *Phytopathology* **68**, 441–445 (1978)

LEVIN, D. A. The chemical defenses of plants to pathogens and herbivores, *Annual Review of Ecology and Systematics* **7**, 121–159 (1976)

LEWIS, A. C. Feeding preference for diseased and wilted sunflower in the grasshopper, *Melanoplus differentialis*, *Entomologia Experimentalis et Applicata* **26**, 202–207 (1979)

LUCKEY, T. D. Insecticide hormoligosis, *Journal of Economic Entomology* **61**, 7–12 (1968)

LYON, F. M. and WOOD, R. K. S. Production of phaseollin, coumestrol and related compounds in bean leaves inoculated with *Pseudomonas* spp., *Physiological Plant Pathology* **6**, 117–124 (1975)

MAJERNIK, O. and MANSFIELD, T. A. Effects of SO_2 pollution on stomatal movements in *Vicia faba*, *Phytopathologische Zeitschrift* **71**, 123–128 (1971)

MALHOTRA, S. S. and BLAUEL, R. A. Diagonosis of air pollutant and natural stress symptoms on forest vegetation in western Canada. Environment Canada, Canadian Forest Service, Northern Forest Research Centre, Information Report NOR-X-228 (1980)

MALHOTRA, S. S. and KHAN, A. A. Effects of sulphur dioxide fumigation on lipid biosynthesis in pine needles, *Phytochemistry* **17**, 241–244 (1978)

MANSOUR, M. H., DIMETRY, N. Z. and ROFAEEL, I. S. The role of coumarin as secondary plant substance in the food specificity of the cow pea aphid *Aphis craccivora* Koch, *Zeitschrift für Angewandte Entomologie* **93**, 151–157 (1982)

MATTSON, W. J. Jr. Herbivory in relation to plant nitrogen content, *Annual Review of Ecology and Systematics* **11**, 119–161 (1980)

McCUNE, D. C., WEINSTEIN, L. H., MANCINI, J. F. and VAN LEUKEN, P. Effects of hydrogen fluoride on plant-pathogen interactions. In *Proceedings of the 3rd International Clean Air Congress*, Düsseldorf, 1973, pp. A146–A149 (1973)

McNARY, T. J., MILCHUNAS, D. G., LEETHAM, J. W., LAUENROTH, W. K. and DODD, J. L. Effect of controlled low levels of SO_2 on grasshopper densities on a northern mixed-grass prairie, *Journal of Economic Entomology* **74**, 91–93 (1981)

McNEIL, S. and SOUTHWOOD, T. R. E. The role of nitrogen in the development of insect/plant relationships. In *Biochemical Aspects of Plant and Animal Coevolution*, (Ed. J. B. HARBORNE), pp. 77–98, Academic Press, Inc., London (1978)

MEUDT, W. J. Interactions of sulfite and manganous ion with peroxidase oxidation products of indole-3-acetic acid, *Phytochemistry* **10**, 2103–2109 (1971)

MITTLER, T. E. Effects of dietary amino acids on the feeding rate of the aphid *Myzus persicae*, *Entomologia Experimentalis et Applicata* **13**, 432–437 (1970)

NATIONAL ACADEMY OF SCIENCES, COMMITTEE ON MEDICAL AND BIOLOGIC EFFECTS OF ENVIRONMENTAL POLLUTANTS. *Nitrogen Oxides*, Washington, DC (1977)

NEUMANN-VISSCHER, S. Plant growth hormone effects on insect growth and reproduction, *Proceedings of the 5th International Symposium on Insect-Plant Relationships*, Pudoc Press, Wageningen, pp. 57–62 (1982)

NYBORG, M. Sulfur pollution and soils. In *Sulfur in the Environment. Part II. Ecological Impacts*, (Ed. J. O. NRIAGU), pp. 359–390, John Wiley and Sons, New York (1978)

PEISER, G. D. and YANG, S. F. Ethylene and ethane production from sulfur dioxide-injured plants, *Plant Physiology* **63**, 142–145 (1979)

PELL, E. J., WEISSBERGER, W. C. and SPERONI, J. J. Impact of ozone on quantity and quality of greenhouse-grown potato plants, *Environmental Science and Technology* **14**, 568–571 (1980)

PFEFFER, A. Destructive insects affecting fir trees in the area of gas emissions [in German], *Zeitschrift für Angewandte Entomologie* **51**, 203–207 (1963)

PORT, G. R. and THOMPSON, J. R. Outbreaks of insect herbivores on plants along motorways in the United Kingdom, *Journal of Applied Ecology* **17**, 649–656 (1980)

PRASAD, B. J. and RAO, D. N. Effects of SO_2 exposure on carbohydrate contents, phytomass and caloric values of wheat plants, *Water, Air, and Soil Pollution* **16**, 287–291 (1981)

PREECE, T. F. Some observations on leaf surfaces during the early stages of infection by fungi. In *Biochemical Aspects of Plant-Parasite Relationships*, (Eds. J. FRIEND and D. R. THRELFALL), pp. 1–10, Academic Press, Inc., London (1976)

PRIEBE, A., KLEIN, H. and JÄGER, H.-J. Role of polyamines in SO_2-polluted pea plants, *Journal of Experimental Botany* **29**, 1045–1050 (1978)

PRZYBYLSKI, Z. Effect of gases and vapors of SO_2, SO_3 and H_2SO_4 on fruit trees and certain harmful insects [in Polish], *Postepy Nauk Rolniczych* **14**, 111–118 (1967)

PRZYBYLSKI, Z. Results of consecutive observations of effect of SO_2, SO_3 and H_2SO_4 gases and vapors on trees, shrubs, and entomofauna of orchards in the vicinity of sulfur mines and a sulfur processing plant in Machow [in Polish], *Postepy Nauk Rolniczych* **15**, 131–138 (1968)

PRZYBYLSKI, Z. Results on observations of the influence of sulfurous gas on fruit trees, shrubs, and arthropods around the sulfur mines and processing plants in the Tarnobrzeg region [in French, English summary], *Environmental Pollution* **6**, 67–74 (1974)

PRZYBYLSKI, Z. The effects of automobile exhaust gases on the arthropods of cultivated plants, meadows and orchards, *Environmental Pollution* **19**, 157–161 (1979)

RENWICK, J. A. A. and POTTER, J. Effects of sulfur dioxide on volatile terpene emission from balsam fir, *Journal of the Air Pollution Control Association* **31**, 65–66 (1981)

ROSENTHAL, G. A. and BELL, E. A. Naturally occurring, toxic nonprotein amino acids. In *Herbivores, Their Interaction with Secondary Plant Metabolites*, (Eds. G. A. ROSENTHAL and D. H. JANZEN) pp. 353–385, Academic Press, Inc., New York (1979)

ROSENTHAL, G. A., DAHLMAN, D. L. and JANZEN, D. H. A novel means for dealing with L-canavanine, a toxic metabolite, *Science* **192**, 256–257 (1976)

ROSENTHAL, G. A. and JANZEN, D. H. (Eds.). *Herbivores, Their Interaction with Secondary Plant Metabolites*, Academic Press, Inc., New York, 718 pp. (1979)

SÁRDI, K. Changes in the soluble protein content of soybean *Glycine max* L. and pea *Pisum sativum* L. under continuous SO_2 and soot pollution, *Environmental Pollution (Series A)* **25**, 181–186 (1981)

SCHOPF, R., MIGNAT, C. and HEDDEN, P. As to the food quality of spruce needles for forest damaging insects. 18. Resorption of secondary plant metabolites by the sawfly *Gilpinia hercyniae* Htg. (Hym., Diprionidae), *Zeitschrift für Angewandte Entomologie* **93**, 244–257 (1982)

SCRIBER, J. M. The effects of larval feeding specialization and plant growth form on the consumption and utilization of plant biomass and nitrogen: An ecological consideration, *Entomologia Experimentalis et Applicata* **24**, 694–710 (1978)

SCRIBER, J. M. and SLANSKY, F. Jr. The nutritional ecology of immature insects, *Annual Review of Entomology* **26**, 183–211 (1981)

SEMTNER, P. J., RASNAKE, M. and TERRILL, T. R. Effect of host-plant nutrition on the occurrence of tobacco hornworms and tobacco flea beetles on different types of tobacco, *Journal of Economic Entomology* **73**, 221–224 (1980)

SEQUEIRA, L. Defenses triggered by the invader: Recognition and compatability phenomena. In *Plant Disease: An Advanced Treatise*, Vol. V, (Eds. J. G. HORSFALL and E. B. COWLING), pp. 179–200, Academic Press, Inc., New York (1980)

SHOREY, H. H. and McKELVEY, J. J. Jr. (Eds.). *Chemical Control of Insect Behavior*, 414 pp., John Wiley and Sons, Inc., New York (1977)

SHRINER, D. S. Vegetation surfaces: A platform for pollutant/parasite interactions. In *Polluted Rain*, (Eds. T. Y. TORIBARA, M. W. MILLER and P. E. MORROW), pp. 259–272, Environmental Science Research, Vol. 17. Plenum Press, New York (1980)

SIERPINSKI, Z. The significance of secondary pine insects in areas of chronic exposure to industrial air pollution [in German], *Mitteilungen der forstlichen Bundesversuchsanstalt Wien* **97**, 609–615 (1972)

SMITH, W. H. Depressed litter decomposition. In *Air Pollution and Metropolitan Woody Vegetation*, (Eds. W. H. SMITH and L. S. DOCHINGER) pp. 23–24, Pinchot Institut, Consortium for Environmental Forestry Studies, USDA Forest Service, Delaware, Ohio (1975)

SOKOLOV, V. S. The influence of certain environmental factors on the formation and accumulation of alkaloids in plants. In *Utilization of Nitrogen and its Compounds by Plants, Symposia of the Society for Experimental Biology* **13**, 230–257 (1959)

SOUTHWOOD, T. R. E. The insect/plant relationship — an evolutionary perspective. In *Insect/Plant Relationships*, (Ed. H. F. VAN EMDEN), pp. 3–30, Symposium of the Royal Entomological Society of London, No. 6. Blackwell Scientific Publications, Oxford (1973)

STÄDLER, E. Host plant stimuli affecting oviposition behavior of the eastern spruce budworm, *Entomologia Experimentalis et Applicata* **17**, 176–188 (1974)

STARK, R. W., MILLER, P. R., COBB, F. W. Jr., WOOD, D. L. and PARMETER, J. R. Jr. Photochemical oxidant injury and bark beetle (Coleoptera:Scolytidae) infestation of ponderosa pine. I. Incidence of bark beetle infestation in injured trees, *Hilgardia* **39**, 121–126 (1968)

STEBBING, A. R. D. An experimental approach to the determinants of biological water quality, *Philosophical Transactions of the Royal Society of London, Series B: Biological Sciences* **286**, 465–481 (1979)

STEWARD, F. C., CRANE, F., MILLAR, K., ZACHARIUS, R. M., RABSON, R. and MARGOLIS, D. Nutritional and environmental effects on the nitrogen metabolism of plants. In *Utilization of Nitrogen and Its Compounds by Plants. Symposia of the Society for Experimental Biology* **13**, 148–176 (1959)

STROJAN, C. L. The impact of zinc smelter emissions on forest litter arthropods, *Oikos* **31**, 41–46 (1978)

SUMIMOTO, M., SHIRAGA, M. and KONDO, T. Ethane in pine needles preventing the feeding of the beetle, *Monochamus alternatus*, *Journal of Insect Physiology* **21**, 713–722 (1975)

THALENHORST, W. Investigations on the influence of fluor containing air pollutants upon the suceptibility of spruce plants to the attack of the gall aphid *Sacchiphantes abietis* (L.), *Zeitschrift für Pflanzenkrankheiten und Pflanzenschutz* **81**, 717–727 (1974)

TINGEY, D. T., STANDLEY, C. and FIELD, R. W. Stress ethylene evolution: A measure of ozone effects on plants, *Atmospheric Environment* **10**, 969–974 (1976)

TINGEY, W. M. and GIBSON, R. W. Feeding and mobility of the potato leafhopper impaired by glandular trichomes of *Solanum berthaultii* and *S. polyadenium*, *Journal of Economic Entomology* **71**, 856–858 (1978)

TODD, J. W., PARKER, M. B. and GAINES, T. P. Populations of Mexican bean beetles in relation to leaf protein of nodulating and non-nodulating soybeans, *Journal of Economic Entomology* **65**, 729–731 (1972)

TOMLINSON, H. and RICH, S. Metabolic changes in free amino acids of bean leaves exposed to ozone, *Phytopathology* **57**, 972–974 (1967)

TROIANO, J. J. and LEONE, I. A. Changes in growth rate and nitrogen content of tomato plants after exposure to NO_2, *Phytopathology* **67**, 1130–1133 (1977)

URS, N. V. R. and DUNLEAVY, J. M. Function of peroxidase in resistance of soybean to bacterial pustule, *Crop Science* **14**, 740–744 (1974a)

URS, N. V. R. and DUNLEAVY, J. M. Bactericidal activity of horseradish peroxidase on *Xanthomonas phaseoli* var. *sojensis*, *Phytopathology* **64**, 542–545 (1974b)

VARGO, R. H., PELL, E. J. and SMITH, S. H. Induced resistance to ozone injury of soybean by tobacco ringspot virus, *Phytopathology* **68**, 715–719 (1978)

WEARING, C. H. and VAN EMDEN, H. F. Studies on the relations of insect and host plant. 1. Effects of water stress in host plants on infestation by *Aphis fabae* Scop., *Myzus persicae* (Sulz.) and *Brevicoryne brassicae* (L.), *Nature* **213**, 1051–1052 (1967)

WEINSTEIN, L. H. Effects of atmospheric fluoride on metabolic constituents of tomato and bean leaves, *Contributions from the Boyce Thompson Institute* **21**, 215–231 (1961)

WEINSTEIN, L. H. and ALSCHER-HERMAN, R. Physiological effects of fluoride on higher plants. In *Effects of Gaseous Air Pollution in Agriculture and Horticulture*, (Ed. M. H. UNSWORTH and D. P. ORMROD), pp. 139–165, Butterworths, London (1982)

WEINSTEIN, L. H., McCUNE, D. C., MANCINI, J. F. and VAN LEUKEN, P. Effects of hydrogen fluoride fumigation of bean plants on the growth, development, and reproduction of the Mexican bean beetle. In *Proceedings of the 3rd International Clean Air Congress, Düsseldorf*, 1973, pp. A150–A153 (1973)

WELLBURN, A. R., CAPRON, T. M., CHAN, H.-S. and HORSMAN, D. C. Biochemical effects of atmospheric pollutants on plants. In *Effects of Air Pollutants on Plants*, (Ed. T. A. MANSFIELD), pp. 105–114, Cambridge University Press, Cambridge (1976)

WHITE, T. C. R. A hypothesis to explain outbreaks of looper caterpillars, with special reference to populations of *Selidosema suavis* in a plantation of *Pinus radiata* in New Zealand, *Oecologia* **16**, 279–301 (1974)

WHITE, T. C. R. Weather, food and plagues of locusts, *Oecologia* **22**, 119–134 (1976)

WILSON, K. G. *Aspects of the physiological ecology of the Mexican bean beetle, Epilachna varivestis Mulsant*. PhD Thesis, North Carolina State University, Raleigh, NC, USA (1981)

WOOD, D. L. The role of pheromones, kairomones, and allomones in the host selection and colonization behavior of bark beetles, *Annual Review of Entomology* **27**, 411–446 (1982)

WOOD, D. L., SILVERSTEIN, R. M. and NAKAJIMA, M. (Eds.). *Control of Insect Behavior by Natural Products*, 345 pp., Academic Press, Inc., New York (1970)

WOOD, T. and BORMANN, F. H. The effects of an artificial acid mist upon the growth of *Betula alleghaniensis* Britt, *Environmental Pollution* **7**, 259–268 (1974)

YANG, S. F. and MILLER, G. W. Biochemical studies on the effect of fluoride on higher plants. 1. Metabolism of carbohydrates, organic acids, and amino acids, *Biochemical Journal* **88**, 505–509 (1963a)

YANG, S. F. and MILLER, G. W. Biochemical studies on the effect of fluoride on higher plants. 3. The effect of fluoride on dark carbon dioxide fixation, *Biochemical Journal* **88**, 517–522 (1963b)

YANG, S. F. and SALEH, M. A. Destruction of indole-3-acetic acid during the aerobic oxidation of sulfite, *Phytochemistry* **12**, 1463–1466 (1973)

ZIEGLER, I. The effect of air-polluting gases on plant metabolism, *Environmental Quality and Safety* **2**, 182–208 (1973)

Part VII

Biochemical mechanisms of pollutant tolerance

Chapter 25

Enzymatic investigations on tolerance in forest trees

L. E. Mejnartowicz

POLSKA AKADEMIA NAUK, INSTYTUT DENDROLOGII, KÓRNIK, POLAND

Introduction

Forest trees provide spectacular evidence for the toxicity of gaseous industrial emissions to plants. Having a very long ontogenetic development, trees can accumulate much greater quantities of toxic substances than herbaceous plants when exposed to a polluted environment. The metabolic disturbances leading initially to injury of leaves or needles can ultimately kill the plant. Literature on this subject is abundant and comprehensive reviews have been provided by Bell (1980), Chang (1975), Horsman and Wellburn (1976), Jeffree (1979), Malhotra and Hocking (1976), Miller and McBridge (1975) and Ziegler (1973, 1975).

On the other hand, there are few papers on the intra- and interpopulational changes occurring in communities of forest trees growing in regions polluted with industrial emissions. If we assume that 'at a given locality the local populations possess particular heritable physiological characteristics that make them better adapted to the local environment than populations from other localities, and that species are composed of different races' (Solbrig, 1971), then we would expect that this ecotypic differentiation will also be reflected in the response of populations to gaseous pollutants.

The existence of racial differentiation among forest trees has been demonstrated in the so-called provenance experiments (Wright, 1976) for polygenic characteristics and in various studies using electrophoretic methods for single genes. Such studies have been done for Scots pine (*Pinus sylvestris*) (Krzakowa, 1979; Mejnartowicz, 1979; Rasmuson and Rudin, 1971; Rudin, Erikson and Rasmuson, 1974; Rudin, 1974), Norway spruce (*Picea abies*) (Bergmann, 1971, 1973; Lundkvist, 1979), European larch (*Larix decidua*) (Mejnartowicz and Bergmann, 1975), Douglas fir (*Pseudotsuga menziesii*) (Mejnartowicz, 1976; Muhs, 1974) and for many other species of North American trees (Brown and Moran, 1979; Conkle, 1979; Hamrick, Mitton and Linhart, 1979; Shaw and Allard, 1979).

Industrial emissions, acting as a drastic agent of selection (Knabe, 1967; Treshow, 1968), can cause a degradation of the existing plant communities, a reduction in their specific composition (Gordon and Gorham, 1963; Karnosky, 1980) and a formation of new associations previously unknown (Mamayev and Shkarlet, 1972; Wolak, 1970, 1979) in which only that part of the original population is maintained which was able to adapt to the polluted environment (Bradshaw, 1976; Sinclair, 1969). Studies on the genetic structure of Scots pine populations that are exposed to the action of SO_2 and fluoride have been conducted by Mejnartowicz (1980).

Genetic determinants of tolerance to air pollution

The scarcity of results from biochemical studies aimed at the understanding of the causes of differentiation in tolerance of individual trees of air pollution may result from underestimating the degree of genetic complexity in forest tree populations. Lack of suitable methods previously hindered the identification of allele frequencies in populations, and consequently the identification of their heterozygosity, and the similarities and genetic distances between ecotypes and races. In recent years this difficulty has been largely overcome following the development of electrophoretic techniques for the analysis of isozymes. Studies of clones, families and provenances indicate that increased tolerance to pollutants in trees may have a genetic basis.

Clonal studies

Clonal studies are widely used in the laboratory testing of trees for sensitivity to industrial gaseous emissions. These studies depend on the selection of phenotypes that are either tolerant or sensitive to a certain pollutant, and their subsequent vegetative propagation by grafting, rooting or tissue culture.

Material obtained in this manner is to a large extent genetically uniform. However, a comparison of results obtained after controlled fumigation with SO_2 of various ramets of the same clone of Scots pine (Lorenc-Plucinska, 1980; Oleksyn, 1980) and an electrophoretic analysis of multiple forms of peroxidase in poplar (*Populus deltoides*) clones (Guzina, 1974) indicate that even in such material a substantial heterogeneity can sometimes occur. For example Dochinger *et al.* (1972) found significant intraclonal variation in SO_2-tolerance of poplar hybrids.

Clones of European larch, Scots pine and many other coniferous trees are usually obtained by grafting scions onto open-pollinated seedling stocks. The stocks represents diversified genotypes, and these may modify both the activity and number of isozymes when the scion and stock are incompatible. In Douglas fir such grafts gave strongly-staining bands of peroxidase and esterase isozymes (Copes, 1978). On the other hand, Bücher-Wallin (1976) compared the activity of peroxidase and its electrophoretic pattern in three clones of Norway spruce grafted onto two different stocks, and observed no influence of the stocks on the enzymatic patterns in the needles of the scions.

To minimize complications, cuttings taken for the study of the effects of gaseous emissions on the metabolic activity of clones should be taken from trees of identical age. Clones propagated from young trees reacted differently to HF treatment than did clones propagated from older trees. In Norway spruce, this difference in clones raised from younger trees is expressed as a higher peroxidase activity, as well as in a series of other differences, such as the ability to regenerate chlorophyll, the response to nitrogen fertilization, and the content of sugars and of phenols (H. Keller, 1976). In hybrid larch (*Latrix × eurolepis* Henry) cuttings from trees up to 15 years old root more readily than those from older trees (John, 1977). Although propagation by root cuttings is easily carried out from tree seedlings there is a risk (Lindgren, 1977) that characters of the clones may change from those of the parents and the cuttings may not respond in the laboratory like the parents in the field (Lampadius, Pelz and Pohl, 1970). However, on the basis of observations of five poplar clones and 11 clones of coniferous trees Dässler (1967) found good agreement between fumigation tests under laboratory conditions and the response observed in the field. The response of hybrid poplar clones to SO_2 fumigation in the studies of Karnosky (1977) depended significantly on the genotype, the duration of fumigation and the interaction between genotype and time of fumigation (see also Dochinger et al., 1972). In studies with clones of eastern white pine (Pinus strobus), Houston and Stairs (1973) have shown a synergistic response to SO_2 and O_3. Clones of Pinus strobus showed a lower tolerance to SO_2 in the presence of O_3. Laboratory testing of clones resulted in substantial progress in the breeding of Norway spruce plants that were more resistant to SO_2 and HF, but such testing was somewhat less effective for Scots pine (Rohmeder, Merz and Schönborn, 1962; Rohmeder and Schonborn, 1965; Wentzel, 1968). Scholz and Knabe (1976) found that clones from Norway spruce trees differing in resistance to SO_2 differ significantly in buffering capacity and this appears to be genetically determined. Clones from trees growing in a region free of air pollution and from trees selected as very resistant to SO_2 show a higher buffering capacity of the cell sap. Norway spruce, silver birch (Betula pendula) and black alder (Alnus glutinosa) clones were exposed to SO_2 and HF for long periods and the effects on phenoloxidase, esterase and phosphatase were measured (Yee-Meiler, 1975, 1978). The esterase activity was found to be an effective indicator of pollution by both SO_2 and HF. Jensen et al. (1976) warn that 'to eliminate environmental responses the preconditioning environment before fumigation must be identical for all clones or progenies. Ramets screened for tolerance should be grown from cuttings taken from about the same position on the tree to avoid topophytic effects'.

Family comparisons

A selection conducted exclusively on the basis of a mean value of the resistance character for half-sib or full-sib progenies is referred to as family selection. In this method, intrafamily variability is ignored and the whole family is either accepted or rejected. This method gives best results when the selected character is of low heritability (Falconer, 1964). Polster, Böritz and Vogl (1965) have proposed family selection for the breeding of forest trees resistant to gaseous pollution. In this way, selections were made of substantially SO_2-tolerant families of Larix decidua and hybrid progenies of Larix decidua × leptolepis; the hybrid progenies were not only more tolerant but were also characterized by a greater growth vigour (Enderlein and Vogl, 1966; Enderlein, Kästner and Heidrich, 1967; Schönbach et al., 1968). Vogl (1970) has found in Scots pine the same degree of resistance to SO_2 in both the progenies and parent trees. Analysing the genotypic and environmental variation in response of Norway spruce families to HF, Scholz, Timmann and Krusche (1980) found that 59% of the observed variation was due to genetic effects and the narrow sense heritability was $h^2 = 0.34$. In the response of Scots pine families to gaseous air pollution, measured as changes in the activity of peroxidase in needles, very significant differences between the families occur (Niemtur, 1979).

New and very significant conclusions for family selection for resistance arise from isozyme studies on the population structure of several species of coniferous trees. Of the high heterozygosity of such populations (Lundkvist, 1979; Mejnartowicz and Bergman, 1975; Mejnartowicz, 1976; Yeh and Layton, 1979) only about 10% of the variation is due to genetic differences between stands, and the remainder is due to genetic variation within local stands. Intrapopulation differences in ponderosa pine (*Pinus ponderosa*) result primarily from interfamily differences (Mitton *et al.*, 1977).

The heterozygosity of trees determined with the help of enzymatic markers appears to have a significant correlation with fitness (Mitton *et al.*, 1979), which is particularly expressed in the greater-than-average heterozygosity of Scots pine trees recognized from a study of four isozyme systems (Krzakowa, Szweykowski and Korczyk, 1977). Also in studies of maize, the effect of SO_2 was more pronounced in homozygous inbred lines than in their heterozygous crosses (Grzesiak, 1979).

On the basis of these data the use of enzymatic markers for the determination of the degree of heterozygosity between families will help in the selection and breeding of trees tolerant to air pollution.

Population studies

Apart from the intrapopulation variability investigated in clonal and family tests there occurs also an interpopulation component of intraspecific variation in the tolerance of trees to gaseous air pollution. Differences between provenances in, say, Norway spruce, can be so large that they may be comparable with differences between species. The most SO_2-tolerant provenances of Norway spruce are comparable in resistance to blue spruce (*Picea pungens*). Here certain geographic correlations can be observed. Provenances of Norway spruce (*Picea abies*) from mountains and the north are usually more resistant than lowland provenances and those from more southern parts of their range (Tzschacksch and Weiss, 1972). The geographic correlation of resistance may also apply to other species, since similar conclusions were reached by Huttunen and Törmälehto (1979) in studies of pollution resistance in some Finnish Scots pine provenances, the more resistant northern provenances being characterized by a greater xeromorphism (Huttunen, 1978). The more northern continental provenances of lodgepole pine (*Pinus contorta*) are also more resistant than southern provenances (Tzschacksch, Vogl and Thümler, 1969).

In ash (*Fraxinus pennsylvanica*) there is a geographic trend in variation that is opposite to the trend described for the pines. Northern provenances were most sensitive to O_3 while the southern ones were most resistant (Karnosky, 1979). In some species, such as larch (*Larix decidua*), Schönbach *et al.* (1964) observed that there were significant interprovenance differences with respect to SO_2 resistance.

Enzymes as markers of tolerance of trees to pollution

When exposed to the gaseous pollutants SO_2, HF and NO_x, the ultrastructural organization of leaf cells is destroyed, which becomes manifest as a swelling and curling of the thylakoids followed by their disintegration, a granulation of the cytoplasmic and plastid matrix, injury to mitochondria and a decrease in the number of ribosomes (Blingy *et al.*, 1973; Godzik and Knabe, 1973; Godzik and Sassen, 1974; Malhotra, 1976; Młodzianowski and Białobok, 1977; Soikkeli and Tuovinnen, 1979). Changes in the ultrastructure of cells caused by the action of HF and SO_2 are similar in the needles of pine and spruce, though HF causes more serious injury (Soikkeli, 1981). Borei (1954) explained the inhibitory action of fluoride on enzymes by its binding with the calcium and/or magnesium needed for their activity and by the possible formation with the fluoride ion of poisonous metabolic analogues. In addition, fluoride substantially reduces the content of free phosphates, and causes the precipitation of calcium.

Studies on the activity of some enzymes in leaves of trees growing in an atmosphere polluted with HF indicate that such an interpretation, though frequently adopted, is very much oversimplified. Peroxidase from the needles of spruce, in spite of the fact that it is an iron-containing (haem) enzyme, is not inhibited; rather its activity increases when the tree is subjected to HF, particularly when the trees are also fertilized with nitrogen (H. Keller, 1976). In addition new isozymes of peroxidase are induced in spruce by H^- ion treatment and the activities of esterases and β-galactosidase increase. However, in beech (*Fagus sylvatica*) the activity of peroxidase does not change in response to HF and in Scots pine there is an increase in the activity of α-mannosidase while the activities of β-galactosidase, α-galactosidase and β-glucosidase do not change significantly (Bucher-Wallin, 1976). In roots of maize (*Zea mays*) the content of ribosomal RNA decreases (Chang, 1970a) and the activity of ribonuclease increases (Chang, 1970b). In the leaves of beech and spruce, the activities of acid phosphatase and ribonuclease did not change (Bücher-Wallin, 1976), though the fluoride ion disturbed the ratios between potassium, magnesium and calcium by precipitating them as fluoride salts, which would have been expected to cause an inhibition of the ribonuclease which is dependent on these cations (Hanson, 1960).

The mechanism of the injurious action of SO_2 on enzymes can be explained in part by the similarity of sulphite with some other anions and in part by its competition with carbon dioxide in metabolic processes (Ziegler, 1973).

Disintegrated organelles liberate the enzymes contained in them or on their surfaces. This may account for an increase in the activity of some enzymes and why there are changes in isozyme patterns sometimes observed in extracts obtained from leaves of trees growing in regions polluted by industrial emissions. Histochemical studies on the distribution and activity of enzymes in leaf tissues indicate that in some varieties and species of trees there are some differences in the subcellular distribution of some enzymes in trees of varying tolerance (H. Keller, 1976; Kieliszewska-Rokicka, 1978; Młodzianowski and Młodzianowska, 1980).

Peroxidase as an indicator of injury caused by gaseous air pollutants

Peroxidase (E.C. 1.11.1.7) is one of the most widely studied enzymes in attempts to develop a method for the early identification of chronic injury from air pollution. This enzyme belongs to the class of oxidoreductases; it has been found in a wide variety of trees and shrubs and appears to be universally distributed. The enzyme decomposes hydrogen peroxide with a simultaneous oxidation of various substrates such as aromatic acids, phenols and heterocyclic compounds (DeKock, Hall and Inkson, 1979; Nicholls, 1962; Saunders, Holmes-Siedle and Stark, 1964).

Peroxidase activity increases in plant cells under various stress conditions, such as the influence of toxic gases, mechanical injuries to plants or attack by parasitic organisms. Farkas *et al.* (1964) suggest that an increase in the activity of enzymes under stress is associated with destruction of cellular membranes and the concomitant release of previously immobilized proteins. Fridovich and Handler (1961) suggested that when hydrogen peroxide is available, sulphite may be oxidized to sulphate in the presence of peroxidase, which would explain the observed increase in peroxidase activity in plant cells subjected to exposure to SO_2.

The increase of peroxidase activity in leaves of trees growing in zones heavily polluted with SO_2, or under conditions of controlled fumigation with SO_2, varied according to the species of tree, the season and the concentration of the gas, as was shown in the studies of Nikolayevskii (1968). At low concentrations of SO_2 a slight increase was observed in the peroxidase activity in leaves of box elder (*Acer negundo*), a very resistant species. In a homogenate of leaves of silver birch (*Betula pendula*), a sensitive species, the changes in levels of peroxidase activity were not significant. At levels of SO_2 which cause acute injuries to leaves, a sixfold increase in the activity of peroxidase was found in both species. The changes in peroxidase activity are accompanied by opposite changes in catalase activity (Nikolayevskii, 1968; Judel, 1972). In both box elder and silver birch (Nikolayevskii, 1979), as well as in larch (*Larix decidua*) (Grill, Esterbauer and Birkner, 1980), the highest peroxidase activity was observed in late-August and in early-September.

In Norway spruce, even low concentrations of SO_2 in the order of 0.01 ppm caused a more than tenfold increase in the peroxidase activity when the trees were exposed for several months (T. Keller, 1976a). Peroxidase activity increased even when the spruce trees were fumigated during the winter dormant season and when no visible signs of injuries appeared on the needles. The increase in peroxidase activity was accompanied by an increased accumulation of sulphur in leaves, a decline in CO_2 assimilation and a reduced content of ascorbic acid (Keller, 1981). A similar increase in peroxidase activity, preceding the appearance of symptoms of leaf injury, were observed in the leaves of bean (*Phaseolus vulgaris*) subjected to fumigation with O_3 (Curtis and Howell, 1971) and in leaves of Norway spruce exposed to HF (Bucher-Wallin, 1976; H. Keller, 1976; Keller and Schwager, 1971).

The increase in peroxidase activity after exposure to O_3 or SO_2 observed in the leaves of the ornamental trees *Weigela florida*, *Weigela hybrida* 'Van Houttei', bean and Scots pine was not accompanied by the synthesis of new isozymes of peroxidase (Curtis and Howell, 1971; Kieliszewska-Rokicka, 1979), though such isozymes were formed in the needles of Norway spruce treated with HF (Bucher-Wallin, 1976). The seasonal increase in peroxidase activity in needles of Norway spruce and European larch are, however, accompanied by the synthesis of new isozymes (Grill, Esterbauer and Birkner, 1980), a fact that should be taken into consideration when analysing electrophoretograms of enzymes extracted from leaves exposed to industrial gaseous pollutants in the field.

On the basis of investigations of peroxidase activity in leaves of box elder and silver birch Nikolayevskii (1979) believes that trees characterized by considerable resistance to the action of SO_2 normally have a high peroxidase activity. However, Kieliszewska-Rokicka (1979) reached the opposite conclusion, on the basis of an analysis of peroxidase activity in the leaves of SO_2-resistant and SO_2-sensitive clones of Scots pine, *Weigela florida* and *W. hybrida* 'Van Houttei'.

According to Bucher-Wallin (1976), Horsman and Wellburn (1975), Keller (1974, 1976b) and other authors, peroxidase may serve as a good indicator of the distribution of gaseous pollution in the vicinity of large towns, because the SO_2, HF and O_3 which accompany such pollution all cause an increase in its activity, which precedes any visible symptoms of leaf injury.

However, peroxidase is inadequate as an indicator when dealing with air pollution with heavy metals, to which it is relatively insensitive (Ernst, 1976) in spite of the fact that after prolonged exposure to high Cd concentrations, for example, its activity can increase (Grünhagel, Klein and Jäger, 1981; Priebe, Klein and Jäger, 1981). It was also found that peroxidase activity is modified by such factors as seasonal climatic changes (Grill, Esterbauer and Birkner, 1980; Nikolayevskii, 1968), leaf age (Esterbauer, Grill and Zotter, 1978; Simola and Sopanen, 1970), the state of the ontogenetic development of the tree (Conkle, 1971; Grill, Esterbauer and Birkner, 1980), the incompatibility of the stock and scion (Copes, 1978), the type of crown morphology and abundance of cone production and its heterozygosity (Grant and Mitton, 1977; Kavac and Ronye, 1975; Linhart *et al.*, 1979) and fertilization with nitrogen compounds (H. Keller, 1976). In addition, an analysis of isozymes in the needles of inbred progeny of Japanese black pine (*Pinus thunbergii*) has shown that the activity of at least some isozymes is under strict genetic control (Kawanobe and Katsuta, 1980). Such a large number of factors influencing the activity of peroxidase prevents any certain determination of which part of the increase in its activity was caused by the action of gaseous air pollution and which was the result of the remaining exo- and endogenous factors listed above.

The contradictory results obtained by various authors in studies on peroxidase activity in leaf tissues of trees growing in areas polluted with industrial fumes may be partly explained by the considerable interclonal variability in trees as regards this character (H. Keller, 1976) and the substantial interfamily variation. Niemtur (1979) observed that trees belonging to various progenies of Scots pine differed substantially in the response of their peroxidase activity to pollution emitted by a zinc smelter. Trees belonging to some progenies showed an increase in peroxidase activity when close to the source of emission, while in other families peroxidase activity was inhibited, or was not affected at all.

Peroxidase may be a good indicator of air pollution only when the measurement of its activity is performed on leaves of trees for which the nature of the response of the enzyme to pollutants has already been determined and whose isoenzymes, separated electrophoretically, have been genetically analysed.

Histochemical localization of peroxidase in Scots pine needles

PEROXIDASE ACTIVITY IN YOUNG AND MATURE NEEDLES

Since difference in the resistance of leaves to gaseous air pollutants has been observed to be related to the age of these organs, the question arises whether the activity and subcellular distribution of peroxidase also changes during ageing of the needles. The comparison of one month old needles with one year old needles of Scots pine performed by Przymusiński (1980) has shown that young needles did not have an active peroxidase in mesophyll cell walls or in the cytoplasm of transfusion tissue cells. Rather, activity appeared to be in the epidermis and walls of sclerenchyma cells. In mature needles, however, there was no peroxidase activity in epidermis and sclerenchyma cell walls, but activity was found in the protoplast.

Simola and Sopanen (1970), when studying the peroxidase activity in box elder cells at four stages of growth in suspension cultures, found an increase in the activity as the tissues aged. This is possibly the result of the accumulation of large quantities of peroxidase in older tissues. This hypothesis is supported by the studies of Sagisaka (1976) on hydrogen peroxide, one substrate for peroxidase, in shoots of the poplar, *Populus 'gelrica'*; large quantities of this compound were found in the bark and in the xylem.

PEROXIDASE ACTIVITY IN TREES GROWING IN POLLUTED AND UNPOLLUTED AIR

The previously described change in peroxidase activity in needles of trees growing in regions polluted with HF and SO_2 is not associated with any change in the localization of the enzyme in needles of Scots pine compared with trees growing in an unpolluted area (Przymusiński, 1980).

LOCALIZATION OF PEROXIDASE ACTIVITY IN NEEDLES OF TREES DIFFERING IN RESISTANCE TO
SO_2 AND FLUORIDE

The localization of peroxidase in needles was studied using an electron microscope in trees differing in sensitivity to SO_2 (Młodzianowski and Młodzianowska, 1980). The enzyme was localized in the cell walls, the endoplasmic reticulum and in the Golgi apparatus. In sensitive trees a higher activity of the enzyme was observed in mesophyll tissue while in resistant trees the peroxidase activity was higher in the epidermis, hypodermis and in the resin canals of needles. A particularly high increase in enzyme activity has been observed in the cells of the hypodermis and the endodermis of trees resistant to SO_2 when they were subjected to the action of this gas, which permitted Młodzianowski and Młodzianowska (1980) to postulate that these tissues in the needle have a particular significance in the process of detoxification.

Influence of gaseous pollutants on the activity of acid phosphatase

DESCRIPTION OF THE ENZYME

Acid phosphatase (APH) is second to peroxidase as the enzyme most commonly studied for possible use as a bio-indicator of air pollution. APH is produced in cell walls and occurs also in the cytosol of higher plants (Hasegawa, Lynn and Brockbank, 1976). In the leaves it occurs primarily in the spongy parenchyma and in minor veins, but it is much less abundant in the pallisade parenchyma (Besford and Syred, 1979). In yeast, APH is a glycoprotein, whose carbohydrate moiety contains mannose, glucosamine and traces of fucose. It has been separated into two components, an 'acid' one and an 'alkaline' one (Wątorek, Morawiecka and Korczak, 1977), and this has also been shown for APH isolated from needles of Scots pine (Jonsson, 1979). The two enzymes have the same molecular weight (in Scots pine, 70 000), but they differ from each other in their isoelectric points, which for the 'acid' enzymes (optimum pH = 4.95) is 3.5 and for the 'alkaline' one (optimum pH = 5.04) is 9.5. Such considerable differences in isoelectric points in two proteins of identical molecular weight probably result from a bonding with some acidic radical. Removal of the carbohydrate component does not lead to deactivation of APH (Boer *et al.*, 1975). The 'acid' form of APH does not occur in the chloroplasts of Scots pine needles (Jonsson, 1979).

An important metabolic function of APH is to decompose phosphate esters (Heredia, Yen and Sols, 1963; Banasik, Rudnicki and Saniewski, 1980) and to release inorganic phosphate in the cells (Heredia, Yen and Sols, 1963). When the level of phosphorus declines below $0.25\%/g$ P 100 g^{-1} dry wt of tissue, the activity of acid phosphatase increases substantially (Besford, 1979).

INFLUENCE OF EXOGENOUS FACTORS ON THE ACTIVITY OF APH

Some ions, including those occurring in air polluted with industrial fumes, have an inhibitory effect on APH activity. *Inter alia* these include Al^{3+}, Fe^{3+}, fluoride, phosphate, arsenate and molybdate (Clancy and Coffey, 1977; Lorenc-Kubis and Morawiecka, 1973; Malhotra and Khan, 1980; Wątorek, Morawiecka and Korczak, 1977) and differences occur in sensitivity to inhibitors between the wall and the cytoplasmic enzymes (Hasegawa, Lynn and Brockbank, 1976). In interspecific hybrids, *Pinus contorta* × *banksiana*, growing in an environment chronically polluted with SO_2 a higher level of manganese was found in the needles and a lower one of iron, potassium and phosphorus; the level of Fe, K and P declined with the age of the needles

(Amundson, 1978). These results differ from those presented by Materna (1962) from studies on the action of SO_2 on the content of inorganic compounds in needles of Norway spruce. Here in needles exposed to SO_2 the content of K, Ca, Mg and Fe increased. Changes in the levels of these elements have an effect on the activity of acid phosphatase which in spruce, a species very sensitive to SO_2 (Guderian, 1977), declines to 80% of the control activity in the tissues of six week old needles after treatment with 0.06 ppm SO_2 48 h^{-1} (150 μg SO_2 m^{-3} 48 h^{-1}) (Rabe and Kreeb, 1976). Similarly in young needles of jack pine (*Pinus banksiana*) there was an inhibition of APH activity to 46% of the control under the influence of 0.35 ppm SO_2 for 24 h and most importantly it declined by a further 10% during the 24 h after fumigation when the seedlings were already returned to pure air and when the first symptoms of needle injury were just beginning to appear externally. These results must lead to alterations in our methods of evaluating the influence of SO_2 on plants in laboratory tests. In old needles of jack pine and in Norway spruce there was a drop in APH activity following treatment with SO_2, even though APH is still much higher than in younger needles (Malhotra and Khan, 1980; Yee-Meiler, 1975). However, Rabe (1978) believes that acid phosphatase is not a very good indicator enzyme since changes in its activity depend to such a large extent on the species of plant and on the concentrations of gases acting on the plant.

Effects of air pollution on structure of Scots pine populations

According to Nei (1972) the genetic structure of a population may be determined by the frequencies of genes and genotypes, by the degree of heterozygosity and by genetic distances from other populations. Mejnartowicz (1978) calculated the frequency of genes and genotypes for five populations of Scots pine, two of which developed in an environment strongly polluted with SO_2 or with both SO_2 and HF and the remaining three in regions free of pollution. The results of this investigation are presented below.

L-*Leucinaminopeptidase* (E.C. 3.4.1.1), referred to below as LAP, has been studied in female gametophyte tissue isolated from seeds collected from these five populations. It appears that LAP is coded at two loci, the most frequent allele being LAP-A1 in locus LAP-A and LAP-B1 in locus LAP-B. These alleles characteristically occur with great frequency in most Scots pine populations in Europe (Mejnartowicz, 1979). Populations from polluted regions had similar frequencies of LAP alleles and similar genotype frequencies to populations growing in regions free of pollution. In another paper, Mejnartowicz, Białobok and Karolewski (1978) analysed LAP in 24 clones of Scots pine and 17 seedling progenies from these clones and showed experimentally that the degree of heterozygosity of these trees in the loci LAP-A and LAP-B was not related in any way to their resistance to SO_2.

Acid phosphatase (E.C. 3.1.3.2) has been analysed in the same populations that were used for the study of LAP and it appears to be coded in four loci, of which only one, locus APH-B, was analysed in detail. Populations from regions polluted with SO_2 and HF were characterized by a significantly lower frequency of genotypes with the allele APH-B6, which suggests that a selection pressure is being exerted against genotypes with this allele. In non-polluted stands differences in APH-B6 frequency are much smaller. In a comparison of two groups of trees from the polluted environment that differed in their degree of injury by the pollutants, it was found that there are significant differences in the frequency of allele APH-B5. Its frequency was much greater in the group of more tolerant trees. Similar results were obtained by Karnosky and Houston (1978), who studied the effect of SO_2 on a stand of eastern white pine. APH-isoenzyme and total activity assays of tissue extracts suggested that population differentiation occurs in response to population stress. Frequencies of three isozymes were significantly dependent on sampling time after fumigation, while two isozyme frequencies varied significantly according to their sensitivity class.

These results appear to have a wider significance in the evaluation of microevolutionary changes caused by air pollution. This point of view is confirmed by the detailed analysis presented below of the frequencies of genes and genotypes in a Scots pine population aged about 80 years and growing from a planting in an environment strongly polluted with HF and SO_2 originating from a phosphate fertilizer plant (Mejnartowicz, 1980). Three seed samples (A, B and C) collected in the stand have been compared with a fourth one (D) collected in a population that is the natural progeny of the old stand. Sample A is a random representation of trees regardless of the degree of injury to their needles, sample B are selected trees with severely injured needles and sample C consists of material from the least affected trees.

An analysis of the genetic parameters has shown that there is a high degree of genetic similarity between the more resistant trees in the parental population and the F_1 progeny, as is shown graphically in *Figure 25.1*. In the F_1 progeny the most frequent genotype was APH-B5B5 and the rarest was APH-B6B6. In an environment polluted with HF and SO_2 a selection pressure appears to exist against genotype APH-B6B6 and in favour of APH-B5B5.

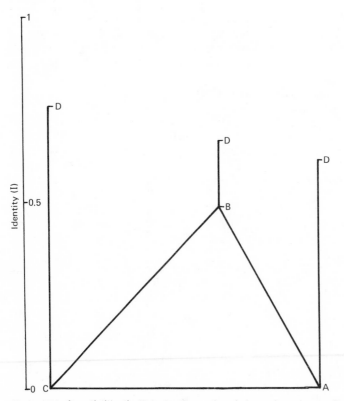

Figure 25.1. Genetic identity (I) between samples of: A, random parents, B, sensitive parents, C, tolerant parents, D. offspring

This method is complicated by the relatively small difference in mobility of the two isoenzymes APH-B5 and B6 (in electrophoresis on starch gel with a borate buffer electrode) and by the usual large frequency of the gene APH-B5 in populations of Scots pine (Mejnartowicz, 1979). We know of no direct link between the presence in a tree of a particular isoenzyme of APH and the resistance of that tree to HF or SO_2, though APH is clearly an important enzyme in nutrition (Hasegawa, Lynn and Brockbank, 1976), in the transport of stored metabolites (Flinn and Smith, 1967) and in stomatal opening (Mishra and Panda, 1970). Disturbance of any of these processes by industrial emissions causes injuries to trees. The presence of some isozymes of acid phosphatase may give an early advantage at the stage of seed germination, when a higher activity of an enzyme may depend on its *de novo* synthesis (Shain and Mayer, 1968) or on the liberation and subsequent activation of partially inactive phosphatases present in various cell fractions (Meyer, Mayer and Harek, 1971). In roots of wheat, Hasegawa, Lynn and Brockbank (1976) have observed five isoenzymes of acid phosphatase differing in sensitivity to various inhibitors, including fluorides, although all were inhibited by NaF. In the presence of 1 mM NaF only 50% of the initial activity was seen in APH-5 while in APH-2 the activity remaining was 81%.

Malate dehydrogenase (E.C. 1.1.1.37) [MDH] may give good marker genes for tolerance of plants to gaseous emissions. Analysis of the genetic structure of the Scots pine population and of its natural progeny described above (Mejnariowicz, 1980) has shown that it is coded in three loci with a relatively large number of alleles, of which in locus MDH-A allele A2 is the most common ($P=0.978$), in locus MDH-B allele B2 is the most common ($P=0.685$) and in locus MDH-C the allele C4 has a large frequency ($P=0.478$).

An analysis of genotype frequencies has shown that the most common genotype is MDH-A2A2 which is almost exclusively homozygous for the whole population, and for this locus no differences were found between the groups of trees differing in tolerance to HF and SO_2. On the other hand significant differences were observed in genotypes for the MDH-B locus, where the group of tolerant trees had the greatest number of genotypes of the type MDH-B2B2 ($P=0.583$), while in the group of sensitive trees its frequency was much lower ($P=0.363$) and the most common genotype was MDH-B2B5. It is characteristic that in the group of resistant trees there were as many as 75% of homozygous trees in the locus MDH-B while in the sensitive group only 36% were homozygous.

In locus MDH-C there occurred seven genotypes in the group of tolerant trees and five in the sensitive group. Genotypes C1C4 and C3C4 have not been found among trees sensitive to HF and the most common genotype is C4C4. In all there were 45% homozygotes in the MDH-C locus in the group of trees sensitive to pollution by HF and only 27% in the tolerant group.

Three years of electrophoretic investigations on isoenzymes of MDH in seed samples collected in consecutive years from the same trees growing in a region polluted with HF and SO_2 have shown that changes occur in the electrophoretic mobility and activity of enzymes, most probably under the influence of these agents. In about 2% of the trees a change in the location of a band on the electrophoregram was observed because of its slower migration, perhaps due to loss of some electrical charge or association of some group increasing the size of the isoenzyme molecule. It was also found that some isoenzymes were losing their activity, so that in a genetic analysis of the electrophoretic pattern they were included in the pool of recessive alleles, that occur in many populations of forest trees, not only in MDH but also in LAP, APH, glutamate/oxalacetate transaminase and others (Mejnartowicz, 1981).

As with APH there was also a substantial reduction in the number of genotypes of MDH observed in the group of trees tolerant to HF and SO_2 compared to the sensitive group.

Peroxidase (E.C. 1.11.1.7) activity in the Scots pine stand described above was examined by Szmidt (1980). There were no significant differences in enzyme activities between the group of more tolerant trees and the more sensitive ones.

No genetic analysis was made of the electrophoretic patterns of this enzyme and it is not possible to say much about the effect of air pollution on the genetic structure of the population. In the group of trees tolerant to HF and SO_2 one of the 19 electrophoretic bands of peroxidase, described as the P-8 band, was more frequent, and its frequency was also greater in the F_1 progeny compared to the sensitive trees. In another stand exposed to SO_2 the frequency of P-8 was greater in the more tolerant trees. On the other hand band P-17 was more common in trees more sensitive to these gases.

Are these permanent mutational changes or are they only modifications that will disappear when air pollution ceases? Are these changes observable only in polluted areas or are they present in regions free of pollution? The studies of enzymes conducted by the author on Scots pine clones growing in regions free of pollution indicate that there is great stability in electrophoretic patterns of isoenzymes, so that the cause of the changes in isoenzyme patterns described above may truly be related to pollution. This would indeed be important in the selection processes that take place in plant populations growing in regions where the environment is polluted by industrial fumes, and where the toxic agents cause an inactivation of some alleles.

Acknowledgements

This project was supported by funds PL-FS-86(J-MOA-USDA-11) made available from the Marie Skłodowska–Curie Fund established by contributions of the United States and Polish Governments, and supported by funds MR.II.16.PAN.

References

AMUNDSON, R. G. The effects of SO_2 on mineral nutrition and gas exchange of pine (*Pinus contorta* × *banksiana*). PhD Dissertation, University of Washington, Seattle, USA (1978)

BANASIK, L., RUDNICKI, R. M. and SANIEWSKI, M. Studies on the physiology of hyacinth bulbs (*Hyacinthus orientalis* L.). XIII. The distribution of amylase and acid phosphatase activities and starch grains in hyacinth bulbs, *Acta Physiologiae Plantarum* 2, 145–156 (1980)

BELL, J. N. Response of plants to sulphur dioxide, *Nature* 284, 399 (1980)

BERGMANN, F. Genetische Untersuchungen bei *Picea abies* mit Hilfe der Isoenzym-Identifizierung. I. Möglichkeiten für genetische Zertifizierung von Forstsaatgut, *Allgemeine Forst- und Jagdzeityng* 11, 278–280 (1971)

BERGMANN, F. Genetische Untersuchungen bei *Picea abies* mit Hilfe der Isoenzym-Identifiezirung. II. Genetische Kontrolle von Esterase- und Leucinaminopeptidase-Isoenzymen im haploiden Endosperm ruhender Samen, *Theoretical and Applied Genetics* 43, 222–225 (1973)

BESFORD, R. T. Quantitative aspects of leaf acid phosphatase activity and phosphorus status of tomato plants, *Annals of Botany* 44, 153–161 (1979)

BESFORD, R. T. and SYRED, A. D. Effect of phosphorus nutrition on the cellular distribution of acid phosphatase in the leaves of *Lycopersicon* L., *Annals of Botany* 43, 431–453 (1979)

BLIGNY, R., BISCH, A. M., GARREC, J. P. and FOURCY, A. Observations morphologigues et structurales des effects du fluor sur les cires epicuticularies et sur les chloroplastes des aiguilles de sapin (*Abies alba* Mill.), *Journal de Microscopie* 17, 207–214 (1973)

BOER, P., VANRIJN, H. J. M., REJNKING, A. and STEYEN-PARVE, E. Biosynthesis of acid phosphatase of baker's yeast. Characterization of a protoplast-bound fraction containing precursor of the exoenzyme, *Biochimica et Biophysica Acta* 377, 331–342 (1975)

BOREI, H. Inhibition of cellular oxidation by fluoride, *Arkiv för kemi, mineralog och geologi*. K. Svenska Vetenskaps-Akademien. Stockholm, Uppsala Almqvist et Wiksells Boktrycker **20A(8)**, 1–215 (1945)

BRADSHAW, A. D. Pollution and evolution. In *Effects of Air Pollution on Plants*, (Ed. T. A. MANSFIELD), pp. 135–160, Cambridge University Press, Cambridge (1976)

BROWN, A. H. D. and MORAN, G. F. Isozymes and the genetic resources of forest trees, *Proceedings of the Symposium on Isozymes of North American Forest Trees and Forest Insects*, July 27, 1979, Berkeley, California, General Technical Report PSW, **48**, 1–10 (1979)

BÜCHER-WALLIN, I. Zur Beeinflussung des physiologischen Blattalters von Waldbäumen durch Fluor-Immissionen, *Mitteilungen Eidgenossiche Anstalt für das Forstliche Versuchswesen, Schweiz* **52**, 101–158 (1976)

CHANG, C. W. Effect of fluoride on ribosomes from corn roots. Changes with growth retardation, *Physiologia Plantarum* **23**, 536–543 (1970a)

CHANG, C. W. Effect of fluoride on ribosomes and ribonuclease from corn roots, *Canadian Journal of Biochemistry* **48**, 450–454 (1970b)

CHANG, C. W. Fluorides. In *Responses of Plants to Air Pollutants*, (Eds. J. B. MUDD and T. T. KOZLOWSKI), pp. 57–95, Physiological Ecology. A series of Monographics, Texts and Treatises, Academic Press, Inc. A subsidiary of Harcourt, Brace, Jovanovich, Publishers (1975)

CLANCY, F. G. and COFFEY, M. D. Acid phosphatase and protease release by the insectivorous plant *Drosera rotundifolia*, *Canadian Journal of Botany* **55**, 480–488 (1977)

CONKLE, M. Th. Isozyme specificity during germination and early growth of knobcone pine, *Forest Science* **17**, 494–498 (1971)

CONKLE, M. Th. Isozyme variation and linkage in six conifer species, *Proceedings of the Symposium on Isozymes of North American Forest Trees and Forest Insects*, July 27, Berkeley, California. General Technical Report PSW **48**, 11–17 (1979)

COPES, D. L. Isoenzyme activities differ in compatible and incompatible Douglas-fir graft unions, *Forest Science* **24**, 297–303 (1978)

CURTIS, C. R. and HOWELL, R. K. Increase in peroxidase isoenzyme activity in bean leaves exposed to low doses of ozone, *Phytopathology* **61**, 1306–1307 (1971)

DÄSSLER, H. G. Zur Aussagekraft experimenteller Resistenzprüfungen, *Archiv fü Forstwesen* **16**, 781–785 (1967)

DeKOCK, P. C., HALL, A. and INKSON, H. E. A study of peroxidase and catalase distribution in the potato tuber, *Annals of Botany* **43**, 295–298 (1979)

DOCHINGER, L. S., TOWNSEND, A. M., SEEGRIST, D. W. and BENDER, F. W. Responses of hybrid poplar trees to sulfur dioxide fumigation, *Journal of the Air Pollution Control Association* **22**, 363–371 (1972)

ENDERLEIN, H., KÄSTNER, W. and HEIDRICH, H. Wie verhält sich auf Rauchhärte geprüftes Pflanzenmaterial der Gattung *Larix* in einem extremen Rauchschadengebiet unter dem Einfluss starker Frost- bzw, Spätfrosteinwirkungen. *Die Sozialistische Forstwirt* **17**, 91–93 (1967)

ENDERLEIN, H. and VOGL, M. Experimentelle Untersuchungen über die SO_2-Empfindlichkeit der Nadeln verschiedener Koniferen, *Archiv für Forstwesen* **15**, 1207–1224 (1966)

ERNST, W. Physiological and biochemical aspects of metal tolerance. In *Effect of Air Pollutants on Plants*, (Ed. T. A. MANSFIELD), pp. 115–133, Cambridge University Press, Cambridge (1976)

ESTERBRAUER, H., GRILL, D. and ZOTTER, M. Peroxidase in Nadeln von *Picea abies* (L.) Karst, *Biochemie und Physiologie der Pflanzen* **172**, 155–159 (1978)

FALCONER, D. S. *Introduction to Quantitative Genetics*, The Ronald Press Company, New York (1964)

FARKAS, G. L., DEZSI, L., HORVATH, M., KISBAN, K. and UDVARDY, J. Common pattern of enzymatic changes in detached leaves and tissues attacked by parasites, *Phytopathologische Zeitschrift* **49**, 343–354 (1964)

FLINN, A. M. and SMITH, D. L. The localization of enzymes in the cotyledons of *Pisum arvense* L. during germination, *Planta* **75**, 10–22 (1967)

FRIDOVICH, I. and HANDLER, P. Detection of free radicals generated during enzymatic oxidations by the initiation of sulphite oxidation, *Journal of Biological Chemistry* **236**, 1834–1840 (1961)

GODZIK, S. and KNABE, W. Vergleichende elektronenmikroskopische Untersuchungen der Feinstruktur von Chloroplasten einiger Pinus Arten aus den Industriegebieten an der Ruhr und in Oberschlesien. In *Proceedings of the Third International Clean Air Congress*, VDI-Verlag GmbH, Düsseldorf, A, 164–170 (1973)

GODZIK, S. and SASSEN, M. M. A. Einwirkung von SO_2 auf die Feinstruktur der Chloroplasten von *Phaseolus vulgaris*, *Phytopathologische Zeitschrift* **79**, 155–159 (1974)

GORDON, T. G. and GORHAM, E. Ecological aspects of air pollution from iron-sintering plant at Wawa Ontario, *Canadian Journal of Botany* **41**, 1063–1078 (1963)

GRANT, M. C. and MITTON, J. B. Genetic differentiation among growth forms of engelmann spruce and subalpine fir at tree line, *Arctic and Alpine Research* **9**, 259–263 (1977)

GRILL, D., ESTERBAUER, H. and BIRKNER, M. Untersuchungen über die Peroxidaseaktivät in Lärchennadel, *Beiträge zur Biologie der Pflanzen*, (Eds. D. VON DENFFER, H. SCHRAUDOLF, and H. WEBER), Duncker und Humblot, Berlin **55**, 67–76 (1980)

GRÜNHAGEL, L., KLEIN, H. and JÄGER, H. J. Langzeiteinwirkungen von Freilandrelevanten SO_2-und Cadmiumkonzentrationen an Pflanzen. Nachweis und Wirkung forstschädlicher Luftverunreinigungen, *XI Internationalen Arbeitstagung forstlicher Rauchsschadenssachverständiger 1–6 September 1980* in Graz, Österreich, pp. 309–317 (1981)

GRZESIAK, S. Influence of sulphur dioxide on the relative rate of photosynthesis in four species of cultivated plants under optimum soil moisture and drought conditions, *Bulletin de L'Academy Polonaise Des Sciences Cl. II. Plant Physiology* **27**, 309–321 (1979)

GUDERIAN, R. *Air Pollution. Phytotoxicity of Acidic Gases and Its Significance in Air Pollution Control*, Springer-Verlag, Berlin (1977)

GUZINA, V. Isoenzimi peroksidaze u genetskim proucavajima topola [*Isoenzymes of peroxidase in genetic provenances of poplar*], *Genetika (Yugoslavia)* **6**, 63–68 (1974)

HAMRICK, J. L., MITTON, J. B. and LINHART, Y. B. Levels of genetic variation in trees: influence of life history characteristics, *Proceedings of the Symposium on Isozymes of North American Forest Trees and Forest Insects*, July 27, 1979, Berkeley, California, General Technical Report PSW, **48**, 42–47 (1979)

HANSON, J. B. Impairment of respiration, ion accumulation and ion retention in root tissue treated with ribonuclease and ethylenediamine tetra-acetic acid, *Plant Physiology* **35**, 372–379 (1960)

HASEGAWA, Y., LYNN, K. R. and BROCKBANK, W. J. Isolation and partial characterization of cytoplasmic and wall-bound acid phosphatases from wheat roots, *Canadian Journal of Botany* **54**, 1163–1169 (1976)

HEREDIA, C. F., YEN, F. and SOLS, A. Role and formation of the acid phosphatase in yeast, *Biochemical and Biophysical Research Communications* **10**, 14–18 (1963)

HORSMAN, D. C. and WELLBURN, A. R. Synergistic effects of SO_2 and NO_2 polluted air upon enzyme activity in pea seedlings, *Environmental Pollution* **8**, 123–133 (1975)

HORSMAN, D. C. and WELLBURN, A. R. Guide to the metabolic and biochemical effects of air pollutants on higher plants. In *Effects of Air Pollutants on Plants*, (Ed. T. A. MANSFIELD), pp. 185–199, Cambridge University Press, Cambridge (1976)

HOUSTON, D. B. and STAIRS, G. R. Genetic control of sulfur dioxide and ozone treatment in eastern white pine, *Forest Science* **19**, 267–271 (1973)

HUTTUNEN, S. The effects of air pollution on provenances of Scots pine and Norway spruce in Northern Finland, *Silvae Fennica* **12**, 1–16 (1978)

HUTTUNEN, S. and TÖRMÄLEHTO, H. Air pollution resistance of some Finnish Scots pine provenances. *Proceedings IUFRO Air Pollution Meeting on the Physiological and Biochemical Effects of Air Pollution on Plants*, 20–24 August, Zabrze, Poland, pp. 27–29 (1979)

JEFFREE, C. E. Plant damage caused by SO_2. Committee on Agricultural Problems, *Timber Committee Symposium on the Effects of Airborne Pollution on Vegetation*, 20–24 August, Warsaw, Poland, pp. 1–26 (1979)

JENSEN, K. F., DOCHINGER, L. S., ROBERTS, B. R. and TOWNSEND, A. M. Pollution Response. In *Modern Methods in Forest Genetics*, (Ed. J. P. MIKSCHE), pp. 189–216, Springer-Verlag, Berlin, Heidelberg (1976)

JOHN, A. Vegetative propagation of hybrid larch (*Larix × eurolepis* Henry) in Scotland. In *Vegetative Propagation of Forest Trees — Physiology and Practice*. Lectures from a Symposium in Uppsala, Sweden, 16–17 February, 1977. The Institute for Forest Improvement and the Department of Forest Genetics, College of Forestry, The Swedish University of Agricultural Sciences, pp. 129–136 (1977)

JONSSON, J. Isolation and partial characterization isoenzymes of acid phosphatase from needles of *Pinus sylvestris* L. *Proceedings of the Conference on Biochemical Genetics of Forest Trees*, Umeå 1978, (Ed. D. RUDIN), pp. 5–14, Report 1. Swedish University of Agricultural Sciences, Department of Forest Genetics and Plant Physiology, Umeå (1979)

JUDEL, G. K. Anderung in der Aktivität der Peroxidase und der Katalase und im Gehalt an Gesamtphenolen in der Blättern der Sonnenblume unter dem Einfluss von Kupfer-und Stickstoffmangel, *Zeitschrift für Pflanzenernährung und Bodenkunde* **133**, 81–92 (1972)

KARNOSKY, D. F. Evidence for genetic control of response to sulfur dioxide and ozone in *Populus tremuloides*, *Canadian Journal of Forest Research* **7**, 437–440 (1977)

KARNOSKY, D. F. Consistency from year to year in the response of *Fraxinus pennsylvanica* provenances to ozone. *Proceedings of the Symposium IUFRO Air Pollution Meeting*. Physiological and biochemical effects of air pollution on plants, Zabrze, Poland, August 27–29 (in press) (1979)

KARNOSKY, D. F. Changes in southern Wisconsin white pine stands related to air pollution sensitivity.

Proceedings of Symposium of Air Pollutants on Mediterranean and Temperate Forest Ecosystems, June 22–27, 1980, Riverside, California, pp. 238 (1980)

KARNOSKY, D. F. and HOUSTON, D. S. Genetics of air pollution tolerance of trees in the northeastern United States. *Proceedings 25th Northeastern Forest Tree Improvement Conference*, Pennsylvania State University, July 25–27, 1978, pp. 161–178 (1978)

KAVAC, Ya. E. and RONYE, V. M. Isoenzimy pyeroksidazy khvoi v populyaciyakh yeli obyknovyennoi. [*Isoenzymes of peroxidases of needles in population of Norway spruce.*] In *Gyenyetichyeskiye isslyedovaniya drevesnykh v Latviiskoi SSR* [*Genetic investigation of trees in Latvian SSR*], Zinatnye, Riga, USSR, pp. 58–62 (1975)

KAWANOBE, K. and KATSUTA, M. Genetic analysis of peroxidase isozymes in *Pinus thunbergii* Parl. self progeny, *Journal of the Japanese Forestry Society* **62**, 357–360 (1980)

KELLER, H. Histologische und physiologische Untersuchungen an Forstpflanzen in einem Fluorschadensgebiet, *Berichte, Eidgenössische Anstalt für das forstliche Versuchswesen* **154**, 1–82 (1976)

KELLER, T. The use of peroxidase activity for monitoring and mapping air pollution areas, *European Journal of Forest Pathology* **4**, 11–19 (1974)

KELLER, T. and SCHWAGER, H. Der Nachweis unsichtbarer ('physiologischer') Fluorimmissionsschädigungen an Waldbäumen durch eine einfache kolorimetrische Bestimmung der Peroxidase-Aktivität, *European Journal of Forest Pathology* **1**, 54–67 (1971)

KELLER, T. Auswirkungen niedriger SO₂-Konzentrationen auf junge Fichten, *Schweizerischen Zeitschrift für Forstwesen* **127**, 237–251 (1976a)

KELLER, T. Der Einfluss von Schwefeldioxid als Luftverunreinigung auf die Assimilation der Fichte, *Berichte, Eidgenössische Anstalt für das forstliche Versuchswesen* **57**, 48–53 (1976b)

KELLER, T. Die beeinflussung physiologischer Prozesse der Fichte durch eine Winterbegasung mit SO₂, *Mitteilungen der forstlichen Bundesversuchstanstalt Wien* **137**, 115–120 (1981)

KIELISZEWSKA-ROKICKA, B. Relation between the activity of peroxidase and the resistance of clones of Scots pine to the action of SO₂. In *Studies on the Effect of Sulphur Dioxide and Ozone on the Respiration and Assimilation of Trees and Shrubs in Order to Select Individuals Resistant to Action of These Gases*, (Ed. S. BIAŁOBOK), pp. 82–84, Fourth Annual Report FG-Po-326. Polish Academy of Sciences Institute of Dendrology, Kórnik, Poland (1978)

KIELISZEWSKA-ROKICKA, B. Peroxidase activity in Varieties of *Weigela* and *Pinus sylvestris* resistant and susceptible to SO₂, *Arboretum Kórnickie* **24**, 313–320 (1979)

KNABE, W. Methoden der Auslese und Züchtung immissionsresistenter Gehölze, *XIV IUFRO-Kongress. München Sektion* **24**, 2–24 (1967)

KRZAKOWA, M. Enzymatyczna zmienność międzypopulacyjna sosny zwyczajnej (*Pinus sylvestris*). [*Enzymatic variability between Scots pine populations.*] *Uniwersytet Adama Mickiewicza w Poznaniu* (*Adam Mikiewicz University, Poznań*], Seria Biologia **17**, 1–44 (1979)

KRZAKOWA, M., SZWEYKOWSKI, J. and KORCZYK, A. Population genetics of Scots pine (*Pinus sylvestris* L.) forest genetic structure of plus-trees in Bolewice near Poznań (West Poland), *Bulletin de l'Académie Polonaise des Sciences. Série des sciences biologiques* **9**, 583–590 (1977)

LAMPADIUS, F., PELZ, E. and POHL, E. Beiträge zum Problem der Beurteilung und des Nachweises der Resistenz von Waldbäumen gegenüber Imissionen, *Biologisches Zentralblatt* **89**, 301–326 (1970)

LINDGREN, D. Possible advantages and risks connected with vegetative propagation for reforestation. In *Vegetative Propagation-Physiology and Practice Lectures from a Symposium in Uppsala, Sweden, 16–17 February, 1977*. The Institute for Forest Improvement and The Department of Forest Genetics, College of Forestry, The Swedish University of Agricultural Science, pp. 9–16 (1977)

LINHART, B., MITTON, J. B., BOWMAN, D. M., STURGEON, K. B. and HAMRICK, J. L. Genetic aspects of fertility differentials in ponderosa pine, *Genetical Research, Cambridge* **33**, 237–242 (1979)

LORENC-KUBIS, J. and MORAWIECKA, B. Phosphatase activity of *Poa pratensis* seeds. I. Preliminary studies on acid phosphatase II, *Acta Societatis Botanicorum Poloniae* **42**, 369–377 (1973)

LORENC-PLUCINSKA, G. Effect of industrial pollutants on the process of CO₂ exchange. In *Studies on the Effect of Sulphur Dioxide and Ozone on the Respiration of Trees and Schrubs in Order to Select Individuals Resistant to Action of These Gases*. Pl-Fs 74 (FG-Po-236), (Ed. S. BIAŁOBOK), pp. 21–37, Polish Academy of Sciences Institute of Dendrology, Kórnik, Poland (1980)

LUNDKVIST, K. Allozyme frequency distributions in four Swedish populations of Norway spruce (*Picea abies* K.). I. Estimation of genetic variation within and among populations, genetic linkage and a mating system parameter, *Hereditas* **90**, 127–143 (1979)

MALHOTRA, S. S. Effects of sulphur dioxide on biochemical activity and ultrastructural organization of pine needle chloroplasts, *New Phytologist* **76**, 239–245 (1976)

MALHOTRA, S. S. and HOCKING, D. Biochemical and cytological effects of sulphur dioxide on plant metabolism, *New Phytologist* **76**, 228–237 (1976)

MALHOTRA, S. S. and KHAN, A. A. Effects of sulphur dioxide and other air pollutants on acid phosphatase activity in pine seedlings, *Biochemie und Physiologie der Pflanzen* **175**, 228–236 (1980)

MAMAYEW, S. A. and SHKARLET, O. D. Effects of air and soil pollution by industrial waste on the fructification of Scots pine in the Urals, *Mitteilungen der Forstlichen Bundes-Versuchsanstalt, Wien* **97**, 443–450 (1972)

MATERNA, J. Vliv kyslicniku siriciteho na mineralni sliżeni smrkoveho jehlici [*Effect of sulphur dioxide on the mineral composition of spruce needles*], *Prace vyzkumnych ustavu lesnickych, CSSR* **24**, 7–36 (1962)

MEJNARTOWICZ, L. Genetic investigations on Douglas-fir (*Pseudotsuga menziesii* (Mirb.) Franco) populations, *Arboretum Kórnickie* **21**, 125–187 (1976)

MEJNARTOWICZ, L. Struktura populacji sosny zwyczajnej w badaniach metodami biochemicznymi [*Population structure of Scots pine in biochemical methods of investigation*]. In *Stan prac nad proweniencjami sosny, świerka i modrzewia w Polsce oraz perspektywa badań populacyjnych tych gatunków* [*The State of Provenance Investigations of Pine, Spruce and Larch in Poland and the Future of Population Investigations of these Trees*], Kórnik, 5–6 June 1978. Polska Akademia Nauk Instytut Dendrologii w Kórniku [*Polish Academy of Sciences, Institute of Dendrology at Kórnik*], pp. 1–10 (1978)

MEJNARTOWICZ, L. Genetic variation in some isoenzyme loci in Scots pine (*Pinus silvestris* L.) populations, *Arboretum Kórnickie* **24**, 91–104 (1979)

MEJNARTOWICZ, L. Changes in genetic structure of Scots pine population affected by industrial emissions of sulphur-oxide and fluorine. In *A Genetic Basis for the Resistance of Forest Trees to Antropopressure, with Special Study of the Effect of Some Toxic Gases*. First Annual Report, (Ed. L. MEJNARTOWICZ), pp. 4–18, Polish Academy of Sciences, Institute of Dendrology, Kórnik, Poland (1980)

MEJNARTOWICZ, L. Application of biochemical genetics in the selection of trees more resistant to the action of sulphur dioxide. In *A Genetic Basis for the Resistance of Forest Trees to Anthropopressure, with Special Study of the Effect of Some Toxic Gases*. First Annual Report, (Ed. L. MAJNARTOWICZ), pp. 19–38, Polish Academy of Sciences, Institute of Dendrology, Kórnik, Poland (1980)

MEJNARTOWICZ, L. and BERGMANN, F. Genetic studies on European larch (*Larix decidua* Mill.) employing isoenzyme polymorphisms, *Genetica Polonica* **16**, 29–36 (1975)

MEJNARTOWICZ, L., BIAŁOBOK, S. and KAROLEWSKI, P. Genetic characteristics of Scots pine specimens resistant and susceptible to SO_2 action, *Arboretum Kórnickie* **23**, 233–239 (1978)

MEYER, H., MAYER, A. M. and HAREK, E. Acid phosphatases in germinating lettuce — evidence for partial activation, *Physiologia Plantarum* **24**, 95–101 (1971)

MILLER, P. R. and McBRIDGE, I. R. Effects of air pollution on forests. In *Response of Plants to Air Pollution*, (Eds. J. B. MUDD and T. T. KOZŁOWSKI), pp. 196–235, Academic Press, New York (1975)

MISHRA, D. and PANDA, K. C. Acid phosphatase of rice leaves showing diurnal variation and its relation to stomatal opening, *Biochemie und Physiologie der Pflanzen* **161**, 532–536 (1970)

MITTON, J. B., LINHART, Y. B., HAMRICK, J. L. and BECKMAN, J. S. Observations on the genetic structure and mating system of ponderosa pine in the Colorado Front Range, *Theoretical and Applied Genetics* **51**, 5–13 (1977)

MITTON, J. B., KNOWLES, P., STURGEON, K. B., LINHART, Y. B. and DAVIS, M. Association between heterozygosity and growth rate variables in three western forest trees. *Proceedings of Symposium on Isozymes of North American Forest Trees and Forest Insects*, July 27, 1979, Berkeley, California, General Technical Report PSW, **48**, 27–34 (1979)

MŁODZIANOWSKI, F. and BIAŁOBOK, S. The effect of sulphur dioxide on ultrastructural organization of larch needles, *Acta Societatis Botanicorum Poloniae* **46**, 629–634 (1977)

MŁODZIANOWSKI, F. and MŁODZIANOWSKA, L. Cytochemical localization of enzyme activity in Scots pine needles treated with SO_2. In *Studies on the Effect of Sulphur Dioxide and Ozone on the Respiration of Trees and Shrubs in Order to Select Individuals Resistant to the Action of These Gases*. Pl-Fs 74 (FG-Po-236) Report, (Ed. S. BIAŁOBOK), pp. 79–81, Polish Academy of Sciences Institute of Dendrology, Kórnik, Poland (1980)

MUHS, H. J. Distinction of Douglas-fir provenances using peroxidase isoenzyme patterns of needles, *Silvae Genetica* **23**, 71–76 (1974)

NEI, M. Genetic distance between populations, *The American Naturalist* **106**, 283–292 (1972)

NICHOLLS, P. Peroxidase as an oxygenase. In *Oxygenase*, (Ed. HAYAISHI), pp. 274–303, Academic Press, New York (1962)

NIEMTUR, S. Influence of zinc smelter emissions on peroxidase activity in Scots pine needles of various families, *European Journal of Forest Pathology* **9**, 142–147 (1979)

NIKOLAYEVSKII, V. S. Aktivnost nekotorikh fyermyentov i gazoustoichivosti rastyenii [*Plants resistance to gaseous emissions and activity of some enzymes*]. *Trudy Instituta Ekologii Rastyenii i Zhyvotnykh. Uralskii Filial AN SSR* **62**, 208–211 (1968)

NIKOLAYEVSKII, V. S. Biologicheskiye osnovy gazoustoichivosti rastyenii [*A biological basis for plant resistance to gaseous emission*], *Akadyemia Nauk SSR, Nauka Sibirskoye Otdyelyeniye*, pp. 1–278 (1979)

OLEKSYN, J. Effect of sulphur dioxide on net photosynthesis and dark respiration of Scots pine individuals differing in susceptibility to this gas. In *Studies on the Effect of Sulphur Dioxide and Ozone on the Respiration of Trees and Shrubs in Order to Select Individuals Resistant to the Action of These Gases*. Pl-Fs 74 (FG-Po-326) Report, (Ed. S. BIAŁOBOK), pp. 37–45, Polish Academy of Sciences, Institute of Dendrology, Kórnik, Poland (1980)

POLSTER, H., BÖRITZ, S. and VOGL, M. Pflanzenphysiologische Untersuchungen im Dienste der Züchtung von Koniferen auf Rauchresistenz, *Sozialistische Forstwirtschaft, Berlin*, pp. 368–370 (1965)

PRIEBE, A., KLEIN, H. and JÄGER, H. J. Cadmiumanreicherung Immissionsbelasteter Fichten. Utteilungen der Forstlichen Bundes-Versuchsanstalt Wien, Nachweis und Wirkung forstschädlicher Luftferunreinigungen. *IUFRO Tagungsbeiträge zur XI Internationalen Arbeitstagung forstlicher Rauchschadenssachverständiger 1–6, September 1980 in Graz, Osterreich*, pp. 319–326 (1981)

PRZYMUSIŃSKI, R. Cytochemical localization of peroxidase activity in needles of Scots pine resistant and susceptible to industrial pollution. In *A Genetic Basis for the Resistance of Forest Trees to Anthropopressure, with Special Study of the Effect of Some Toxic Gases*, (Ed. L. MEJNARTOWICZ), pp. 38–49, Polish Academy of Sciences, Institute of Dendrology, Kórnik, Poland (1980)

RABE, R. Bioindication von Luftverunreinigungen auf Grund der Änderung von Enzymaktivität und Chlorophyllgehalt von Testpflanzen, *Disertationes Botanicae*, (Ed. J. CRAMER), A. R. Gantner Verlag **45**, 1–219 (1978)

RABE, R., and KREEB, L. Eine Methode zur Laborbegasung von Testpflanzen mit Schwefeldioxide und ihre Anwendung bein Untersuchungen zur Enzymaktivität, *Angewandte Botanik* **50**, 71–78 (1976)

RASMUSON, B. and RUDIN, D. Variations in esterase zymogram patterns in needles of *Pinus silvestris* from provenances in northern Sweden, *Silvae Genetica* **20**, 1–52 (1971)

ROHMEDER, E., MERZ, W. and von SCHÖNBORN, A. Zuchtung von gegen Industrialgase relative resistenten Fichten -und Kiefernsorten, *Forstwissenschaftliches Zentralblatt* **81**, 321–332 (1962)

ROHMEDER, E. and von SCHÖNBORN, A. Der Einflus von Umwelt und Erbgut auf die Widerstandsfähigkeit der Waldbäume gegenüber Luftverunreinigungen durch Industrieabgase. Ein Beitrag zur Züchtung einer relativ rauchresistenten Fichtensorte, *Forstwissenschaftliches Zentralblatt, Hamburg und Berlin* **84**, 1–68 (1965)

RUDIN, D. Gene and genotype frequencies in Swedish Scots pine populations a study by the aid of the isozyme technique, *Proceedings of Joint IUFRO Meeting*, Section S.02.04.1–3, Special Reports, 465–467 (1974)

RUDIN, D., ERIKSSON, G. and RASMUSON, M. Studies of allele frequencies and inbreeding in Scots pine populations by the aid of the isozyme technique, *Silvae Genetica* **23**, 10–13 (1974)

SAGISAKA, S. The occurrence of peroxide in a perennial plant, *Populus gelrica, Plant Physiology* **57**, 308–312 (1976)

SAUNDERS, B. C., HOLMES-SIEDLE, A. G. and STARK, B. P. *Peroxidase. The Properties and Use of a Versatile Enzyme and of Some Related Catalysts*, pp. 271, Butterworths, London (1964)

SCHOLZ, F. and KNABE, W. Investigations on buffering capacity in spruce clones of different resistance to air pollution, *XVI IUFRO World-Congress, Oslo 1976*, Section S.2.09.04, pp. 1–6 (1976)

SCHOLZ, F., TIMMANN, T. and KRUSCHE, D. Genotypic and environmental variance in the response of Norway spruce families to HF-fumigation. In *Papers Presented to the Symposium on the Effects of Airborne Pollution on Vegetables*, pp. 277, United Nations Economic Commission for Europe, Warsaw, Poland (1980)

SCHÖNBACH, H., DÄSSLER, R. G., ENDERLEIN, H., BELLMAN, E. and KÄSTNER, W. Über den unterschiedlichen Einfluss von Schwefeldioxide auf die Nadeln verschiedener zweijährigen Lärchekreuzungen, *Der Züchter* **34**, 312–316 (1964)

SCHÖNDBACH, H., DÄSSLER, H. G., POLSTER, H., BÖRITZ, S., ENDERLEIN, H., LUX, H., RANFT, H., STEIN, G. and VOGL, M. How to increase forest productivity. Present international scientific findings, (Ed. G. VINCENT), pp. 437–484, *Statni zemedelske nakladatelstvi* (1968)

SHAIN, A. and MAYER, A. M. Activation of enzymes during germination — trypsinlike enzyme in lettuce, *Phytochemistry* **7**, 1491–1498 (1968)

SHAW, D. V. and ALLARD, R. W. Analysis of mating system parameters and populations structure in Douglas-fir using single locus and multilocus methods. In *Proceedings of the Symposium on Isozymes of North American Forest Trees and Forest Insects*, July 27, 1979, Berkeley, California. General Technical Report PSW, **48**, 18–22 (1979)

SIMOLA, L. K. and SOPANEN, T. Changes in the activity of certain enzymes of *Acer pseudoplatanus* L. cells of four stages of growth in suspension cultures, *Physiologia Plantarum* **23**, 1212–1222 (1970)

SINCLAIR, W. A. Polluted air: potent new selective force in forest, *Journal of Forestry* **67**, 305–309 (1969)

SOIKKELI, S. The types of ultrastructural injuries in conifer needles of northern industrial environments, *Silvae Fennica* **15**, 339–404 (1981)

SOIKKELI, S. and TUOVINNEN, T. Damage in mesophyll ultrastructure of needle of Norway spruce in two industrial environments in central Finland, *Annales Botanici Fennici* **16**, 50–64 (1979)

SOLBRIG, O. T. *Principle and Methods of Plant Biosystematics.* The Macmillan Biology Series, (Eds. N. H. GILES and J. G. TOREY), The Macmillan Company, New York (1971)

SZMIDT, A. E. Peroxidase variation and activity in Scots pine (*Pinus sylvestris* L.) needle from polluted areas. In *A Genetic Basis for the Resistance of Forest Trees to Anthropopressure, with Special Study of the Effect of Some Toxic Gases.* J-MOA-USDA-11, (Ed. L. MEDNARTOWICZ), pp. 20–23, Polish Academy of Sciences Institute of Dendrology, Kórnik, Poland (1980)

TRESHOW, M. The impact air pollutants on plant populations, *Phytopathology* **58**, 1108–1113 (1968)

TZSCHACKSCH, O., VOGL, M. and THÜMLER, K. Vorselektion geeigneter Provenienzen von *Pinus Contorta* Douglas (*Pinus murrayana* Belf.) für den Anbau in den Rauchschadgebieten des oberen Erzgebirges, *Archiv für Forstwesen* **18**, 979–982 (1969)

TZSCHACKSCH, O. and WEISS, M. Die Variation der SO$_2$ Resistenz von Provenienzen der Baumart Fichte (*Picea abies* (L.) Karst.), *Beitrage für die Forstwirtschaft* **3**, 21–23 (1972)

VOGL, M. Untersuchungen über Unterschiede in der relativen SO$_2$-Resistenz bei Keifern-Nachkommenschaften, *Archiv für Forstwesen* **19**, 3–12 (1970)

WATOREK, W., MORAWIECKA, B. and KORCZAK, B. Acid phosphatase of the yeast *Rhodotorula rubra.* Purification and properties of the enzyme, *Acta Biochemica Polonica* **24**, 153–162 (1977)

WENTZEL, K. F. Empfindlichkeit und Resistenzunterschiede der Pflanzen gegenüber Luftverunreinigungen, *Forstarchiv* **39**, 189–194 (1968)

WOOD, F. A. and COPPOLINO, I. B. The influence of ozone on deciduous tree species, *Mitteilungen der Forstlichen Bundesversuchsanstalt Wien* **97**, 233–253 (1972)

WRIGHT, I. W. *Introduction to Forest Genetics*, Academic Press, New York (1976)

WOLAK, J. Modyfikacje sosny (*Pinus sylvestris*) pod wplnwem zanieczyszczenia powietrza [*Modification in pine under the impact of industrial air pollution*], *Sylwan* **114**, 33–39 (1970)

WOLAK, J. Reaction of ecosystems to sub-necrotic pollution. *United Nations Committee on Agricultural Problems, Timber Committee Symposium on the Effects of Air-Borne Pollution on Vegetation, Warsaw (Poland).* 20–24 August 1979 (1979)

YEE-MEILER, D. Über die Eignung von Phosphatase und Esteraseaktivitätbestimmungen an Fichten-nadeln und Birkenblättern zur Nachweis, unsichtbarer physiologischer Fluorimmissionschädigungen, *European Journal of Forest Pathology* **5**, 329–338 (1975)

YEE-MEILER, D. Der Einfluss von kontinuierlichen, niedrigen SO$_2$-Begasungen auf den Phenolgehalt und die Phenoloxidase-Activität in Blättern einiger Waldbaumarten, *European Journal of Forestry Pathology* **8**, 14–20 (1978)

YEH, F. C. and LAYTON, C. The organization of genetic variability in central and marginal populations of lodgepole pine (*Pinus contorta* spp. *Latifolia*), *Canadian Journal of Genetics and Cytology* **21**, 487–503 (1979)

ZIEGLER, I. The effect of air-polluting gases on plant metabolism. In *Environmental Quality and Safety, Global Aspects of Chemistry Toxicology and Technology as Applied to the Environment*, (Eds. F. COULSTON and F. KORTE), pp. 181–208, Georg Thieme Publishers, Stuttgart and Academic Press, Inc., New York (1973)

ZIEGLER, I. The effect of SO$_2$ pollution on plant metabolism *Residue Reviews* **56**, 79–105 (1975)

Chapter 26

Emissions of volatiles from plants under air pollution stress

J. B. Bucher

EIDGENÖSSISCHE ANSTALT FÜR DAS FORSTLICHE VERSUCHSWESEN, BIRMENSDORF, SWITZERLAND

Introduction

Although the detrimental effects of air pollutants on vegetation are widely accepted nowadays, damage can be difficult to prove in a concrete case, as the reactions of the plant are unspecific and resemble those caused by other stresses. The recognition of air pollution damage or injury may therefore often be subject to a differential diagnosis, which includes physiological and biochemical parameters (Tingey, Wilhour and Taylor, 1979). Hence, air pollution research needed basic knowledge on the mechanisms of the pollutant-plant interactions (Heath, 1980; Mudd, 1982; Tingey and Taylor, 1982; Weinstein and Alscher-Herman, 1982; Wellburn, 1982)*. As it might be more effective to avoid pollution damage than to repair it, physiological and biochemical parameters have also been used for preventive measures, for example to evaluate acceptable pollution levels (Bucher and Keller, 1978) and the selection of resistant cultivars (Bressan *et al.*, 1979). In the physiological and biochemical methods of assessing the plant responses the main focus has been on metabolic products in plants, and little attention has been given to substances emitted by vegetation. Some of these, such as H_2S, may be regarded as products of an elevated metabolic activity or a detoxifying process (de Cormis, 1968; Wilson, Bressan and Filner, 1978); others, such as ethylene, may be an expression of a general stress (Abeles, 1973), while ethane (Konze and Elstner, 1976) or monoterpenes (Rasmussen and Went, 1965) may be the result of cellular disturbance. There are many plant volatiles which may also be affected by pollutants (e.g. acetaldehyde and ethanol, Kimmerer and Kozlowski, 1982) but this contribution will mainly focus on ethylene and monoterpenes, because these substances, being of general interest in plant physiology and pathology, have been studied in most detail (Abeles, 1973; Hogsett, Raba and Tingey, 1981; Kuc and Lisker, 1978; Lieberman, 1979; Loomis and Croteau, 1980; Yang and Pratt, 1978).

Ethylene, monoterpenes and other volatiles released from vegetation are not only the consequence of a pollution stress, they can be regarded as pollutants themselves. However, the ambiguous character of these compounds is only mentioned here. Forests especially contribute considerable amounts of monoterpenes, isoprene and perhaps ethylene by the normal metabolism of their trees or by release due to fire incidents (Rasmussen, 1972, 1981; Tingey and Burns, 1980). But the role of the natural

---

* No attempt is made to review the literature on general topics; therefore only recent reviews are cited.

organics in air pollution is still in discussion and the significance of elevated O_3 concentrations in urban and in rural areas might be a minor one (Bufalini, 1981; Dimitriades, 1981; Dodge, 1980), with the exception of fire incidents (Westberg, Sexton and Flyckt, 1981).

Ethylene

Elevated ethylene emissions from plants as a reaction to different abiotic or biotic stresses are a widely reported phenomenon (Hogsett, Raba and Tingey, 1981; Lieberman, 1979; Yang and Pratt, 1978). According to Abeles (1973) the primary function of stress ethylene is to accelerate abscission of organs damaged by disease, insect, drought and temperature extremes; it may also play a role in growth regulation and disease-resistance mechanisms. Hence, ethylene might also be responsible for some of the symptoms of pollutant stress, e.g. reduced growth, epinasty or necrosis, and be of potential value in the assessment of plant response and the selection of resistant individuals.

Craker (1971) was the first to demonstrate a great increase in ethylene production in tomato (*Lycopersicon esculentum*), tobacco (*Nicotiana tabacum*) and bean (*Phaseolus vulgaris*) plants in response to O_3 treatment (0.25 ppm for 2 h). Later, ethylene emissions were also shown to increase in various plant species, including forest trees, after treatment with SO_2 (Bressan *et al.*, 1978a, 1979; Bucher, 1979, 1981; Bucher and Keller, 1978; Kimmerer and Kozlowski, 1981; Peiser and Yang, 1979), sulphur dust (Recalde-Manrique and Diaz-Miguel, 1981), chlorine (Tingey, Pettit and Bard, 1978), cadmium (Fuhrer, Geballe and Fries, 1981; Rodecap, Tingey and Tibbs, 1981) and automobile exhaust fumes (Flückiger *et al.*, 1979).

Tingey's group (Tingey, Standley and Field, 1976; Hogsett, Raba and Tingey, 1981) used headspace sampling techniques of previously exposed plants to document the O_3-induced ethylene release for a large number of plant species. They established, based on a log-linear relationship, a close correlation between O_3 concentration (up to 0.75 ppm for 4 h) and ethylene level and suggested that the production of stress ethylene be used as a test for effects of O_3 on plants. This parameter seemed to be more sensitive, more reproducible and less subjective than assessing visible foliar injury, especially as it could be measured even before the appearance, or in the absence of leaf necrosis. Stan, Schicker and Kassner (1981) successfully correlated O_3 concentrations (up to 0.5 ppm for 15 h) and the production of stress ethylene in bean plants. Similarly, close linear correlations between SO_2 concentrations (up to 5.5 ppm for 16 h) and the amount of ethylene and ethane generated by fumigated cucumber plants (*Cucumis sativa*) have been found (Bressan *et al.*, 1979). However, in all these fumigation experiments with O_3 or SO_2 concentrations well above the ambient had been used. A relationship between chlorine concentrations (up to 1 ppm for 2 h) and stress ethylene was also established by Tingey, Pettit and Bard (1978), but this was curvilinear and different from the response elicited by O_3 and SO_2, and depended on the plant species tested. Recently, Craker, Fillatti and Grant (1982), using wheat (*Triticum aestivum*) seedlings enclosed in a test-tube and stress ethylene as an indicator, developed a quick, quantitative biotest for detection of phytotoxicants in air samples. Unfortunately rather high concentrations of pollutants, up to 20 times those normally shown to injure plants, were needed to produce a reaction under the test conditions used; within the range used, an increase in concentration of all the gases tested (namely NO_2, H_2S, SO_2 and chlorine) resulted in a linear increase in ethylene production. However, our own experiments on forest trees fumigated in close to field conditions (up to 0.225 ppm SO_2 for several weeks), suggested that stress ethylene was unsuitable as indicator and measure of plant response in the field, because the ethylene emissions fluctuated strongly under the persistent stress (Bucher, 1979, 1981). Problems with the great variations in ethylene emissions were also reported for bean plants by Stan, Schicker, and Kassner (1981); they were able to use only one primary leaf for the fumigation, with the other primary leaf from the same plant serving as the control. As normal ethylene production as well as that induced by stress depends among other factors on the position of the leaf on the stem (Lavee and Martin, 1981) or the leaf age (Roberts and Osborne, 1981), a great variation in pollution-induced ethylene emission can be expected. Evidence for this has been found in ponderosa pines (*Pinus ponderosa*) exposed to O_2 in which ethylene emissions are greater in the summer just after elongation of the needles, when the plants are known to be most sensitive, than in experiments conducted in the fall (Tingey, Standley and Field, 1976).

Figure 26.1 presents selected ethylene responses from plants to different pollutants and shows some of the problems discussed in measuring plant injury on a biochemical basis. Apart from the different relationships between pollutant levels, concentrations and fumigation times, and stress ethylene production, the heterogeneity in the test conditions of the various laboratories can be seen. There is a lack of standardization in the exposures. However, such standardization is difficult to achieve because the fumigation conditions depend greatly on plant species and test used. Nevertheless, it might be concluded that stress ethylene can serve as a good indicator or measure of air pollution effects under standardized conditions with high pollutant concentrations.

The emission of volatiles, especially ethylene, may direct our interest to the question why these volatiles are produced so rapidly upon injury. Are they merely a consequence of disorganization when cell structures are destroyed? Or is there a direct influence upon their biosynthesis? Do they act as messengers for other processes leading to necrosis, or are they simply indicators of injury, having no physiological effects themselves? These questions, which have not yet been satisfactorily answered, were partly discussed in recent reviews (e.g. Dörfling, 1980). The whole problem is complicated by the fact that ethylene acts autocatalytically to induce its own production (Kende and Baumgartner, 1974) and works in close relationship with other hormones (Liebermann, 1979). It is well documented that stress ethylene is formed through the same biochemical pathway as normal ethylene in the plant:

methionine → S-adenosylmethionine → aminocyclopropane-carboxylic acid (ACC) → ethylene

(Adams and Yang, 1979; Boller, Herner and Kende, 1979; Hogsett, Raba and Tingey, 1981; Jones and Kende, 1979; Lürssen and Naumann, 1979; Yu and Yang, 1979). The ability to produce ethylene is restricted to living cells (Elstner and Konze, 1976); when most of the leaf has died (Kobayashi, Fuchigami and Brainerd, 1981; Peiser and Yang, 1979) or aged, as in late senescence (Roberts and Osborne, 1981), the production of ethylene ceases.

Roberts and Osborne (1981) suggested that disruption in the permeability of the cell membrane due to stress would release cofactors activating enzymes involved in ethylene biosynthesis. However, in water-stressed plants it was found that ethylene synthesis increased prior to any change in electrolytic leakage and decreased as the leakage increased, which implies that membrane integrity must be maintained for the biosynthesis of ethylene (Kobayashi, Fuchigami and Brainerd, 1981). According to the work of several groups, it seems that stress ethylene is likely to be produced through a

402

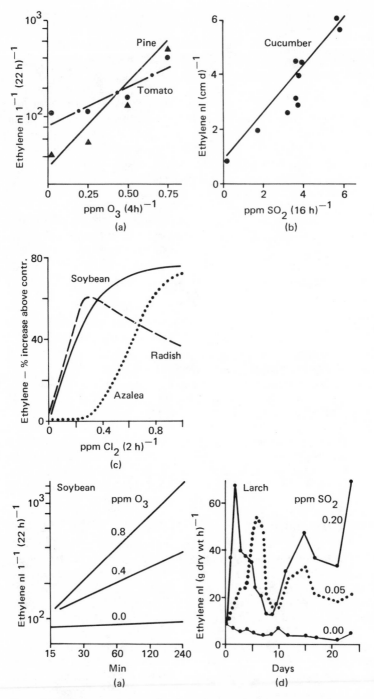

Figure 26.1. Ethylene emissions of various plant species subjected to different pollutants. Selected data after (a) Tingey *et al.* (1976), (b) Bressan *et al.* (1979), (c) Tingey *et al.* (1978) and (d) Bucher (1981)

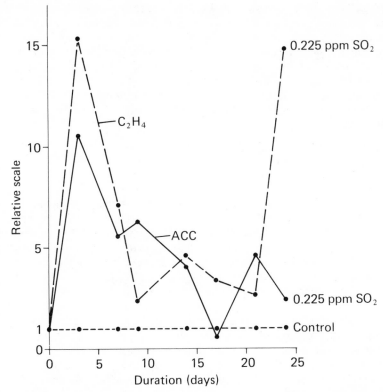

Figure 26.2. ACC content and ethylene emission of larch foliage under the influence of a continuous SO$_2$ exposure (Bucher, 1981)

de novo synthesis, rather than activation of ACC-synthase (Boller and Kende, 1980; Wang and Adams, 1982; Yu and Yang, 1980). Recent work shows that under certain conditions the conversion of ACC to ethylene may also be a rate-limiting step (Hoffman and Yang, 1982; de Laat and van Loon, 1982). However, the biochemical mechanism by which stress causes the induction of the ethylene pathway remains unknown (Yu and Yang, 1980). As a result of their experiments on the effect of sulphur dust on bean plants Recalde-Manrique and Diaz-Miguel (1981) suggested that the sulphur dust absorbed as SO$_2$ would increase the ethylene precursor methionine in the cells and so induce the biosynthesis of ethylene. According to other investigations a stress influence on methionine or S-adenosylmethionine seems less probable (de Laat and van Loon, 1982). However, the direct precursor of ethylene, ACC, is affected by air pollutants and biotic stress. Sulphur dioxide fumigation at an ambient level increased the ACC content in the foliage of larch trees (*Figure 26.2*), and similar effects were found in virus infected tobacco leaves (de Laat and van Loon, 1982).

Contrary to the observations from the tobacco mosaic virus experiment, where the increase in ACC appeared prior to the visible injury and the emission of the ethylene afterwards, pollutants have been shown to cause elevated ethylene emissions before visible foliar symptoms appeared (Bucher, 1979, 1981; Bucher and Keller, 1978; Stan, Schicker and Kassner, 1981; Tingey, Standley and Field, 1976). Saltveit and Dilley (1978) concluded from their experiments with excised segments of aetiolated pea

(*Pisum sativum*) seedlings that wound ethylene was rapidly induced after the transmission of an unknown stimulus. From waterlogging experiments it seems that ACC could be this signal, since the ACC concentration was higher in the xylem sap from waterlogged tomato plants than from aerated controls (Bradford and Yang, 1980). The appearance of ACC in the sap preceded the onset of epinastic curvatures of the petioles, and its concentration correlated with the production of ethylene in the sheet. One may therefore conclude that air pollutants and other stresses influence ethylene synthesis and that not ethylene itself but rather its precursor plays a hormonal role (Dörfling, 1980; Yang, 1980).

Ethane

In contrast to ethylene, ethane production increases with increasing membrane damage and is mostly connected with visible injury and dying cells (Kobayashi, Fuchigami and Brainerd, 1981; Konze and Elstner, 1976). For SO_2 effects (1.6 ppm for 8 h) Peiser and Yang (1979) suggested that the tissue damage in alfalfa (*Medicago sativa*) plants was caused by free radicals generated through oxidation of the pollutant, which would lead to lipid peroxidation and ethane formation (Yu *et al.*, 1982). The small amount of stress ethylene still formed in dying cells might be synthesized in the same way as ethane (Konze, 1977).

The emission of ethylene and ethane has also been tested as a method for assessing resistance to air pollutants. Bressan and co-workers (1978, 1979) used ethane emissions from leaf discs of field grown cucurbit (*Cucumis* and *Cucurbita*) varieties floated on bisulphite solutions for a simple test for SO_2 resistance. Sensitive plants emitted more ethane than resistant ones, so that a rank order could be established. It was concluded that the reason for the differences in resistance among these cultivars was the relative rate of absorption of the gas (Bressan, Wilson and Filner, 1978). The resistance was not attributed to stomatal differences but to another barrier which could be the plasma membrane. This interpretation was supported by the fact that the rank order remained stable when influences of other stresses as chilling, heating, alcohols, mechanical injury and desiccation were tested (Bressan *et al.*, 1981). Yet, the physiological role of ethylene or ethane in regard to this resistance is unknown and the emissions of these volatiles are therefore considered to be biochemical markers only. However, as the method is rapid and needs only small amounts of plant material it was suggested to be superior to conventional phenotypic classification (Bressan *et al.*, 1979).

Monoterpenes

The usefulness of ethylene and ethane emissions from plants as an indicator for resistance to air pollutants and other stresses has only recently been proposed. For other volatile plant compounds, the monoterpenes, this was postulated some 20 years ago (Cvrkal, 1959). It is currently accepted that terpenes can be used as biochemical markers in genetic studies and in the breeding of plants resistant to various pollutants (Anderson *et al.*, 1980; Muhs, 1981; Squillace, 1977). Cvrkal (1959) observed that spruce (*Picea abies*) which proved to be sensitive to industrial smoke, contained only small amounts of camphene and limonene and/or much β-pinene, whereas the volatile oil of resistant trees was high in camphene. He did not suspect his findings to be the result of any pollutant effect on the plant, but saw in the different monoterpene contents a sign of individual smoke resistance which could be used in selection breeding. However, Dässler (1964), who designed a fumigation experiment with young spruce trees on these resistance criteria, could not confirm Cvrkal's work, although he observed a distinct visible response to the pollutant. Unfortunately the experiment was conducted for only a few hours or days with unrealistically high SO_2 exposures (up to 100 ppm and 100 h). In addition, a recent long-term SO_2-exposure at concentrations closer to ambient (0.15 ppm for 60 days, followed by 0.5 ppm for seven days) showed no correlation between the monoterpene pattern and the rank order of affected individuals in a Scots (*Pinus sylvestris*) and eastern white pine (*Pinus strobus*) population (*Figure 26.3*).

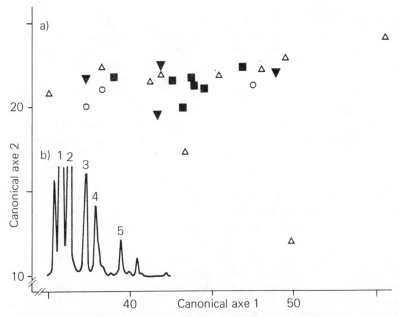

Figure 26.3. Susceptibility of Scots pine individuals to SO_2 in relation to their monoterpene pattern. (a) Biochemical distances based on the monoterpene pattern (MANOVA, Canonical Analysis) and foliar symptom classes according to the degree of needles affected: ○ 0%, △ 1–33%, ■ 34–66%, ▼ 67–100%. (b) GC-chromatogram of the extracted monoterpenes: 1 α-pinene, 2 camphene, 3 β-pinene, 4 myrcene, 5 limonene (Bucher and McCune, unpublished)

However, keeping in mind that the volatile monoterpenes play an important role in the host-parasite relationship of tree colonization by insects (Hanover, 1975), pollutant effects on the terpene metabolism would be likely to influence infestation, and indeed there are several field observations of catastrophic insect infestations in polluted areas (Berge, 1973; Bösener, 1969; Charles and Villemant, 1977; Novakova, 1969; Pfeffer, 1978; Stark *et al.*, 1968; Wentzel and Ohnesorge, 1961). The research done is limited and only in one investigation was an increased monoterpene production shown. According to Lehtiö (1981) the emissions of a fertilizer plant led to higher contents of camphene, β-pinene, myrcene and tricyclene in Scots pine needles depending on their degree of injury. In contrast, SO_2-fumigation experiments with spruce (Dässler, 1964) and Scots pine (Bucher, 1982) revealed a general slight decrease in the oil content of the needles. Similarly the ambient oxidant smog of the Los Angeles basin seems to be responsible for the production of smaller amounts of total volatile oils in sensitive ponderosa pines, although the amounts of monoterpenes themselves were unchanged (Miller, Cobb and Zavarin, 1968). Hence, it seems that pollutant stress hardly influences the monoterpene production and may only lead to a decreased production in severely affected plants. So far the reported information is based on steam distillation of foliage or investigations of resin sap, and what might be true for the volatile oil content within the plant may not necessarily be so for the air surrounding the plant, and indeed the latter might be more relevant to insects (Hanover, 1972).

There is some evidence that SO_2, at least, influences the halo of conifers and can lead to drastic changes in the absolute and relative amount of monoterpenes. Renwick and Potter (1981), fumigating young balsam fir (*Abies balsamea*) trees with rather high SO_2 concentrations for some days (2 ppm for 5 h on three consecutive days) and collecting the emitted volatiles after the experiment on Porapak traps, showed that the fumigated trees released much more β-pinene, camphene and α-pinene than the control trees. The SO_2-induced increase in the emissions of α-thujene, β-phellandrene (+ limonene) and bornylacetate was smaller but still significant. In a somewhat differently designed experiment, collecting the volatile emissions of mature Scots pine twigs every three weeks during a long-lasting fumigation with low SO_2-concentrations (0.225 ppm up to 19 weeks) on active charcoal traps, we were able to demonstrate that the fumigated trees released significantly more α-pinene, camphene and limonene, whereas the monoterpene content in the needles showed tendency to a decrease (Bucher, 1982). However, there was a great variation in the release of the components in the individual trees of the same clone (*Figure 26.4*). Contrary to the findings of Renwick and Potter, where no visible symptoms were seen immediately after the fumigation, our effects were significant only when necrosis appeared.

As the release of volatiles is normally via the stomatal pathway, and the cuticle probably represents a great barrier to emissions (*cf.* Hanover, 1972), the occurrence of necrosis, breaking open this barrier, may therefore explain the large air pollutant-induced monoterpene emissions. There is also evidence that the volatilization of the monoterpenes is primarily a physical process depending upon the vapour pressure of the terpenoids, as the emission rates from dead, but intact, and live slash pine (*Pinus elliottii*) needles and black sage (*Salvia mellifera*) leaves are similar to each other, and the release is temperature but not light-dependent (Tingey *et al.*, 1980; Tingey and Burns, 1980). Also Rasmussen and Went (1965) found that the emissions of volatiles from a forest canopy strongly depends upon the meteorological conditions, and that terpenes are not released until the leaves become older or when their cells die. The difference in the relative amounts of the stress-induced monoterpenes might result from a different composition of the compounds in the secretory cells within the foliage (Carde, 1976; Hanover, 1975), and therefore depend on the part of needle affected. The physiological role of the terpenes is still unclear and one can hardly assume that their qualitative and quantitative changes are a response to a specific stress. Rather they seem to be a general expression of disturbance, and related to wound repair processes or disease resistance mechanism (Kuc and Lisker, 1978). Whereas the role of sesquiterpenoids and phytoalexins is more obvious, the volatile monoterpenes may, apart from the fungistatic effects and the ecological function in insect attraction already discussed, have a messenger function within the plant itself. It has been shown that β-pinene behaves as a potent inhibitor of photosynthesis and respiration in pine (Pauly,

Control trees

Trees of treatment
0.225 ppm SO$_2$

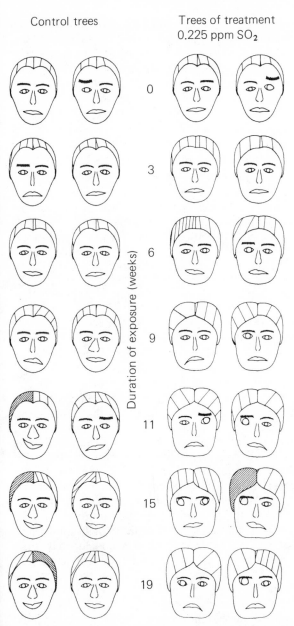

Figure 26.4. Monoterpene emission and content of Scots pine foliage under the influence of a continuous SO$_2$ exposure. The multivariate data of eight trees is represented in four so-called CHERNOFF-faces (one tree per each face side) per each time point. The computer program, ASYM (Flury, 1980), sets the normalized means of the variables to neutral face parameters. The following monoterpene variables have been investigated and were attributed to various face parameters: emission of α-pinene to size and vertical position of eye, emission of camphene to slant and horizontal position of eye; emission of β-pinene to curvature and density of eyebrow; emission of myrcene to horizontal and vertical position of eyebrow; emission of limonene to size and position of pupil; content of α-pinene to curvature of mouth; content of camphene to nose line; content of β-pinene to size of mouth; content of myrcene to darkness of hair; content of limonene to hair shading slant; and content of sulphur to upper and lower hair line as well as face line (Bucher, 1982)

Douce and Carde, 1981). One may speculate that plants suffering from environmental stress would be able to slow down their metabolism in this way and so partly avoid the stress. Monoterpenes have also been suggested as a sink for ozone pollution, and may therefore have a preventive effect. However, only terpinolene, α-phellandrene and α-terpinene consumed significant amounts of O_3, suggesting that these terpenoids, which are normally minor components in the foliage of trees, are insignificant scavengers of O_3 (Tingey and Taylor, 1982).

Conclusions

Although much is known about the wounding mechanism and the biosynthesis of stress volatiles, their physiological significance and consequences still remain unclear. For ethylene, which is emitted by damaged yet viable cells, the physiological role seems to be an active one, leading to various pollution symptoms as epinasty, premature senescence or leaf abscission, and to metabolic activities associated with wound healing processes or increased disease resistance. Unlike ethylene, ethane and monoterpene emissions may only be the indirect result of leaf necrosis. Although it is tempting to speculate on the ecological impact of a pollution-induced release of monoterpenes resulting in a changed, insect-favourable host–parasite relationship, such an effect has still to be proven.

Ethylene and ethane emissions from a great variety of plant species have been successfully used for assessing plant response to pollutant stress and for screening resistant cultivars before visible injury appeared. However, in such tests the plants were subjected to pollutant concentrations which were well above the ambient levels. There is some evidence that under lower but persistent pollutant concentrations biochemical stress indicators may not be as useful as under acute exposure. In Keller's laboratory where we worked with moderate SO_2-fumigations, yet still far above the annual mean concentrations of central Europe, many stress-induced metabolic changes occurred with or only shortly before the appearance of necrosis (Bucher and Keller, 1978). The biochemical indicators investigated included several enzyme activities (Bucher, 1978; Bucher-Wallin, Bernhard and Bucher, 1978), early products of photosynthesis (Landolt, 1982) and volatiles (Bucher, 1979, 1981, 1982). Additionally Garsed and Rutter (1982) could show that for populations of pine tree species the rank order of resistance was inverted when the pollutant concentrations and durations of the exposure were changed from chronic (0.08 ppm $SO_2 = 200$ μg m^{-3} for 11 months) to acute (3 ppm $SO_2 = 8000$ μg m^{-3} for 6 h) levels. It therefore can be concluded that the potential value of biochemical indicators should also be verified under ambient levels. Furthermore, biochemical indicators should be judged according to Arndt's criteria (Arndt, 1970) as recently emphasized by Tingey, Wilhour and Taylor (1979). This would mean that biochemical indicators should be rapid and unambiguous in their response to low pollutant concentrations, and that they should also have a very specific reaction with different pollutants and be detectable by simple methods with a high degree of reproducibility in the results.

References

ABELES, F. B. *Ethylene in Plant Biology*, Academic Press, New York and London, 302 pp. (1973)
ADAMS, D. O. and YANG, S. F. Ethylene biosynthesis: identification of 1-aminocyclopropane-1-carboxylic acid as an intermediate in the conversion of methionine to ethylene, *Proceedings of the National Academy Sciences of the USA* **76**, 170–174 (1979)

ANDERSSON, B. A., HOLMAN, R. T., LUNDGREN, L. and STENHAGEN, G. Capillary gas chromatograms of leaf volatiles. A possible aid to breeders for pest and disease resistance, *Journal of Agricultural and Food Chemistry* **28**, 985–989 (1980)

ARNDT, U. Konzentrationsänderungen bei freien Aminosäuren in Pflanzen unter dem Einfluss von Fluorwasserstoff und Schwefeldioxid, *Staub-Reinhalt. Luft* **30**, 256–259 (1970)

BERGE, H. Beziehungen zwischen Baumschädlingen und Immissionen, *Anzeiger Schädlingskunde* **46**, 155–156 (1973)

BOLLER, Th., HERNER, R. C. and KENDE, H. Assay for and enzymatic formation of an ethylene precursor, 1-aminocyclopropane-1-carboxylic acid, *Planta* **145**, 293–303 (1979)

BOLLER, Th. and KENDE, H. Regulation of wound ethylene synthesis in plants, *Nature* **286**, 259–260 (1980)

BÖSENER, R. Zum Vorkommen rindenbrütender Schadinsekten in rauchgeschädigten Kiefern- und Fichtenbeständen, *Archiv für Forstwesen* **18**, 1021–1026 (1969)

BRADFORD, K. J. and YANG, S. F. Xylem transport of 1-aminocyclopropane-1-carboxylic acid, an ethylene precursor, in waterlogged tomato plants, *Plant Physiology* **65**, 322–326 (1980)

BRESSAN, R. A., LECUREUX, L., WILSON, L. G. and FILNER, P. Emission of ethylene and ethane by leaf tissue exposed to injurious concentrations of sulfur dioxide or bisulfite ion, *Plant Physiology* **63**, 924–930 (1979)

BRESSAN, R. A., LECUREUX, L., WILSON, L. G., FILNER, P. and BAKER, L. R. Inheritance of resistance to sulfur dioxide in cucumber, *Hort Science* **16**, 332–333 (1981)

BRESSAN, R. A., WILSON, L. G. and FILNER, P. Mechanisms of resistance to sulfur dioxide in the *Cucurbitaceae*, *Plant Physiology* **61**, 761–767 (1978)

BRESSAN, R. A., WILSON, L. G., LECUREUX, L. and FILNER, P. Use of ethylene and ethane emissions to assay injury by SO_2, *Plant Physiology* **61** (Supplement), 93, Abstract no. 509 (1978)

BUCHER, J. B. SO_2-induziertes Stress-Aethylen in den Assimilationsorganen von Waldbäumen, *Zbornik (Mitteilungen) Institut Forst- und Holzwirtschaft Ljubljana (YU), IUFRO-Bericht der X. Fachtagung in Ljubljana*, Sept. 1978, Fachgruppe S2.09, pp. 93–101 (1979)

BUCHER, J. B. SO_2-induced ethylene evolution of forest tree foliage and its potential use as stress-indicator, *European Journal of Forest Pathology* **11**, 369–373 (1981)

BUCHER, J. B. Einfluss von SO_2 auf Terpen-Emissionen von Kiefern (*P. sylvestris* L.). XII. *International Meeting for Specialists in Air Pollution Damage in Forests*. IUFRO S2.09, Oulu (Finland), 23–29 August 1982. Poster (1982)

BUCHER, J. B. and KELLER, Th. Einwirkungen niedriger SO_2-Konzentrationen im mehrwöchigen Begasungsversuch auf Waldbäume, *VDI-Bericht* **314**, 237–242 (1978)

BUCHER-WALLIN, I. K., BERNHARD, L. and BUCHER, J. B. Einfluss niedriger SO_2-Konzentrationen auf die Aktivitäten einiger Glykosidasen der Assimilationsorgane verklonter Waldbäume, *European Journal of Forest Pathology* **9**, 6–15 (1978)

BUFALINI, J. J. Summary comments. In Atmospheric Biogenic Hydrocarbons. *Ambient Concentrations and Atmospheric Chemistry* (Vol. 2), pp. 211–212, (Eds. J. J. BUFALINI and R. R. ARNTS), Ann Arbor Science (1981)
Société Botanique France, Coll. Sécrét. Vég. 181–189 (1976)

CHARLES, P. J. and VILLEMANT, C. Modifications des niveaux de population d'insectes dans les jeunes plantations de pins sylvestres de la forêt de Roumare (Seine Maritimes) soumises à la pollution atmosphérique, *Comptes Rendus Académie Agriculture France* **63**, 502–510 (1977)

CRAKER, L. E. Ethylene production from ozone injured plants, *Environmental Pollution* **1**, 299–304 (1971)

CRAKER, L. E., FILLATTI, J. J. and GRANT, L. A rapid, quantitative bioassay for detecting phytotoxic gases using stress-ethylene, *Atmospheric Environment* **16**, 371–374 (1982)

CVRKAL, H. Biochemische Diagnose an Fichten in Rauchgebieten, *Berichte Tschechoslovakischen Akademie Landwirtschaftswissenschaft, Sektor Forstwesen* **5** (XXXII), 1033–1048 (1959)

DÄSSLER, H.-G. Der Einfluss des Schwefeldioxides auf den Terpengehalt von Fichtennadeln, *Flora* **154**, 376–382 (1964)

DE CORMIS, L. Dégagement d'hydrogène sulfuré par des plantes soumises à une atmosphère contenant de l'anhydride sulfureux, *Comptes Rendus Académie Sciences Série D* **266**, 683–685 (1968)

DE LAAT, A. M. M. and VAN LOON, L. C. Regulation of ethylene biosynthesis in virus-infected tobacco leaves, *Plant Physiology* **69**, 240–245 (1982)

DIMITRIADES, B. The role of natural organics in photochemical air pollution. Issues and research needs, *Journal of the Air Pollution Control Association* **31**, 229–235 (1981)

DODGE, M. C. A modeling study of the effect of biogenic hydrocarbons on rural ozone formation, *Journal of Environmental Science and Health* **A15**, 601–612 (1980)

DÖRFLING, K. IV. Growth. In *Progress in Botany* **42**, 111–125, (Eds. ELLENBERG *et al.*), Springer-Verlag, Berlin (1980)

ELSTNER, E. F. and KONZE, J. R. Effect of point freezing on ethylene and ethane production by sugar beet leaf disks, *Nature* **263**, 351–352 (1976)

FLÜCKIGER, W., OERTLI, J. J., FLÜCKIGER-KELLER, H. and BRAUN, S. Premature senescence in plants along a motorway, *Environmental Pollution* **13**, 171–176 (1979)

FLURY, B. Construction of an asymetrical face to represent multivariate data graphically. *Technischer Bericht* No. 3. Institut Mathematische Statistik + Versicherungslehre, Universität Bern (CH), pp. 10 (1980)

FUHRER, J., GEBALLE, G. T. and FRIES, C. Cadmium-induced change in water economy of beans: involvement of ethylene formation, *Plant Physiology* **67** (Supplement) 55, Abstract No. 307 (1981)

GARSED, S. G. and RUTTER, A. J. The relative sensitivities of conifer populations to SO₂ in screening tests with different concentrations of sulphur dioxide. In *Effects of Gaseous Air Pollution in Agriculture and Horticulture*, (Eds. M. H. UNSWORTH and D. P. ORMROD), pp. 474–475, Butterworths, London (1982)

HANOVER, J. W. Factors affecting the release of volatile chemicals by forest trees, *Mitteilungen aus der Forstlichen Bundesversuchsanstalt, Wien* **97**, 625–644 (1972)

HANOVER, J. W. Physiology of tree resistance to insects, *Annual Review of Entomology* **20**, 75–95 (1975)

HEATH, R. L. Initial events in injury to plants by air pollutants, *Annual Review of Plant Physiology* **31**, 395–431 (1980)

HOFFMAN, N. E. and YANG, S. F. Enhancement of wound-induced ethylene synthesis by ethylene in preclimacteric cantaloupe, *Plant Physiology* **69**, 317–322 (1982)

HOGSETT, W. E., RABA, R. M. and TINGEY, D. T. Biosynthesis of stress ethylene in soybean seedlings: Similarities to endogenous ethylene biosynthesis, *Physiologia Plantarum* **53**, 307–314 (1981)

JONES, J. F. and KENDE, H. Auxin-induced ethylene biosynthesis in subapical stem sections of etiolated seedlings of *Pisum sativum* L., *Planta* **146**, 649–656 (1979)

KENDE, H. and BAUMGARTNER, B. Regulation of aging flowers of *Ipomoea tricolor* by ethylene, *Planta* **116**, 279–289 (1974)

KIMMERER, T. W. and KOZLOWSKI, T. T. Ethylene, ethane and ethanol production by SO₂-stressed woody plants, *Plant Physiology* **67** (Supplement), 56, Abstract No. 312 (1981)

KIMMERER, T. W. and KOZLOWSKI, T. T. Ethylene, ethane, acetaldehyde, and ethanol production by plants under stress, *Plant Physiology* **69**, 840–847 (1982)

KOBAYASHI, K., FUCHIGAMI, L. H. and BRAINERD, K. E. Ethylene and ethane production and electrolyte leakage of water-stressed 'Pixy' plum leaves, *Hort Science* **16**, 57–59 (1981)

KONZE, J. R. Bildung von Aethylen und Aethan in Geweben höherer Pflanzen und in Modellreaktionen. Dissertation Universität Bochum, pp. 69 (1977)

KONZE, J. R. and ELSTNER, E. F. Ethylene- and ethane formation in leaf disks, plastids and mitochondria, *Berichte der Deutschen Botanischen Gesellschaft* **89**, 547–553 (1976)

KUC, J. and LISKER, N. Terpenoids and their role in wounded and infected plant storage tissue. In *Biochemistry of Wounded Plant Tissues*, (Ed. G. KAHL), pp. 203–242, de Gruyter, Berlin (1978)

LANDOLT, W. Der Einfluss einer praxisnahen SO₂-Begasung auf das ¹⁴CO₂-Fixierungsmuster von Buchen (*Fagus sylvatica* L.), *European Journal of Forest Pathology* **12**, 331–339 (1982)

LAVEE, S. and MARTIN, G. C. Ethylene evolution from various developing organs of olive (*Olea europea*) after excision, *Physiologia Plantarum* **51**, 33–38 (1981)

LEHTIÖ, H. Effect of air pollution on the volatile oil in needles of Scots pine (*Pinus sylvestris* L.), *Silva Fennica* **15**, 122–129 (1981)

LIEBERMANN, M. Biosynthesis and action of ethylene, *Annual Review of Plant Physiology* **30**, 533–591 (1979)

LOOMIS, W. D. and CROTEAU, R. Biochemistry of terpenoids. In *The Biochemistry of Plants*, (Eds. P. K. STUMPF and E. E. CONN), Academic Press, New York **4**, 363–415 (1980)

LÜRSSEN, K. and NAUMANN, K. 1-Aminocyclopropane-1-carboxylic acid — a new intermediate of ethylene biosynthesis, *Naturwissenschaften* **66**, 264–265 (1979)

MILLER, P. R., COBB, F. W. Jr. and ZAVARIN, E. III. Effect of injury upon oleoresin composition, phloem carbohydrates and phloem pH, *Hilgardia* **39**, 135–140 (1968)

MUDD, J. B. Effects of oxidants on metabolic function. In *Effects of Gaseous Air Pollution in Agriculture and Horticulture*, (Eds. M. H. UNSWORTH and D. P. ORMROD), pp. 189–203, Butterworths, London (1982)

MUHS, H-J. Terpenes and some other substances as biochemical markers in forestry. A review of European studies since 1976. IUFRO Congress, Kyoto, Japan, 6–10 Sept. 1981, WP S.2-04-5 *Biochemical Genetics and Cytology* (voluntary paper) (1981)

NOVAKOVA, E. Influence des pollutions industrielles sur les communautés animales et l'utilisation des animaux comme bioindicateurs. *Proceedings First European Congress Influence of Air Pollutants on Plants and Animals, Wageningen 1968*, pp. 41–48, Pudoc, Wageningen, Netherlands (1969)

PAULY, G., DOUCE, R. and CARDE, J.-P. Effects of β-pinene on spinach chloroplast photosynthesis,

Zeitschrift für Pflanzenphysiologie **104**, 199–206 (1981)

PEISER, G. D. and YANG, S. F. Ethylene and ethane production from sulfur dioxide-injured plants, *Plant Physiology* **63**, 142–145 (1979)

PFEFFER, A. Wirkungen von Luftverunreinigungen auf die freilebende Tierwelt, *Schweizerische Zeitschrift für Forstwesen* **129**, 362–367 (1978)

RASMUSSEN, R. A. What do the hydrocarbons from trees contribute to air pollution? *Journal of the Air Pollution Control Association* **22**, 537–543 (1972)

RASMUSSEN, R. A. A review of the natural hydrocarbon issue. In *Atmospheric Biogenic Hydrocarbons. Emissions* (Vol. 1), (Eds. J. J. BUFALINI and R. R. ARNTS), pp. 3–14, Ann Arbor Science (1981)

RASMUSSEN, R. A. and WENT, F. W. Volatile organic material of plant origin in the atmosphere, *Proceedings of the National Academy Sciences of the USA* **53**, 215–220 (1965)

RECALDE-MANRIQUE, L. and DIAZ-MIGUEL, M. Evolution of ethylene by sulphur dust addition, *Physiologia Plantarum* **53**, 462–467 (1981)

RENWICK, J. A. A. and POTTER, J. Effects of sulfur dioxide on volatile terpene emission from balsam fir, *Journal of the Air Pollution Control Association* **31**, 65–66 (1981)

ROBERTS, J. A. and OSBORNE, D. J. Auxin and the control of ethylene production during the development and senescene of leaves and fruits, *Journal of Experimental Botany* **32**, 875–887 (1981)

RODECAP, K. D., TINGEY, D. T. and TIBBS, J. H. Cadmium-induced ethylene production in bean plants, *Zeitschrift für Pflanzenphysiologie* **105**, 65–74 (1981)

SALTVEIT, M. E. and DILLEY, D. R. Rapidly induced wound ethylene from excised segments of etiolated *Pisum sativum* L., cv. Alaska, *Plant Physiology* **62**, 710–712 (1978)

SQUILLACE, A. E. Use of monoterpene composition in forest genetics research with slash pine. *Proceedings Fourteenth Southern Forest Tree Improvement Conference*, pp. 227–238, Gainesville, Florida, June 1977 (1977)

STAN, H.-J., SCHICKER, S. and KASSNER, H. Stress ethylene evolution of bean plants — a parmeter indicating ozone pollution, *Atmospheric Environment* **15**, 391–395 (1981)

STARK, R. W., MILLER, P. R., COBB, F. W. Jr., WOOD, D. L. and PARMETER, J. R. Jr. I. Incidence of bark beetle infestation in injured trees, *Hilgardia* **39**, 121–126 (1968)

TINGEY, D. T. and BURNS, W. F. Hydrocarbon emissions from vegetation. *Proceedings of the Symposium on the Effects of Air Pollutants on Mediterranean and Temperate Forest Ecosystems*, pp. 24–30, Riverside, California, June 22–27, 1980 (1980)

TINGEY, D. T., MANNING, M., GROTHAUS, L. C. and BURNS, W. F. Influence of light and temperature on monoterpene emission rates from slash pine, *Plant Physiology* **65**, 797–801 (1980)

TINGEY, D. T., PETTIT, N. and BARD, L. Effect of chlorine on stress ethylene production, *Environmental and Experimental Botany* **18**, 61–66 (1978)

TINGEY, D. T., STANDLEY, C. and FIELD, R. W. Stress ethylene evolution: a measure of ozone effects on plants, *Atmospheric Environment* **10**, 969–974 (1976)

TINGEY, D. T., WILHOUR, R. G. and TAYLOR, O. C. The measurement of plant responses. In *Handbook of Methodology for the Assessment of Air Pollution Effects on Vegetation*, (Eds. W. W. HECK, S. V. KRUPA and S. N. LINZON), pp. 7-1-7-35, Upper Midwest Section, Air Pollution Control Association, April 1978 (1979)

TINGEY, D. T. and TAYLOR, G. E. Jr. Variation in plant response to ozone: a conceptual model of physiological events. In *Effects of Gaseous Air Pollution in Agriculture and Horticulture*, (Eds. M. H. UNSWORTH and D. P. ORMROD), pp. 113–138, Butterworths, London (1982)

WANG, C. Y. and ADAMS, D. O. Chilling-induced ethylene production in cucumbers (*Cucumis sativus* L.), *Plant Physiology* **69**, 424–427 (1982)

WEINSTEIN, L. H. and ALSCHER-HERMAN, R. Physiological responses of plants to fluorine. In *Effects of Gaseous Air Pollution in Agriculture and Horticulture*, (Eds. M. H. UNSWORTH and D. P. ORMROD), pp. 139–167, Butterworths, London (1982)

WELLBURN, A. R. Effects of SO_2 and NO_2 on metabolic function. In *Effects of Gaseous Air Pollution in Agriculture and Horticulture*, (Eds. M. H. UNSWORTH and D. P. ORMROD), pp. 169–187, Butterworths, London (1982)

WENTZEL, K. F. and OHNESORGE, B. Zum Auftreten von Schadinsekten bei Luftverunreinigung, *Forstarchiv* **32**, 177–186 (1961)

WESTBERG, H., SEXTON, K. and FLYCKT, D. Hydrocarbon production and photochemical ozone formation in forest burn plumes, *Journal of the Air Pollution Control Association* **31**, 661–664 (1981)

WILSON, L. G., BRESSAN, R. A. and FILNER, P. Light-dependent emission of hydrogen sulfide from plants, *Plant Physiology* **61**, 184–189 (1978)

YANG, S. F. Regulation of ethylene biosynthesis, *Hort Science* **15**, 238–243 (1980)

YANG, S. F. and PRATT, H. K. The physiology of ethylene in wounded plant tissues. In *Biochemistry of Wounded Plant Tissues*, (Ed. G. KAHL), pp. 595–622, de Gruyter, Berlin, New York (1978)

YU, S.-W., LIU, Y., LI, Z.-G., TAN, C. and YU, Z.-W. Studies on the mechanism of SO_2 injury in plants. In *Effects of Gaseous Air Pollution in Agriculture and Horticulture*, (Eds. M. H. UNSWORTH and D. P. ORMROD), pp. 507–508, Butterworths, London (1982)

YU, Y. B. and YANG, S. F. Auxin-induced ethylene production and its inhibition by aminoethoxyvinylglycine and cobalt ion, *Plant Physiology* **64**, 1074–1077 (1979)

YU, Y. B. and YANG, S. F. Biosynthesis of wound ethylene, *Plant Physiology* **66**, 281–285 (1980)

Chapter 27

Biochemical aspects of plant tolerance to ozone and oxyradicals: superoxide dismutase

Jesse H. Bennett, E. H. Lee and H. E. Heggestad

PLANT PHYSIOLOGY INSTITUTE, AGRICULTURAL RESEARCH SERVICE, US DEPARTMENT OF AGRICULTURE, BELTSVILLE, MD, USA

Superoxide dismutases: general properties and functions

Superoxide dismutase (SOD) activity in biological systems was discovered in the late 1960s during the course of research using an oxidase enzyme (xanthine oxidase) to generate superoxide radicals ($\cdot O_2^-$) in experimental fractions obtained from bovine blood (Fridovich, 1981). Actually, a hemocuprein protein, which was in fact a copper-containing SOD, was isolated 30 years earlier and subsequently studied rather extensively without its enzymatic role being understood. Three classes of SOD metalloproteins are now known to occur, separated on the basis of the specific metal cofactor (iron, manganese or copper) required for activity. This family of oxyradical-scavenging enzymes has been shown to be present in all aerobic and aerotolerant cells studied to date.

Table 27.1 summarizes general biochemical properties and characteristics of the Fe-, Mn- and CuZn-containing SOD types along with information about their expected occurrences within the plant kingdom and sites of cellular localization. The 'active center' metal cofactor apparently undergoes alternate reduction and reoxidation in the catalytic cycle during successive encounters with the superoxide radical. Amino acid analyses indicate a close evolutionary relationship between the Mn- and Fe-SOD types but the CuZn SODs seem to have evolved along an independent line (Asada *et al.*, 1980). Zinc in the CuZn-containing SOD plays a secondary structural role not directly related to the catalytic center of the enzyme. Zinc can be replaced by a variety of other metals without the loss of enzymatic activity (Fridovich, 1981).

The widespread occurrence of superoxide dismutases in aerobic organisms has led to the acceptance of SODs as important cellular protective enzymes. Since the enzyme substrate, $\cdot O_2^-$, is of relatively low toxicity compared to other potential secondary oxyradicals that may be formed in cells, the protective action is believed to be largely indirect, i.e. by removing $\cdot O_2^-$ radicals which may lead to the generation of more toxic chemical species. The comparatively low reactivity of the superoxide anion radical gives it the unique capability of transmitting radical properties over wider distances in the cell. Superoxide dismutases along with catalase and the peroxidases that act on the end product (H_2O_2) of SOD activity can interact to regulate injurious oxyradical and peroxyl concentrations in cells and organelles and determine equilibrium rates.

Superoxide dismutases are found in tissues both as water-soluble enzymes and as

TABLE 27.1. General biochemical properties of superoxide dismutases

General
  Reaction catalyzed[a]: $\cdot O_2^- + \cdot O_2^- + 2H^+ \rightarrow H_2O_2 + O_2$
  Molecular weight: Dimer (enzyme) $\sim 32,000$
                       Monomer subunit $\sim 16,000$

Specific types and occurrence

Types	Occurrence	Localization
(A) Iron-containing SOD	Anaerobic cells and primitive aerobic cells; a few eucaryotes	Intracellular and periplasmic sites
(B) Manganese-containing SOD	Increasingly aerobic cells; eucaryotes	Mitochondria; chloroplast membranes
(C) Copper-zinc-containing SOD	Aerobic cells; higher plants	Cytosol; chloroplast stroma

[a] Mechanism: Catalytic cycle involves alternate reduction and oxidation of the active site metal during successive encounters with $\cdot O_2^-$.

metalloproteins bound to cell membranes. When associated with cell membranes, SODs appear to increase the order (decrease the entropy) of the lipid layer fluidity, resulting in membrane structural alterations. This is believed to decrease the susceptibility of membranes to chemical attack.

Reactions of superoxide and other oxyradicals at activated membrane surfaces can be highly complex. Indeed, it is left to the reader to conceive of the many possible reactions, reaction products, and effects involving both organic and inorganic constituents. *Figure 27.1* schematically illustrates some pertinent mechanisms involving $\cdot O_2^-$ and its protonated conjugate acid, enzymatically-catalyzed processes, and O_3 interactions. The most damaging potential reaction products include the extremely potent $\cdot OH$ free radical and various organic radicals and peroxides. Reactions induced by hydroxyl free radicals can lead to the formation of chain-propagating organic radicals which mimic $\cdot OH$ effects but tend to be more specific in their actions leading to delayed toxic responses. Such chain-propagating mechanisms combined with peroxidation reactions are thought — based on kinetic considerations — to more nearly account for cellular injury as observed in biological systems.

Should oxyradicals lead to phytotoxicity in cells, cell defenses that evolve for

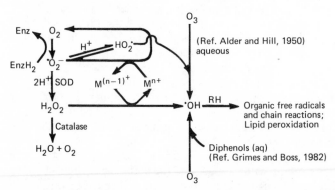

Figure 27.1. Possible reaction mechanisms in aqueous media involving the superoxide anion radical ($\cdot O_2^-$) and conjugate acid ($HO_2\cdot$), SOD dismutation product (H_2O_2), and O_3 leading to injurious hydroxyl ($\cdot OH$) free radicals and organic reaction products. [M = Mn or Fe]

protection would be expected to act either to prevent their generation and/or to minimize the damage caused by those produced. Superoxide dismutases and the catalase/peroxidase enzyme systems serve as interlinked primary protection mechanisms in reducing the potential for cellular injury. Common metabolic reductants such as α-tocopherol and ascorbic acid would further minimize damage due to oxidants present in spite of these primary protective enzyme systems.

Ozone generated oxyradicals

Photochemical oxidants often build up to phytotoxic concentrations downwind of urban-industrial centers during periods of aggravated air pollution. More than 90% of the damage to vegetation in the USA due to air pollution has been attributed to photochemical oxidants of which O_3 is the most important component (Heck, Taylor and Heggestad, 1973). Ozone stress of vegetation, however, is a worldwide problem which deserves greater international concern and study. Although O_3 is troublesome in elevated concentrations, it is also a natural component of the atmosphere to which there has been evolutionary adaptation. Consequently, all green plants show some tolerance to O_3 and varieties vary in their relative susceptibilities or tolerances. Indeed, every crop species we have studied has exhibited some genotypes that are more sensitive than others to O_3 exposure (Bennett et al., 1981).

The biological activity of ozone (O_3) probably involves oxyradicals, at least in part. Recent spin-trapping experiments by Grimes and colleagues at North Carolina State University using electron spin resonance spectroscopy (personal communication) indicate that O_3 produces $\cdot$OH radicals in buffered solutions containing phenolic compounds found in plant tissues (e.g. caffeic acid). The acidity of the chemical microenvironment affects $\cdot$OH radical production as O_3 is more stable in ($\cdot O_2H$-free) aqueous solutions at low pH. There have been conflicting experimental results by different authors regarding the mechanism of O_3 decomposition in aqueous solution but much chemical evidence indicates that $\cdot$OH and $\cdot O_2H$ probably take part in the mechanism (Ardon, 1965; Alder and Hill, 1950). In this regard, O_3 solutions may have properties similar to those caused by radiation of low linear energy transfer (such as electrons produced by the interaction of gamma radiation with water). Free radical scavengers protect against irradiation effects. A point of interest moreover is that radiation damage is potentiated in the presence of oxygen. Protection from irradiation has been shown to be afforded by SOD in mice, bacteria, cell cultures and model membranes (Bors, Saran and Czapski, 1980), implicating the superoxide radical or reaction products in the mechanism of injury. Bors and coworkers postulated that $\cdot$OH radicals may attack membranes with subsequent fixation of the damage by the formation of peroxy radicals. This involves membrane-derived organic oxygen radicals; $\cdot O_2^-$ serves only as a chain propagator. Sugars and alcohols are among cellular products in large quantities that readily scavenge the initiating $\cdot$OH radicals. Carbohydrates in plant tissues including free sugars, starch, and other cellular components may be important stabilizers of cells and organelles exposed to $\cdot$OH radicals.

Because SOD and catalase did not apparently reduce substantially the amount of $\cdot$OH formed in experiments by Grimes and Boss (1982), $\cdot O_2^-$ and H_2O_2 were not considered to be major intermediates in the O_3-induced $\cdot$OH radical production measured. The greater complexities of biological systems — particularly when cellular reactions involving metal-catalyzed one-electron transfers are encountered — and inconsistency found in the chemical literature, however, demand more research into this problem. Reactions occurring with illuminated mesophyll tissue where structured electron-rich chloroplasts are found adpressed to the cell membranes must be better understood. These reactions and processes may be quite different from those for isolated protoplasts in solution with swollen organelles and for respiratory metabolism. Cell surface properties and activities might be very different and transmission of effects to internal organelles might be altered, particularly at low oxidant levels where O_3 stress in older leaves causes primarily premature leaf chlorosis.

Oxyradicals may be especially prevalent in chloroplasts where their concentrations may be enhanced during illumination, resulting in light-induced injury to susceptible chloroplasts. Starch and sugars in carbohydrate-loaded chloroplasts might mitigate the injury.

Ozone and secondary products can react with a wide range of bioorganic molecules and inorganic cell constituents. High O_3 concentrations cause general destruction to the cells. Chronic lower-level concentrations most often found in ambient air can have more specific effects. It has been commonly supposed that O_3 would probably effect damage at the first available sites encountered when diffusing into the leaf mesophyll tissues. This may be an oversimplification that should be addressed in terms of the specific tissue microenvironments encountered and the integrated activities of cellular processes occurring both before and during exposure.

Oxidant injury and aging in leaves

Recent research using selective chemical treatments that alter plant metabolic processes prove certain plant varieties which are naturally quite sensitive to oxidant injury can be transformed into highly tolerant plants by such treatment (Lee and Bennett, 1982). The most effective chemical tool found to date that enhances plant tolerance to oxidants is N-[2-(2-oxo-1-imidazolidinyl)ethyl]-N'-phenylurea (EDU) developed by E. I. DuPont de Nemours Company and made available for study in the mid 1970s. Up to a 30-fold enhancement in the foliar tolerance to O_3 has been reported for plants treated systemically by EDU soil applications or applied as foliar sprays (Jenner, Carnahan and Wat, 1978). Physiological and toxicological investigations using this compound have sought to characterize O_3 stress and to determine mechanisms by which O_3 tolerance is induced in plants.

Is plant tolerance to ozone controlled by biophysical or biochemical processes?

Concern was expressed during early investigations into the mechanism of EDU-induced plant tolerance to O_3 that EDU might cause stomatal closure and hence protect mesophyll tissue by reducing O_3 absorption rates of the sensitive cells. Subsequent experiments showed this was not the case (Bennett, Lee and Heggestad, 1978; Lee and Bennett, 1979). Non-EDU treated control plants did not differ measurably in their stomatal conductances from plants treated with EDU (*Figure 27.2*). Moreover, apparent photosynthesis rates were also shown to be equivalent in these studies. Ozone fumigation trials, nevertheless, demonstrated that the control plants were extensively injured by O_3 exposures that caused no injury to EDU-treated plants.

Stomata in plants made tolerant to O_3 because of EDU treatment stay open even longer during O_3 fumigation than in leaves of O_3-stressed control plants which

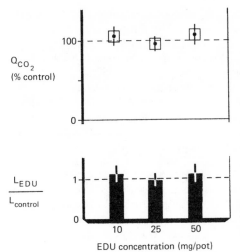

Figure 27.2. Relative stomatal conductances ($L_{EDU}/L_{Control}$) and apparent photosynthesis rates (Q_{CO_2}) for EDU-treated leaves vs control leaves of *Phaseolus vulgaris* L. 'Bush Blue Lake 290'. EDU soil applications are shown on the abscissa. The data represent means and standard deviations for all data taken during a two-week period following EDU application. Paired trifoliate leaves were examined daily in matched Physiological Activity and Diagnostic Chambers under steady-state experimental conditions. (Bennett, Lee and Heggestad, 1978)

progressively close with time of exposure. This leads to an increasing disparity between the actual doses absorbed by the mesophyll tissues compared to those predicted from exposure × time data if the progressive increases in stomatal resistances were not taken into account. These and other data indicate that EDU-induced plant tolerance to O_3 is biochemical in nature and not primarily biophysical due to stomatal closure and restricted O_3 diffusion from the air to sensitive leaf mesophyll sites.

Stomatal conductance can, of course, play an important role in plant damage by gaseous air pollutants. Stomata are the first physiologically sensitive leaf cells encountered by ambient O_3 as it diffuses into the leaf, and the highest O_3 tissue concentrations occur there. Any effect of O_3 or its reaction products on the membrane permeabilities and function of guard cells might be expected to affect stomatal activity. Stomatal closure is also, however, coupled to mesophyll tissue stress due to the suppression of net photosynthesis or water stress. The resulting enhancement of mesophyll CO_2 concentrations and/or developing tissue dehydration caused by O_3 may lead secondarily to stomatal closure.

The environmental conditions prior to and during O_3 fumigation of the plants affect enzyme and metabolite levels in the cells as well as stomatal conductance (and O_3 uptake) at the time of fumigation. Leaves of snap bean (*Phaseolus vulgaris*), soybean (*Glycine max*) and clover (*Trifolium pratense*) plants preconditioned under full sunlight in the greenhouse before being fumigated in controlled fumigation chambers were shown to have greater stomatal conductances and be injured more than plants preconditioned in growth chambers under typical chamber lighting conditions (ca. 300 μE m^{-1} s^{-1} PAR) (Lee and Bennett, 1979). Plants preconditioned for only one day showed more injury than those given longer pre-exposure times in the chambers. Leaf stomatal conductances decreased during the first two to three days in the chamber from one-half to one-third of those exhibited by the greenhouse plants when measured under the same fumigation conditions. Chamber preconditioned plants absorbed correspondingly less O_3 than the sunlight pre-exposed plants. Photosynthates and chemical energy supplies were undoubtedly reduced in chamber-pretreated plants at the lower PAR intensities which could affect both mesophyll metabolism and stomatal function.

Ozone absorption studies conducted in 'physiological activity and diagnostic chambers' show that healthy leaf mesophyll tissue can quantitatively absorb O_3 at typical leaf permeability rates when exposed to subinjurious concentrations (Bennett, 1978). The rates of absorption appear to be more important than the total doses absorbed. If oxidoreductase enzymes and other active-oxygen detoxifying constituents function in the uptake process, limited tissue protection may be expected when saturation conditions are not exceeded. In addition to the SOD-catalase-peroxidase interacting systems for primary protection against oxyradicals and peroxides, various reducing enzymes and constituents such as glutathione reductase and single-electron transfer reactions involving quinones or iron redox agents might be important in the overall process. The minimum absorbed doses of O_3 required to cause any detectable cellular necrosis (trace amount) to highly sensitive snap bean leaves were shown to be above $55\,\mu g\ O_3\ h^{-1}\ dm^{-2}$ (Bennett, 1978). Twice this foliar absorbed dose was required if absorbed over 3-h periods. Leaf cells tolerate subthreshold absorption rates over extended periods of time while maintaining their health. This permits leaves to develop and grow in air when exposed to background O_3 concentrations.

Possible involvement of SOD in O_3 and oxyradical stress tolerance

Relatively little is known about the biochemical mechanisms of action by which leaf mesophyll tissues are able to resist oxidant exposures. However, certain chemical compounds that protect plants against acute and chronic O_3 injury can sustain cellular integrity in physiologically stressed leaves and retard senescence (Gilbert, Elfving and Lisk, 1977; Lee, Bennett and Heggestad, 1981; Ormrod and Adedipe, 1974; Pellissier, Lacasse and Cole, 1972). Chronic, low-level O_3 stress leads to premature senescence in older leaves. Acute doses tend to cause necrosis and chlorosis. Comparing the enzyme and metabolite status of one O_3-tolerant plant with another 'less tolerant' one involves many inherent biological and environmental factors that cannot be easily controlled or standardized. To minimize these difficulties, leaf tissues from a normally O_3-sensitive plant variety made highly tolerant to O_3 in a short period of time (less than one day) by EDU treatment were compared with control plant tissues not treated with EDU. This technique minimizes many difficulties due to plant age and development, nutritional differences associated with mineral efficient vs inefficient plants, and other metabolite- and environmentally-influenced enzyme-induction factors.

At O_3 test concentrations causing acute foliar injury to non-EDU-treated control plants (0.30–0.45 ppm O_3 for 2 to 4 h), tissues treated with EDU in every case sustained reduced O_3 injury depending upon the concentration applied (*Figures 27.3* and *27.4*). EDU at the optimal dose that promoted maximum O_3 tolerance and retardation of senescence (*Table 27.2* and *Figure 27.5*) was also the most effective in inducing SOD and catalase enhancement in normally sensitive (untreated) snap bean trifoliate leaves (*Phaseolus vulgaris* L. 'Bush Blue Lake 290'). It is not known to what extent EDU treatment affects other cellular enzymes such as glutathione peroxidase, which has been associated with enzymatic detoxification and found to respond to oxidant exposure in animal tissues (Mustafa and Lee, 1977); although EDU treatment does, however, generally enhance RNA and protein concentrations in stressed leaf discs (Lee, Bennett and Heggestad, 1981). EDU-treated leaf discs held in the dark (under starvation conditions) or exposed to low PAR levels near the CO_2 compensation point retain their chlorophyll and vigor over much longer periods of time than control discs. The cellular structure was better preserved and energy supplies may have been utilized more effectively. EDU treatment has been observed to increase soluble carbohydrate levels

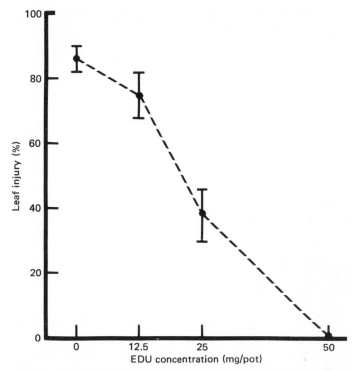

Figure 27.3. Effects of various concentrations of EDU soil applications on leaf injury induced by O₃ fumigation (0.45 ppm for 4 h). Plant tissue: fully expanded BBL 290 snap bean trifoliate leaves (*see Figure 27.5*)

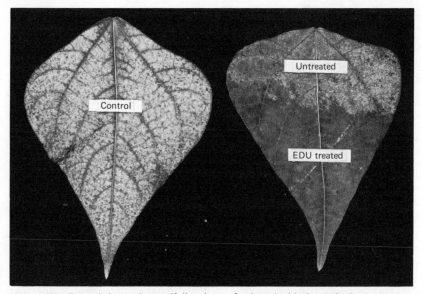

Figure 27.4. Expanded snap bean trifoliate leaves fumigated with O₃. *Left*: Control (untreated with EDU); *Right*: Terminal portion of leaf dipped in 500 ppm EDU aqueous solution one day prior to fumigation

TABLE 27.2. Effects of EDU soil applications on superoxide dismutase and catalase activities of expanded Bush Blue Lake 290 snap bean trifoliate leaves (*Phaseolus vulgaris* L. 'BBL 290'). **Measured two days after treatment**[a]

EDU dose (mg/pot)	Enzyme activities		
	(units/g dry wt)	(units/mg protein)	
	SOD	SOD	Catalase
Control	282	129	18
25	340	139	21
50 (optimal dose)	863	298	36
100	430	170	29

[a] Source: Lee and Bennett, 1982.

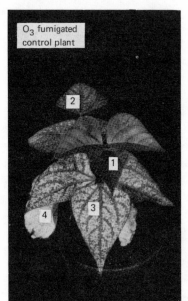

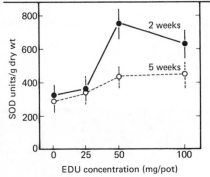

Stage of development	SOD activity		%
	Control	EDU-treated	
	(units/mg protein)		
Younger leaves (less than 40% expanded)	305	307	101
Expanded leaves (50–70% expanded)	142	270	190
Mature leaves (fully expanded)	125	297	238

Figure 27.5. Bush Blue Lake 290 snap bean plant (not treated with EDU) two days after exposure to 0.45 ppm O_3 for 4 h. Youngest trifoliate leaves (1) were not injured. Expanding leaves (2) were injured less than the fully expanded trifoliates (3). Leaf 4 represents a primary leaf. *Right*: (Top) SOD activity for leaves at different stages of development taken from EDU-treated (50 mg/pot) and control plants. Plants were treated two days before sampling. (Bottom) Retention of SOD activity induced in BBL 290 trifoliates by EDU treatment determined two weeks and five weeks after soil application. Compare data in *Figure 27.5* with *Table 27.2*. (Lee and Bennett, 1982)

in Bush Blue Lake 290 snap bean trifoliate leaves which were transformed into a tolerant state by the treatment (Lee, Wang and Bennett, 1981). The induced transformation from sensitive into tolerant leaves required less than 24 h for maximum development of tolerance.

Dhindsa, Plumb-Dhindsa and Reid (1982) recently reported that during the development of senescence in oat (*Avena sativa*) leaf segments and *Rumex* leaf discs, SOD and catalase activities decreased while lipid peroxidation increased. Treatment with kinetin inhibited the decline in SOD and catalase enzyme activities and delayed senescence. Treatment with ethanol or α-tocopherol (free radical scavengers) or with diphenylisobenzofuran (singlet oxygen scavenger) resulted in a strong inhibition of lipid peroxidation and senescence but did not substantially affect the decline in enzyme activities. Kinetin was believed to inhibit senescence through the modulation of membrane lipid peroxidation by maintaining higher levels of cellular scavengers such as SOD, catalase and other free radical counteractants. Damaged membranes lead to cell organelle leakage and disruption with mixing of partitioned metabolites with degradative enzymes. Loss of vigor and senescence ensues. Preservation of cell and membrane integrity is essential to the structure-function interrelations required for cellular vitality. Genetic activity, particular (constitutive or induced) enzymic systems present, the availability of cell metabolites and turnover rates, and energy inputs and reserves constitute the control systems, components, activators and power to maintain and repair the cells.

Oxyradicals and other active oxygens should not be considered as being only injurious to cells. A number of vital cell processes require them as reaction intermediates and regulators. This, however, requires their controlled utilization and prevention of oxyradical amplification (by chain reaction mechanisms) in cells. Two particularly significant cell processes involving superoxide and hydroxyl radicals are the so-called pseudocyclic photophosphorylation process in photosynthesizing chloroplasts, and hydroxylation reactions leading to the formation of diphenols. During pseudocyclic photsphorylation, electrons are transferred from ferredoxin to molecular O_2 (eventually forming H_2O) instead of to NADP as the electron acceptor (Simmons and Urbach, 1973). ATP is generated during the overall process. More recent research (Robinson and Gibbs, 1982) suggests, however, that the univalent reduction of oxygen by ferredoxin forming the superoxide anion radical may be to poise the Photosystem I and Photosystem II processes to prevent overreduction of the photosystems.

Important diphenolic acids produced by enzymatically-controlled hydroxylation of phenolics may be involved in important inducible biological reductive reactions, such as the reduction of ferric iron to active ferrous iron making it more available for metabolic reactions (Olsen, Clark and Bennett, 1981). Free radical polymerization of diphenolic acids furthermore leads to the production of lignans, lignins and other compounds. Lignins constitute the second largest class of plant structural components found in nature.

Conclusion

This treatise discussed the possible importance of SOD and associated oxidoreductase enzymes in the protection of aerobic cells from active-oxygen and free radical toxicity. A number of researchers present evidence to support such a mechanism (Asada, 1980; Dhindsa, Plumb-Dhindsa and Reid, 1982; Lee and Bennett, 1982; Rabinowitch, Sklan and Budowski, 1982; Ream and Wilson, 1982; Tanaka and Sugahara, 1980). Some investigators, however, are skeptical as they have not been able to show increases in SOD and related activity that were proportional to enhanced tolerance of leaves to oxidant stress and injury (McKersie, Beversdorf and Hud, in press; Runeckles, personal communication). These latter studies employed crude enzyme extracts which are subject to a number of chemical interferences. Nevertheless, these results should be carefully considered as we too have found similar results when using crude extracts for enzyme activity determinations. Some suggestions are given below to aid the future researcher in investigating the problem.

Difficulties encountered in the analysis of oxyradical generation and reactions using crude extract preparations can be multifaceted. Prepurification (and perhaps reconstitution) of the enzyme preparations are required to minimize some of these problems. Also, since SOD exists in cells both in soluble and membrane-bound insoluble forms, extracts used for analysis must be verified for quantitative extraction of the active enzyme. This is further complicated by the possibility that the metal redox cofactor can be lost during the process leading to inactivation and may need to be reintroduced. It is also important to know to what extent the protein subunits encountered in the isolated fractions occur as the dimer or monomer forms when evaluating SOD induction in the tissues. Electrophoretic screening of the extracted protein components should be routinely employed to provide this information.

Many questions are left unanswered relating to the enzyme scavenging hypothesis for protection against exogeneous oxidants. Perhaps the most significant of these relate to tissue location factors and the primary sites of reaction. Most information to date indicates that O_3 and oxyradical stress and injury is initiated at cell membrane sites. Solute leakage and chemiosmotic perturbations can cause secondary effects leading to cell injury and senescence. The location and activation of extracellular protective enzymes and reducing substances need to be better researched. The interaction of inductive stresses and preconditioning should be given greater emphasis in such studies. Transmission of the cellular stresses to internal organelles and cytoplasmic structure–function processes need to be given more attention. It would be improper at this time to single out one apparent tolerance-enhancing process to the exclusion of others. All protective mechanisms that serve to mitigate lipid peroxidation in membranes and scavenge damaging free radicals should be given close scrutiny.

Most biochemical research into O_3 and oxyradical phytotoxicity has dealt with reactions involving primary metabolites and components associated with protoplast structure. Secondary metabolism and effects deserve greater attention than has been previously given in the assessment of plant protection and detoxification mechanisms. Secondary metabolism can also affect plant sensitization, as indicated by the work of Grimes and coworkers, who showed the potential importance of common plant phenolics in generating hydroxyl radicals in O_3 solutions. Neish (1965) suggested that secondary waste products in plant organs and in the cell walls may serve as evolutionary detoxification processes and physical protection in certain cases. Superoxide dismutases, hydroxylases and many other cell enzymes influence these processes. A careful study of plant evolutionary detoxification mechanisms related to active-oxygen and free radical stress may provide clues as to how tolerant plant varieties resist oxidant and oxyradical stresses today.

References

ALDER, M. G. and HILL, G. R. The kinetics and mechanism of hydroxide in catalyzed ozone decomposition in aqueous solution, *American Chemical Association Journal* 72, 1884–1886 (1950)

ARDON, M, *Oxygen: Elementary forms and Hydrogen Peroxide*, W. A. Benjamin Press, New York, 106 pp. (1965)

ASADA, K. Formation and scavenging of superoxide in chloroplasts, with relation to injury by sulfur dioxide, *Research Report, National Institute for Environmental Studies, Japan* 11, 165–179 (1980)

ASADA, K., KANEMATSU, S., OKADA, S. and HAYAKAWA, T. Phylogenetic distribution of three types of superoxide dismutase in organisms and in cell organelles. In *Chemical and Biochemical Aspects of Superoxide and Superoxide Dismutase*, (Eds. J. V. BANISTER and H. A. O. HILL), pp. 136–153, Elsevier, North Holland, Inc., New York (1980)

BENNETT, J. H. Foliar exchange of gases. In *Handbook of Methodology for the Assessment of Air Pollution*

Effects on Vegetation, (Eds. W. W. HECK, S. V. KRUPA and S. N. LINZON), pp. 10.1–10.29, Air Pollution Control Association, Minneapolis, MN, USA (1978)

BENNETT, J. H., LEE, E. H. and HEGGESTAD, H. H. Apparent photosynthesis and leaf stomatal diffusion in EDU treated ozone-sensitive bean plants. In *Proceedings Fifth Annual Meeting of the Plant Growth Regulator Working Group,* (Ed. M. ABDEL-RAHMAN), pp. 242–246, Agway, Inc., Syracuse, NY, USA (1978)

BENNETT, J. H., LEE, E. H., HEGGESTAD, H. E., OLSEN, R. A. and BROWN, J. C. Ozone injury and aging in leaves: Protection by EDU. In *Oxygen and Oxy-Radicals in Chemistry and Biology,* (Eds. M. A. J. RODGERS and E. L. POWERS), pp. 604–605, Academic Press, New York (1981)

BORS, W., SARAN, M. and CZAPSKI, G. The nature of intermediates during biological oxygen activation. In *Biological and Clinical Aspects of Superoxide and Superoxide Dismutase,* (Eds. W. W. BANNISTER and J. V. BANNISTER), pp. 1–31, Elsevier, North Holland, Inc., New York (1980)

DHINDSA, R. S., PLUMB-DHINDSA, P. L. and REID, D. M. Leaf senescence and lipid peroxidation: Effects of some phytohormones and free radical scavengers, *Plant Physiology* 69 (Supplement), 10, Abstract No. 48 (1982)

FRIDOVICH, I. Role and toxicity of superoxide in cellular systems. In *Oxygen and Oxy-Radicals in Chemistry and Biology,* (Eds. M. A. J. RODGERS and E. L. POWERS), pp. 197–239, Academic Press, New York (1981)

GILBERT, M. D., ELFVING, D. C. and LISK, D. J. Protection of plants against ozone injury using the antiozonant, N-(1,3-di-methylbutyl)-N′-phenyl-p-phenylenediamine, *Bulletin Environmental Contamination and Toxicology* 18, 783–786 (1977)

GRIMES, H. D. and BOSS, W. F. Ozone-induced HO· effects on plasma membranes *in vivo, Plant Physiology Supplement* 69 (Supplement), 49 (1982)

HECK, W. W., TAYLOR, O. C. and HEGGESTAD, H. E. Air pollution research needs: Herbaceous and ornamental plants and agriculturally generated pollutants, *Journal Air Pollution Control Association* 23, 257–266 (1973)

JENNER, E. L., CARNAHAN, J. E. and WAT, E. K. W. A new antiozone-plant protectant. In *Proceedings Fifth Annual Meeting of the Plant Growth Regulator Working Group,* (Ed. M. ABDEL-RAHMAN), pp. 238–241, Agway, Inc., Syracuse, NY, USA (1978)

LEE, E. H. and BENNETT, J. H. Comparative studies of foliar protection from ozone induced by EDU in greenhouse and growth chamber preconditioned plants. In *Proceedings of the Sixth Annual Meeting of the Plant Growth Regulator Working Group,* (Ed. ABDEL-RAHMAN), p. 218, Agway, Inc., Syracuse, NY, USA (1979)

LEE, E. H. and BENNETT, J. H. Superoxide dismutase: A possible protective enzyme against ozone injury in snap beans (*Phaseolus vulgaris* L.), *Plant Physiology* 69, 1444–1449 (1982)

LEE, E. H., BENNETT, J. H. and HEGGESTAD, H. E. Retardation of senescence in red clover leaf discs by a new antiozonant N-[2-(2-oxo-1-imidazolidinyl)ethyl]-N′-phenylurea, *Plant Physiology* 67, 347–350 (1981)

LEE, E. H., WANG, C. Y. and BENNETT, J. H. Soluble carbohydrates in bean leaves transformed into oxidant-tolerant tissues by EDU treatment, *Chemosphere* 10, 889–896 (1981)

McKERSIE, B. D., BEVERSDORF, W. D. and HUCL, P. The relationship between ozone insensitivity, lipid soluble antioxidants, and superoxide dismutase in *Phaseolus vulgaris* L., *Canadian Journal of Botany* 60, 2686–2691 (1982)

MUSTAFA, M. G. and LEE, S. D. Biological effects of environmental pollutants: Methods for assessing biochemical changes. In *Assessing Toxic Effects of Environmental Pollutants,* (Eds. S. D. LEE and J. B. MUDD), pp. 105–120, Ann Arbor Science Publishers, Ann Arbor, MI, USA (1977)

NEISH, A. C. Coumarins, phenylpropanes, and lignin. In *Plant Biochemistry,* (Eds. J. BONNER and J. E. VARNER), pp. 581–617, Academic Press, New York (1965)

OLSEN, R. A., CLARK, R. B. and BENNETT, J. H. The enhancement of soil fertility by plant roots, *American Scientist* 69, 378–384 (1981)

ORMROD, D. P. and ADEDIPE, N. O. Protecting horticultural plants from atmospheric pollutants, *Hort Science* 9, 108–111 (1974)

PELLISSIER, M., LACASSE, N. L. and COLE, H. Jr. Effectiveness of benomyl and benomyl-folicote treatments in reducing ozone injury to pinto beans, *Journal of the Air Pollution Control Association* 22, 722–725 (1972)

RABINOSWITCH, H. D., SKLAN, D. and BUDOWSKI, P. Photo-oxidative damage in the ripening tomato fruit: Protective role of superoxide dismutase, *Physiologia Plantarum* 54, 369–374 (1982)

REAM, J. E. and WILSON, L. G. Possible significance to SO_2 injury of dark and light dependent bisulfite oxidizing activities in *Cucumis sativa* leaves, *Plant Physiology* 69 (Supplement): 16, Abstract No. 85 (1982)

ROBINSON, J. M. and GIBBS, M. Hydrogen peroxide synthesis in isolated spinach chloroplast lamellae:

An analysis of the Mehler Reaction in the presence of NADP reduction and ATP formation, *Plant Physiology* **70**, 1249–1254 (1982)

SIMONIS, W. and URBACK, W. Photophosphorylation *in vivo*, *Annual Review of Plant Physiology* **24**, 89–114 (1973)

TANAKA, K. and SUGAHARA, K. Role of superoxide dismutase in the defense against SO_2 toxicity and induction of superoxide dismutase with SO_2 fumigation, *Research Report National Institute for Environmental Studies, Yatabe, Japan* **11**, 155–164 (1980)

Chapter 28

Epilogue: a biochemical overview

Michael Treshow

DEPARTMENT OF BIOLOGY, UNIVERSITY OF UTAH, SALT LAKE CITY, UT, USA

An epilogue should strive to 'assimilate and integrate the material that has gone on before and, from this, draw conclusions'. In attempting to do this, I will do my best to assimilate the excellent material that has been presented in this volume and ideally draw from this a simple common denominator expressing a 'unified' theory of air pollution effects. Although this does not seem possible, we may be narrowing the possibilities as the basic stress mechanisms become better understood. Many agents — pollutants included — cause stress; all affect plant health, all can cause disease or disorders. But how similar are the fundamental stress mechanisms at the biochemical level? Is each pollutant unique? Or do they share some common properties?

For this epilogue I shall do my best to draw together the state of the art at this time, and attempt to present a general theory of air pollution damage to plants.

I should like to attempt to draw together what we have learned as holistically as I can, considering the major pollutants collectively. Biochemically, are there mechanisms common to all?

Initially each pollutant interacts biochemically with the most sensitive receptor; ultimately this becomes reflected in some loss to production. I shall attempt to generalize how the biochemical mechanisms of pollutants may influence natural or agricultural ecosystems and production. What mechanisms common among all pollutants exist that interfere with the intervening metabolic processes? How can biochemical effects be translated into reduced growth and reproduction? I shall be approaching the problem largely through the eyes of a pathologist and ecologist.

Plants evolved in a gaseous environment in which low concentrations of certain chemicals were natural components. In higher amounts, they would be pollutants. Small amounts of ozone, and nitrogen and sulphur oxides and, in the soil, even fluorides, are natural to most environments. But when the composition of pollutants exceed the critical limits of adaptation and tolerance, stress is imposed, and the most sensitive components of the plant system begin to malfunction. Any gas can cause such stress when a threshold concentration is exceeded. The impact to the ecological system, regardless of whether natural or manipulated, is reflected not only on the system but on the organism and is felt first at the molecular level of organization. Regardless of the air pollutant, the impact invariably involves interactions with one or more biochemical metabolic processes. Let us follow the pathway of the gaseous pollutant into the plant to the site or sites of reaction.

Pollutant receptors

Cuticle and stomata

The cuticle and stomata are the initial receptors or targets — the first structures the pollutant encounters. We need look no further for the initial responses. It is natural that most thought is given to the stomata since these provide the path by which gases enter the leaf, but the direct impact on the cuticle must also be considered. The effects of SO_2 and acid deposition on the cuticular waxes are well documented (Fowler et al., 1980; Godzik, and Sassen, 1978). The degradation of the cuticular wax structure is seen in coniferous needles such as those of Scots pine (Huttunen, Mäkelä and Laine, 1982). Studies on Norway spruce have shown that degradation of wax microfibers begins soon after needle flusing, and clear signs of erosion and wax degradation can be found within a few months of emergence. The normal weathering of needle cuticles may not be extensive the first year, but under air pollution or acid deposition conditions it may be many times faster than in unpolluted forest areas. Similar observations of surface degradation have been described in other forest species including lichens and mosses (Huttunen, Mäkelä and Laine, 1982). The consequences of this are not clear but may well render the tissues more sensitive to invasion by pathogenic organisms. More immediately, evapotranspiration would be greater which could be critical in arid environments.

Pollutants absorbed by the guard cells and subsidiary cells may initially affect the stomatal aperture. Sulfur dioxide has a notable effect in stimulating stomatal opening (Majernik and Mansfield, 1970), interacting with CO_2 and atmospheric moisture. Low concentrations of SO_2 may stimulate stomatal conductance within 15 min of exposure when the relative humidity is high, and the effect may persist for several days (Black and Unsworth, 1980). However, this could be due to the destruction of adjacent, subsidiary, epidermal cells (Black and Black, 1979). Wider stomata naturally influence the rate of pollutant uptake and indirectly therefore, sensitivity. However, wider stomata would also facilitate an increased CO_2 concentration which then induces stomatal closure or might enhance photosynthesis and conceivably production. On the negative side, the increased potential for water loss could limit growth and production in arid environments. However, under such conditions, the humidity would be low and the stomata less responsive to SO_2.

The rate at which SO_2 is absorbed into plants varies among species and with previous exposure to SO_2 (Jensen and Kozlowski, 1975). These differences in leaf resistance to uptake have been found to be associated with differences in SO_2 tolerance (Winner and Mooney, 1980). Sulfur dioxide may increase or decrease resistance depending apparently on its concentration, thus affecting CO_2 uptake and potential photosynthetic performance (Hällgren, 1978). Hällgren (this volume) pointed out how the atmospheric concentration alone is not enough to consider. The other environmental parameters are at least as important: for example as RH increases from 30–70%, O_2 uptake was increased fourfold and SO_2 uptake, threefold.

Once within the leaf, SO_2 diffusion is slowed and the stomata may be induced to remain open for longer periods, especially when the relative humidity is above 40% (Mansfield and Majernik, 1970). It is important to note that a given concentration of SO_2 could cause the stomata of one species to open, while causing stomata of another to close (Mudd, 1975). In fact, different plant species can respond oppositely when exposed to the same SO_2 concentrations (Biggs and Davis, 1980).

Ozone also influences the stomatal aperture, and again there is considerable disagreement how this mechanism works. As with SO_2, concentrations of O_3 appear to be critical as well as the O_3-tolerance of the plant. When the exposure concentration is high enough to cause the early 'waterlogging' symptoms of visible damage, the stomata were found to close (Mansfield, 1973). At still higher O_3 concentrations, leaf resistance steadily decreases (Evans and Ting, 1974). The mechanisms involved suggest that at low concentrations of O_3, a flow of potassium ions into the guard cells is followed by water which increases turgidity, thus causing the guard cells to open. But should water loss be excessive, the guard cell turgidity would decrease and the stomata would close. It is interesting to note that membrane permeability disruption may be the underlying cause of the changes in stomatal resistance.

Intercellular areas

Once the gaseous pollutants pass through the stomata, they enter the substomatal, intercellular spaces where they dissolve in water on the moist cellular surfaces. Sulfur dioxide, for example, forms sulfite (SO_3^{2-}) and bisulfite (HSO_3^-). Cellular pH would be influenced by the generation of protons which may produce secondary effects. An increase in the hydrogen ion concentration, for instance, could cause leakage of K^+, Cl^- and malate (Smith and Raven, 1979). In K^+-regulated stomatal opening, anions such as malate and Cl^- play important roles. Both SO_2 and O_3 have been found to cause an immense loss of K^+ from cells.

Fluoride is largely dissolved in the water phase and translocated toward the marginal and apical parts of the leaves. Garrec (personal communication) has shown that F accumulates closest to the pathway of entry — near the stomata and transfusion parenchyma between the conducting vessels. The photochemical pollutants, O_3 and PAN, also presumably would be absorbed in the water phase.

Ozone is about ten times more soluble in water than oxygen, and roughly half is absorbed depending in part on temperature and acidity. In alkaline solution, O_3 decomposes rapidly releasing molecular oxygen. The concentration of O_3 in the tissues is affected by its solubility, rate of decomposition, and pH at the site of absorption.

Ozone can produce a number of free radicals (e.g. $HO_2^\cdot$, $^\cdot O_2^-$ and $^\cdot OH$) which can oxidize various cellular metabolites and affect a number of membrane constituents such as SH groups, amino acids, proteins and unsaturated fatty acids (Heath, 1975). While a certain amount of free radicals is normal, pollutants can cause the system to become overloaded. When this happens, and the radicals are not neutralized, the surplus may attack other substrates. Some of these, at least, are associated with photosynthetic pathways (Bennett, this volume), very likely the thylakoid membranes, or phosphorylation processes.

Cell wall

Pollutants that have not been totally absorbed in the intercellular spaces, in other words the aqueous phase, next contact the cell walls. Presumably the walls would not provide a particularly reactive site. There is no evidence, for instance, that SO_2 reacts with either the cell wall or its components. Ozone similarly is unlikely to react with the cell wall (Mudd, this volume).

Membranes (plasmalemma)

Most likely, the primary critical and permanent impact of a gaseous pollutant in the cell is on the plasmalemma, the first major barrier a pollutant contacts. Unfortunately, the reported effects are not always consistent. The greatest influence of pollutants on membranes has been argued to be most likely on the lipid component. Decreases in lipid concentrations following SO_2 exposure may be brought about by either reduced synthesis, increased lipase activity, peroxidation of fatty acid chains or a combination of these. The magnitude of the inhibition of synthesis depends on the concentration and duration of exposure to SO_2 (Malhotra and Khan, 1978). However, the inhibition is reversed upon removal of the plants from the SO_2 atmosphere.

Mudd (this volume) provided some convincing arguments that the membrane protein component may be more sensitive than the lipid component. His experiments demonstrated that polypeptide chains exposed outside lipid bilayers of the plasma membrane are susceptible to attacks by O_3, and that this reaction takes place before oxidation of membrane lipid.

Digestion of oxidized lysozyme and analysis of the peptides showed the oxidized residue to be consistent with the loss of the capability of the enzyme to bind substrate. Ozone has also been shown to alter a number of amino acids, all of which are found in the proteins of the cell membranes.

Fluoride caused early rupturing of the tonoplast while leaving the plasmalemma of soybean leaves intact (Wei and Miller, 1972). Fundamentally, the capacity for selective absorption by membranes is lost, which may trigger a rapid reduction in photosynthesis, respiration and other processes as secondary effects. This is not unlike many other stress responses. ATP and sulfhydryls also decrease rapidly following O_3 damage to the membranes while the peroxide content increases. Permeability may not return to normal for several hours after exposure which causes a further upset in water balance.

We should keep in mind that, as pointed out by Pell and Weisberger (1976), O_3 may react differently on different membranes. An attack on the plasmalemma and/or tonoplast may lead to death whereas attacks on other membranes only limit function.

Cellular components

Having passed through the plasmalemma, the gaseous pollutants that have not already reacted remain free to attack the cytoplasmic components including the various organelles and their membranes. The endoplasmic reticulum being a network of exceedingly fine double membranes that often traverses all regions of the cytoplasm, would be most readily attacked. The plasmalemma, endoplasmic reticulum and nuclear envelope, for convenience, may be considered collectively as being equally susceptible to pollutant attack although this has not been established experimentally.

Wei and Miller (1972) found that the first changes in HF-fumigated soybean leaves involved an increase in the amount and aggregation of endoplasmic reticulum (ER); this preceded the appearance of any visible symptoms of injury. Once injury became apparent, the amount of ER decreased and small vacuoles and lipid-droplets appeared in the cytoplasm.

CHLOROPLASTS

The chloroplast envelope is presumably the first part of this organelle to be subjected to a pollutant. Godzik and Knabe (1973) described invaginations extending from the inner membrane of the envelope, and doubling of the envelopes in needles of some pines collected near S-polluted industrial areas.

Similarly in high F areas, the envelope became stretched and later ruptured in otherwise healthy-appearing needles in an early stage of injury. The stretching was accompanied by an increase in the staining density and accumulation of electron-dense material between the two membranes of the envelope.

While it is convenient to generalize about chloroplasts, we should keep in mind that they all may not react the same. In addition to the genetic variations in sensitivity, plants with C_4-type photosynthesis may have different mechanisms for converting S either in normal metabolism or in detoxification reactions (Gerwick, Ku and Black, 1980).

Following exposure of cells to SO_2, the chloroplasts can be seen to change shape from ellipsoidal to round. Hydrogen fluoride has also been shown to have this effect (Wei and Miller, 1972); furthermore the chloroplasts were smaller than normal. Ozone has a generally similar effect although the affected chloroplasts become more irregular in shape. Nitrogen dioxide causes a swelling of chloroplast membranes (Wellburn, Majernik and Wellburn, 1972). This may occur if the ammonia produced from NO_2 is not rapidly incorporated into amino acids. Ammonia also has been shown to impair photosynthesis by uncoupling electron transport (Avron, 1960) and inducing structural alterations.

The effects on ultrastructure are naturally related to the biochemical effects, and the next step is to try to integrate these. Whether the swelling is a cause of impaired photosynthesis, or a result of it, or both a function of a third factor (e.g. some free radical or a membrane disruption) remains an unanswered question, but certainly many photosynthetic reactions are affected by SO_2 and other gaseous pollutants, and the process is impaired.

It has been suggested (Hampp and Ziegler, 1977) that both SO_3^{2-} and SO_4^{2-} are transported to the inner chloroplast membranes by phosphate translocators and that light modulates this process. Ziegler (1977) showed that sulfur from either SO_2 or sulfite (SO_3^{2-}) could be incorporated in the chloroplast thylakoids to a much greater extent than sulfate (SO_4^{2-}) sulfur. She suggested that SO_3^{2-} was either directly incorporated into the sulfuric groups of the sulfolipids (as reported by Benson, 1963) or was taken up at the binding sites in the thylakoids (Schwenn, Depka and Hennies, 1976).

Some excellent, further insight on this process was developed by Ruth Alscher-Herman (this volume). Conversion of light modulation of key enzymes to their activated state involves a reduction of crucial disulfide bonds and a subsequent shift of the enzyme to a more active form. The source of reductant for the change seems to be dithiol groups produced in the chloroplast as a result of photosynthetic electron transport. SO_2 and SO_3^{2-} have been shown to affect the extent to which the light-dependent transformation takes place on several light-modulated enzymes. For example, SO_2 may interact with disulfide bonds to produce a sulfonate group and a sulfhydryl group. Enzymes on the thylakoid membranes are apparently especially effective in providing sites capable of picking up free SO_2. Work with SO_2-sensitive and tolerant soybean varieties provided good evidence that the thylakoids contained the most sulfur-susceptible factor and are an important site of pollutant activity — at least for SO_2.

Changes to the thylakoids show up early in the ultrastructure studies. Exposure to SO_2 include swelling of the lamella and reduction of the grana. Grana were often reduced even in visibly healthy material (Soikkeli and Tuovinen, 1979). Hydrogen fluoride has been shown to produce this same effect, and injury may develop within 24 h of exposure (Wei and Miller, 1972). Soikkeli and Tuovinen (1979) reported the occurrence of two types of lamellar injuries in healthy appearing pine needles in a fluoride-polluted area. One was the smooth swelling of thylakoids beginning from the stroma lamella; the second, a curling of the thylakoids with no apparent swelling.

Looking at photosynthesis, according to Sugahara (this volume), SO_2 preferentially inhibited Photosystem II. He suggested that the site of SO_2 inhibition was the primary electron donor or even the reaction center itself. While he did not emphasize the point, his work showed that the oxygen evolved after SO_2 fumigations decreased sharply after 1 h of exposure. This suggests that a 1 h air quality standard might be considered to fully protect plants.

Less is understood how other pollutants act on the chloroplast stroma or thylakoids and the biochemical reactions. However, some parallels might be drawn between responses to O_3 and SO_2. Thomson, Dugger and Palmer (1966) speculated that the granulation of chloroplast stroma seen in electron micrographs of O_3-treated tissues was due to oxidation of SH groups in ribulose bisphosphate carboxylase. Fumigation with ozone reduced the activity of RuDP carboxylase (Nakamura and Saka, 1978).

Ozone also has been found to inhibit electron transport in both Photosystems I and II, Photosystem I being the more sensitive (Coulson and Heath, 1974). Schreiber *et al.* (1978) suggested that O_3 acted within the photosynthetic apparatus with the initial damage at the donor site of Photosystem II (H_2O-splitting system). Increasing exposure concentrations of O_3 inhibited electron transport from Photosystem II to Photosystem I.

Ozone also reduces the chlorophyll content of plants (Leffler and Cherry, 1974), but this occurs late in the injury process and essentially accompanies the appearance of visible symptoms.

Dugger *et al.* (1965) showed that exposure to PAN for even brief exposures inhibited O_2 evolution in chloroplasts. Treatment of isolated chloroplasts with PAN inhibited electron transport, photophosphorylation and CO_2-fixation. The inhibitory effect on enzymes has been attributed to its ability to oxidize SH groups in proteins, and metabolites including cysteine. PAN can also oxidize NADH and NADPH.

Effects of fluoride on the two photosystems is even more nebulous, especially at F concentrations not causing visible symptoms of injury. Most photosynthetic studies had dealt with gas exchange measurements of apparent photosynthesis until Ballantyne (1972) demonstrated that the Hill reaction was inhibited in bean (*Phaseolus vulgaris*), a relatively F-tolerant species.

Chlorophyll synthesis has been found to be reduced by F. The mechanism still is not known, although fluoride tends to accumulate in chloroplast and dilates the internal membranes (Weinstein, 1961). Fluoride may render Mg^{2+} (essential for thylakoid function) physiologically inactive by precipitating if from solution.

The report by Wellburn (this volume) was especially enlightening in suggesting ways in which pollutants may disrupt energy flow. Numerous previous reports as well as that by Wellburn have borne out that ATP formation is depressed by SO_4^{2-}, SO_3^{2-} or SO_2. Such suppression alone may well limit normal growth and production. But how might the mechanism of ATP formation work? It was suggested that the competitive inhibition between orthophosphate and either SO_3^{2-} or SO_2 for both PGA-dependent O_2 evolution and photophosphorylation could be critical.

Wellburn explained how such pollutants as SO_2, NO_x, O_3 and PAN all can interact in the photoreduction of NADP and nitrite and so may further relate to an overall breakdown in energy flow.

Since chlorophyll and other pigments are necessary to harnessing light energy by Photosystems I and II, it is clear that the effects of SO_2 on these processes may indirectly impair photosynthesis. High concentrations of SO_2 (5 ppm; 24 h) may degrade the chlorophyll molecule to phaeophytin $+ Mg^{2+}$, but the same effect has been reported to be induced by much lower concentrations (0.34 ppm) (Khan and Malhotra, 1982).

MITOCHONDRIA

Mitochondria, at least ultrastructurally, appear far less responsive to either SO_2 or fluoride than the chloroplasts (Soikkeli and Tuovinen, 1979). Nevertheless, biochemical reactions in the mitochondria also may be adversely affected by gaseous pollutants through various mechanisms. Inhibition of photorespiration is suggested by the inhibition of glycine and serine synthesis, although CO_2 fixation was stimulated (Koziol and Cowling, 1978).

Ozone seems to cause mixed responses in respiration and can either stimulate or inhibit the process. The effects on photorespiration and related processes is not certain. More is known about the effects of fluoride on respiration. Mitochondrial phosphorylation was severely inhibited whether tissue respiration was inhibited or stimulated. Since fluoride also caused mitochondrial swelling and leakage of proteins, it is likely that membranes are the primary site of fluoride action. Several metabolic processes of respiration have been demonstrated to be affected by fluoride. Succinate, malate and NADH oxidation in mitochondria are all inhibited.

CYTOPLASM, RIBOSOMES, AMINO ACIDS AND PROTEINS

Gaseous pollutants not reacting directly with the cell walls, membranes or organelles could react as readily with any number of other cellular components including the cytoplasm, ribosomes, amino acids, proteins, carbohydrates, fatty acids and lipids.

Sulfur dioxide can cause an increase in the free amino acid content in a number of plant species when the supply of sulfur is adequate (Malhotra and Sarkar, 1979; Ziegler, 1975); in sulfur-deficient plants though amino acid content decreased (Cowling and Bristow, 1979). Heath (this volume) pointed out that O_3 also tends to cause an increase in the amount of free amino acids. Increases in the concentrations of glycine, alanine, threonine, lysine and methionine in pine needles (*Pinus banksiana*) treated with SO_2 were thought by Malhotra to be due mostly to the breakdown of proteins (Malhotra and Sarkar, 1979). Along with a greater concentration of amino acids, SO_2 can cause an increase in the polyamines derived from them (Pribe, Klein and Jäger, 1978).

Several enzymes involved in amino acid metabolism have also been shown to be affected by SO_2. However, the effects are inconsistent, with the activity of some apparently inhibited and others stimulated (Horsman and Wellburn, 1975).

Ozone also can cause either an increase or decrease in amino acid content (Tingey, Fites and Wickliff, 1973). Likewise, protein content may be either reduced or increased depending on the O_3 concentration (Ting and Mukeriji, 1971; Constantinidou and Kozlowski, 1979). As with SO_2, one way O_3 may act is to break down existing proteins or prevent synthesis of the new ones.

Fluoride has an especially strong effect on amino acid metabolism. As with O_3 and SO_2, increases in free amino acids and amines have been reported (Jäger and Grill, 1975). It was not certain whether the increase was due to increased protein breakdown or inhibition of protein synthesis or both.

Lipids are particularly significant components of membranes; in chloroplasts, glycolipids constitute about 50% by weight of thyalkoid membranes (James and Nichols, 1966). Khan and Malhotra (1977) have shown that when lodgepole pine (*Pinus contorta*) needles were exposed to SO_2 there was a marked reduction in the concentration of many glycolipids. Since these are involved in the structure and function of chloroplasts, it was suggested that the structural alterations in chloroplasts in lodgepole pine were due to changes in galactolipid concentration (Khan and Malhotra, 1977).

Ozone can affect polyunsaturated fatty acids by oxidative mechanisms that can change the properties of the membranes. The biosynthesis of galactolipids in chloroplasts was inhibited by O_3 and other agents that bind SH groups (Mudd *et al.*, 1971). Heath (this volume) found that O_3 attacked the double bonds of lipids to produce various compounds including aldehydes and peroxide. But he considered the change to be small and perhaps secondary to changes in the protein component of membranes.

Results concerning reducing sugars are inconsistent. Ozone also can either increase or decrease carbohydrate concentrations. Koziol (this volume) summarized the carbohydrate alterations, pointing out that HF, O_3, SO_2 and NH_3 each can cause an increase in reducing sugars, but that conflicting results have been found, especially with SO_2. Ozone can cause either an increase or decrease in starch content, presumably depending on pollutant concentrations and interacting environmental parameters and predisposition.

The significance of carbohydrate effects may lie mostly in translocation and storage. The limited work available in this field suggest that phloem loading is reduced, at least by SO_2, leading to less storage in the roots which could have serious consequences on production, and in pasture crops, on the success of the next year's growth.

Another impact suggested was that changes in carbohydrate metabolism could influence the attractiveness to insects. This could be particularly critical if terpenes or other natural compounds were involved that can mediate sensitivity of plants to insect attack.

Fluoride causes changes in sugars, polysaccharides and organic acids (McCune *et al.*, 1964). As with other pollutants, some of the changes can be explained by the toxic effects on various enzymes.

Defense mechanisms

Various postulates have been proposed to explain the differences in sensitivity to pollutants and ways by which plants might get rid of waste products. The role of free radical scavengers is one. If free radicals take part in membrane lipid peroxidation, and this is important to injury, enzymes such as superoxide dismutase that can metabolize and hence detoxify free radicals could provide an important defense.

A mechanism discussed by Filner (this volume), however, sounded at least equally appealing. Cucurbit plants (*Cucumis sativa* and *Cucurbita pepo*) exposed to SO_2 were found to release H_2S. In fact, sensitive gas measurements disclosed that H_2S was released even in the absence of SO_2, but about 1000 times more was released after exposure to SO_2. Young leaves absorbed more SO_2 per unit area than the old leaves, but were more resistant to injury. This suggested that the young leaves possessed a biochemical resistance mechanism. They produced H_2S as much as 100 times more rapidly than the old leaves. A cycle of cysteine being desulfhydrated to sulfide, and then oxidized to SO_4, then reduced to sulfide again may enable the plant to expel excess sulfur and prevent accumulation of excess thiol groups.

Along the same line, Bell described some research he did with broadbean ten years ago. Plants were fumigated at only 0.01–0.06 ppm SO_2, and the evolution of significant amounts of H_2S was detected. Hällgren described similar findings with Scots pine whereby a correlation was found between exposures to 50 to 400 μg SO_2 m^{-3} and the amount of H_2S evolved. The fact that plants exposed to SO_2, especially resistant plants, emit considerable amounts of H_2S would support this mechanism of defense and tolerance (Sekiya, Wilson and Filner, 1980).

The emission of ethylene and ethane has also been mentioned in response to SO_2 exposure, but this appears to be more of a general stress response than a defense mechanism.

Integration of biochemical effects and production

Production losses in agriculture, and biomass reduction in natural ecological systems, have often been equated to the visible chlorosis and necrosis that characterizes the effects of gaseous air pollutants. However, we see that physiological responses may well occur long before the appearance of such symptoms. Indeed, it is not difficult to envisage the existence of prolonged adverse biochemical effects occurring indefinitely in the absence of any apparent leaf markings. Reduced yields in the absence of chlorosis or necrosis are well documented for O_3 and SO_2 where exposures are continuous although less definite for other pollutants.

The effects of sustained, high concentrations of a pollutant as were most common in past decades, can still be important. But in the relatively cleaner air of today, the significance of gaseous pollutants lies not so much in the myriad of metabolic, chemical reactions caused by exposures to high concentrations of the pollutant, as in the insidious effect of lower, but sometimes frequent, exposures that may exert sustained chronic effects on even one metabolic process.

The influence of very low concentrations may be minimal or even temporary if the pollutant is absorbed on the plant surfaces or reacts with the relatively inert cell walls. However, the effects on reducing stomatal resistance would aggravate moisture loss in areas where water stress was a factor in growth and production. Furthermore, it would facilitate entry of other pollutants.

Membranes appear to be most vulnerable to many pollutants and their dysfunction could lead to a number of secondary effects, including disrupted permeability. The inadequacy of regulation of selective ion absorption, together with the leakage of certain ions could produce a number of adverse responses. This leakage could include even changes in permeability to water as described following O_3 exposures (Evans and Ting, 1974). Even if this were the only consequence, such an effect could upset the water balance of the plant and limit productivity. The inner thylakoid membrane of the chloroplasts may be most sensitive of all. Thus, we see early and serious effects on photosynthesis giving rise to the potential for significant losses in productivity.

Together with effects on membranes, pollutants often initiate the formation of free

TABLE 28.1. Generalized responses of plants to air pollution

	SO_2	F	O_3	PAN	NO_2
Stomatal conductance	+/−		+/−		
Membrane disruption oxidize SH groups	+	+	+	+	
Chloroplast and photosynthesis	+	+	+	+	+
Respiration inhibition	+/−	+	+/−		
Enzymes	+/−	−	+/−		
Amino acid increase	+/−	+	+/−		+
Lipid synthesis	−		+		
Free radical formation	+	+	+		
Carbohydrate synthesis	+/−	−	+/−		
e- transport	−		−		

radicals which may well be toxic. If there is enough of a pollutant to reach the organelles, especially the chloroplasts, the adverse effects become highly significant, often lethal, and result in visible symptoms of injury. Furthermore, SO_2 and HF may act as poisons at specific metabolic sites associated with many processes. Ozone and SO_2 may impair photosynthetic pathways in the absence of visible symptoms, and thereby restrict productivity. In a natural ecosystem the impact may be sufficient to reduce the ability of the affected plants to compete successfully.

While it is tempting to draw analogies between all the pollutants, we must accept that each has its own characteristic properties; each acts in its own distinctive manner, and it is not possible to draw valid generalizations about the mode of action of the major gaseous air pollutants (*Table 28.1*). Likewise, the concentrations of a pollutant to which a given plant first responds adversely varies with the pollutant. Thus we see injury to sensitive plants when atmospheric SO_2 concentrations are in the 500–1000 μg range for a few hours, while fluoride can be injurious at concentrations below 10 μg over the same time periods.

Finally, it is important to remember that the rate of entry of a pollutant into a plant is particularly critical to toxicity. Plants have an inherent capacity to absorb, detoxify or, in some cases, metabolically incorporate or expel the pollutant, as in the case of SO_2. This varies not only among species but among varieties and individuals in a population and is further influenced by the varying environmental conditions during, before and even following exposure. All interact to determine the plant's response to a pollutant exposure.

Conclusions

So what does all this come down to?

(1) The impact of gaseous pollutants on the cuticle is well defined.
(2) Stomata are affected, but the mechanism is uncertain.
(3) By direct effects on membrane proteins, permeability is disrupted followed by ion leakage leading to other adverse subsequent reactions.
(4) The pollutant enters the intercellular spaces and reaches the cell interior. Free radicals in superabundant amounts may be formed which produce further effects, perhaps including disruption of the membranes, chloroplasts, or bioenergetic pathways.
(5) Through the direct action of pollutants on disulfide bonds, notably on light-modulated enzymes, photosynthetic processes may be disrupted directly, or indirectly by effects on proteins.

Finally, then, let us look at some practical aspects.

Gaseous pollutant concentrations in the atmosphere are not significant in themselves. They do not tell us what is actually reaching the cell, or even less, reaching the metabolic site.

More research is needed on the actual exposure, or dose. In laboratory experiments, this would be based on the input minus the output from the exposure chambers. Perhaps this is not practical, but some of the chapters in this volume attest to its importance. Only this type of dose measurement can be validly related to effects of pollutants on plant productivity.

We should also keep in mind that the findings revealed in laboratory studies at the chemical level do not necessarily reflect how a given system, or plant, may respond in the field. The most sensitive metabolic processes as found *in vitro* may not be the most sensitive *in vivo*, although it is most likely that they will be.

We must further bear in mind that our ultimate goal is to understand how gaseous pollutants act so that we may prevent or minimize production losses and maintain the integrity of both natural and agricultural plant communities.

References

AVRON, M. Photophosphorylation by swiss chard chloroplasts, *Biochimica et Biophysica Acta* **40**, 257–272 (1960)

BALLANTYNE, D. J. Fluoride inhibition of the Hill reaction in bean chloroplasts, *Atmospheric Environment* **6**, 267–273 (1972)

BENSON, A. A. The plant sulfolipids, *Advances in Lipid Research* **1**, 387–394 (1963)

BIGGS, A. R. and DAVIS, D. D. Stomatal response of three birch species exposed to varying acute doses of SO_2, *Journal of the American Society of Horticultural Science* **105**, 514–516 (1980)

BLACK, C. R. and BLACK, V. J. The effect of low concentrations of sulphur dioxide on stomatal conductance and epidermal cell survival in field bean (*Vicia faba* L.), *Journal of Experimental Botany* **30**, 291–298 (1979)

BLACK, V. J. and UNSWORTH, M. H. Stomatal responses to sulphur dioxide and vapour pressure deficit, *Journal of Experimental Botany* **31**, 667–677 (1980)

CONSTANTINIDOU, H. A. and KOZLOWSKI, T. T. Effects of sulphur dioxide and ozone on *Ulmus americana* seedlings. II. Carbohydrates, proteins and lipids, *Canadian Journal of Botany* **57**, 176–184 (1979)

COULSON, C. and HEATH, R. L. Inhibition of the photosynthetic capacity of isolated chloroplasts by ozone, *Plant Physiology* **53**, 32–38 (1974)

COWLING, D. W. and BRISTOW, A. W. Effects of SO_2 on sulphur and nitrogen fractions and on free amino acids in perennial ryegrass, *Journal of the Science of Food and Agriculture* **30**, 354–360 (1979)

DUGGER, M., MUDD, J. B. and KOUKOL, J. Effect of PAN on certain photosynthetic reactions, *Archives of Environmental Health* **10**, 195–200 (1965)

EVANS, L. S. and TING, I. P. Ozone sensitivity of leaves: Relationship to leaf water content, gas transfer resistance, and anatomical characters, *American Journal of Botany* **61**, 592–597 (1974)

FOWLER, D., CAPE, J. N., NICHOLSON, I. A., KINNARID, L. W. and PATERSON, I. S. The influence of a polluted atmosphere on cuticle degradation in Scots pine (*Pinus sylvestris*), *Proceedings of an International Conference on the Ecological Impact of Acid Precipitation*, Norway, March 11–14, 1980 (1980)

GERWICK, B. C., KU, S. B. and BLACK, C. C. Initiation of sulfate activation. A variation in C_4 photosynthesis plants, *Science* **209**, 513–515 (1980)

GODZIK, S. and KNABE, W. Vergleichende elektronmikroscopische Untersuchungen der Feinstruktur von Chloroplasten einiger *Pinus* – Arten aus den Industriegebieten an der Ruhr und in Oberschlesien. In *Proceedings of the Third International Clean Air Congress*, Düsseldorf, UDI-Verlag, Düsseldorf, pp. A164–A170 (1973)

GODZIK, S. and SASSEN, M. M. A. A scanning electron microscope examination on *Aesculus hippocastanum* L. leaves from control and air polluted areas, *Environmental Pollution* **17**, 13–18 (1978)

HÄLLGREN, J. E. Physiological and Biochemical Effects of Sulphur Dioxide on Plants. In *Sulphur in the Environment*, (Ed. J. O. NRIAGU), pp. 163–210, John Wiley and Sons, New York (1978)

HAMPP, R. and ZIEGLER, I. Sulfate and sulfite translocation via the phosphate translocator of the inner envelope membrane of chloroplasts, *Planta* **137**, 309–312 (1977)

HEATH, R. L. Ozone. In *Responses of Plants to Air Pollution*, (Eds. J. B. MUDD and T. T. KOZLOWSKI), pp. 23–25, Academic Press, New York (1975)

HORSMAN, D. C. and WELLBURN, A. R. Synergistic effect of SO_2 and NO_2 polluted air upon enzyme activity in pea seedlings, *Environmental Pollution* **8**, 123–133 (1975)

HUTTUNEN, S., MÄKELÄ, M. and LAINE, K. Quality and structure of surface waxes in pine needles and long term effects of air pollutants. In *XII International Meeting for Specialists in Air Pollution Damage in Forests, August 23–29, 1982, Oulu, Finland*, p. 27 (1982)

JÄGER, H. S. and GRILL, D. Einfluss von SO_2 und HF auf freier Aminosauren der Fichte (*Picea abies* L. Karsten), *European Journal of Forest Pathology* **5**, 279–286 (1975)

JAMES, A. T. and NICHOLS, B. W. Lipids of photosynthetic systems, *Nature* **210**, 372–375 (1966)

JENSEN, K. F. and KOZLOWSKI, T. T. Absorption and translocation of sulfur dioxide by seedlings of four forest tree species, *Journal of Environmental Quality* **4**, 379–382

KHAN, A. A. and MALHOTRA, S. S. Effects of aqueous sulphur dioxide on pine needle glycolipids, *Phytochemistry* **16**, 539–543 (1977)

KHAN, A. A. and MALHOTRA, S. S. Ribulose bisphosphate carboxylase and glycolate oxidase from jack pine: effects of sulphur dioxide fumigation, *Phytochemistry* **21**, 2607–2612 (1982)

KOZIOL, M. J. and COWLING, D. W. Growth of ryegrass (*Lolium perenne* L.) exposed to SO_2, *Journal of Environmental Botany* **29**, 1431–1439 (1978)

LEFFLER, H. R. and CHERRY, J. H. Destruction of enzymatic activities of corn and soybean leaves exposed to ozone, *Canadian Journal of Botany* **43**, 677–685 (1974)

McCUNE, D. C., WEINSTEIN, L. H., JACOBSON, J. S. and HITCHCOCK, A. E. Some effects of fluoride on plant metabolism, *Journal of the Air Pollution Control Association* **14**, 465–468 (1964)

MAJERNIK, O. and MANSFIELD, T. A. Direct effect of SO_2 on the degree of opening of stomata, *Nature* **227**, 377–378 (1970)

MALHOTRA, S. S. and KHAN, A. A. Effects of sulphur dioxide fumigation on lipid biosynthesis in pine needles, *Phytochemistry* **17**, 241–244 (1978)

MALHOTRA, S. S. and SARKAR, S. K. Effects of sulphur dioxide on sugar and free amino acid content of pine seedlings, *Physiologia Plantarum* **47**, 223–228 (1979)

MANSFIELD, T. A. The role of stomata in determining the response of plants to air pollutants, *Current Advances in Plant Science* **2**, 11–20 (1973)

MANSFIELD, T. A. and MAJERNIK, O. Can stomata play a part in protecting plants against air pollution? *Environmental Pollution* **1**, 149–154 (1970)

MUDD, J. B. Sulphur dioxide. In *Responses of Plants to Air Pollution*, (Eds. J. B. MUDD and T. T. KOZLOWSKI), pp. 9–22, Academic Press, New York (1975)

MUDD, J. B., McMANUS, T. T., ONGUN, A. and McCULLOGH, T. E. Inhibition of glycolipid biosynthesis in chloroplasts by ozone and sulfhydryl reagent, *Plant Physiology* **48**, 335–339 (1971)

NAKAMURA, H. and SAKA, H. Photochemical oxidants injury in rice plants. III. Effect of ozone on physiological activities in rice plants, *Japanese Journal of Crop Science* **47**, 707–714 (in Japanese with English summary) (1978)

PELL, E. J. and WEISSBERGER, W. C. Histopathological characterization of ozone injury to soybean foliage, *Phytopathology* **66**, 856–861 (1976)

PRIEBE, A., KLEIN, H. and JÄGER, H. J. Role of polyamines in SO_2-polluted pea plants, *Journal of Experimental Botany* **29**, 1045–1050 (1978)

SCHREIBER, U., VIDAVER, W., RUNECKLES, V. C. and ROSEN, P. Chlorophyll fluorescence assay for ozone injury in intact plants, *Plant Physiology* **61**, 80–84 (1978)

SCHWENN, J. O., DEPKA, B. and HENNIES, H. H. Assimilatory sulfate reduction in chloroplasts: Evidence for the participation of both stromal and membrane-bound enzymes, *Plant and Cell Physiology* **17**, 165–171 (1976)

SEKIYA, J., WILSON, L. G. and FILNER, P. Positive correlation between H_2S emission and SO_2 resistance in cucumber, *Plant Physiology* **65** (Supplement): 74, Abstract No. 407 (1980)

SMITH, F. A. and RAVEN, J. A. Intracellular pH and its regulation, *Annual Review of Plant Physiology* **30**, 289–311 (1979)

SOIKKELI, S. and TOUVINEN, T. Damage in mesophyll ultrastructure of needles of Norway spruce in two industrial environments in central Finland, *Annales Botanici Fennici* **16**, 50–64 (1979)

THOMSON, W. W., DUGGER, W. M. Jr. and PALMER, R. L. Effect of ozone on the fine structure of the palisade parenchyma cells of bean leaves, *Canadian Journal of Botany* **44**, 1677–1682 (1966)

TING, I. P. and MUKERJI, S. K. Leaf ontogeny as a factor in susceptibility to zone: Amino acid and carbohydrate changes during expansion, *American Journal of Botany* **58**, 497–504 (1971)

TINGEY, D. T., FITES, R. C. and WICKLIFF, C. Ozone alteration of nitrate reductase in soybean, *Physiologia Plantarum* **29**, 33–38 (1973)

WEI, L. L. and MILLER, G. W. Effect of HF on the fine structure of mesophyll cells from *Glycine max*. Merr, *Fluoride* **5**, 67–73 (1972)

WEINSTEIN, L. H. Effects of atmospheric fluoride on metabolic constituents of tomato and bean leaves, *Contributions from the Boyce Thompson Institute* **21**, 215–231 (1961)

WELLBURN, A. R., MAJERNIK, O. and WELLBURN, F. A. M. Effects of SO_2 and NO_2 polluted air upon the ultrastructure of chloroplasts, *Environmental Pollution* **3**, 37–49 (1972)

WINNER, W. E. and MOONEY, H. A. Ecology of SO_2 resistance. I. Effects of fumigations on gas exchange of deciduous and evergreen shrubs, *Oecologia* (*Berlin*) **44**, 290–295 (1980)

ZIEGLER, I. The effects of SO_2 pollution on plant metabolism, *Residue Reviews* **56**, 79–105 (1975)

ZIEGLER, I. Subcellular distribution of ^{35}S-sulphur in spinach leaves after application of $^{35}SO_3^{2-}$ and $^{35}SO_2$, *Planta* **131**, 25–32 (1977)

Latin and common names of plants and insects cited in the text

Plants

Abies   fir
Abies alba   common silver fir
Abies balsamea   balsam fir
Acer   maple, sycamore
Acer negundo   ash-leafed maple, box-elder
Acer platanoides   Norway maple
Acer saccharum   sugar or rock maple
Aesculus hippocastanum   horse chestnut
Agropyron smithii   western wheatgrass
Ailanthus altissima   tree of heaven
alder, black   *Alnus nigra*
Alectoria capillaris   a lichen
alfalfa   *Medicago sativa*
Alnus nigra   black alder
Anacystis   a blue-green alga
apple *Malus* sp.
Arachis hypogaea   peanut
Armillaria mellea   non-obligate pathogenic fungus (honey or root fungus)
ash, red   *Fraxinus pennsylvanica*
Atriplex sabulose   a C_4 'salt bush' (Australian)
Atriplex triangularis   a C_3 'salt bush (Australian)
Avena sativa   oat
Azalea   ornamental plant

barley   *Hordeum* spp.
bean (pinto, red kidney, French, bush, snap)   *Phaseolus vulgaris*
bean, broad or field   *Vicia faba*
beech   *Fagus* spp.
beech, common   *Fagus sylvatica*
beet   *Beta vulgaris*
Beta vulgaris   beet
Beta vulgaris var. *Cicla*   Swiss chard
Betula   birch

439

Betula alba   silver birch, sometimes also a hybrid of *pendula* × *pubescens*
Betula papyrifera   canoe or paper-bark birch
Betula pendula   silver birch
Betula verrucosa   silver birch
birch, canoe or paper-bark   *Betula papyrifera*
birch, silver   *Betula pendula*
Botrytis   a nonobligate pathogenic fungus (neck and leaf rot, chocolate spot)
box elder   *Acer negundo*
Brassica napus   rape, cole, swedish turnip or swede
Brassica oleracea   collards (kale)
Brassica oleracea (*capita*)   cabbage
buckwheat   *Fagopyrum* spp.

cabbage   *Brassica oleracea* (*capita*)
Canavalia ensiformis   jack-bean
cantaloupe   *Cucumis meto annuum*
Capsicum   pepper
carrot   *Daucus carota*
Cassia occodentalis   senna
Chara   a green alga
Chenopodium murale   nettle-leaved goosefoot
Chlamydomonas reinhardtii   a green alga
Chlorella pyrenoidosa   a green alga
Chlorella sorokiniana   a green alga
Chlorella vulgaris   a green alga
Chrysanthemum indicum   chrysanthemum
Citrullus vulgaris   water melon
Citrus aurantiifolia   lime
Citrus aurantium   orange
Citrus limon   lemon
Cladonia rangiferina   a lichen
clover, red   *Trifolium pratense*
clover, white   *Trifolium repens*
cocksfoot   *Dactylis glomerata*
Coleus   an ornamental plant
collards (kale)   *Brassica oleracea*
Colocasia antiquorum   taro
Commelina communis   an ornamental plant
corn   *Zea mays*
cotton   *Gossypium hirsutum*
cottonwood   *Populus deltoides*
cowberry   *Vaccinium vitis-idaea*
Crataegus monogyna   hawthorn
crowberry   *Empetrum nigrum*
cucumber   *Cucumis sativus*
Cucumis melo   cantaloupe
Cucumis sativus   cucumber
cucurbit   referring to *Cucurbitaceae*, i.e. cucumber, squash, pumpkin
Cucurbita pepo   pumpkin, squash

Dactylis glomerata   cocksfoot, orchard grass
Daucus carota   carrot
Diplacus aurantius (= *Mimulus*)   monkey flower
dock, curly   *Rumex crispus*

elder, common   *Sambucus nigra*
elm, American white   *Ulmus americana*
Empetrum nigrum   crowberry
Escherichia   enteric bacteria
Euglena gracilis   a green alga
Euphorbia pulcherrima   poinsettia

Fagopyrum cymosum   a buckwheat
Fagus   beech
Fagus sylvatica   common beech
fir   *Abies* spp.
fir, balsam   *Abies balsamea*
fir, common silver   *Abies alba*
fir, Doublas   *Pseudotsuga menziesii, P. douglasii*
Fomes annosus   a nonobligate pathogenic fungus
Fragaria   strawberry
Fragaria chiloensis   strawberry
Fraxinus pennsylvanica   red ash

Gladiolus   an ornamental species
Glycine max   soyabean
goosefoot, nettle-leaved   *Chenopodium murale*
Gossypium hirsutum   cotton
grape   *Vitis vinifera*

Habbacterium   bacterium requiring high salt concentrations
hawthorn   *Crataegus monogyna*
Helianthus annuus   sunflower
Heteromeles arbutifolia   tollon, toyon, Christmas berry
Hibiscus esculentus   okra
Hordeum distichon   a barley
Hordeum vulgare   a barley
horse chestnut   *Aesculus hippocastanum*
Hydrodictyon africanum   a green alga
Hypogymnia physodes   a lichen

Ipomoea batatas   sweet potato
Ipomoea sp.   morning glory

jack-bean   *Canavalia ensiformis*

Kalanchoë   a Crassulacean Acid Metabolism (CAM) plant
Kochia trichophylla   a 'salt bush' (Australian)
kudzu   *Pueraria lobata*

Lactuca sativa   lettuce
larch   *Larix* spp.
larch, Dunkeld (hybrid)   *Larix × eurolepis*
larch, European   *Larix decidua*
Larix   larch
Larix decidua   European larch
Larix decidua × eurolepis   a hybrid larch
Larix × eurolepis   Dunkeld (hybrid) larch
Larix leptolepis   Japanese larch
Lemna gibba   gibbous duckweed
lemon   *Citrus limon*
Lens culinaris   lentil
lentil   *Lens culinaris*
lettuce   *Lactuca sativa*
lilac   *Syruiga vulgaris*
lime   *Citrus aurantifolia*
Liquidambar styraciflua   sweet gum
Lolium multiflorum   Italian ryegrass
Lolium perenne   perennial ryegrass
lucerne   *Medicago sativa*
Lycopersicon esculentum   tomato

maize   *Zea mays*
maple, ash-leafed   *Acer negundo*
maple, Norway   *Acer platanoides*
maple, sugar   *Acer saccharum*
Medicago sativa   alfalfa, lucerne
milo maize   *Sorghum vulgare*
Mimulus cardinale   monkey flower, musk
morning glory   *Pharbitis nil* or *Ipomoea* sp.
Morus nigra   mulberry
mulberry   *Morus nigra*
Myrothamnus flabellifolia   resurrection plant

Nicotiana affinis   a tobacco
Nicotiana tabacum   a tobacco
Nitella translucens   a green alga

oak, English   *Quercus robur*
oak, sessile or durmast   *Quercus petraea*
oat   *Avena sativa*
okra   *Hibiscus esculentum*
orange   *Citrus aurantium*
orchard grass   *Dactylis glomerata*
Oryza sativa   rice

pansy   *Viola tricolor*
Papaver somniferum   opium poppy
Parmelia bolliana   a lichen
Parmelia sulcata   a lichen

pea　*Pisum sativum*
peanut　*Arachis hypogaea*
pear　*Pyrus communis*
pepper　*Capsicum annuum*
Phaeoceros　a hornwort
Pharbitis nil　Japanese morning glory
Phaseolus vulgaris　bean (pinto, red kidney, French, bush, snap)
Physica millegrana　a lichen
Phleum pratense　timothy
Picea　spruce
Picea abies ⎫
Picea excelsa ⎬　Norway spruce
Picea pungens　Colorado spruce
pine　*Pinus* spp.
pine, eastern white　*Pinus strobus*
pine, jack　*Pinus banksiana*
pine, Japanese black　*Pinus thunbergii*
pine, loblolly　*Pinus taeda*
pine, lodgepole　*Pinus contorta*
pine, Monterey　*Pinus radiata*
pine, pond　*Pinus serotina*
pine, ponderosa　*Pinus ponderosa*
pine, red　*Pinus resinosa*
pine, Scots　*Pinus sylvestris*
pine, shortleaf　*Pinus echinata*
pine, slash　*Pinus elliottii*
pine, western yellow　*Pinus ponderosa*
pine, Weymouth　*Pinus strobus*
Pinus　pine
Pinus banksiana　jack pine
Pinus contorta　lodgepole pine
Pinus contorta × *banksiana*　a hybrid pine
Pinus echinata　shortleaf pine
Pinus elliottii　slash pine
Pinus ponderosa　ponderosa pine, western yellow pine
Pinus radiata　Monterey pine
Pinus resinosa　red pine
Pinus serotina　pond pine
Pinus strobus　eastern white pine, Weymouth pine
Pinus sylvestris　Scots pine
Pinus taeda　loblolly pine
Pinus thunbergii　Japanese black pine
Pisum sativum　pea
poinsettia　*Euphorbia pulcherrima*
Polygonum orientale　a knotweed
poplar　*Populus* spp.
poplar, balsam　*Populus balsamifera*
poplar, hybrid black　*Populus canadensis*
poppy, opium　*Papaver somniferum*
Populus　poplar

Populus balsamifera   balsam poplar
Populus canadensis   hybrid black poplar
Populus deltoides   cottonwood
Populus gelrica   a poplar
Populus × *interamericana*   a hybrid poplar
potato   *Solanum tuberosum*
Pseudomonas glycinea   a bacterium infecting soyabean leaves
Pseudomonas phaseolicola   a bacterium infecting *Phaseolus* spp.
Pseudotsuga douglasii   ⎫
Pseudotsuga menziesii   ⎬ Douglas fir
Pueraria lobata   kudzu
pumpkin   *Cucurbita pepo*
Pyrus communis   pear

Quercus petraea   sessile or durmast oak
Quercus robur   English oak

radish   *Raphanus sativus*
Raphanus sativus   radish
rice   *Oryza sativa*
Rosa rugosa   a rose
Rumex crispus   curly dock
rye   *Secale cereale*
ryegrass, Italian   *Lolium multiflorum* (formerly *L. italicum*)
ryegrass, perennial   *Lolium perenne*

Saccharomyces cerevisiae   yeast
Saccharum officinale   sugar cane
sage, black   *Salvia mellifera*
Salvia mellifera   black sage
Sambucus nigra   common elder
Scenedesmus obtusiusculus   a green alga
Secale cereale   rye
sesame   *Sesamum orientale*
Sesamum orientale   sesame
Solanum pseudocapsicum   a ornamental plant
Solanum tuberosum   potato
sorghum   *Sorghum vulgare*
Sorghum vulgare   sorghum, milo maize
soyabean   *Glycine max*
Sphagnum fimbriatum   a bog or moorland moss
spinach   *Spinacia oleracea*
Spinacia oleracea   spinach
spruce   *Picea* spp.
spruce, Colorado   *Picea pungens*
spruce, Norway   *Picea abies, P. excelsa*
squash   *Cucurbita pepo*
strawberry   *Fragaria* spp.
sunflower   *Helianthus annuus*
sweet gum   *Liquidambar styraciflua*

sweet potato   *Ipomoea batatas*
Swiss chard   *Beta vulgaris* var. *Cicla*
Syringa vulgaris   lilac

taro   *Colocasia antiquorum*
Taxus baccata   common yew
Tidestromia sp.   a C_4 plant
timothy   *Phleum pratense*
tobacco   *Nicotiana* spp.
tomato   *Lycopersicon esculentum*
tree of heaven   *Ailanthus altissima*
Trifolium pratense   red clover
Trifolium repens   white clover
Triticum aestivum   wheat
tulip   *Tulipa* spp.
Tulipa   tulip
Tulipa gesneriana   a tulip

Ulmus americana   American white elm

Vaccinium vitris-idaea   cowberry
Valerianella olitoria   lamb's lettuce, corn salad
Vicia faba   broad or field bean
Viola tricolor   pansy
Vitis vinifera   grape

water melon   *Citrullus vulgaris*
Weigela florida   an ornamental tree
Weigela hybrida   an ornamental tree
western wheatgrass   *Agropyron smithii*
wheat   *Triticum aestivum*

Xanthomonas phaseoli   leaf spot on *Phaseolus* and *Glycine*
Xanthoria parietina   a lichen

yeast   *Saccharomyces cerevisiae*
yew, common   *Taxus baccata*

Zea mays   (sweet) corn, maize
Zygosaccharomyces acidifaciens   a yeast

Insects

Epilachna varivestis   Mexican bean beetle
Monochamus alternatus   a wood-boring beetle
Sacchiphantes abietis   gall aphid

Appendix 2

Conversion of air pollutant units

Concentrations of gaseous pollutants are generally expressed either on a volume/volume basis such as parts per million (10^6) or parts per billion (10^9), ppm or ppb, respectively, or on a mass/volume basis such as microgrammes per cubic metre ($\mu g\ m^{-3}$). The expression of concentrations of gaseous pollutants on a volume/volume basis has the benefit of being essentially independent of differences in temperature and pressure, assuming the gases obey the ideal gas law ($P_1 V_1/T_1 = P_2 V_2/T_2$). Units such as ppm or ppb are perhaps best suited to descriptions of ambient pollutant concentrations in monitoring studies or of pollutant exposure concentrations in field or laboratory growth studies. In physiological or biochemical studies aimed at describing plant response to specific amounts of gaseous pollutants, or for evaluating pollutant dosages and fluxes, expression of pollutant concentrations on a mass/volume basis is preferred. Units such as $\mu g\ m^{-3}$ are, however, affected by changes in temperature and pressure, which should be borne in mind when comparing pollutant concentrations between studies and when interconverting mass/volume and volume/volume units.

The conversion of values from ppb to $\mu g\ m^{-3}$ or vice versa depend upon temperature (T) and pressure (P). The equations below have been generalized for use with all pollutant gases by including the factor ($M/22.4$), where M is the weight in mg of 1 mmol of pollutant gas and 22.4 is the volume in ml occupied by 1 mmol of gas:

$$\mu g\ m^{-3} = (\text{ppb})\left(\frac{M}{22.4}\right)\left(\frac{T_0}{T}\right)\left(\frac{P}{P_0}\right)$$

$$\text{ppb} = (\mu g\ m^{-3})\left(\frac{22.4}{M}\right)\left(\frac{T}{T_0}\right)\left(\frac{P_0}{P}\right)$$

where T_0 and P_0 are the standard temperature and pressure, i.e. 273 K and 1 atmosphere, respectively. *Table A.1* lists the molecular weights of nine common gaseous pollutants.

TABLE A.1. Molecular weights of some gaseous pollutants

Gaseous pollutant	mg mmol^{-1}	Gaseous pollutant	mg mmol^{-1}
Ammonia (NH_3)	17.03	Nitrogen dioxide (NO_2)	46.01
Hydrogen chloride (HCl)	36.46	Ozone (O_3)	48.00
Hydrogen fluoride (HF)	20.01	Peroxyacetyl nitrate (PAN)	105.05
Hydrogen sulphide (H_2S)	34.08	Sulphur dioxide (SO_2)	64.06
Nitric oxide (NO)	30.01		

Air pollutant conversions: A computer program

A. R. Wellburn, *Department of Biological Sciences, University of Lancaster, Lancaster, UK*

The following BASIC program may be useful to those who wish to convert concentrations of a number of pollutants (HF, NH$_3$, NO, NO$_2$, O$_3$, PAN and SO$_2$) from ppm or ppb into μg m^{-3} or vice versa. The program allows for the correction of temperatures and pressures in a variety of different units. Further, the gaseous to aqueous conversions of Nieboer *et al.* (1977) applicable to SO$_2$ have been included and expanded to include those for NO$_2$ as well. Other gaseous to aqueous conversions may be added if the appropriate factor is calculated and inserted on line 1430 and appropriate adjustments made to lines 1080, 1220 and 1280 allowing for rearrangement of the data order in lines 1410 and 1420.

The program is written in Applesoft, but only minor changes (e.g. to line 1100) are required to allow the use of this program on any microcomputer.

Reference

NIEBOER, E., TOMASSINI, F. D., PUCKETT, K. J. and RICHARDSON, D. H. S. A model for the relationship between gaseous and aqueous concentrations of sulphur dioxide in lichen exposure studies *New Phytologist* **79**, 157–162 (1977)

Program

```
1000  REM   **AIR POLLUTANT CONC CONVERSIONS**
1010  REM   *WRITTEN BY A.R.WELLBURN*
1020  REM   *UNIV.LANCASTER,U.K.1982*
1030  REM   *INCORPORATING GAS)SOLN CONVERSION*
1040  REM   *OF NIEBOER ET AL.NEW PHYT.79(1977)157 *
1050  REM   *WHEN PROGRAM STOPS ANY KEY RESTARTS*
1060  FOR I = 1 TO 7: READ A$(I): NEXT I
1070  FOR I = 1 TO 7: READ A(I): NEXT I
1080  FOR I = 1 TO 2: READ B(I): NEXT I: RESTORE
1090  B$ = "**************************":C$ = "AIR POLLUTANT CONVERSIONS"
1100  HOME : VTAB (5): HTAB (8): PRINT  LEFT$ (B$, LEN (C$)): PRINT : HT
AB (8): INVERSE : PRINT C$: NORMAL : PRINT : HTAB (8): PRINT  LEFT$ (B$,
 LEN (C$))
1110  PRINT : INPUT "Temperature (Degrees Celsius) ? ";T
1120  PRINT : INPUT "Atmospheric pressure in Atmospheres (1),kPa (2) or
mmHg (3)  ? ";E
1130  IF E < 1 OR E > 3 THEN 1120
1140  IF E = 1 THEN F = 1
1150  IF E = 2 THEN F = 101.3
1160  IF E = 3 THEN F = 760
1170  PRINT : INPUT "What is the pressure in the chosen units ? ";K
1180  PRINT : INPUT "Range required - PPM (1) or PPB (2) ? ";A
1190  IF A = 1 THEN A$ = " PPM":G = 1: GOTO 1210
1200  A$ = " PPB":G = .001: IF A < 1 OR A > 2 THEN  GOTO 1180
1210  PRINT : PRINT "Do you want to convert";A$;" to micrograms per cubi
c meter (1) or the reverse (2) ": INPUT B
```

```
1220  IF B < 1 OR B > 2 THEN  GOTO 1100
1230  FOR I = 1 TO 7
1240  PRINT : PRINT : PRINT A$(I); SPC( 25 -  LEN (A$(I)));I
1250  NEXT I
1260  PRINT : PRINT : INPUT "Which pollutant(1 - 6) ? ";C
1270  PRINT : PRINT A$(C): PRINT  LEFT$ (B$, LEN (A$(C)))
1280  IF B = 2 THEN  GOTO 1340
1290  PRINT : PRINT "Amount of pollutant in";A$: INPUT P
1300  Q = ((P * A(C) * G) / .0224) * (273.15 / (273.15 + T)) * (K / F):Q1
   = ( INT (Q * 1000 + .5)) / 1000
1310  PRINT : PRINT "Is equivalent to ";Q1;" micrograms  per cubic meter
..........................."
1320  GET T$: IF C < 0 OR C > 2 THEN  GOTO 1100
1330  GOTO 1380
1340  PRINT : INPUT "Amount of pollutant in micrograms per cubic meter ?
  ";Q
1350  P = ((Q * .0224 * G) / A(C)) * ((273.15 + T) / 273.15) * (F / K):P1
   = ( INT (P * 1000 + .5)) / 1000
1360  PRINT : PRINT "Is equivalent to ";P1;A$
1370  GET T$: IF C < 1 OR C > 2 THEN  GOTO 1100
1380  R = ( INT (B(C) *  SQR ((P * G)) * 1000 + .5) / 1000):S = ( INT ((R
   / A(C)) * 10000 + .5) / 10000)
1390  PRINT "or alternatively ";R;" PPM or            ";S;" millimoles/li
tre in solution."
1400  GET T$: GOTO 1100
1410  DATA    "Sulphur dioxide","Nitrogen dioxide","Nitric oxide","Ammoni
a","Ozone","PAN","Hydrogen fluoride"
1420  DATA    64.066,46.008,30.008,17.029,48,105.051,20.008
1430  DATA    10.3,358.9
```

List of participants

Alscher, Dr R.
Boyce Thompson Institute for Plant Research at Cornell University, Tower Road, Ithaca, NY 14853, USA

Ballantyne, Dr D. J.
Department of Biology, University of Victoria, PO Box 1700, Victoria, BC, Canada V8W 2Y2

Bell, Dr J. N. B.
Department of Pure and Applied Biology, Imperial College of Science and Technology, Silwood Park, Ascot, Berkshire SL5 7PY, UK

Bennett, Dr J. H.
Plant Physiology Institute, Agricultural Research Service, United States Department of Agriculture, Beltsville, MD 20705, USA

Black, Dr V. J.
Ecology Section, Department of Human Sciences, Loughborough University of Technology, Loughborough, Leicestershire LE11 3TU, UK

Bonte, Dr J.
INRA, Laboratoire d'Étude de la Pollution Atmosphérique, Montardon, 64121 Serres Castet, France

Bucher, Dr J. B.
Eidgenössische Anstalt für das forstliche Versuchswesen, CH-8903 Birmensdorf ZH, Switzerland

Buckenham, Miss A. H.
University of Nottingham, School of Agriculture, Sutton Bonnington, Loughborough, Leicestershire LE12 5RD, UK

Connelly, Mr P. R.
Department of Biology, St Michael's College, Winooski, VT 05404, USA

Darrall, Dr N. M.
CEGB, Central Electricity Research Laboratory, Kelvin Avenue, Leatherhead, Surrey KT22 7SE, UK

Deeley, Ms S.
Butterworth and Company (Publishers) Ltd, Borough Green, Sevenoaks, Kent TN15 8PH, UK

Filner, Professor P.
ARCO Plant Cell Research Institute, 6560 Trinity Court, Dublin, CA 94566, USA

Furukawa, Dr A.
Department of Environmental Biology, National Institute for Environmental Studies, Japan Environment Agency, Yatabe, Tsukuba, Ibaraki 305, Japan

Garrec, Dr J.-P.
Commissarait à l'Énergie Atomique, Centre d'Études Nucléaires de Grenoble, Laboratoire de Biologie Végétale, 85X-38041 Grenoble CEDEX, France

Garsed, Dr S. G.	Imperial College Field Station, Silwood Park, Ascot, Berkshire SL5 7PY, UK
Godzik, Dr S.	Polska Akademia Nauk, Institut Podstaw Inżynierii Środowiska, ul. M. Skłodowskiej-Curie 34, 41-800 Zabrze, Poland
Guillot, Dr P.	Commission of the European Communities, Directorate-General for Science, Research and Development, XII/G-Environment, Raw Materials and Materials Technologies, Rue de la Loi 200, B-1049 Bruxelles, Belgium
Hällgren, Dr J.-E.	Swedish University of Agricultural Sciences, The Faculty of Forestry, Department of Forest Genetics and Plant Physiology, S-901 83 Umeå, Sweden
Heath, Professor R. L.	Department of Botany and Plant Sciences, University of California, Riverside, CA 92521, USA
Heck, Dr W. W.	Agricultural Research Service, US Department of Agriculture and Botany Department, North Carolina State University, PO Box 5186, Raleigh, NC 27650, USA
Hughes, Dr P. R.	Boyce Thompson Institute for Plant Research at Cornell University, Tower Road, Ithaca, NY 14853, USA
Huttunen, Dr S.	Kasvitieteen laitos, Oulun yliopisto, Pl 191, SF-90101 Oulu 10, Finland
Jakobsson, Ms C.	Sveriges lantbruksuniversitet, Institutionen för växt- och skogsskydd, Box 7044, S-750 07 Uppsala 7, Sweden
Khan, Dr A. A.	Northern Forest Research Centre, Canadian Forestry Service, Environment Canada, 5320 122nd Street, Edmonton, ALTA, Canada T6H 3S5
Kliffen, Mr C.	Research Institute for Plant Protection, Binnenhaven 12, 6709 Wageningen, The Netherlands
Kozioł, Dr M. J.	Botany School, University of Oxford, South Parks Road, Oxford OX1 3RA, UK
Kvist, Dr K.	Sveriges lantbruksuniversitet, Institutionen för växt- och skogsskydd, Box 7044, S-750 07 Uppsala 7, Sweden
Last, Professor F. T.	Institute of Terrestrial Ecology, Bush Estate, Penicuik, Midlothian EH26 0QB, Scotland
Lendzian, Dr K. J.	Lehrstuhl für Botanik, Institut für Botanik und Mikrobiologie der Technischen Universität München, Arcisstrasse 21, D-8000 München 2, FRG
Lex, Dr M.	Natural Environment Research Council, Polaris House, North Star Avenue, Swindon, Wiltshire SN2 1EU, UK
Lorenzini, Dr G.	Istituto di Patologia Vegetale, Università di Pisa, Via del Borghetto 80, 56100 Pisa, Italy
Louguet, Professor P.	Université Paris Val de Marne, U.E.R. de Sciences, Service de Physiologie Végétale, Avenue du Géneral de Gaulle, 94010 Créteil CEDEX, France
Mansfield, Professor T. A.	Department of Biological Sciences, University of Lancaster, Bailrigg, Lancaster, Lancashire LA1 4YQ, UK
Mejnartowicz, Dr L. E.	Polska Akademia Nauk, Instytut Dendrologii, ul. Parkowa 5, 63-120 Kórnik, Poland
Mudd, Dr J. B.	ARCO Plant Cell Research Institute, 6560 Trinity Court, Dublin, CA 94566, USA

Murray, Dr F.	Department of Biological Sciences, University of Newcastle, New South Wales 2308, Australia
Nieboer, Professor E.	McMaster University, Department of Biochemistry, Health Services Centre, 1200 Main Street West, Hamilton, ONT, Canada L8N 3Z5
Parry, Mr M. A. J.	Botany Department, Rothamsted Experimental Station, Harpenden, Hertfordshire AL5 2JQ, UK
Pell, Dr E. J.	The Pennsylvania State University, Department of Plant Physiology, 211 Buckhout Laboratory, University Park, PA 16802, USA
Pierre, Dr M.	Centre National de la Recherche Scientifique, Laboratoire du Phytotron, 91190 Gif sur Yvette, France
Posthumus, Dr A. C.	Research Institute for Plant Protection, Binnenhaven 12, 6709 PD Wageningen, The Netherlands
Rennenberg, Dr H.	Botanisches Institut der Universität Köln, Gyrohofstrasse 15, D-5000 Köln 41, FRG
Richardson, Professor D. H. S.	Botany School, University of Dublin, Trinity College, Dublin 2, Eire
Roberts, Dr T. M.	CEGB, Central Electricity Research Laboratory, Kelvin Avenue, Leatherhead, Surrey KT22 7SE, UK
Ro-Poulsen, Dr H.	University of Copenhagen, Institute of Plant Ecology, Øster Farimagsgade 2D, DK-1353 Copenhagen K, Denmark
Runeckles, Professor V. C.	Department of Plant Science, The University of British Columbia, Suite 248, 2357 Main Hall, Vancouver, BC, Canada V6T 2A2
Sajdl, Dr V.	INEP, Laboratorija za Biofizicku i Analiticku Hemiju, Ranatska 31b, 11080 Zemun, Yugoslavia
Satoh, Dr S.	Bio-Environment Laboratory, Central Research Institute of Electric Power Industry, 1646 Abiko, Abiko City, Chiba, Japan
Saunders, Dr P. J. W.	Natural Environment Research Council, Polaris House, North Star Avenue, Swindon, Wiltshire SN2 1EU, UK
Schenone, Dr G.	ENEL (Italian National Electricity Generating Board), Thermal and Nuclear Research Centre, Via Robattino 54, Milano, Italy
Singh, Dr S. P.	Centre of Advanced Study in Botany, Department of Botany, Banaras Hindu University, Varanasi 221005, India
Skärby, Dr L.	Swedish Water and Air Pollution Research Institute, Box 5207, S-402 24 Göteborg, Sweden
Steinkamp, Mr R.	Botanisches Institut der Universität Köln, Gyrohofstrasse 15, D-5000 Köln 41, FRG
Sugahara, Dr K.	Department of Environmental Biology, National Institute for Environmental Studies, Japan Environment Agency, Yatabe, Tsukuba, Ibaraki 305, Japan
Treshow, Professor M.	The University of Utah, Department of Biology, 201 Biology Building, Salt Lake City, UT 84112, USA
Ulbricht, Dr T. L. V.	Agricultural Research Council, 160 Great Portland Street, London W1N 6DT, UK

Wellburn, Dr A. R.   Department of Biological Sciences, University of Lancaster, Bailrigg, Lancaster, Lancashire LA1 4YQ, UK

Whatley,
    Professor F. R.   Botany School, University of Oxford, South Parks Road, Oxford OX1 3RA, UK

Whittingham, Dr C. P.   Botany Department, Rothamsted Experimental Station, Harpenden, Hertfordshire AL5 2JQ, UK

Yu, Professor S.-W.   Shanghai Institute of Plant Physiology, Academia Sinica, 300 Fonglin Road, Shanghai, China 200032

Species index

Subject index

For convenient reference, the pollutants are listed first in the subheadings, with the remaining descriptive entries following. No distinction has been made between pollutants and their possible chemically-active species.